D1451927

Microbial Evolution

GENE ESTABLISHMENT, SURVIVAL, AND EXCHANGE

Microbial Evolution

GENE ESTABLISHMENT, SURVIVAL, AND EXCHANGE

edited by

ROBERT V. MILLER
MARTIN J. DAY

ASM
PRESS WASHINGTON, D.C.

Library of Congress Cataloging-in-Publication Data

Microbial evolution: gene establishment, survival, and exchange/
edited by Robert V. Miller and Martin J. Day.
        p. ; cm.
    Includes bibliographical references and index.
    ISBN 1-55581-271-6 (alk. paper)
    1. Microorganisms—Evolution.
 [DNLM: 1. Genetics, Microbial. 2. Chromosomes, Bacterial—genetics.
QW 51 M62625 2004] I. Miller, Robert V. (Robert Verne), 1945- II. Day,
Martin J.

QR13.M525 2004
579′.138—dc22

                         2003024010

Address editorial correspondence to ASM Press, 1752 N St., N.W., Washington, DC
20036-2904, U.S.A.

Send orders to: ASM Press, P.O. Box 605, Herndon, VA 20172, U.S.A.
Phone: 800-546-2416; 703-661-1593
Fax: 703-661-1501
E-mail: books@asmusa.org
Online: www.asmpress.org

CONTENTS

CONTRIBUTORS

Mark J. Bailey
CEH Oxford
Oxford OX1 3SR, United Kingdom

David L. Balkwill
Department of Biomedical Sciences
Florida State University College of Medicine
Tallahassee, FL 32306

John R. Battista
Department of Biological Sciences
Louisiana State University and A & M College
Baton Rouge, LA 70803

Michael Chandler
Laboratoire de Microbiologie et de Génétique Moléculaire
Toulouse, France

Madhusudan Choudhary
Department of Microbiology and Molecular Genetics
University of Texas Medical School—Houston
Houston, TX 77030

François Cornet
Laboratoire de Microbiologie et de Génétique Moléculaire
Toulouse, France

Martin J. Day
Cardiff School of Biosciences
Cardiff University
Cardiff CF10 3TL, Wales, United Kingdom

Angela E. Douglas
Department of Biology, University of York, York YO10 5YW,
United Kingdom

Ashlee M. Earl
Department of Biological Sciences, Louisiana State University and A & M College
Baton Rouge, LA 70803

Renato Fani
Dipartimento di Biologia Animale e Genetica, Università degli Studi di Firenze
Florence, Italy

Susse Kirkelund Hansen
Center for Microbial Interactions, BioCentrum-DTU
Lyngby, Denmark

I. King Jordan
National Center for Biotechnology Information, National Library of Medicine
National Institutes of Health, Bethesda, MD 20894

Samuel Kaplan
Department of Microbiology and Molecular Genetics
University of Texas Medical School-Houston
Houston, TX 77030

Eugene V. Koonin
National Center for Biotechnology Information, National Library of Medicine
National Institutes of Health, Bethesda, MD 20894

Günther Koraimann
Institut für Molekularbiologie, Biochemie und Mikrobiologie (IMBM)
Karl-Franzens-Universität Graz, Graz, Austria

Andrew K. Lilley
CEH Oxford, Oxford OX1 3SR, United Kingdom

Chris Mackenzie
Department of Microbiology and Molecular Genetics
University of Texas Medical School-Houston
Houston, TX 77030

Eshwar Mahenthiralingam
Cardiff School of Biosciences, Cardiff University
Cardiff CF10 3TL, Wales, United Kingdom

Mike Manefield
CEH Oxford, Oxford OX1 3SR, United Kingdom

Roger Milkman
Professor Emeritus, University of Iowa
Iowa City, IA

Robert V. Miller
Department of Microbiology and Molecular Genetics
Oklahoma State University, Stillwater, OK 74074

Søren Molin
Center for Microbial Interactions, BioCentrum-DTU
Lyngby, Denmark

Klaus Nüsslein
Department of Microbiology, University of Massachusetts
Amherst, MA 01003

Mario Pedraza-Reyes
Institute of Investigation in Experimental Biology
Faculty of Chemistry, University of Guanajuato
Noria Alta, Guanajuato 36050
Mexico

Abigail A. Salyers
Department of Microbiology, University of Illinois-UC
Urbana, IL 61801

Steven J. Sandler
Department of Microbiology, University of Massachusetts
Amherst, MA 01003

Nadja B. Shoemaker
Department of Microbiology, University of Illinois-UC
Urbana, IL 61801

Sarah L. Turner
CEH Oxford, Oxford OX1 3SR, United Kingdom

Lorraine G. van Waasbergen
Department of Biology, University of Texas at Arlington
Arlington, TX 76019

Gabrielle Whittle
Department of Microbiology, University of Illinois-UC
Urbana, IL 61801

Ronald E. Yasbin
College of Sciences, University of Nevada, Las Vegas
Las Vegas, NV 89154

PREFACE

Microbes have been around for about four billion years. Their environments have undergone vast changes over this time. They have had to adapt to these environmental changes to survive, and they have done an excellent job so far. Plants and animals only started to evolve about one billion years ago, and so bacteria and their ancestors had a huge head start in their evolutionary history (almost three billion years). They have conserved in their genomes clues to how they started and traveled this evolutionary road. These clues do not lie solely in the gene sequences they carry, but also in the genetic strategies they have evolved for evolutionary change.

This book developed from a desire to place what we know in a context that may stimulate others to see the many disparate themes as a whole. To understand evolution we need an overview of all the topics that culminate to produce novel sequences and allow individual cells to produce phenotypes better suited to survival and growth. In this book we have addressed the natural processes used in bacteria genome evolution. Processes of anthropogenic intervention and manipulation are not addressed. We will leave these to texts on biotechnology. We will never really understand the consequences of our genetic manipulation of microorganisms until we understand how bacterial communities interact and evolve naturally. All the concerns raised and expressed in the GEM debate are those that many microbiologists have asked about normal bacterial populations. Answer one, and you have the answer to the other.

This book is meant to be a learning book. It is not merely a monograph on the subject of bacterial evolution. We hope that it will help you to begin your journey into the study of this fascinating subject. We hope that you will use it to gain a basic understanding so that you can more effectively pose and answer questions to move us toward greater insights into bacterial evolution. To help you on your way, we have asked the contributors to address questions, identify important evolutionary points, and differentiate what we understand from what we do not. We asked contributors not to review the literature on the subject but to draw insight and inference from it. Therefore, you will not find a cacophony of references in this text. Two types of references will be listed at the end of each chapter. There will be some general references and books for further reading on the subject. These will include some historical references important in understanding the history of the subject. In addition, there will be a list of specific references. These will include papers that back up specific facts in the chapter. They should be consulted to gain a fuller understanding of the experiments that led to the conclusions outlined in the text. These references are those that the authors feel will be most illustrative of the subjects discussed. In an attempt to allow discussion and speculation about the subject, we have also asked the authors to provide a series of thought-provoking questions that address the academic issues raised in their chapters. Following each section of the book, we have summarized the highlights and themes presented in the section in a short chapter. We hope that you will find these summaries useful.

The knowledge base in microbiology is growing rapidly. As it becomes larger, its various subdisciplines, including genetics, taxonomy, and physiology, among others, are diverging so rapidly that it is becoming harder and harder to see the links among them. If we are to under-stand how bacteria "see" the world they live in, we must have ways to bridge these gaps. This book is an attempt to begin this process. Understanding how bacteria evolved and have adapted to their environments underlies the genetics, taxonomy, and physiology of bacteria. To understand the processes of genetic diversity is to appreciate the diversity we see in the microbial world. This is not the end of a journey, but the beginning. The end is far from known. You, the reader, will help to uncover the mysteries that wait along the path. To help you and us on this journey to understanding, we invited a few colleagues to share their expertise in research and teaching of the various topics covered in this book. We extend our thanks to all of them. Finally, we owe a huge thank you to Greg Payne of ASM Press. Without him pushing, we might never have put our feet on the road at all.

We hope you enjoy reading the book. Let it stimulate you to continue down the trail of exploring, through your own research, the ways bacteria adapt and evolve and continue their successful life story.

BOB MILLER AND MARTIN DAY
Stillwater, Oklahoma and Cardiff, Wales

INTRACELLULAR MECHANISMS FOR GENERATING DIVERSITY

I

MUTAGENESIS AND DNA REPAIR: THE CONSEQUENCES OF ERROR AND MECHANISMS FOR REMAINING THE SAME

John R. Battista and Ashlee M. Earl

I

The evolution of any species is defined by an accumulation of hereditary changes over time. At the molecular level, these genetic changes (mutations) occur randomly and are passed on to the offspring, increasing genetic variability within the species. Mutagenesis, therefore, plays an obvious and fundamental role in the evolutionary process. Put simply, a species cannot evolve without mutagenesis. Despite this fact, all characterized species maintain efficient mechanisms to prevent mutation, because mutations are a potential threat to an individual cell's viability. Muta-

John R. Battista and Ashlee M. Earl, Department of Biological Sciences, Louisiana State University and A & M College, Baton Rouge, LA 70803.

tional inactivation of an essential protein will result in cell death, and, if this inactivation occurs in a unicellular organism, that individual is lost from the population, reducing the species' genetic variability. In this chapter we will learn about a subset of the processes that contribute to change and stability in the prokaryotic genome (Fig. 1) to consider the necessary balance that exists between the two processes and to discuss circumstances under which a high mutation rate may be necessary for adaptive evolution.

ORIGINS OF CHANGE

Spontaneous mutation rates are remarkably similar among DNA-based microbes. In 1991, Jan Drake made a survey comparing rates derived from phage, bacteria, and fungi and found that the spontaneous mutation frequency was the same for all the species examined, approximately 0.003 mutations per genome per DNA replication, suggesting that these rates were optimized during evolution. As discussed below, the processes that initiate genetic change are inexorable. They are inherent to an organism's genetic material and to the enzymatic machinery that replicates the genetic material. All cells are, therefore, under constant pressure to change, but there are mechanisms that restrain the tendency toward change, enzymatic processes that appear to have evolved to limit mutation and provide

Microbial Evolution: Gene Establishment, Survival, and Exchange
Edited by Robert V. Miller and Martin J. Day, ©2004 ASM Press, Washington, D.C.

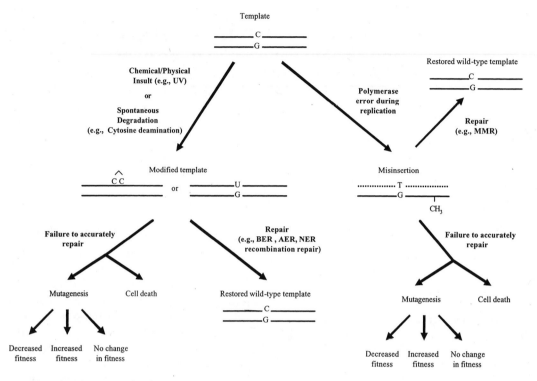

FIGURE 1 Overview of mutagenesis and DNA repair. Native DNA is subjected to endogenous and exogenous sources of DNA damage. The nature of the damage determines the repair pathways activated to counteract these lesions. Failure to accurately restore the wild-type sequence has potentially lethal or mutagenic consequences. MMR, methyl-directed mismatch repair.

genetic stability. Although these processes are efficient, they are not foolproof, and genetic change occurs because not all potentially mutagenic events are identified and corrected. Historically, these "errors" in genome maintenance have been considered the principal source of the measurable spontaneous mutation frequency, and it has been assumed that this genetic variation, accumulating slowly, is the foundation of the evolutionary process.

Friedberg et al. (1995) have published detailed descriptions of mutagenesis and the mechanisms for avoiding mutation, and we do not intend to discuss these topics exhaustively. Instead, we will focus only on the events that give rise to and prevent the formation of base substitution mutations (Fig. 1), because these opposing processes nicely illustrate the balance that has evolved to limit mutagenesis without entirely eliminating a species' ability to accumulate genetic variation.

Errors That Occur during DNA Replication

DNA replication is the most significant source of genome stability in an organism (Kunkel and Bebenek, 2000). Faithful replication ensures accurate hereditary transmission of essential characteristics between parent and offspring, but each time the genome is duplicated, there is the potential for the misincorporation of bases, and, therefore, the potential for heritable changes in the nucleotide sequence. Even the most accurate DNA polymerases make mistakes during DNA synthesis, and these mistakes, in part, govern the spontaneous mutation frequency (Radman et al., 1981). Table 1 illustrates the error rate for sev-

TABLE 1 Reported error rates for purified prokaryotic DNA polymerases

Organism	Error rate (errors/bp)
Thermus aquaticus	2.0×10^{-4}
Thermococcus litoralis	5.0×10^{-5}
Pyrococcus furiosus	2.0×10^{-6}
Thermus flavis	1.0×10^{-4}
E. coli DNA polymerase III holoenzyme	6.0×10^{-7}
E. coli DNA polymerase I Klenow fragment	6.0×10^{-6}

eral prokaryotic DNA polymerases in vitro. Each purified polymerase appears to have a characteristic error rate, and these rates vary by over two orders of magnitude. The differences in error rate reported in Table 1 reflect the enzyme's accuracy in the absence of any accessory proteins that may influence polymerase fidelity. In general, the higher fidelity polymerases, for example, those from *Escherichia coli* and *Pyrococcus furiosus*, have an associated $3' \rightarrow 5'$ exonuclease activity that proofreads the terminal 3′ nucleotide of the growing polymer and removes mispaired bases.

Errors may also arise during DNA replication when the polymerase encounters a modified base in the template strand that misdirects nucleotide incorporation in the nascent strand. A large number of modified bases capable of causing this type of mispairing have been described by Friedberg et al. (1995) and they can arise by one of two mechanisms in vivo: being generated as the primary structure of DNA decays or as heteroatoms of the nucleobases reacting with a wide variety of chemical and physical agents. These chemical alterations of the DNA can lead to (i) heritable mutations by changing the base-pairing properties of the affected nucleobase or (ii) cell death by disrupting normal cellular processes such as DNA replication or transcription. Many base modifications elicit both outcomes.

The replicative DNA polymerase can also contribute to the error rate by incorporating modified or damaged bases found within the nucleotide pool. Free nucleotides are subject

to the same modifications as bases found in the double helix, and some will serve as substrates for the replicative polymerase. If these damaged bases are not recognized and removed before the next round of DNA synthesis, a mutation may be fixed in the resulting daughter cell.

Endogenous Sources of DNA Damage

SPONTANEOUS DECOMPOSITION OF DNA

Given the fundamental role of DNA in information storage and hereditary transmission of that information, it seems counterintuitive that this molecule is relatively unstable chemically. Two processes associated with the decay of DNA in vivo are of particular concern when considering spontaneous mutagenesis: depurination and deamination.

The *N*-glycosyl bond (the base sugar bond) of deoxyribonucleosides is susceptible to hydrolytic attack at 37°C and pH 7.4, releasing the base and leaving an apurinic/apyrimidinic (AP) site in the DNA. Depurination is greatly favored over depyrimidination, cytosine and thymine being released at about 5% the rate of adenine and guanine. Release of an individual base does not appear to be affected by the sequence context in which it is found, and bases in single-stranded DNA (ssDNA) depurinate at a rate only fourfold higher than those found in double-stranded DNA (dsDNA). AP sites are strong blocks to DNA replication and, as such, will lead to cell death unless repaired. When AP sites are incorporated into coliphage, the major effect observed is lethality, presumably because phage replication cannot proceed. By inference, it is assumed that AP sites are also capable of effectively blocking DNA replication when incorporated into the bacterial chromosome.

AP sites are noninstructional lesions in that they offer no base for the replicative polymerase to pair with during DNA replication (Loeb and Preston, 1986). Infrequently, the DNA polymerase will bypass an AP site that has stopped DNA synthesis, and, when this occurs, the polymerase may, in principle, insert any one of the four nucleotides. In *E. coli*, adenine (A) is preferentially inserted opposite the AP site. The AP/A pair is thermodynamically more stable than the pairs that form between an AP site and the other bases. In fact, the T_m of a sequence of B-form DNA changes only slightly when an AP/A pair is substituted for an A/T at a given position in a DNA strand. Since the rate of depurination exceeds that of depyrimidination, the insertion of A opposite an AP site should most frequently result in GC → TA or AT → TA transversion mutations. Lindahl and Nyberg (1972) estimated that for an individual *E. coli* cell grown at 37°C, only 0.5 purines are lost during spontaneous depurination per chromosome per generation. In other words, it is predicted that, on average, the cell will lose only one purine every other generation. However, GC → TA and AT → TA transversions constitute a significant part of the mutational spectrum in *E. coli* growing at 37°C. Schaaper and Dunn (1991) found that transversions make up approximately 25% of the spontaneous mutations detected in this species, GC → TA and AT → TA being half of those transversions.

Deamination occurs when the exocyclic amino groups of deoxyribonucleotides are hydrolyzed. If alkaline solutions of DNA are heated, cytosine, 5-methylcytosine, adenine, and guanine readily deaminate, forming uracil, thymine, hypoxanthine, and xanthine, respectively. The modifications that arise, if not repaired, may give rise to mutations. The newly formed uracil and thymine will pair with adenine during replication, ultimately resulting in a GC → AT transition mutation. Hypoxanthine pairs with cytosine and gives rise to an AT → GC transition. Xanthine, if it forms, pairs with cytosine and does not give rise to a mutation. The rate of deamination is extremely low in neutral solutions at 37°C, however. Under these conditions, the half-life of a cytosine residue in ssDNA is about 200 years, and that number is approximately 150-fold higher than the half-life of cytosine in dsDNA, indicating that the exocyclic amino group is protected by the double helix. 5-Methylcytosine deaminates

at a rate 25% higher than that of cytosine, but deamination of adenine and guanine is only 2% that of cytosine. The rate of base deamination is increased dramatically in alkaline solutions and at elevated temperatures. The problems associated with deamination in *E. coli* are presumably magnified in thermophilic and/or alkalophilic organisms.

Despite the low levels of deamination observed in DNA in solution, there is evidence that deamination contributes significantly to the overall mutation frequency. In *E. coli*, the 3′-cytosine in the sequence CC(A/T)GG is methylated by the *dcm* gene product (Palmer and Marinus, 1994). These methylation sites are hotspots for spontaneous GC → AT transitions. In the *lacI* gene of *E. coli*, methylated cytosines generate transitions at a rate at least 10-fold higher than that observed in unmethylated cytosines. The pattern of hotspots observed in dcm^+ strains disappears in *E. coli* strains that lack the *dcm* methylase. The mutations presumably arise because of the deamination of 5-methylcytosine and the formation of a G/T mismatch that is not repaired prior to DNA replication. Unmethylated cytosines are not hotspots for base substitution mutation even though they are expected to deaminate at a rate only three- to fourfold lower than 5-methylcytosine. Deamination of an unmethylated cytosine produces uracil. Unlike the thymine generated from 5-methylcytosine, uracil is rapidly excised from the DNA by the uracil-*N*-glycosylase expressed in *E. coli*. The hotspot found at *dcm* methylation sites, therefore, reflects differences in the rate at which the lesion formed is repaired.

INTRACELLULAR GENERATION OF DNA-DAMAGING AGENTS

In general, those chemical species capable of covalently modifying DNA are electrophiles that react with nucleophilic centers (primarily heteroatoms) found in the DNA. The modifications that result from the reaction of these compounds with DNA may be innocuous, or the modification may significantly alter the chemical and/or biological properties of the nucleic acid, especially the ability to pair with its complementary base. A number of biological molecules produced during normal metabolism are highly reactive and can modify DNA. In 1993, Lindahl described evidence that two of these molecules, S-adenosylmethionine and hydrogen peroxide, have the potential to contribute to spontaneous mutagenesis.

S-Adenosylmethionine (SAM), a cofactor in many transmethylation reactions, will nonenzymatically methylate DNA in vitro. At 37°C, 7-methylguanine and 3-methyladenine form linearly as a function of time when a neutral solution of DNA is exposed to 10 μM SAM. SAM is a weak electrophile, however, being 1,000- to 3,000 times less reactive than methyl methanesulfonate, a compound frequently used to induce mutagenesis in *E. coli*. Modification of the N7 position in guanine does not directly interfere with DNA metabolism. This lesion is not mutagenic, and it does not block movement of the DNA polymerase. The N7 modification does, however, increase the lability of the *N*-glycosyl bond, and, therefore, the propensity of 7-methylguanine residues to depurinate. As discussed, the resulting AP site is potentially mutagenic. 3-Methyladenine (3mA) is cytotoxic, strongly blocking DNA replication. The potential threat from 3mA in the DNA is underscored by the fact that *E. coli* expresses two glycosylases (Table 2) that target and remove 3mA from DNA. 3mA-DNA glycosylase I selectively excises 3mA, whereas 3mA-DNA glycosylase II has a broader substrate specificity, removing 7-methylguanine and 3-methylguanine in addition to 3mA. SAM may contribute to spontaneous DNA damage and mutagenesis in vivo, since, even though SAM is a relatively weak electrophile, the constant presence of this compound provides an opportunity for nonenzymatic methylation of DNA.

In aerobic organisms, the major reduction product of dioxygen is water, and this process, which is catalyzed by the respiratory chain, can be illustrated as a sequence of four one-electron reductions summarized in the following equation.

$$\overset{1e^-}{\tfrac{1}{2}O_2 \rightarrow} \overset{1e^-}{\tfrac{1}{2}O_2^{\cdot -} \rightarrow} \overset{1e^-}{\tfrac{1}{2}H_2O_2 \rightarrow} \overset{1e^-}{OH^{\cdot} \rightarrow H_2O}$$

dioxygen · · · superoxide · · · hydrogen peroxide · · · hydroxyl radical · · · water

TABLE 2 Characterized DNA glycosylases and their substrates[a]

Enzyme	Substrate	Products
Ura-DNA glycosylase	DNA containing uracil	Uracil + AP site
Hmu-DNA glycosylase	DNA containing hydroxymethyluracil	Hydroxymethyluracil + AP site
Thymine mismatch-DNA glycosylase	DNA containing G-T mispairs	Thymine + AP sites
MutY-DNA glycosylase	DNA containing G-A mispairs	Adenine + AP sites
3mA-DNA glycosylase I	DNA containing 3-methyladenine	3-Methyladenine + AP sites
3mA-DNA glycosylase II	DNA containing 3-methyladenine, 7-methylguanine, or 3-methylguanine	3-Methyladenine + 7-methylguanine, or 3-methylguanine + AP sites
FaPy-DNA glycosylase	DNA containing formamidopyrimidine moieties, 8-hydroxyguanine or hypoxanthine	2, 6-Diamino-4-hydroxy-5-*N*-methylformamidopyrimidine, or hypoxanthine + AP sites
5,6-HTDNA glycosylase (endonuclease III)	DNA containing 5,6-hydrated thymine moieties	5,6-Dihydroxydihydrothymine or 5,6-dihydrothymine + AP sites
PD-DNA glycosylase	DNA-containing pyrimidine dimers	Pyrimidine dimers in DNA with hydrolyzed 5′-glycosyl bonds + AP sites

[a] Data are from Friedberg et al. (1995).

The three intermediates formed ($O_2^{·-}$, H_2O_2, OH·) are potentially genotoxic. Superoxide and H_2O_2 do not directly react with DNA, but they can be converted to OH·. The hydroxyl radical is highly electrophilic, and it appears to be the species that generates most oxidative DNA damage. Aerobic organisms have enzymatic and nonenzymatic defenses that combat all three intermediates. Superoxide dismutase converts two molecules of $O_2^{·-}$ to H_2O_2, and catalase uses H_2O_2 in a bimolecular dismutation to form water and dioxygen. Cells also contain a number of compounds (e.g., glutathione and ascorbic acid) that serve as radical scavengers limiting the potential for hydroxyl radical-induced DNA damage.

Overall, the reduction of dioxygen to water by the respiratory chain is very efficient, but electrons do leak from chain components. In *E. coli*, for example, there is a significant release of electrons from NADH dehydrogenase and ubiquinone with concomitant $O_2^{·-}$ and H_2O_2 production. Gonzales-Flecha and Demple (1997) established that the steady-state concentration of H_2O_2 in *E. coli* is approximately 0.15 μM. Apparently, catalase does not remove all the H_2O_2 as it is formed. Hydrogen peroxide is, therefore, constantly available to serve as a source of OH·. In vitro, Fe^{2+} or Cu^+ ions will serve as an electron donor to H_2O_2.

$$Fe^{2+} + H_2O_2 \rightarrow OH· + OH^- + Fe^{3+}$$
$$Cu^+ + H_2O_2 \rightarrow OH· + OH^- + Cu^{2+}$$

It has been suggested that H_2O_2 reacts with transition metals complexed with DNA in vivo, so obviously, if this occurs, the OH· is formed adjacent to the DNA, increasing the probability of DNA damage.

Hydroxyl radical is capable of abstracting a proton from any position within the DNA, and dsDNA, ssDNA, and free nucleotides are susceptible to attack. In general, OH· reacts preferentially with the nucleobase, but 10 to 20% of the time a sugar proton will be removed. A wide variety of base damage is generated (far more than can be covered here),

and the biological significance of the majority of these lesions has yet to be determined. There are, for example, at least 40 species of cytosine adduct formed when DNA is exposed to OH·-generating agents.

The best-characterized hydroxyl radical-induced DNA lesions are *cis*-thymine glycol and 8-oxoguanosine (8-oxo-dG). The hydroxyl radical is highly reactive toward the 5,6 double bond of thymine residues, producing an abundance of thymine glycol in DNA. Thymine glycol is mutagenic in ssDNA, causing AT → GC transitions, and toxic in dsDNA, where this lesion may block DNA replication. The hydroxyl radical also forms lesions at the C-8 position of purines, and 8-oxo-dG is the predominant species formed in vitro and in vivo. 8-Oxo-dG can mispair with adenine during DNA replication. If this mismatch is not repaired, it leads to a GC → TA transversion mutation. Further, the imidazole ring of the 8-oxopurines is labile, opening to a 5-formamidopyrimidine adduct that strongly blocks DNA replication.

Environmental Influences: Exogenous Sources of DNA Damage

Prokaryotes inhabit an impressive range of environments. Members of the *Archaea* and *Bacteria* have been isolated from hot springs, abyssal hydrothermal vents, highly saline lakes, acidic solfatara fields, alkaline hot springs, soda lakes, and extremely dry and/or cold deserts. A series of references to papers describing these isolations can be found at the end of this chapter (Billi and Potts, 2002; Brock, 1986; L'Harison et al., 1995; Stetter et al., 1990). Since the deleterious physical effects of acidic or alkaline pH, elevated temperature, UV light, ionizing radiation, and desiccation on isolated DNA are well documented (Friedberg et al., 1995), it is not unreasonable to assume that an organism living in an extreme environment would be under greater pressure to change (exhibit a higher spontaneous mutation frequency) than a neutrophilic, mesophilic organism like *E. coli*. The rates of depurination and deamination, for example, increase dramatically with decreasing pH and with increasing temperature, suggesting that organisms found in low-pH and/or high-temperature environments should be more susceptible to this type of DNA damage. Available evidence argues against this assumption, however. By selecting for 5-fluoro-orotic acid-resistant mutants, Jacobs and Grogan (1997) estimated the spontaneous mutation frequency to be approximately 2×10^{-7} mutational events per generation at the *pyrE* and *pyrF* genes of *Sulfolobus acidocaldarius,* a value similar to the spontaneous mutation frequency reported for individual loci in *E. coli.* Thus, the genomic DNA of *S. acidocaldarius* growing at 75°C and pH 3.6 is as genetically stable as that of *E. coli* growing at 37°C and neutral pH. This result indicates that *S. acidocaldarius,* at least the cultured strain used in these studies, has evolved specific mechanisms that limit the lethal and mutagenic effects that high temperature and low pH have on DNA.

Section Summary

- The replicative DNA polymerase may contribute to spontaneous mutagenesis.
- DNA is inherently unstable and this instability may contribute to spontaneous mutagenesis.
- Reactive molecules, generated intracellularly, induce DNA damage that may result in mutation.
- There are many exogenous sources of DNA damage.

MAINTAINING GENOME INTEGRITY

As discussed above, numerous mechanisms can contribute to genetic change by altering the DNA molecule. These modifications, if left uncorrected, can direct the replicative polymerase to rewrite the genome, creating a novel draft of the original sequence. More often than not, however, the resulting changes are neutral or deleterious. When deleterious, the change weakens the offspring so that it cannot compete effectively with its parent. In other words, the cost of unrestricted mutagenesis is high. Given this outcome, it is not surprising that a number of mechanisms for avoiding mutation have evolved. Cellular strategies for maintaining genome integrity can be classified as mechanisms that maintain the sequence of the template and mechanisms that improve the fidelity of the replicative polymerase.

Template Maintenance: Mechanisms of DNA Repair

The effect that a DNA-damaging agent has on the cell is determined by the interaction between the lesions it forms and the cell's replicative DNA polymerase (Fig. 1). When the polymerase encounters DNA damage, it will either continue to synthesize DNA, stop at the site of the lesion unable to bypass the injury, or stop transiently at the site of the lesion before bypassing the adduct. If DNA synthesis continues, the lesion has the potential to permanently alter the genome of that cell's progeny by directing the misincorporation of a base into the daughter strand. If DNA synthesis is completely blocked by the damage, replication of the chromosome cannot be completed, and the cell dies.

By avoiding interactions between the replication fork and DNA damage, cells avoid a circumstance that may result in mutation, and it appears that the "purpose" of most DNA repair proteins is to facilitate this avoidance. An impressive arsenal of proteins is dedicated to maintaining the template for DNA replication, repairing the duplex by (i) directly reversing base damage or (ii) excising the damage and using the undamaged complementary strand to faithfully restore the original sequence. DNA repair proteins are, therefore, most effective when repairs are made prior to a replication fork reaching a lesion, allowing replication to proceed uneventfully.

Despite the efficiency of repair proteins, blocked replication forks will occur. When this happens, cells literally "go around the problem" taking advantage of their ability to exchange homologous DNA between sister duplexes formed during replication. In doing so, replication is reinitiated downstream of the blocking lesion, rescuing the cell from death, while creating a duplex that can be repaired.

DIRECT REVERSION OF DNA DAMAGE

There are two well-characterized mechanisms for DNA damage repair that involve direct reversion of the offending lesion: photoreversal of a specific UV photoproduct and transalkylation of alkylated bases. DNA photolyases are enzymes that catalyze a light-dependent direct monomerization of cyclobutyl pyrimidine dimers, the most abundant lesion formed following the cell's exposure to UV light. They appear to be widely distributed in nature, being found in many, but not all, members of the *Archaea*, *Bacteria*, and *Eukarya*. The *E. coli* protein (the *phr* gene product) is present in low abundance (10 to 20 copies per cell), and it has two associated chromophores, FADH$_2$ (1,5-dihydroflavin adenine dinucleotide) and a reduced, conjugated pterin (5,10-methylene tetrahydrofolyl-polyglutamate). FADH$_2$ acts as a photocatalyst initiating the monomerization of the cyclobutyl pyrimidine dimer. Light at wavelengths between 300 and 500 nm excites the reduced pterin, which in turn transfers energy to FADH$_2$, generating a singlet state. The excited FADH$_2$ transfers an electron to monomerize the pyrimidine dimer.

Alkyltransferases are proteins that recognize and remove alkyl groups from the phosphate backbone (alkylphosphotriesters), the O^6 position of guanine, and the O^4 position of thymine. These lesions form when bacterial cells are exposed to monofunctional alkylating agents such as *N*-methyl-*N*′-nitro-*N*-nitrosoguanidine. Alkylphosphotriesters, which are formed in abundance, are innocuous, but O^6-alkylguanine and O^4-alkylthymine are potentially mutagenic, pairing with thymine and guanine, respectively, during chromosomal DNA synthesis. *E. coli* encodes two alkyltransferases, named O^6-methylguanine DNA methyltransferase I and II (O^6-MGT I and O^6-MGT II), which specifically repair all these lesions. O^6-MGT I (the *ada* gene product) is

an alkyltransferase and a regulatory protein. Normally, cellular levels of O^6-MGT I are extremely low (2 to 3 molecules/cell), but levels rise dramatically (to >3,000 molecules/cell) when the cell is exposed to an alkylating agent. O^6-MGT I catalyzes the direct transfer of offending alkyl groups to one of two cysteine residues within itself. Residue 69 accepts transfer from alkylphosphotriesters, and once this site is alkylated the protein undergoes a conformational change that permits O^6-MGT I to function as a transcriptional activator that increases expression of the *ada* regulon. In other words, alkylation of Cys-69 by the numerous but harmless phosphotriesters induces expression of O^6-MGT I. Residue 321 of O^6-MGT I accepts transfer of alkyl groups from O^6-alkylguanine and O^4-alkylthymine, restoring the normal base-pairing properties to these sites. Even though their substrates overlap, O^6-MGT II (the *ogt* gene product) is distinct from O^6-MGT I. This enzyme is expressed constitutively (30 molecules/cell), does not have a regulatory function, and is roughly half the size of O^6-MGT I. Both enzymes appear to play a role in maintaining genome integrity. *ada ogt* strains have a higher spontaneous mutation frequency than *ada$^+$ ogt* or *ada ogt$^+$* strains.

REMOVAL OF DNA DAMAGE: EXCISION REPAIR

As the name implies, excision-repair processes remove a modified base or nucleotide from the DNA duplex. The proteins that catalyze excision repair recognize perturbations in the DNA molecule and then initiate a sequence of events that expel the damaged base (Sancar, 1996). Removal coincides with new DNA synthesis by a high-fidelity DNA polymerase that restores the original sequence by using the complementary strand as the template for new synthesis. Three types of excision repair have been described. They are base-excision repair, nucleotide-excision repair, and alternative-excision repair (Fig. 2).

Base-excision repair (BER) refers to a repair process that is initiated by a large number of DNA-repair proteins with varying substrate specificity (Wilson et al., 1998). The DNA damage-specific DNA glycosylase (Table 2) recognizes an inappropriate base in dsDNA and catalyzes cleavage of the *N*-glycosyl bond, separating that base from the deoxyribose, and creating an AP site. In *E. coli*, AP sites are very efficiently repaired by an AP endonuclease that binds exclusively to abasic sites, incising the 5′ phosphodiester bond leaving a 3′-OH group and a 5′ abasic deoxyribose phosphate. The 3′-OH terminus will serve as a primer for a DNA polymerase, but the abasic sugar must be removed. This is accomplished by a DNA deoxyribophosphodiesterase (dRPase), a protein that removes this sugar leaving a 5′ phosphate. A DNA polymerase then fills in the resulting one

base pair gap. AP sites formed by spontaneous depurination are repaired in an identical manner.

Some DNA glycosylases exhibit an AP lyase activity that is distinct from the activity of the AP endonuclease. In addition to cleaving the *N*-glycosyl bond of their substrate, these glycosylases cut on the 3′ side of the abasic deoxyribose leaving a 5′-phosphate and a 3′ unsaturated aldehyde terminus that is removed by the activity of an AP endonuclease.

Nucleotide-excision repair (NER) refers to a set of enzymatic reactions initiated by the action of a single DNA-repair protein, the UvrABC endonuclease. Unlike the DNA glycosylases that mediate BER, the UvrABC endonuclease repairs almost all forms of DNA damage, including most of the damage recognized by these gly-

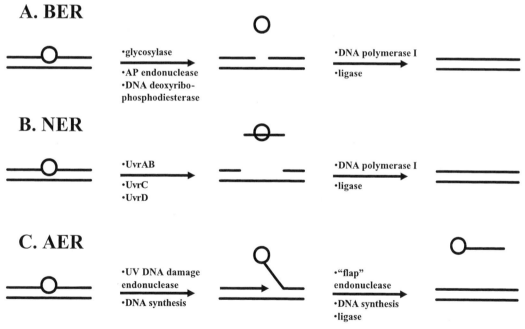

FIGURE 2 Simplified diagrammatic representation of three types of excision repair. (A) Base-excision repair (BER). The modified base is removed by a lesion specific glycosylase and an AP site created by the action of AP endonuclease and DNA deoxyribophosphodiesterase. This 1-bp gap is filled in by DNA polymerase I. (B) Nucleotide-excision repair (NER). A UvrAB complex recognizes distortions in the DNA helix and recruits the UvrC protein to the damaged base. UvrC initiates a bimodal incision. The UvrD helicase releases the damaged fragment leaving a 12- to 13-bp gap. The gap is filled in by DNA polymerase I. (C) Alternative excision repair (AER). A UV DNA damage endonuclease recognizes the lesion and cleaves the phosphate backbone directly 5′ to the lesion. The remaining repair is carried out either by enzymes that unwind the damaged substrate, resulting in a "flap" that is acted on by a flap endonuclease, or by exonucleases that degrade the damaged fragment. DNA synthesis fills the resulting gap.

cosylases. As described by Van Houten (1990), the UvrABC complex does not appear to recognize specific chemical modifications of the bases; instead this endonuclease seems to key on the conformational changes induced by damage to the DNA duplex explaining the broad substrate specificity associated with this protein complex. A dimer of the UvrA protein associates with UvrB, and this complex tracks along the DNA in search of damage-induced deformity. Once DNA damage is identified, the UvrAB complex recruits UvrC to that site. This event triggers an incision on both sides of the damaged base. DNA helicase II (the UvrD protein of *E. coli*) removes the damage-containing fragment. The 12- to 13-bp gap generated is filled in by DNA polymerase I and sealed by DNA ligase.

In *E. coli*, NER operates on transcriptionally active and inactive DNA, but the transcribed strands are preferentially repaired. This preference, which presumably enhances the survival of the cell by promoting the expression of proteins necessary to maintain viability, is mediated by a protein called transcription-repair coupling factor (TRCF). The current model for TRCF action has TRCF specifically interacting with an RNA polymerase installed at the site of a lesion in the DNA, displacing that polymerase. TRCF remains bound to the site of damage, however, recruiting the UvrAB complex to this site, which in turn displaces TRCF, before effecting repair.

The third excision repair system, alternative-excision repair (AER), has been described in a limited number of microorganisms. Repairs are catalyzed by a UV DNA damage endonuclease that binds to a wide spectrum of DNA lesions, including UV-induced photoproducts, apurinic/apyrimidinic sites, and base-base mismatches. This protein introduces a nick immediately 5′ to a lesion, leaving a 5′-phosphate and a 3′-OH. The nicks become the focus of other repair proteins, which first digest the strand containing the damage and then fill in the resulting gapped heteroduplex. In *Schizosaccharomyces pombe* and *Deinococcus radiodurans*, AER-proficient cells are fully resistant to UV light, even in the absence of NER.

HOMOLOGOUS RECOMBINATION AND DAUGHTER STRAND GAP REPAIR

In *E. coli*, the replication of a DNA strand containing a lesion that blocks the DNA polymerase is fatal unless the polymerase can bypass the adduct, and one strategy the cell uses to circumvent the lesion requires homologous recombination. By an unknown mechanism, replication restarts downstream from blocking lesions leaving a single-strand gap in the daughter strand containing the adduct. This gap is filled using the sister duplex by homologous recombination. In brief, there is an exchange of isologous strands (strands from each sister with the same orientation) between the sister DNA molecules (Fig. 3). The daughter strand whose synthesis is blocked by the lesion is displaced so that it anneals to its complement in the sister duplex, and the isolog of the blocked strand anneals to its complement (the strand containing the lesion). Through a process known as branch migration, an exchange of strands, which can cover several thousand base pairs, occurs. This exchange of homologous strands has two effects: (i) the invading DNA strand primes DNA synthesis downstream of the lesion that has blocked replication, permitting the gap that formed to be filled, and (ii) the lesion becomes part of a DNA duplex and is now subject to excision-repair proteins.

Twenty-two *E. coli* proteins have been identified as being involved in homologous recombination. Of these, the RecA protein, which facilitates homologous pairing and strand exchange, is best studied (Roca and Cox, 1997). Without RecA, *E. coli* strains are extremely sensitive to many types of DNA-damaging agents, emphasizing the critical role that homologous recombination plays in DNA repair in *E. coli*. In the presence of ATP, RecA binds ssDNA, forming a nucleoprotein filament extending in the 5′ → 3′ direction as it covers the DNA. This filament has the ability to pair with naked DNA from a homologous duplex.

Improving Polymerase Fidelity

The numbers presented in Table 1 do not necessarily reflect the fidelity of DNA replication

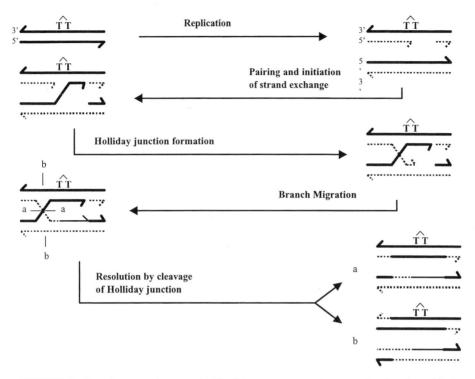

FIGURE 3 Daughter-strand gap repair. The lesion, represented here as a thymine dimer, blocks DNA replication in one strand of the duplex. Replication proceeds in the sister strand. DNA replication is reinitiated downstream of the lesion creating a single-strand gap. This gap is filled by a multistep process involving RecA-mediated strand exchange. The intermediates formed are identical with those observed during homologous recombination. The thick black lines represent the parental strands including the damaged strand. The dotted black lines represent daughter-strand DNA synthesized after damage. Branch migration followed by incision at either a or b resolves the structure, yielding two intact DNA molecules. The dimer can now be corrected by excision repair. (Adapted from Friedberg et al., 1995.)

in vivo. The overall accuracy of cellular DNA replication is determined by a number of factors, including: the complementarity of base pairing, the polymerase's ability to distinguish between appropriately and inappropriately paired bases (the polymerase's ability to proofread), and the presence of a mismatch-correction system that functions post-replication by identifying mismatched bases and removing the misincorporated base from the daughter strand. For *E. coli*, the error frequency during DNA replication has been estimated to be 10^{-10} mutations per base pair per generation. The mutation frequency increases dramatically when any of the factors involved in keeping this rate low are inactivated. For

example, genetic inactivation of the proofreading subunit of DNA polymerase III (the replicative polymerase of *E. coli*) results in a mutator phenotype, and the spontaneous mutation frequency increases by about 100-fold when compared with the wild-type rate (Schaaper and Radman, 1989).

E. coli also encodes a protein that removes 8-oxo-dGTP, a potentially mutagenic form of deoxyguanosine from the nucleotide pool. 8-Oxo-dGTP is readily incorporated into DNA by DNA polymerase III and can mispair with adenine creating the potential for an AT →CG transversion mutation. The MutT protein is a phosphohydrolase that converts 8-oxo-dGTP to the monophosphate, which cannot be added to the growing DNA chain, thus decreasing DNA polymerase's opportunity to contribute to mutagenesis.

Mismatch Repair

As indicated above, mismatched bases arise either through base misincorporation during DNA synthesis or as the result of hydrolytic deamination of cytosine. In *E. coli*, there are several well characterized systems that deal with mismatched bases (Harfe and Jinks-Robertson, 2000). Long-patch repair (also referred to as methyl-directed mismatch repair in *E. coli*) corrects mismatches that arise during semiconservative DNA synthesis (Fig. 4), selectively removing the mismatched base from the newly formed strand. The DNA in wild-type strains of *E. coli* is methylated at the N^6 position of adenine in GATC sequences, and prior to replication both strands of DNA are methylated. The duplex formed immediately following DNA synthesis, however, is hemimethylated. The rate at which GATC sites are methylated in the new strand is slower than the rate of DNA synthesis, creating a transient difference in the state of methylation between the parental and daughter strands. Long-patch repair takes advantage of this difference. The MutS protein detects the mismatch and recruits MutL and MutH to the site. The interaction of these three proteins is believed to facilitate a bending of the DNA that brings the nearest GATC sequence into contact with this protein complex. MutH is an endonuclease that nicks the unmethylated DNA strand at the GATC site. Thus, the state of methylation determines the specificity of long-patch repair. DNA helicase II unwinds the helix, and the segment of newly synthesized DNA between the GATC site and the mismatch is removed by an exonuclease. The GATC site may be 3′ or 5′ to the mismatch, exonucleases with different directionality being responsible for the removal of the unwanted sequence. In *E. coli*, when the nick is made 5′ to the mismatch, either exonuclease VII or the RecJ protein catalyzes the digestion, whereas exonuclease I removes the segment created by a nick made 3′ to the mismatch. The resulting gap, which may be several thousand base pairs long, is filled in by DNA polymerase III.

The mechanism of short-patch repair is quite different from that of long-patch repair, even though the MutS and MutL proteins are also required for this type of mismatch repair. The short-patch system functions on fully methylated DNA, recognizing either G/T or G/U mismatches, removing the thymine or uracil from this pair, and leaving an AP site. The AP site is corrected by the mechanism discussed previously and the single-nucleotide gap filled in by DNA polymerase I. This type of mismatch will appear in fully methylated DNA following deamination of 5-methylcytosine or cytosine, and they are not substrates for long-patch mismatch repair, which can only function effectively on a hemimethylated substrate. The short-patch system, therefore, appears to deal exclusively with spontaneous DNA damage brought about by deamination.

Section Summary

- Cells express a variety of DNA-repair proteins that limit mutation.
- DNA-repair proteins accurately correct DNA damage by either reversing the chemical modification or excising the altered base.
- When DNA damage blocks the replicative polymerase, this potentially fatal event can be circumvented by exchange of homologous strands between sister DNA molecules.
- The accuracy of the replicative polymerase is increased by a proofreading subunit that removes misincorporated bases during DNA synthesis.
- A mismatch correction system functioning after replication improves the accuracy of DNA replication by two orders of magnitude.

ACCELERATING THE PACE OF EVOLUTION AS NEEDED

For many years bacterial evolution was thought of as a process whereby a population of cells

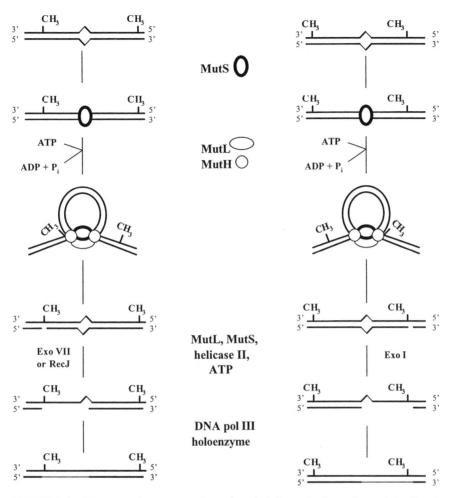

FIGURE 4 Diagrammatic representation of methyl-directed mismatch repair in *E. coli*. Black lines represent newly synthesized DNA duplex. Dotted black lines represent synthesis by DNA polymerase III after mismatch correction. The DNA-damage surveillance protein, MutS, recruits MutL and MutH to the site of the mismatch. In the presence of ATP these proteins catalyze the loop formation, which brings the nearest methylated GATC site in contact with the endonuclease MutH. MutH discriminates between template and newly synthesized daughter strands and nicks the strand opposite the methylated site. The nicked DNA is then acted on by the helicase, UvrD, and directional exonucleases that unwind and degrade the unmethylated daughter strand. Resynthesis is carried out by DNA polymerase III. (Adapted from Friedberg et al., 1995.)

would, over time, accumulate a sufficient number of genetic variants so that, on a shift in environmental conditions, at least one of these variants would contain a suitable combination of mutations to allow it to compete successfully in the new environment. The selected strain would then take over as the progenitor of the resulting population and the evolutionary process would

be repeated in preparation for the next environmental change. However, given the fidelity of the replicative polymerase and the myriad of DNA-damage surveillance and repair enzymes described, it is difficult to imagine that there would be sufficient genetic diversity already present within individuals in the population to achieve this feat, especially if the environmental change were abrupt and catastrophic for the majority of the population. The rates of mutation established under laboratory conditions are sufficiently low to suggest that the adaptive genetic diversity within a given population may be quite small. Assuming 0.003 mutations per genome per DNA replication, and that only 1 in 10,000 mutations are beneficial, an adaptive mutation is expected to arise in a single step within a population at a frequency of approximately 10^{-7}. If only one adaptive change per cell is necessary to overcome the adverse conditions, then a sufficiently large population (>10^7 cells) should survive a strongly selective event. It is, however, unlikely that modifying a single locus will be sufficient to protect a cell from a catastrophic environmental change. The frequency of cells carrying two adaptive mutations simultaneously is the product of the frequency at which the individual mutations appear. If each arose at a frequency of 10^{-7}, the expected frequency of both mutations appearing in a single cell would be 10^{-14}, excluding the possibility that a strain with both mutations will be found in a reasonably sized population. Assuming that a stationary-phase culture of E. coli contains 10^8 cells/ml of growth medium, a culture with 10^{14} cells would occupy 1,000 liters of culture medium. As odd as it may sound, the processes of faithful DNA replication and the mechanisms of repair that limit the deleterious effects of mutation may themselves be detrimental to the population's survival because they limit the cell's capacity to adapt quickly when necessary.

Despite these theoretical concerns, it is apparent that bacterial populations do adapt rapidly while undergoing selection, suggesting that cells have the ability to generate more genetic diversity than is predicted from the spontaneous mutation frequency. Radman (2001) has described two mechanisms that seem to account for the rapid evolution observed in bacterial populations. The first involves the fortuitous inactivation of part of the cellular machinery that mediates the faithful replication of DNA. The resulting mutator strain rapidly generates genetic variation that is subject to selection. The second mechanism requires the synthesis of a specialized group of low-fidelity DNA polymerases that in the appropriate context introduce genetic variation. In bacteria, these polymerases are inducible in response to environmental stresses that introduce DNA damage, suggesting they are part of a process that facilitates long-term survival of species encountering these stresses.

The Mutator Phenotype

Mutators are cells that exhibit a higher mutation rate than normal cells, because they lack one or more of the systems that facilitate high-fidelity DNA replication. In E. coli, for example, inactivation of the mutH, mutL, or mutS gene products of the mismatch-repair system increases the forward mutation rate by as much as 200-fold relative to the wild-type organism (Schaaper and Dunn, 1991). Mutators are, therefore, a potential source of genetic variation within a population. It is assumed that a subpopulation of mutators is always present in bacterial cultures, arising as a consequence of random spontaneous mutagenesis. In the absence of selection, the mutator subpopulation does not compete well with nonmutators. Recall that adaptive mutations only arise at a frequency of about 10^{-4}. Under strong selective pressure, however, a mutator will be more competitive because the best opportunity for generating an adaptive mutation that ameliorates the negative effects of the selective pressure resides within the subpopulation of mutators. It is well documented that, under selection, bacteria exhibiting a mutator phenotype have a selective advantage over nonmutator strains and that there is an enrichment of mutator strains during selection (Mao et al., 1997; Miller et al., 1999). Presumably this enrichment occurs because the mutator carries an adaptive mutation that allows it to survive as the rest of the population is lost. Even though the mutator produces deleterious mutations, the cost of these mutations is not as onerous, in the short term, as might be assumed intuitively. The mutator's progeny survive only because they carry the beneficial adaptation. After the selection is removed, the mutational load created in the mutator will lower overall fitness, and the number of mutators found within the population is reduced.

Specialized DNA Polymerases

When *E. coli* cultures are exposed to DNA-damaging agents that create lesions capable of inhibiting DNA replication, there is a rapid increase in the transcription of approximately 30 loci in what has been labeled the SOS response. Many of the genes transcribed encode proteins involved in DNA repair, including UvrA, UvrB, and RecA. It is believed that blocked replication forks generate a signal that modifies the conformation of the RecA protein, which is constitutively expressed at low levels. Modified RecA, also known as activated RecA, can interact with the LexA protein, this interaction resulting in the proteolytic cleavage of LexA. LexA is a transcriptional repressor that can no longer bind to its operator when cleaved. The genes induced during the SOS response are under LexA control. As the DNA damage that initiated the response is removed, the level of activated RecA drops. LexA is no longer cleaved, and transcription from the SOS genes stops when intact LexA binds to its operator.

The SOS response is associated with a transient increase in mutation frequency, referred to as SOS mutagenesis. Evelyn Witkin (1976) noted that this increase was eliminated in some bacterial mutants, indicating that cellular proteins facilitate SOS mutagenesis. Among the genes regulated by LexA in *E. coli* are three loci that encode members of a distinctive class of DNA polymerase known as Y-family polymerases (Friedberg et al., 2002). They are DNA polymerase II, DNA polymerase IV, and DNA polymerase V. These polymerases perform a highly specialized function in that they rescue stalled replication forks by a process known as translesion synthesis (TLS). As the name implies, TLS is the extension of the DNA strand through a lesion. These polymerases minimize cell death by alleviating blocks to cell replication; this is believed to be their principal function.

These specialized DNA polymerases have two unusual properties that account for SOS mutagenesis. (i) They can insert the "correct" base opposite a blocking lesion. For example, preparations of DNA polymerase V, the *umuDC* gene product, will carry out TLS on UV-induced thymine dimers, inserting the appropriate complementary nucleotides in the newly synthesized strand. This fidelity does not, however, necessarily extend to all lesions the polymerase may encounter. DNA polymerase V frequently inserts the wrong nucleotide opposite the less abundant UV-induced dimer, the [6-4] photoproduct. (ii) They lack the $3' \rightarrow 5'$ exonucleolytic proofreading subunit characteristic of high-fidelity polymerases. As a consequence, these polymerases exhibit error rates as much as two orders of magnitude higher than a replicative polymerase, when replicating on normal undamaged templates. Mutations can, therefore, arise specifically at the sites of DNA damage or be untargeted, appearing anywhere one of these specialized DNA polymerases replicates undamaged DNA. Since these polymerases appear only during the SOS response, their ability to elevate the mutation frequency is only observed during this time. SOS mutagenesis provides the stressed cell with additional genetic variation at a time when it is most needed.

Section Summary

- The accuracy of DNA replication and repair prevents the individual cell from succumbing to the high mutational load that would exist in the absence of this accuracy.
- The accuracy of DNA replication and repair limits genetic diversity within a population.
- Populations rely on mutators, cells that exhibit a higher mutation rate than normal cells, because they lack one or more of the systems that facilitate high-fidelity DNA replication, to provide the genetic variability needed for rapid evolution.
- Some cells experiencing environmental stress synthesize a family of error-prone DNA polymerases that transiently in-

crease the mutation frequency, providing a means of increasing genetic variability directly in response to the stress.

- These error-prone polymerases may be increasing the number of mutators within the population.

ACHIEVING BALANCE

It is axiomatic that modification of DNA has the potential to alter the informational content of an organism's genome. Intrinsic and extrinsic processes are constantly altering the DNA molecule in vivo, creating the possibility for genetic change. In a population well adapted to its environment these changes are to be avoided, as evidence indicates that only 1 in 10,000 mutations gives rise to an adaptive or beneficial mutation. Unrestricted mutagenesis places a cell growing under "normal" conditions at a tremendous disadvantage, the high-mutation load decreasing the fitness of this cell. DNA-damage repair appears to have evolved to limit mutation and prevent the negative outcomes of DNA modification. The DNA-repair processes provide balance. The cell avoids most mutation, but continues to generate, at a low level, genetic variation. It is as if nature is playing a percentage and taking the safe bet. In the vast majority of cases, preventing mutation does not alter the cell's fitness relative to the rest of the population. The cell sacrifices genetic variation in deference to genetic stability.

By limiting genetic variation, however, a population puts itself at risk. If that population is exposed to an abrupt, unpredictable environmental change that threatens its survival, it needs to adapt rapidly to that threat. This selective force will begin to reshape the population, allowing the most fit members to out-compete all others. Eventually genetic variants incapable of long-term survival are lost. Under these circumstances, the success of the population depends on its genetic variability, or, perhaps more appropriately, on its ability to generate genetic variability. Cells need a means to circumvent the restraints on mutagenesis and provide a source of genetic variability that might allow the population to survive through the

uniqueness of some of its members. Recent experiments described by Radman (2001) suggest that there are two systems for increasing genetic variability in *E. coli*. (i) The fate of a population depends on a subpopulation of mutators, cells with genetic defects that dramatically increase their spontaneous mutation frequency. Mutators presumably arise as a random mutation and are part of the genetic variation found in any population. During abrupt environmental change, the mutators have a selective advantage because they are more likely to generate adaptive mutations. The mutators survive and propagate rescuing the at-risk population. Eventually, the fitness of the mutators decreases because of their high-mutational load, and there is a return to a nonmutator phenotype. (ii) Under sustained stress, cells will contribute to the level of mutagenesis by synthesizing specialized error-prone polymerases through a well characterized and tightly regulated stress response called the SOS response. These polymerases copy undamaged DNA with very low fidelity and cause a substantial but transient increase in mutation frequency. The increase in mutation may be adaptive, but it also may serve to generate additional mutators within the population. Thus, as an environment becomes hostile, mutators could be created in direct response to the stress.

QUESTIONS

1. The journal *Science* published a study [R. J. Cano and M. K. Boracki, May 19, 1995, **268**(5213):1060–1064] that claimed that bacterial spores could be revived and cultured from the abdomens of bees preserved for 25 to 40 million years in amber. This report was very controversial; critics argued that no organism could persist for this period of time in a dormant state. Based on the content of this chapter, what arguments did these critics use to support their position?

2. Assume that a strain of *E. coli* evolved that exhibited no evidence of spontaneous mutagenesis. How do you think this strain would fare over evolutionary time, if (a) the culture was kept in a chemostat, and (b) the culture was released into a natural environment?

3. In his movie *Unbreakable*, M. Night Shyamalan told the story of a human being who has a supernatural ability to resist being harmed when confronted with situations that kill other humans. Could a bacterium be engineered that completely resists the lethal effects of DNA-damaging agents?

4. Very little is known about how hyperthermophiles (e.g., *Pyrococcus* spp.) protect their genome from the increases in deamination and depurination observed when purified DNA is exposed to these temperatures. Generate two hypotheses to explain this phenomenon.

5. Human immunodeficiency virus (HIV) is a ssRNA virus that on entering a cell is converted to dsDNA by the HIV's reverse transcriptase, before it is integrated into the host genome. Progeny viruses are produced by the host cell's RNA polymerase. Clinical treatment of HIV is complicated by the fact that HIV's reverse transcriptase and the RNA polymerase lack a proofreading subunit. Based on the content of this chapter, explain the previous sentence and develop an explanation for why the most effective therapies for treating HIV utilize multiple chemotherapeutics that target different proteins that the virus expresses.

REFERENCES

Billi, D., and M. Potts. 2002. Life and death of dried prokaryotes. *Res. Microbiol.* **153**(1):7–12.

Brock, T. D. 1986. *Thermophilic Microorganisms and Life at High Temperatures.* Springer-Verlag, New York, N.Y.

Friedberg, E. C., R. Wagner, and M. Radman. 2002. Specialized DNA polymerases, cellular survival, and the genesis of mutations. *Science* **296**:1627–1630.

Gonzalez-Flecha, B., and B. Demple. 1997. Homeostatic regulation of intracellular hydrogen peroxide concentration in aerobically growing *Escherichia coli. J. Bacteriol.* **179**:382–388.

Harfe, B. D., and S. Jinks-Robertson. 2000. DNA mismatch repair and genetic instability. *Annu. Rev. Genet.* **34**:359–399.

Jacobs, K. L., and D. W. Grogan. 1997. Rates of spontaneous mutation in an archaeon from geothermal environments. *J. Bacteriol.* **179**:3298–3303.

Kunkel, T. A., and K. Bebenek. 2000. DNA replication fidelity. *Annu. Rev. Biochem.* **69**:497–529.

L'Harison, S., A. L. Reyenbach, P. Glenat, D. Prieur, and C. Jeanthon. 1995. Hot subterranean biosphere in a continental oil reservoir. *Nature* **377**:223–224.

Loeb, L. A., and B. D. Preston. 1986. Mutagenesis by apurinic/apyrimidinic sites. *Annu. Rev. Genet.* **20**:201–230.

Lyndahl, T., and B. Nyberg. 1972. Rate of depurination of native deoxyribonucleic acid in neutral solution. *Biochemistry* **11**:3610–3618.

Mao, E. F., L. Lane, J. Lee, and J. H. Miller. 1997. Proliferation of mutators in a cell population. *J. Bacteriol.* **179**(2):417–422.

Miller, J. H., A. Suthar, J. Tai, A. Yeung, C. Troung, and J. L. Stewart. 1999. Direct selection for mutators in *Escherichia coli. J. Bacteriol.* **181**(5):1576–1584.

Norton, C. F. 1992. Rediscovering the ecology of halobacteria. *ASM News* **58**:363-367.

Palmer, B. R., and M. G. Marinus. 1994. The dam and dcm strains of *Escherichia coli*: a review. *Gene* **143**:1–12.

Radman, M., C. Dohet, M.-F. Bourgingnon, O. P. Doubleday, and P. Lecomte. 1981. High fidelity devices in the reproduction of DNA, p. 431–445. *In* E. Seeberg and K. Kleppe (ed.), *Chromosome Damage and Repair.* Plenum Publishing Company, New York, N.Y.

Roca, A. I., and M. M. Cox. 1997. RecA protein: structure, function, and role in recombinational DNA repair. *Prog. Nucleic Acid Res. Mol. Biol.* **56**:129–223.

Sancar, A. 1996. DNA excision repair. *Annu. Rev. Biochem.* **65**:43–81.

Schaaper, R. M., and R. L. Dunn. 1991. Spontaneous mutation in the *Escherichia coli* lacI gene. *Genetics* **129**:317–326.

Schaaper, R. M., and M. Radman. 1989. The extreme mutator effect of Escherichia coli mutD5 results form saturation of mismatch repair by excessive DNA replication errors. *EMBO J.* **8**(11):3511–3516.

Stetter, K. O., G. Fiala, R. Huber, and A. Segerer. 1990. Hyperthermophilic microorganisms. *FEMS Microbiol. Rev.* **75**:117–124.

Van Houten, B. 1990. Nucleotide excision repair in *Escherichia coli. Microbiol. Rev.* **54**:18–51.

Wilson, D. M., B. P. Engelward, and L. Samson. 1998. Prokaryotic base excision repair, p. 29–64. *In* J. A. Nickoloff and M. F. Hoekstra (ed.), *DNA Damage and Repair*, vol. I. DNA repair in prokaryotes and lower eukaryotes. Humana Press, Totowa, N.J.

FURTHER READING

Coulondre, C., J. H. Miller, P. J. Farabaugh, and W. Gilbert. 1978. Molecular basis of base substitution hotspots in *Escherichia coli*. *Nature* **274:**775–780.

Drake, J. W. 1991. A constant rate of spontaneous mutation in DNA-based microbes. *Proc. Natl. Acad. Sci. USA* **88:**7160–7164.

Friedberg, E. C., G. C. Walker, and W. Siede. 1995. *DNA Repair and Mutagenesis.* ASM Press, Washington, D.C.

Lindahl, T. 1982. DNA repair enzymes. *Annu. Rev. Biochem.* **51:**61–87.

Lindahl, T. 1993. Instability and decay of the primary structure of DNA. *Nature* **362:**709–715.

Miller, J. H. 1996. Spontaneous mutators in bacteria: insights into pathways of mutagenesis and repair. *Annu. Rev. Microbiol.* **50:**625-643.

Radman, M. 2001. Fidelity and infidelity. *Nature* **413:**115.

Samson, L., and J. Cairns. 1977. A new pathway for DNA repair in *Escherichia coli. Nature* **267:**281-283.

Sutton, M. D., B. T. Smith, et al. 2000. The SOS response: recent insights into umuDC-dependent mutagenesis and DNA damage tolerance. *Annu. Rev. Genet.* **34:**479–497.

Taddei, F., I. Matic, et al. 1997. To be a mutator, or how pathogenic and commensal bacteria can evolve rapidly. *Trends Microbiol.* **5:**427-428. (Discussion, 428-429.)

Witkin, E. M. 1976. Ultraviolet mutagenesis and inducible DNA repair in *Escherichia coli. Bacteriol. Rev.* **40:**869–907.

RecA-DEPENDENT MECHANISMS FOR THE GENERATION OF GENETIC DIVERSITY

Steven J. Sandler and Klaus Nüsslein

2

Recombination and mutagenesis are processes that create new combinations of genes. These genes can be existing within the genome, or they can be newly imported into the cell by mechanisms of horizontal gene transfer. Recombination between regions of DNA with similar sequences is termed homologous, legitimate, or RecA-dependent recombination, because it requires, among other proteins, a functional single-strand exchange protein, RecA. Current models are consistent with the notion that recombination is substrate limited and occurs more often than previously thought. Even nonhomologous DNA can be integrated into a recipient chromosome if this DNA is flanked by homologous regions. Although it is now thought that homologous recombination evolved to maintain genomic integrity by repairing double-stranded breaks in DNA and fixing stalled DNA-replication forks, homologous recombination also has an important role in the creation of genetic diversity. Although the RecA protein is perhaps the most important contributor to recombination-mediated roles in the cell, it also contributes to the regulation of the SOS DNA damage-inducible response and mutagenesis. While the roles of RecA and other proteins in homologous recombination will be stressed in this chapter, RecA's other roles will also be discussed, as these also contribute to increases in genetic diversity.

The process of homologous recombination and the creation of genetic diversity are often discussed together because these two processes go hand in hand. For instance, in 1965, Clark and Margulies used the inability to inherit genetic markers as their assay to discover the first mutants of the *recA* gene. Soon after, in 1966, Howard-Flanders and Theriot found that deficiencies in recombination also led to deficiencies in the repair of DNA. This suggested that these three processes were closely related. Now, thirty-seven years later, it is recognized that another function of recombination in the cell is the repair of broken replication forks. While it is arguable that the evolutionarily

Steven J. Sandler and Klaus Nüsslein, Department of Microbiology, University of Massachusetts, Amherst, MA 01003.

Microbial Evolution: Gene Establishment, Survival, and Exchange
Edited by Robert V. Miller and Martin J. Day, ©2004 ASM Press, Washington, D.C.

important function of recombination in the cell lies in the repair of replication forks because it is a function that occurs during every cell cycle, the creation of genetic diversity by homologous recombination also remains an important factor in evolution. In this chapter, we will discuss how homologous recombination occurs at the molecular level and how this can lead to an increase in genetic diversity. To focus this chapter, we will summarize the current understanding of *recA*'s crucial and essential role in homologous recombination, its other roles in the regulation of damage-inducible DNA-repair pathways, and its participation in mutagenic events that can lead to increases in genetic diversity. Since much research on this topic has been done with the *Escherichia coli* recombination system, we will use the *E. coli* system as a paradigm. Other systems will be included as space allows. Many good reviews on homologous recombination have been written. References to several of these can be found at the end of this chapter.

RecA IN RECOMBINATION

Recombination: An Amazing Process

When thinking about recombination, one of the first questions to ask is how often does recombination occur? Much research over the years has led to the notion that recombination in the cell is substrate limited. Even so, if the cell is presented with a recombination substrate, how often will it recombine? This has been measured for the type of recombination that takes place during conjugation. A typical frequency of a recipient cell inheriting a selective marker after conjugation by recombination in the laboratory is 30% or 30 per 100 donors. The frequency of inheriting selective markers by recombination after P1 transduction is about 1% to 2% (of the bacteria that receive a particular marker). This may not seem like a lot of recombination or it may seem that the process of recombination is not very efficient because it is not 90% or 100%. However, given some of the following considerations and a detailed explanation of the

mechanism (see below), we believe that recombination is actually highly efficient and it is amazing that it occurs as often as it does. To begin with, the recombination frequencies mentioned above are underestimates because they refer to the production of colonies that have the selected phenotype. Presumably, much recombination is likely to be occurring in the other cells that do not acquire the selected phenotype. Second, one has to remember that substrates that are introduced into the cell must be compared with every segment of DNA in the recipient cell to find the homologous sequence. If one were to look for a 1,000-bp sequence in a nonoverlapping search pattern for a typical prokaryotic genome of 4,000,000 bp, one would have only 4,000 chances to find the proper sequence. This approach is likely to only work 1 in 1,000 times since a sequence could only be considered correct if every base is aligned. If an alignment is off by one base, it will look like a completely different sequence. Third, this search for the correct sequence takes place in a very small volume where the DNA is compacted on the order of 1,000-fold. Compounding this problem is the observation that the DNA to be searched is not naked, but is covered with proteins that regulate genes, transcribe the DNA, and maintain the shape of the nucleoid and its superhelicity. Lastly, only when the correct DNA is found, is recombination attempted. The process itself requires the accurate cutting and religation of strands of DNA. If this is not accomplished precisely, the organism may lose genetic information or its chromosome may be in a tangle from which it might not be able to recover. Given these considerations, it is amazing that recombination occurs successfully as often as it does.

Recombination in cells may occur even more often than previously thought. In the examples above, recombination was initiated by the presentation of substrates to recipient cells by conjugation or transduction. Do cells that are not in the process of bringing in new DNA from the outside recombine? If so, what are the DNA substrates? Many prokaryotic cells can have more than one copy of their chromosome during the later stages of their cell cycle. This is easily thought of as sister chromosomes that are made as the products of semiconservative DNA replication. Can these two circular chromosomes then recombine? In 1977, Konrad constructed a strain of *E. coli* that had two *lac* operons with nonoverlapping deletions in the *lacZ* gene, such that the cell was phenotypically Lac⁻ but could recombine these two operons such that, in theory, $lacZ^+$ genes could be formed. It was shown that these cells could give rise to Lac⁺ cells, and this was dependent on standard recombination functions (see below) and was shown to occur by recombination between two sister chromosomes. Additionally, hyper-rec mutants could be isolated in these strains. Some of these mutants included mutations in *polA* or *lig* (DNA ligase). These gene products help to heal Okazaki fragments that are formed on the lagging strand during DNA replication. Failure for this processing to occur efficiently leads to a cell with more nicks and gaps. Therefore, nicks and gaps in the DNA are recombinogenic. More recently, it has been shown that recombination between sister chromosomes occurs in about 15% of the cells in a simple log-phased culture. It is thought that the substrates for these recombination events are broken replication forks.

The take-home lesson for this section is that recombination occurs often in cells whether or not DNA is introduced that would allow the creation of genetic diversity. If the recombination enzymes and substrates are present, the enzymes will recombine any DNA that is available with great accuracy, precision, and efficiency.

Substrates for Recombination

What does a recombination substrate look like? This is very important because the form of the DNA substrate will dictate the genetic requirements for recombination. Its form, vis-à-vis standard duplex B form, will also distinguish itself from the vast excess of nonrecombinational DNA substrate in cells. Basically, there are two broad categories of recombinational DNA substrates (Fig. 1). One type of substrate is a linear DNA duplex with an available end (as opposed to a teleomeric sequence found at the ends of eukaryotic chromosomes). This type of substrate is generated during the processes of conjugation and transduction. During conjugation, a single strand of DNA is passed from one cell to another (5′ end first). This single-stranded DNA (ssDNA) is then partially replicated to yield a linear duplex of DNA with a double-strand end. It is thought that recombination

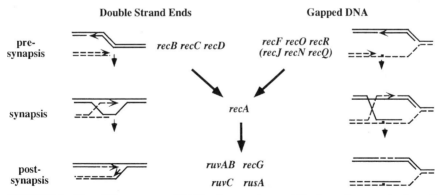

FIGURE 1 Two different types of DNA substrates that could be formed as a result of a replication fork running into a nick (formation of a double-strand end) or a noncoding lesion (square on the DNA) to form a gap. The gene products and the particular stage of recombination at which they are predicted to function are indicated.

with these types of substrates begins by processing the DNA at the ends of the DNA molecules. A second category or type of DNA substrate is the middle of DNA molecules. The actual sites of the DNA where recombination is initiated may be at sites containing nicks and gaps that may have formed during the replication of the single strand to the double strand.

It is important to realize, however, that understanding the DNA substrates is only half the battle. One must also know what recombination enzymes are available to work on the DNA. Figuring out just what recombination enzymes are available is sometimes confusing because the cell has many enzymes that can metabolize DNA and participate in recombination. These can include topoisomerases, endo- and/or exonucleases, helicases, and DNA polymerases. These enzymes may work to change one type of substrate into another. For instance, two close gaps in the DNA on opposite strands could lead to a double-strand break through the action of a helicase or exonuclease. What was the middle of a DNA molecule is now an end.

A useful analogy to recombination in the cell is a wood-working shop. There are many ways in which one can begin to build a piece of furniture and many tools to use. While the raw material with which one is initially presented may make one method perhaps easier or more efficient or lead to a different or better outcome, it is certain that once started, one is committed to a particular plan or pathway. To accomplish this plan, one must have the proper tools to do it. In the next section we will discuss the main tool of recombination: the RecA protein.

A Brief History of RecA

Mutations in the *recA* gene of *E. coli* K-12 were first discovered by Clark and Margulies in 1965. In 1970, the identity of an *E. coli* protein was sought whose expression was highly induced after DNA damage. This protein, infamously called Protein X, was found to be the RecA protein. Further work showed that

RecA, besides having an enzymatic role in recombination and DNA repair, was also a regulatory protein governing the SOS response. The purification and characterization of the RecA protein has shown that it has ATPase and strand-transfer activities that are important for its role in recombination. In addition, it acts as a co-protease, helping to increase the rate of autoproteolysis of LexA, UmuD, and λ cI proteins. This latter activity accounts for its role in the regulation of the SOS response, SOS mutagenesis, and prophage induction, respectively.

The hallmark of the RecA protein is its ability to bind to ssDNA and form a protein nucleic acid helical filament (Fig. 2). It is thought that the helical filament is crucial for all its activities. Thus, RecA acts as a polymer in its roles in the cell, not as a monomer. It is possible, however, that there may be some roles for RecA in the cell yet to be defined that require only the monomer. The filament is formed by the monomers in a head-to-tail arrangement (Fig. 3).

Recombination involves the interactions of RecA with two molecules of DNA. For RecA to perform its task, it must somehow bind the two molecules and bring them into exact alignment. While the actual physical details of the interactions are still unknown, it is thought that the ssDNA interacts with RecA initially to create the RecA-ssDNA helical filament and that the second DNA (double-stranded DNA [dsDNA]) interacts with the RecA-ssDNA helical filament by binding in its major groove (Fig. 2). It is also thought that, when the RecA coprotease activity is manifested by LexA binding the protein-DNA helical filament, this site may overlap or preclude the binding of the second DNA molecule. Competition between LexA and a second molecule of DNA (ssDNA or dsDNA) for binding to the RecA-DNA helical filament has been demonstrated.

Homologs of RecA have been found in almost every organism where they have been sought except one, the pea-aphid endosymbiont *Buchnera aphidicola*. The reason for this

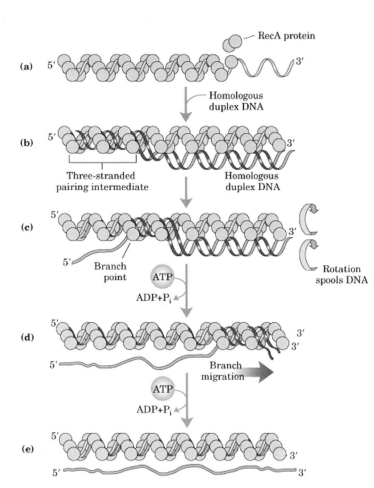

FIGURE 2 A three-stranded reaction. (a) RecA initially binds the ssDNA, creating a protein DNA helical filament. (b) A duplex of DNA then interacts in the major groove of the filament. (c) One strand is then exchanged for the other. (d and e) Completion of the reaction. (This figure is reprinted from Nelson and Cox [2000] with permission of the authors.)

oddity is not yet clear, but it may have to do with this organism's close symbiosis with its host. There are two main families of RecA-like proteins: the bacterial RecA family and the Rad51-like family. Sequence identities between bacterial RecA proteins from different genera of bacteria are usually on the order of more than 65% identity. The RAD51-like family is more diverse. It includes both eukaryotic RAD51 and DMC1 and the archaeal RadA subfamilies. The identity within each subfamily is about 50%, and the identity between subfamilies is about 40%. Although the overall identity between the bacterial RecA and the Rad51-like families is low, on the order of 20%, it is hypothesized that they share a very similar core structure. A physical

difference between the two main families relative to this core structure is that the RAD51-like family has an amino-terminal extension, and the bacterial RecA family has a carboxy-terminal extension. It is thought that these extensions occupy the same space in the respective 3-D structure, and serve the same function. However, this remains to be shown. The protein-DNA filaments and enzymatic activities of the RecA and RAD51-like proteins are very similar, but not identical.

How many *recA*-like genes does an organism have or need? Although most eukaryotic organisms have multiple *recA*-like genes, most prokaryotic organisms have a single *recA* gene. Examples of the RAD51-like proteins in humans include RAD51, RAD51B, RAD

FIGURE 3 Model of a filament of RecA at the molecular level. Twenty-four monomers of the RecA crystal structure have been assembled in a filament. One monomer is shown in darker shading. (This figure is reprinted from Nelson and Cox [2000] with permission of the authors.)

51C, RAD51D, XRCC2, and DMC1. The unraveling of the roles of these many proteins has just begun. In prokaryotic organisms, *Myxococcus xanthus* is the only organism known to date to have two *recA* genes. These two genes are differentially expressed. Another feature of the diversity is seen in the actual *recA* gene structure. The RecA protein of *Mycobacterium tuberculosis* has an intein. This intein codes for an endonuclease that is similar to other intron homing endonucleases. Additionally, a self-splicing Group I intron has been identified in the *recA* gene of *Bacillus anthracis*.

The X-ray crystal structure of the *E. coli* RecA protein bound with ADP has been solved. The structure suggests that the protein can be divided into amino-terminal, central core, and carboxy-terminal domains. The protein as it is found in the crystal forms a helical filament (without DNA) with an open central core (Fig. 3). Several regions of the protein are unordered in the crystal. These include part of the amino- and carboxy-terminal sequences, and two sequences in the middle of the protein called Loop 1 and Loop 2. These latter two structures are each about 12 to 15 amino acids in length and are proposed to bind the molecules of DNA. Modeling suggests that Loop 1 and Loop 2 are in or near the central cavity of the helix in the crystal structure. When all currently known *recA* sequences are compared, computer programs identify many more highly conserved residues on the surface of the protein facing the DNA than on the

exterior surface of the filament (unpublished result). It has been hypothesized that this outer side may be a place for other proteins to interact with the RecA-helical filament. Thus, these outer-surface amino acid residues could evolve at a higher rate and be specific for proteins in each species. It is possible that these proteins could then, in turn, modulate RecA activities.

There is practically no information of where in the cell RecA is located. There is thought to be on the order of 5,000 to 10,000 molecules of RecA per cell. A priori, one would imagine that, in cells that replicate their DNA at replication factories, RecA may be stored close by to be readily available for repair of collapsed replication forks. Alternately, recombination, such as the type that occurs after conjugation, can occur anywhere on the chromosome. Therefore, one might also think that RecA may be generally available to bathe the nucleoid. There has been at least one report of RecA found in the membrane of a naturally competent bacterium (*Bacillus subtilis*) and that RecA is induced during competence of a second bacterium. Hence, a third view might be that some RecA protein is stored in the membranes and becomes associated with the ssDNA as it comes into the cell. Lastly, storage structures for RecA in the cytoplasm have been proposed. Electron micrographs of cells that overproduce RecA show array-like structures that contain some RecA protein. The physiological importance of these structures remains to be determined. Thus, the location of RecA in the cell is varied and may be important for its different roles.

Mechanistic Overview of Recombination: The Three Stages of Recombination

It is most convenient to think of recombination as a three-stage process (Fig. 1). The three stages are called presynapsis, synapsis, and postsynapsis. We will use these stages to discuss the processing of the DNA substrates and the enzymes that act on them. Presynapsis describes the tailoring of the DNA substrates so that they can interact with the RecA protein. Synapsis is the central step in homologous recombination. This step is catalyzed by the RecA protein and involves the search for the homologous sequence followed by the exchange of strands of DNA. The final step is postsynapsis, which usually involves movement and cleavage of the crossover or Holliday junction.

What we score as a recombinant in the laboratory depends on the assay chosen. If the assay targets Thr$^+$ Leu$^+$ recombinants, then we score for the production of a viable cell that is able to grow on a certain type of medium. Other criteria for recombination do not require viability. These include the production of a particular PCR product or a recombinant restriction fragment detected by Southern blotting. Thus, the former assay may require additional steps and gene products that the latter does not. It is conceivable that a cell may be able to do the initial set of reactions in recombination but not be able to do the latter. For instance, one reason why *priA* is required for conjugal and transductional recombination may be that it is necessary to initiate replication at a recombinational intermediate to produce a viable recombinant. Thus, *priA* mutants are likely to do presynapsis, synapsis, and postsynapsis, physically manipulate the DNA as described below, and then fail in a step (initiation of replication just after recombination) to produce viable recombinants. Therefore, when or how one measures or tests for recombination is also critical.

Presynapsis: At the Ends of DNA Molecules

The first stage of recombination usually involves the creation of regions of ssDNA that RecA can bind. Since this can be accomplished by many different groups of enzymes, this has been one of the most confusing areas of recombination for the beginner to learn. Starting with a linear DNA substrate with a flush double-strand end, helicase and exonucleases can liberate regions of ssDNA by either spreading the strands of DNA apart or selectively degrading one strand or another. The archetypal presynaptic enzyme that functions in *E. coli* on the ends of DNA is the RecBCD enzyme. This heterotrimeric enzyme complex has a very potent ATP-dependent helicase/exonuclease activity that can degrade DNA at the rate of 1,000 bp/sec. RecBCD binds to the ends of DNA molecules and then processively melts the strands apart and degrades them. One may reasonably question why such a destructive enzyme would be involved in recombination. The answer seems to be that this exonuclease activity can be attenuated. The RecBCD enzyme will degrade linear duplex DNA until it encounters a properly oriented *chi* site. *chi* is an 8-bp asymmetric sequence that "turns off" the exonuclease activity but allows the helicase activity to continue unwinding the DNA for at least several more kilobases. The RecB subunit also has a special activity in which it helps RecA bind to the ssDNA generated by the helicase. The *chi* sequence at the end of the ssDNA may also aid in RecA binding since it has been shown that RecA binds *chi*-like sequences preferentially. The signal for RecBCD to stop its helicase activity and dissociate from the DNA is not yet known.

Presynapsis: In the Middle of DNA Molecules

As stated above, the other broad type of substrate vis-à-vis the ends of DNA molecules is the middle of DNA molecules. It is hard, however, to conceive of how or where recombination can begin with a perfectly sealed duplex of DNA. Therefore, thinking about recombination that initiates in the middle of DNA molecules has focused on gaps or nicks in the DNA. Those irregularities may be localized there because an Okazaki fragment was not properly processed during DNA replication or the cell may be in the process of base-excision repair or nucleotide-excision repair of an oxidized or otherwise modified base. All these processes may create a nick or small gap in one of the strands. A second presumption about DNA involved with replication or DNA in the middle of chromosomes is that any available ssDNA will be coated with ssDNA-binding proteins (SSB). The presence of SSB in recombination is a double-edged sword since it both aids (in small amounts) and hinders (in large amounts) the binding of RecA. SSB may aid in RecA binding by possibly smoothing out secondary structure in the ssDNA or by protecting the ssDNA from attack by nucleases. SSB hinders recombination in large amounts because it competes with RecA for binding the ssDNA. Therefore, when thinking about the presynaptic steps for initiating recombination in the middle of DNA molecules, an important element is to coordinate the replacement of SSB with RecA. This can be framed in a more formal way as the transition from a replicative intermediate (SSB-coated ssDNA) to a recombinogenic intermediate (RecA-coated ssDNA) (Sandler, 2001). The proteins that are thought to catalyze this transition are the RecF, RecO, and RecR (RecFOR) proteins.

The *recFOR* genes are found in many different genera of bacteria. It is interesting, however, that while, for instance, *E. coli* and *Neisseria gonorrhoeae* have homologs of all three proteins, other bacteria may only have homologs of one or two of them. It is hypothesized that in bacteria where the set is incomplete, other nonhomologous proteins with analogous functions fill in the "gaps." An extreme example of this is in the budding yeast, *Saccharomyces cerevisiae*. In this organism there are no evolutionary homologs of the *recFOR* proteins, yet the cell has *recFOR*-like activities helping the eukaryotic SSB protein (called RPA) off the DNA and helping the eukaryotic RecA (called RAD51) onto the DNA. These *recFOR*-like activities are encoded by the RAD52, RAD55, and RAD57 genes.

The exact function of the RecFOR proteins in *E. coli* remains to determined. In vivo, genetic studies indicate that the three proteins contribute equally to the same process (reviewed in Sandler, 2001). In vitro, however, no reaction has yet been found that requires all three proteins. In brief, two of the more informative in vitro reactions to date

identified are the following. (i) The RecOR proteins help to nucleate and stabilize RecA-DNA helical filaments. (ii) Once these filaments have formed on a gapped DNA substrate and have polymerized from the ssDNA region into the dsDNA, the RecFR proteins stop the polymerization process in the dsDNA region. If you are interested in learning more about the activities of individual proteins and pairs of proteins, check out the 2001 review by Sandler.

Presynapsis: a Thought on Other Pathways and Other Gene Products

Historically, the main pathway of homologous recombination was thought to be the RecBCD pathway. We now know that this view was prejudiced, because the DNA substrate used to assay recombination was a linear substrate. Had a circular plasmid substrate been used, the textbooks might read that the RecF pathway formed the main pathway of homologous recombination in *E. coli*. With this in mind, it is interesting to review additional enzymes that have contributed to our thinking about recombination.

Clark and Margulies (1965) found two types of suppressors to the Rec⁻ UV^S phenotype of *recBC* double mutants. The first was called *sbcA* (for suppressor to *recBC*) and the second was called *sbcB*. These were assumed to turn on the RecE and RecF pathways, respectively. The *sbcA* mutation turned on the *recE* and *recT* gene products (RecET) that have a potent exonuclease activity that is coupled to a strand-annealing activity. RecET activities are almost identical with the activities described for bacteriophage lambda proteins *exo* and *bet*. These proteins degrade a single strand of duplex DNA from an end (5′ to 3′) and liberate the other strand for the single-strand annealing. *sbcB* mutations inactivate Exonuclease I (*xonA*) in a very specific way that is still not fully understood. These mutant *xonA* proteins have lost most of their exonuclease activity. The remaining activity, however, appears to be essential for the suppression because null mutants of *xonA* do not fully suppress both the

Rec$^-$ and UV^S phenotype of the *recBC* double mutants. Full suppression of the *recBC* mutants by *sbcB* also requires a mutation in either *sbcC* or *sbcD*. SbcCD form an enzyme that cleaves hairpin structures in DNA. The RecF pathway was historically important because it helped to identify and characterize a number of other gene products important for recombination. These gene products include RecJ (potent single-strand exonuclease), RecN (no known activity), RecO, RecQ (helicase), and RecR.

The take-home message from this section is that there are many enzymes in the cell that can tailor DNA so that it becomes more readily available for binding by the RecA proteins. These enzymes involve combinations of helicases and exonucleases on one hand and proteins that remodel other proteins bound on the DNA on the other. The exact proteins required for recombination are dictated by the form of the initial DNA substrate and the complement of "recombination" proteins in the cell.

Synapsis: RecA Magic

Synapsis is the central step in homologous recombination. The search for homology is done here. In vitro, this reaction can be catalyzed solely by the RecA protein. As stated above, during presynapsis, the RecA protein binds to the ssDNA after it is liberated from SSB, creating a protein DNA helical filament. The binding occurs in the 5′ to 3′ direction and prefers to start from close to a 3′ end. For this reason, RecA loading can be discontinuous. This may reflect the fact that the 3′ end of the DNA may become available first and, then, as other regions are liberated by the actions of helicases and/or exonucleases, RecA can bind it. It is this polymer, this filament, that is thought to be the active agent searching the duplex for a homologous sequence of DNA. Once a homologous sequence has been found, the RecA protein then catalyzes the strand exchange between the two interacting DNAs. The resulting structure is called a D loop (for displacement loop; Fig. 4), and the activity that RecA uses

to make this structure is called its Strand-Exchange Activity. RecA also has a potent ATPase activity when bound to DNA. The activity is not required for making the D loop and the initial strand exchange reaction in vitro. Mutations that selectively remove the ATPase activity are Rec$^-$ UV^S. Therefore, the ATPase activity is required for in vivo function. It should be pointed out that although models in most textbooks will show *E. coli* RecA binding to ssDNA first and then interacting with a duplex of DNA (since this has been the paradigm for many years; Fig. 2), it has been very recently shown that the RecA protein from *Deinococcus radiodurans* binds to dsDNA first and then ssDNA in vitro. Thus, the biochemistry of strand-exchange proteins may be more varied than previously thought.

What is the structure of the RecA-DNA when it forms the D loop? The invading strand is presumably base paired with its complement in the duplex. The exact structure and composition of this three-stranded structure in the RecA filament is unknown. However, biochemical attempts have been made recently to characterize this structure. How RecA uses the ssDNA to search a duplex for its exact sequence is also unknown. Likewise, there have been recent attempts to characterize this process. Presumably, this is a very fast, efficient reaction with high fidelity since RecA bound to a small oligonucleotide can be used to search the human genome (three billion base pairs) to find the correct sequence and to make a D loop. Although in vitro RecA is able to catalyze these reactions by itself and the filament formation is independent of any other proteins, it is an attractive idea that other proteins may decorate the RecA-DNA filaments and modify or regulate its activity. These proteins could include RecF, RecO, RecR, or SSB.

Postsynapsis: Movement and Cleavage of the Holliday Junction

The D loop gives rise to the formation of a DNA structure called a Holliday junction. In some ways, one can think of a D loop as half of

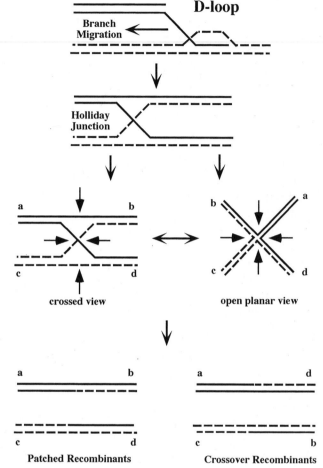

FIGURE 4 Formation of a Holliday junction, its isomerization between the crossover and open-planar forms, and its resolution into patched and crossover recombinants. The arrows pointing to the strands in the Holliday junction are indicative of the strands cleaved by RuvC. Note that pairs of strands are cleaved to form the recombinants. The letters provide orientation for how the arms of the structure are rotated in space. The double arrows indicate isomerization.

a Holliday junction (Fig. 4). The Holliday junction is the so-called crossover intermediate that occurs during recombination. The D loop gives rise to a full Holliday junction when the initial invading strand branch migrates from a region of three strands of DNA to a region of four strands of DNA. Authors take many liberties when drawing recombinational intermediates that may be confusing to the uninitiated reader. First, many authors draw the two interacting duplexes with a Holliday junction between them with a sizable distance between the duplexes of several duplex diameters for clarity's sake (Fig. 1). In reality however, the two duplexes are right next to one another. They are able to form the crossover between adjacent nucleotides without any

torsional stress or loss of base pairs between the two interacting duplexes. Second, the two duplexes are often drawn with the two exchange strands closest to one another when the two duplexes lie side by side. However, only identical strands (same sequence and polarity) can be exchanged, top with top or bottom with bottom. It is convenient (conventional) to explain the Holliday junctions in this way to simplify them. This works if one remembers there is no top or bottom strand to DNA, because the two strands of a duplex are wrapped around one another.

During postsynapsis, the crossover can undergo two processes. The first process is called branch migration. This is when the Holliday junction diffuses up and down the

DNA molecule. This process can create a larger region of heteroduplex or strand exchange and, thus, is important in the creation of genetic diversity. In vitro, this reaction occurs spontaneously in either direction. In vivo, however, the direction of branch migration is likely not to be random but determined by the cell. It is catalyzed by a variety of the enzymes including RuvA and RuvB, RecG, and RecA.

The process of RuvABC action has been reviewed elsewhere in detail. In brief, a tetramer of RuvA proteins binds first to the Holliday junction stabilizing it in the open, plane configuration with four "acidic pins." Then, two hexamers of RuvB proteins will load onto a duplex on opposite sides of the Holliday junction. The binding produces an asymmetry in the structure that then determines the direction of branch migration. As will be considered below, the RuvC protein cleaves the Holliday junction.

The RecG protein can also catalyze branch migration. The mechanism by which RecG catalyzes branch migration may be much different than the one catalyzed by RuvABC for several reasons. First, RecG catalyzes this reaction presumably as a monomer, and RuvAB, as explained above, involves four RuvA and twelve RuvB proteins. Second, genetic evidence suggests that *recG* mutations are additive with *ruvA, ruvB,* or *ruvC* mutations for recombination and UV sensitivity. This last observation is interesting because it suggests that RecG also has a Holliday junction resolution activity and no Holliday junction resolution activity has yet been described for RecG. This reaction must be efficient in *ruvC* mutants since they are fairly proficient for recombination. Recent work supports the idea that the main role for RecG in the cell may be in catalyzing replication fork reversal (Fig. 5). A Reversed Replication Fork is a structure that may form at broken replication forks where the newly synthesized strands anneal with one another. As this duplex increases in size, the replication fork moves backward and the structure formed is essentially a Holliday junc-

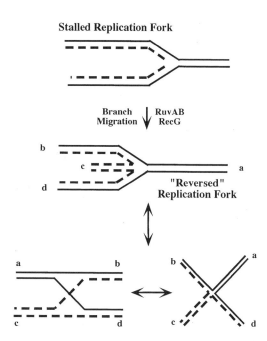

FIGURE 5 Replication fork reversal. The two differently shaded strands indicate parental and newly synthesized DNA. The enzymes that are known to catalyze this reaction are shown at the sides. Although only RecG is mentioned in the text as an enzyme that can perform this reaction, there is evidence that RuvAB also catalyzes this reaction. The letters provide orientation for how the arms of the structure are rotated in space. The double arrows indicate isomerization.

tion. Hence, the two roles proposed for RecG in recombination and replication fork reversal are compatible.

The final step in postsynapsis is the cleavage of the Holliday junction, which dictates whether any recombinants will be formed. The Holliday junction is a symmetrical intermediate. The resolution of the Holliday junction requires cleavage of two strands of DNA. As stated above, the Holliday junction can be thought to exist in two conformations: standard crossover, where the duplexes are side by side, and the open, plane conformation (stabilized by RuvA) (Fig. 4). When a Holliday junction is in the open, plane confirmation, the two strands targeted to be cleaved are opposite to one another. When the Holliday junction is in the crossover conformation, the two strands that are cleaved are either the two that are crossed or the two that are not crossed. Resolution of one pair of strands leads to the patch recombinants, while the resolution of the other pair of strands leads to the displaced or crossover recombinants. The cleavage of the Holliday junction is of critical importance because it dictates whether any recombinants will be formed. For instance, resolution to the patched recombinant would only give rise to a recombinant phenotype if the mutant alleles were contained within the patched region. If, however, resolution allowed the exchange of flanking markers (a crossover), then recombination would be apparent with any markers on either side of the Holliday junction along the length of the chromosome (Fig. 4).

The enzyme that cleaves the Holliday junction is called RuvC. There is little sequence specificity in its recognition of the Holliday junction. Theoretically, there is a 50% chance of the Holliday junction being resolved in one orientation or the other. Historically, this thought pervaded the thinking of researchers in the field. Recent biochemical work, however, has shown that the way in which RuvAB loads on to the DNA will dictate how RuvC cleaves the Holliday junction. In vivo, recombinational repair of replication forks favors formation of the noncrossover resolution of the Holliday junction. This may be because crossovers between sister chromosomes lead to dimer formation and that these dimers then need to be resolved before segregation of the chromosomes can occur at cell division. Although cells come equipped with a special set of genes (*xerC*, *xerD*, and *ftsK*) and sites (*dif*) to accomplish this task, not having to do this would certainly give a selective advantage to the cells. Thus resolution of Holliday junctions at replication forks and other places of recombination elsewhere in the chromosome may respond to different and yet unknown structural clues. A second Holliday junction resolvase has been identified in *E. coli*. This is called *rusA*, and is encoded on a cryptic prophage.

Section Summary

- Recombination events occur frequently.
- RecA plays an essential role in recombination; research on it began in the 1960s.
- RecA's mode of action is complex.
- There are three stages of RecA-dependent recombination.
- Numerous enzymes have to tailor DNA to make it available for RecA.
- The mechanistic resolution of recombinational intermediates is a complex process.

OTHER FUNCTIONS OF RecA

Other Activities of RecA That Can Lead to Increases in Genetic Diversity

RecA has two other activities that can aid in the creation of genetic diversity. The first of these activities is the ability to regulate the SOS response in *E. coli*. The SOS response is a coordinated response to damage in the cells of DNA. More than 30 genes are turned on at the level of transcription as part of this response. The role of these genes is to repair and mutagenize the DNA and to inhibit cell division. LexA is the protein that directly regulates the SOS response. It is a transcriptional repressor that binds in the promoters of SOS-regulated genes. The SOS response is initiated by a RecA binding to ssDNA. This is apparently a convenient measure of the amount of repair occurring in a cell at any given time, because ssDNA is often a by-product of the repair process. RecA binding to ssDNA creates the RecA-DNA helical filament needed for recombinational repair of the DNA. The LexA protein interacts with this filament and increases its rate of autoproteolysis. Thus, the RecA DNA filament is an allosteric effector of the LexA autoproteolysis reaction. In other words, RecA is a coprotease for the LexA cleavage reaction. In addition to increasing the level of expression of certain genes after DNA damage, it has been shown recently that many genes are also turned down as part of the SOS response. Traditionally, this has been more difficult to study, but with the advent of gene chips, it is now possible to see these. Other genes are apparently RecA regulated without being part of the SOS response (e.g., *uspA*). These genes may somehow rely on sensing the process or products of recombination (as in repair of broken replication forks causing dimer formation) for their regulation.

The second role of RecA is in the process of mutagenic bypass of stalled replication forks. More specifically, these replication forks are thought to be stalled at noncoding lesions. These are damaged bases (and not strand breaks) that DNA polymerase III cannot recognize. This process requires RecA's participation in the induction of the SOS response, a coprotease activity on a new protein (in addition to LexA) and a new activity, which has yet to be defined. Using its role as moderator of the SOS response, RecA induces the expression of the *umuC* and *umuD* genes in response to DNA damage. UmuC and a proteolyzed form of UmuD combine to make the DNA polymerase V. DNA polymerase V is an error-prone (mutagenic) DNA polymerase that can replicate (add bases) across certain types of DNA lesions that normally inhibit the progress of DNA polymerase III. DNA polymerase V bypasses the damaged region by inserting bases essentially at random opposite the damaged bases. It is thought that DNA polymerase V inserts only a few bases in a mutagenic manner, because this polymerase is nonprocessive. In a mechanism yet to be discovered, this highly distributive polymerase comes off the template and the processive, normal DNA polymerase III is reloaded. Thus, this aspect of the SOS response has been called SOS mutagenesis.

RecA plays two important roles in this process. The first is similar to that of initiating the SOS response; RecA interacts with UmuD protein (like the LexA protein) as a coprotease. This allows the UmuD protein to be proteolyzed to an active form called UmuD′. A dimer of UmuD′ then interacts with UmuC to form the active form of DNA polymerase V. RecA then plays another role in helping DNA polymerase V replicate across the lesion. Although the exact mechanism of RecA in

this process is unknown, Goodman (2000) has suggested that RecA interacts with UmuD′$_2$C at a template lesion.

The mechanisms above represent what are now considered to be fairly classical ways in which the RecA protein can increase genetic diversity. Recently, a new activity for RecA has been discovered. This is the ability of RecA to bind to single-stranded RNA (ssRNA), and to use this RNA in search for homologous regions of duplex DNA. When it finds a homologous region, it can then perform a strand-exchange reaction (as it would with ssDNA) to create an R loop. This is thought to be very similar to the reaction that it performs above with ssDNA and dsDNA. Such a process may lead to an increase in genetic diversity by providing an additional way to initiate DNA replication that may in turn lead to a mutation. This has been suggested to be one possible explanation for part of the phenomenon called adaptive mutagenesis.

Regulation of Homologous Recombination

The prevailing thought in this chapter so far is that once recombination substrates are presented to a cell, given that the cell has the proper enzymes, recombination will occur. This may be true, but we know of at least three systems that can moderate RecA's activity in homologous recombination and SOS mutagenesis. These include *dinI*, *recX*, and the mismatch repair system. These systems and their effect on RecA are described below.

dinI is a LexA-regulated gene that encodes a small 81-amino-acid protein. When overproduced from a multicopy plasmid, this protein inhibits processing of LexA and UmuD after DNA damage, thus blocking recombination, and cells become very sensitive to DNA damage. Null mutations in *dinI* allow faster processing of UmuD to UmuD′ during an SOS response. For these reasons, it has been suggested that DinI functions by inhibiting the action of RecA. There are currently seemingly conflicting reports on the exact mechanism by which DinI mediates this inhibition. One report suggests that DinI binds to the RecA-DNA helical filament inhibiting RecA activities. The other report suggests that DinI competes with ssDNA for binding to RecA.

recX is a gene that is often found just downstream of the *recA* gene in many organisms. *recX* is often cotranscribed with *recA* and induced during the SOS response. It therefore has been hypothesized to have some function related either to RecA or to recombination. In *N. gonorrhoeae*, RecX appears to help RecA recombine the DNA necessary for antigenic variation. In *Streptomyces lividans* overproduction of RecA is toxic in a *recX* mutant, suggesting that RecX may regulate RecA activity. In *Xanthomonas oryza,* RecX is required for normal RecA levels.

Another modifier of recombination in vivo is the mismatch repair system. This was discovered when researchers tried to measure interspecies recombination. In this study, it was suggested that, during crosses between closely related species, where some mismatches would be generated during homologous recombination, the gene products involved in the methyl-directed mismatch repair system, *mutH*, *mutL*, and *mutS*, serve to inhibit recombination.

Section Summary

- RecA-mediated homologous recombination rearranges genes or parts of genes both within and between replicons.
- RecA limits the divergence of repeated DNA sequences.
- RecA repairs stalled replication forks.
- RecA aids in the repair of double-strand breaks in DNA.

QUESTIONS

1. What are the important properties (catalytic activities) of the protein RecA, and how are these activities used in the case of (a) homologous recombination and (b) recombinational repair?

2. Discuss the activities of RecA that can lead to increases in genetic diversity.

3. How can you separate RecA-dependent recombination into different stages? Does that cover the entire spectrum of RecA's abilities?

4. Describe in your own words how RecA contributes to the fidelity of DNA replication or the repair of potentially mutagenic lesions in DNA.

5. Why should RecA be a filamentous structure rather than a single protein?

REFERENCE

Goodman, M. F. 2000. Coping with replication 'train wrecks' in *Escherichia coli* using PolV, PolII and RecA proteins. *Trends Biochem. Sci.* **25:**189–195.

FURTHER READING

Clark, A. J. 1996. *recA* mutants of *E. coli* K12: a personal turning point. *BioEssays* **18:**767–772.

Clark, A. J., and A. D. Margulies. 1965. Isolation and characterization of recombination deficient mutants of *Escherichia coli* K-12. *Proc. Natl. Acad. Sci. USA* **53:**451–459.

Cox, M. M. 2001. Historical overview: searching for replication help in all of the *rec* places. *Proc. Natl. Acad. Sci. USA* **98:**8173–8180.

Howard-Flanders, P., and L. Theriot. 1966. Mutants of *Escherichia coli* K-12 defective in DNA repair and in genetic recombination. *Genetics* **53:** 1137–1150.

Konrad, E. B. 1977. Method for the isolation of *Escherichia coli* mutants with enhanced recombination between chromosomal duplications. *J. Bacteriol.* **130:**167–172.

Lovett, S. T., and A. J. Clark. 1983. Genetic analysis of regulation of the RecF pathway of recombination in *Escherichia coli* K-12. *J. Bacteriol.* **153:**1471–1478.

Lusetti, S. L., and M. M. Cox. 2002. The bacterial RecA protein and the recombinational DNA repair of stalled replication forks. *Annu. Rev. Biochem.* **71:**71–100.

Masters, M. 1996. Generalized transduction, p. 2421–2441. *In* F. C. Neidhardt (ed.), *Escherichia coli and Salmonella: Cellular and Molecular Biology,* vol. 2. ASM Press, Washington, D.C.

Nelson, D. L., and M. M. Cox. 2000. *Lehninger Principles of Biochemistry.* Worth Publishers, New York.

Sandler, S. J. 2001. Post-replication repair: a new perspective that focuses on the coordination between recombination and DNA replication,

p. 21–42. *In* M. F. Hoekstra and J. A. Nickoloff (ed.), *DNA Damage and Repair: Advances from Phage to Humans*, vol. 3. Humana Press, Totowa, N.J.

Walker, G. 1996. The SOS response of *Escherichia coli*, p. 1100–1116. *In* F. C. Neidhardt, R. Curtiss III, J. L. Ingraham, E. C. C. Lin, K. B. Low, B. Magasanik, W. S. Reznikoff, M. Riley, M. Schaechter, and H. E. Umbarger (ed.), *Escherichia coli and Salmonella, Molecular and Cellular Biology*, 2nd ed., vol. 1. American Society for Microbiology, Washington, D.C.

NONHOMOLOGOUS RECOMBINATION

François Cornet and Michael Chandler

3

Nonhomologous recombination is a generic term covering a wide variety of genetic rearrangements in both prokaryotes and eukaryotes. These types of events range from highly accurate and frequent programmed DNA rearrangements involving sequence-specific recombination to non-sequence-specific events that result in a large range of DNA rearrangements. The essential feature of these types of processes is that they do not require large regions of homology between the recombining DNA partner segments. Many of the genetic elements that promote or undergo nonhomologous recombination have been generally known for some time. However, the general importance of the large-scale genetic changes they induce in shaping bacterial genomes has become evident only during the past decade. This appreciation has resulted from the development of genetic systems for previously intractable bacterial species and from the wealth of data being extracted from the increasing number of available bacterial genome sequences. This chapter focuses on the two principal and best-studied types of nonhomologous recombination event: site-specific recombination and transposition.

A general view of the bacterial genome is that

François Cornet and Michael Chandler, Laboratoire de Microbiologie et de Génétique Moléculaire, 118, route de Narbonne, F-31062 Toulouse Cedex, France.

Microbial Evolution: Gene Establishment, Survival, and Exchange
Edited by Robert V. Miller and Martin J. Day, ©2004 ASM Press, Washington, D.C.

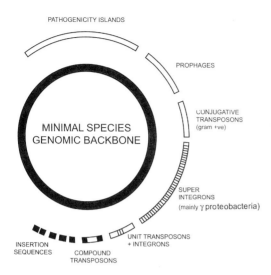

PATHOGENICITY ISLANDS

PROPHAGES

MINIMAL SPECIES
GENOMIC BACKBONE

CONJUGATIVE
TRANSPOSONS
(gram +ve)

SUPER
INTEGRONS
(mainly γ proteobacteria)

INSERTION
SEQUENCES

COMPOUND
TRANSPOSONS

UNIT TRANSPOSONS
+ INTEGRONS

FIGURE 1 Components of the horizontal gene pool.

it is composed of a core "species genomic backbone" decorated with a variety of additional elements (Fig. 1). These include bacteriophages, conjugative transposons, integrons (and integron cassettes), unit transposons, composite transposons, insertion sequences, and extrachromosomal elements such as plasmids. These elements can represent a significant fraction of the genome. For example, in the two *Escherichia coli* strains, K-12 and O157, the minimal common species backbone is approximately 4 Mbp. However, K-12 carries about 0.5 Mbp and O157 carries about 1.3 Mbp of species-specific DNA. The elements can be found in different combinations in different bacterial species or in different isolates of one species and are perceived to form part of an extensive horizontal gene pool. Many bacterial genomes can be considered as having a mosaic structure made up of different assemblages of these modules. Bacterial plasmids, which can themselves also exhibit a complicated modular structure composed of many such elements (e.g., the virulence plasmid, pWR100, resident in *Shigella flexneri*), are thought to play a major role in horizontal transmission.

In addition to horizontal transfer, bacteria also use related nonhomologous recombina-

tion processes as part of their normal lifestyle. These processes include genetic switches, such as the DNA-inversion systems, which can determine surface properties of some bacteria (e.g., the Hin system in *Salmonella*), and switches that determine the recognition of these surfaces by certain bacteriophages (e.g., the Gin and Pin systems of phage Mu and P1, respectively). They also include the machinery involved in assuring separation of chromosome or plasmid dimers.

Nonhomologous recombination reactions can be divided loosely into two major classes: *site-specific recombination* and *transposition*. Site-specific recombination occurs between short specific DNA sequences and leads to reciprocal exchange of DNA adjacent to these recombination sites. A classical example of this type of process is the integration and excision of bacteriophage λ in *E. coli*. Transposition was originally defined as the capacity of a defined genetic entity (a transposon) to insert as discrete DNA segments at many different sites in a genome. Although this definition of transposition remains valid, it has become increasingly clear that the degree of sequence specificity at the insertion site can vary significantly depending on the particular transposon.

Despite the number and diversity of site-specific recombination and transposition systems, a limited number of catalytic mechanisms are used by these elements to accomplish their particular recombination steps. This is reflected in the small number of protein families associated with these systems.

In site-specific recombination, two alternative types of enzyme, the tyrosine (Y) and serine (S) recombinases, are involved. The names derive from the amino acid located in the active site. These recombinases form two distinct families based on their sequence similarities and on their mechanisms of DNA-strand exchange. In most cases, a single recombinase protein is involved for a given element. A major characteristic of the reactions accomplished by these enzymes is the formation of a transient covalent bond with the DNA substrate. No repair or further DNA processing is required. This contrasts with all other recombination systems including transposition and homologous recombination.

The enzymes involved in transposition reactions are known as transposases (Tpases). The majority are in a class generally called DDE transposases because of the presence of three key amino acids, two aspartate residues and a glutamate, as part of the active site. The distinctive characteristic of these enzymes is that the recombination reaction they catalyze does not involve the formation of a covalent enzyme–substrate intermediate. Although DDE enzymes represent most known transposases, three additional types of transposases have been identified. One class is similar to the rolling–circle replicases used by certain bacteriophages and plasmids. Members of this class are known as Y2 transposases since their catalytic mechanism involves two highly conserved tyrosine residues. The other two classes are members of the site-specific recombinase families. However, while these transposases show site specificity for the ends of the donor transposon, they appear to be more flexible than classic site-specific recombinases in the DNA target sequences they recognize and use. They are known as Y-transposases and S-transposases to distinguish them from tyrosine and serine recombinases.

In this chapter we present a short overview of nonhomologous prokaryotic recombination systems and of the mechanisms involved in their respective recombination reactions. We first describe the protein and DNA components involved and consider the various levels of control in these systems. We also try to present the reader with a glimpse of the multiplicity and diversity of these systems. This chapter does not pretend to be an exhaustive review. For more information the reader is referred to a recent American Society for Microbiology publication (*Mobile DNA II*) that describes most of these systems.

SITE-SPECIFIC RECOMBINATION: NUTS AND BOLTS

Site-specific recombination generally occurs within DNA–protein complexes, which include a tetramer of recombinase and two short DNA sequences (of about 30 bp) containing recombinase binding sites. The sites are composed of short inverted repeats flanking a 2-bp (S recombinases) or 6- to 8-bp (Y recombinases) sequence at which DNA-strand exchange takes place (Fig. 2). However, efficient recombination often requires additional sequence elements and proteins (Fig. 3). For the sake of clarity, we will call the minimal DNA sequences "core sequences" and the internal sequence separating the recombinase binding sites "central regions." Supplementary sequences and proteins required in addition to the core sequences and the catalytic tetramer of recombinase are referred to here as "accessory elements." Some of these are indicated for different systems in Fig. 3. Although we use these definitions as general generic terms, it should be noted that other nomenclatures have been used in the literature, depending on the particular recombination system considered. For instance, the central region is called the "coupling sequence" in conjugative transposons and the "overlap region" in bacteriophage λ.

S Recombinases

The serine recombinases form a rather large and homogenous group of related proteins (about 100 members reported to date). The Res proteins of transposons γδ and Tn*3* and the Gin and Hin invertases, from bacteriophage Mu and *Salmonella enterica* serotype Typhimurium, respectively, are the paradigms of this family. In general, serine recombinases are about 200 amino acids in length and consist of an N-terminal catalytic and dimerization domain and a C-terminal domain containing a helix-turn-helix DNA-binding motif (Fig. 2A). The N-terminal domain contains two patches of nearly perfect identity that include amino acids directly involved in catalysis (RS and DRR in Fig. 2A). The first patch contains the catalytic serine residue (S10 in Hin and Res). Three conserved arginine residues, one in the first patch and the two others in the second patch, together with an aspartate residue of the second patch are thought to be directly

A S recombinase driven systems

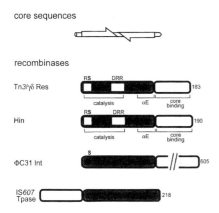

B Y recombinase driven systems

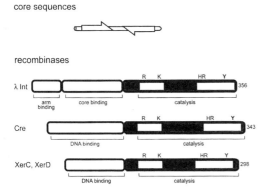

FIGURE 2 Nuts and bolts of site-specific recombination. The core sequences and domain organization of the recombinases are shown for S recombinase-driven systems (A) and Y recombinase-driven systems (B). The core se-quences are represented with their two inverted recombinase recognition sequences symbolized by the inverted arrows. The lengths of the central region separating these binding sites are indicated in base pairs. The major domain organization of the recombinases is symbolized for different members of the two families. These represent the major subclasses within each family. The names of the proteins and the assigned function of the domains are indicated. Catalytic domains are shown in gray with patches of residues important for catalysis indicated by white squares. The number of amino acid residues comprising each protein is also indicated. (A) For resolvase (Res) and invertase (Hin), the catalytic serine (S) that makes covalent bonds with DNA is shown together with additional important residues defining

involved in catalysis by activating the phosphodiester bond (the "scissile" phosphodiester bond), which will be cleaved in the recombining DNA (Fig. 2A).

The S recombinase family may be divided into three major classes based on their functions and sequence similarities: the invertases, of which Hin and Gin are the archetypes; the resolvases, of which the Res proteins of Tn3 and γδ are the archetypes; and the "resolvase/invertases" which include the β-recombinases encoded by certain plasmids from grampositive bacteria. Other members of the family are involved in integration-excision of phages or transposition and differ from these three classes by their unusual length or domain organization (Fig. 2A; see below).

The core sequences recognized by S recombinases are about 26 bp long and contain two binding sites for the recombinases placed as inverted repeats on each side of a 2-bp central region (Fig. 2A). S recombinases bind to the core sequences as dimers with strong cooperativity.

The structure of a dimer of the γδ resolvase bound to a single core sequence has been determined by crystallography. The two C-terminal DNA-binding domains are located on one side of the DNA and connected through a long α-helical linker (αE in Fig. 2A) to the N-terminal domains located on the other side (Fig. 4A). Both the C-terminal domain and the linker participate in DNA binding, whereas residues inferred from mutational analysis to be involved in dimerization and synapsis lie in the N-terminal domain.

the two conserved patches of catalytic residues. These are highly conserved throughout the subfamilies. The position of the α-helix E that links the catalysis and the DNA-binding domain in the Tn3/γδ resolvase is also indicated (see the text). For the less well described φC31 Int and IS607 Tpase, only the global domain organization is shown. (B) The two patches that contain the main catalytic residues are shown with the most important residues indicated. The catalytic tyrosine (Y) is shown in bold. The XerC and XerD recombinases have the same number of residues.

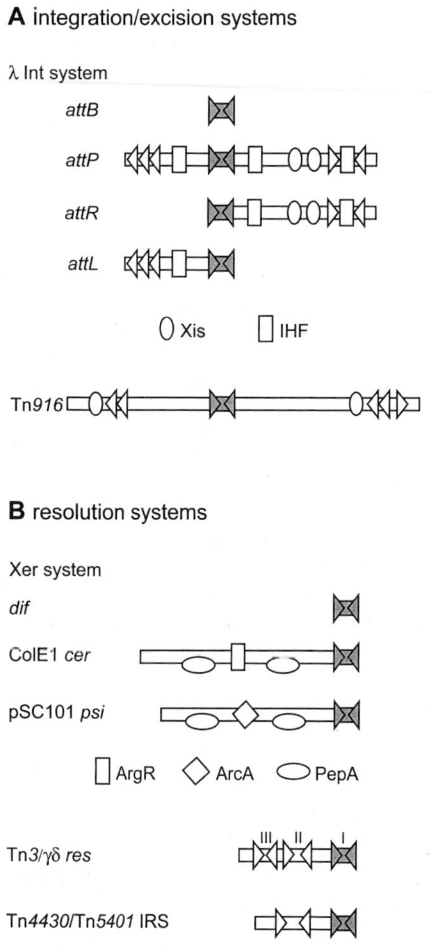

A integration/excision systems

B resolution systems

FIGURE 3 Accessory elements in model systems. Schematic representation of various recombination sites with the position and orientation of binding sites for the different accessory elements. The arrowheads represent recombinase-binding sites and their relative orientations. The core sequences are shown in gray and the accessory binding sites by open symbols. In the Tn*916*, the sequence arrangement shown represents that carried by the excised circular element. I, II, and III are sites I, II, and III. See text for details.

Catalysis by S Recombinases

Cleavage and strand transfer by S enzymes are mediated by successive transesterification reactions that occur within a catalytically competent synaptic complex. The exact architecture of this complex, which may involve additional accessory sequences and diverse host proteins, varies with the specific recombination system. The basic structure includes the two partner core sequences bound by a tetramer of recombinase. The nucleophile hydroxyl groups of the catalytic serine residues initiate recombination by attacking specific phosphodiester bonds (the scissile phosphates) of the DNA backbone (Fig. 4B, step i). Cleavage results in covalent attachment of the proteins to the 5' ends of the cleaved DNA strands and production of 3'-OH free ends (Fig. 4B, step ii). S enzymes cleave the four strands of the paired core sites concomitantly at both edges of the 2-bp central region. This generates an intermediate containing 2 bp staggered DNA ends held together by interactions between the bound recombinase tetramer (Fig. 4B, steps ii and iii). Strand transfer occurs via nucleophilic attacks of the covalent DNA–enzyme bonds by the 3'-hydroxyl OH ends generated by cleavage (Fig. 4B, step iii). The two pairs of strands are exchanged concomitantly. This implies a rotation of 180° between the two pairs of half-sites. It is unclear at present how this is achieved without breaking the protein–protein interface that holds the complex together. The catalytic tetramer is instantly reset to a configuration competent for additional rounds of recombination with further 180° rotations that occur without dissociation of the complex (Fig. 4B, step iv).

Y Recombinases

Tyrosine recombinases, of which bacteriophage λ integrase is the paradigm, form a larger and more heterogeneous family than serine recombinases. The most detailed structural data have been obtained for the Cre/*loxP* system of bacteriophage P1 and the Flp/FRT system of the yeast 2μ plasmid. Y recombinases share a catalytic C-terminal domain, also found in type IB topoisomerases (Fig. 2B). This domain contains two conserved blocks involved in catalysis. One includes the conserved tyrosine residue together with an arginine, a lysine, and a histidine residue and the other carries a second arginine (Fig. 2B). The tyrosine acts as the nucleophile in DNA cleavage, and the other residues are known to be

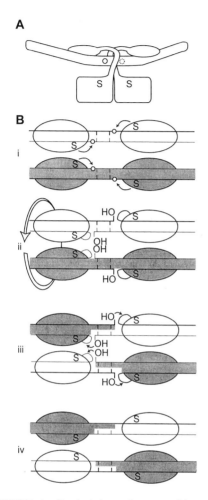

FIGURE 4 Catalysis by serine recombinases. (A) The essential features of recombinase binding to a single core sequence (site I in Fig. 3B). The catalytic domain is shown as a square including the catalytic serine, S, and is connected to the DNA-binding domain that contacts the DNA on the opposite side of the helix (oval), by α-helix E. Open circles indicate the scissile phosphates: one above and the other (dotted line) below the DNA. (B) The catalytic mechanism of S recombinase. (i) The synapse. The two core sequences are shown in a parallel configuration. A tetramer of recombinase is bound at the core sequences. Open circles indicate the scissile phosphates and vertical lines indicate the 2 base pairs of the central region. S indicates the catalytic serine that attacks the scissile phosphodiester bonds (symbolized by the curved arrows). (ii) Cleaved intermediate before rotation. Cleavage generates four covalent DNA-recombinase 5′-phosphoserine bonds and liberates four 3′-OH ends. Strand exchange involves a 180° rotation of the two left-half sites with respect to the two right-half

directly involved in activation of the scissile phosphodiester bonds. The N-terminal regions of these recombinases are rather heterogeneous in length and may contain one of a variety of different types of DNA-binding domain (in some cases two) (Fig. 2B). Y recombinases display low overall conservation, except for their catalytic centers. Some members possess sequence peculiarities related to their functions. For instance, the integrase of phage λ and related recombinases of other phages and conjugative transposons display a long N-terminal domain containing two different DNA-binding domains (Fig. 2B). One of these recognizes the core sequence and the other the accessory (arm) sequences.

Y recombinases bind as dimers with strong cooperativity to core sequences of about 30 bp. These contain two recombinase-binding sites in inverse repetition separated by a 6- to 8-bp central region (Fig. 2B). The structure of several Y recombinases has been solved and, despite their limited homology, they share a common general fold. The Cre (bacteriophage P1) and Flp (yeast 2μ plasmid) recombinases were cocrystallized bound to their core sequences. They form C-shaped clamps with N-terminal domains lying on one side and the C-terminal catalytic domains on the opposite side of the DNA (shown schematically in Fig. 5). The catalytic synaptic complex contains four monomers of recombinase bound to two core sequences and is arranged so that interactions between monomers bound to the same or different core sequences are equivalent (Fig. 5, step i). To achieve this cyclic interaction, a region of each recombinase monomer, generally the extreme C-terminal part, extends and docks into a recipient pocket inside the adjacent monomer.

sites (symbolized by the large circular arrow). (iii) Cleaved intermediate after rotation. After the rotation step, the four 5′-OH ends perform a nucleophilic attack (symbolized by the curved black arrows) on the four 3′-phosphoserine bonds, thereby closing the four DNA nicks and liberating the recombinase monomers from their covalent attachment to the DNA substrate. (iv) Recombination products.

Catalysis by Y Recombinases

Recombination is catalyzed by a tetramer of recombinase bound at the two partner core sequences. The two partner DNA core sites are brought together in an antiparallel configuration. As for the S recombinases, correct synaptic complex formation may require accessory elements. The structure of the basic complex is such that only two opposing monomers at a time are in a configuration competent for cleavage (Fig. 5, step i). The first pair of strands is cleaved in a concerted manner at one edge of the central region by a nucleophilic attack of the hydroxyl group of the conserved tyrosine (Fig. 5, steps i and ii). This results in a 3′-phosphotyrosyl bond with the cleaved strand liberating a corresponding 5′-OH (Fig. 5, step ii). Note that the polarity of attack and cleavage is opposite to that found with S recombinases. The 5′-OH then attacks the phosphotyrosyl bond created during cleavage in the partner core sequence to produce a four-way "Holliday" junction (HJ) intermediate (Fig. 5, step iii). The first cleavage and/or strand exchange allows the reaction to proceed to the second pair of strand exchanges that occurs at the other end of the central region of the core site (Fig. 5, step iv). Exchange of the second pair of strands is thus temporally and spatially separated from the first. This, together with the inverted polarity of cleavage, represents the major differences from S recombinase-driven catalysis.

Reaction Control in Site-Specific Recombination: Adapting Catalysis to Biology

Uncontrolled DNA cleavage and recombination may be detrimental to cells. Site-specific recombination must thus be tightly controlled to recombine only the cognate sites, and only when appropriate. This control requires prevention of cleavage except when the DNA sites are included in a proper (productive) synaptic complex. Such regulation of recombination activity involves both sequence-specific recognition of the sites and a mechanism that senses their location and orientation with respect to each other.

Control of Cleavage: a Requirement for Recombinase Multimers

Structural and functional analyses suggest that assembly of tetramers is essential to obtain a recombinationally active complex. Isolated monomers or dimers of several S or Y recombinases appear inactive, because the attacking nucleophile is too far from the other catalytic residues or from the DNA. This has been observed in structures of the integrases of phages λ and HP1, the XerD recombinase (Y recombinases), and the γδ resolvase (S recombinase). A special case is that of the Flp recombinase, for which two adjacent monomers participate in the formation of the active site of the protein; one monomer provides the attacking tyrosine and the second the activating residues.

Structures with a conformation competent for cleavage have only been observed for the Y recombinases Cre and FLP. In both cases, the structure contained a tetramer recombinase bound to a pair of core sequences. In these cases, two nonadjacent monomers are in active conformation, while the two others are inactive. This arrangement ensures the temporal separation of cleavage of the two pairs of strands (Fig. 5). Thus, in most cases, cleavage is controlled and coordinated to prevent unwanted DNA lesions and occurs only within an appropriate synaptic complex.

Specificity and Orientation of the Core Sequences: the Role of Homology

In the Y recombinases, the primary determinant in the recognition of site orientation is the central region of the core sequences. This region is also an important determinant of the specificity of the recombination partners, especially in systems that may recombine several different pairs of sites (e.g., the Xer system, see "Flexibility in site-specific recombination: example of the Xer system" below). Homology in the central region is required for efficient recombination by Y recombinases. It is believed that during strand exchange (Fig. 5), the attacking 5′-OH extremity anneals with the complementary strand of the partner before ligation. A

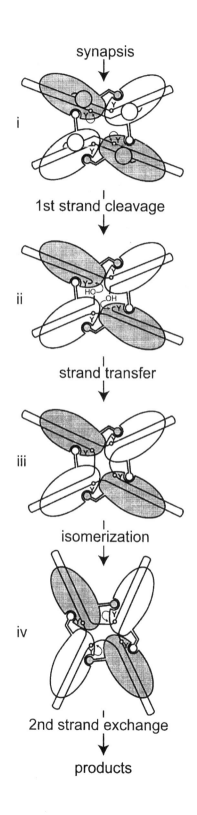

synapsis

↓

1st strand cleavage

↓

strand transfer

↓

isomerization

↓

2nd strand exchange

↓

products

FIGURE 5 Structure of the synaptic complex and catalytic mechanism of Y recombinases. (i) Synapse. The two core sequences are in an antiparallel configuration bound with a recombinase tetramer. The small open circles on the DNA represent the scissile phosphates that border the central region. The large circles in front of the DNA represent the N-terminal DNA-binding domain (not shown in the rest of the figure). A flexible linker connects this domain to the catalytic C-terminal domain (the large ellipse with the catalytic tyrosine residue shown) to form a C-shaped clamp around the DNA. The main interaction domains between the recombinases (the hairpin) extend from the C-terminal domains to dock into specific recipient pockets into the adjacent monomer. The main DNA deformation (the pronounced kink at the uncleaved edge of the central region) is shown. The gray recombinases are in an active conformation for cleavage and perform a nucleophilic attack on the scissile phosphodiester bonds (symbolized by the small curved arrows). (ii) Transfer of the first pair of strands. Cleavage generates two DNA-recombinase 3'-phosphotyrosine bonds (symbolized by the link between the DNA and the catalytic tyrosines of the gray monomers) and liberates two 5'-OH ends. The 5'-OH ends then attack the phosphotyrosine bond of the partner core sequence, thereby liberating the gray recombinase monomers from their covalent bond with the DNA and creating a four-way Holliday junction intermediate. (iii) The Holliday junction intermediate before isomerization. In the four-way junction, generated by exchange of the first pair of strands, the angles formed between the four DNA arms are not equivalent. The interactions between adjacent recombinase monomers are also not equivalent (here two C-terminal extensions are represented kinked and the two others are straight). This Holliday junction intermediate must isomerize to allow exchange of the second pair of strands. (iv) The Holliday junction intermediate after isomerization. Isomerization inverts the small and large angles between the DNA arms. Similarly, the short (here kinked) C-terminal interaction domains extend and the previously extended (here straight) domains contract. This leads to catalytic activation of the previously inactive (white) pair of recombinase monomers and inactivation of the previously active (gray) monomers and allows exchange of the second pair of strands by a mechanism similar to that involved in exchange of the first pair of strands.

notable exception for this requirement for homology is found in Y transposons, which can integrate into a variety of DNA-target sequences (see "Y and S transposases" below).

The 2-bp central region of S recombinase core sequences is usually symmetrical (i.e., identical whatever the orientation of the recombining sites) so that homology in this central region cannot be a determinant of site orientation. Accessory elements are the main determinants of orientation in these systems. However, homology does participate in the reaction and is required for efficient recombination. It is thought that it is involved at a stage following complex assembly and that pairing of these 2 bp after the 180° rotation facilitates the final ligation step (Fig. 4B, step iii). This requirement for homology would therefore tend to preclude successful recombination between nonpartner sites.

Control of the Order of Strand Exchange in Y Recombinase-Driven Systems

In Y recombinase catalysis, exchange of the two pairs of strands is separated in time. A body of functional data obtained from different systems together with structural data of the Cre/loxP system underline the versatile way this control can be achieved. The order of strand exchange is thought to be encoded by the asymmetric conformation of the synaptic complex (Fig. 5, step i). The two monomers that will cleave and exchange the first pair of strands (gray monomers in Fig. 5, step i) extend their C-terminal interaction domains to the monomers bound at the same core sequence, whereas the two "noncleaving" monomers extend this domain to the cleaving monomers bound at the partner core sequence. The conformations of the cleaving and noncleaving recombinase monomers are therefore not equivalent. In addition, the DNA itself exhibits asymmetry since the core sequences are kinked at the noncleaved edge of the central region (Fig. 5, step i).

Minor changes in the core sequences, introduced into the central region or at its immediate boundaries, can lead to inversion of the order of strand exchanges. This has been shown for λInt, Flp, Cre, and XerCD. It is unlikely that this effect is due to direct interactions between the cognate Y recombinases and the central sequences since the enzymes make only minor contacts with these regions. It seems more likely that inversion of the order of strand exchanges is due to changes in intrinsic structural properties (curvature or deformability) of the DNA that may influence the configuration or assembly of the synaptic complex. The order of strand exchange may also be imposed in certain cases by the recombinase itself (e.g., when asymmetry is introduced by composite recombinases such as XerCD acting on the plasmid-borne psi or cer sites; see below) or by accessory elements that can influence the conformation of the synapse.

Once the first pair of strands is exchanged, the Holliday junction intermediate isomerizes so that the two inactive monomers adopt an active conformation to catalyze exchange of the second pair of strands (Fig. 5, steps iii and iv). This isomerization step may also represent an important checkpoint of the reaction. The conformation of the Holliday junction and its ability to isomerize may be influenced by the same factors that influence the conformation of the initial synapse. Depending on the system these can include the sequence of the central region and the intrinsic properties of the recombinase tetramer and accessory elements.

The Nature and Role of Accessory Elements

Accessory elements can play a crucial role in the recombination process. Although they are not directly involved in catalysis, they may influence all steps of the reaction. They do so by different strategies, usually involving the formation of a highly organized nucleoprotein complex. Accessory sequences may be adjacent to the core site (in resolution systems) or distant (in inversion systems) and may be recognized by the recombinase itself (e.g., the Tn3 and γδ resolution systems), by other DNA-binding proteins (e.g., the Xer system, the Hin and Gin inversion systems) or both (e.g., the λInt system). We will briefly describe the best-known examples involving three kinds of rearrangement: programmed insertion into and excision from the host chromosome of a temperate bacteriophage, resolution of replicon fusions, and genetic switches involving inversions.

INTEGRATION-EXCISION OF BACTERIOPHAGE λ

The integration/excision system of bacteriophage λ is a paradigm not only for temperate phage but also for Y recombinase-based systems in general. λDNA is injected into *E. coli* cells as a linear molecule, which then circularizes and, in following the lysogenic pathway, integrates into the *E. coli* chromosome. Integration occurs primarily by recombination between the phage "attachment" site, *attP*, and a unique chromosomal site, *attB* (Fig. 3A). Much more rarely, integration may occur at related secondary sites. *attB* consists only of a core sequence, while *attP* is much more complex and includes binding sites for the accessory proteins, IHF and Fis (two host-encoded proteins that also play an architectural role, determining a precise three-dimensional structure of DNA protein complexes in other DNA-processing reactions), the λ-encoded protein Xis, and the integrase itself, Int, which recognizes its accessory-specific sites (*arm* sites) more efficiently than the core sequences (Fig. 3A). Recombination between *attP* and *attB* leads to integration of the phage genome into the *E. coli* chromosome. This results in a λ prophage flanked by the hybrid *attR* and *attL* sites, whose organization differs significantly due to the asymmetry of the *attP* site (Fig. 3A).

When a lytic cycle is induced, recombination between *attR* and *attL* leads to excision of the prophage and restores the *attP* and *attB* sites. Integrative recombination requires Int and IHF to form a preintegration complex with *attP*. This complex captures the naked *attB* site to form the integrative synaptic complex. Excisive recombination requires Xis in addition to Int and IHF and is stimulated by Fis. *attR-attL* synapsis requires the formation of a higher-order nucleoprotein complex within which Int bridges the two sites via its two DNA-binding specificities to arm and core sites. Thus integration and excision occur via clearly different pathways.

The major determinant of excision control is the highly complex regulation of *int* and *xis* gene expression, but it may also be influenced by the intracellular concentrations of IHF and Fis, which vary significantly, depending on growth conditions.

TOPOLOGICAL FILTERS IN MODEL S RECOMBINASE SYSTEMS

Since, in most site-specific S recombinase systems, the two partner sites are identical, a key issue is then to differentiate the recombination substrates from products. This has been intensively studied in two model S recombinase-driven systems: (i) resolution of cointegrates resulting from transposition of the Tn*3* and γδ transposons and (ii) inversion of specific DNA fragments by the Hin and Gin systems of *S. enterica* serotype Typhimurium and bacteriophage Mu, respectively. In these systems, accessory elements use DNA topology to restrict recombination to a higher-order synaptic complex that is conditioned by the respective position of the sites on DNA.

Transposition of Tn*3*-like elements generates fused donor and target molecules, or cointegrates, in which the donor-target junctions are each separated by a directly repeated copy of the transposable element (see "Catalysis by DDE transposases and the importance of second strand processing," below). Donor and target replicons must subsequently be separated. This process is called resolution. Resolution of Tn*3* cointegrates is catalyzed by the Tn*3* resolvase between *res* sites composed of a core sequence plus a 120-bp adjacent accessory sequence that contains four binding sites for the resolvase arranged as two "core-sequences sequence-like" units (Fig. 6A). Both the correct position and the distance of the accessory sequence with respect to the core sequence are important for recombination. No accessory protein is required in addition to the resolvase. It is thought that 12 monomers of resolvase form a filament-like structure around

A

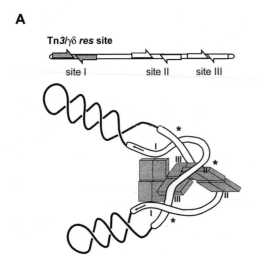

Tn3/γδ *res* site

site I   site II   site III

B

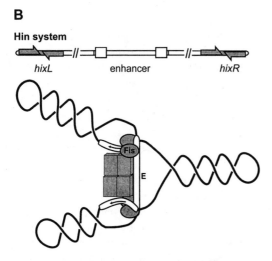

Hin system

hixL   enhancer   hixR

which the two *res* sites wrap in a precise manner (Fig. 6A). In this complex, the two *res* sites cross each other three times to form three negative nodes (*, in Fig. 6A). Wrapping of the *res* sites around the resolvase complex requires negative supercoiling and is highly favored when the two sites are placed in direct repetition on the same plectonemically supercoiled DNA molecule as shown in Fig. 6. Achieving the same synaptic complex between independent DNA molecules or between inverted *res* sites would require the introduction of positive nodes between or inside the DNA molecules, respectively. This is highly disfavored on negatively supercoiled DNA. For this reason, this type of control is referred to as a "topological filter."

The Hin invertase catalyzes inversions between two *hix* sites. *hix* sites consist only of a core sequence. Recombination requires the presence on the same DNA molecule of a 63-bp enhancer sequence located between the *hix* sites. The enhancer carries binding sites for two dimers of the host protein Fis. A DNA loop must be generated between the enhancer and its closest recombination site to promote specific contacts between Fis bound at the enhancer and invertase bound at the *hix* sites (Fig. 6B). The nonspecific DNA-binding protein HU stimulates the reaction by stabilizing this DNA loop. The orientation and distance of the accessory elements with respect to the recombination sites are less important than in the resolution systems. Recombination occurs

FIGURE 6 Topological filters in S-recombinase-driven systems. A linear representation of the recombination sites is shown (top panels) with the arrows representing the binding sites for the recombinases. The core sequences are shown in gray. The enhancer sequence of the Hin system is shown as an open bar edged with the two open squares that represent the binding sites for the Fis protein. The drawing at the bottom of each section represents the organization of the respective synaptic complexes as postulated by recent models (see text). The open curved bars symbolize the recombination sites, and the enclosed small black arrows indicate their relative orientation (in direct repetition in the *res* sites and inverted in the *hix* sites). The black lines represent the rest of the DNA of the closed, negatively supercoiled, circular DNA molecules that carry the recombination sites. The cubes symbolize the recombi-

nases (12 monomers in the Res system and 4 monomers in the Hin system). (A) Synapse between two directly repeated Tn*3 res* sites. The 12 monomers are thought to form a filament-like structure around which the two *res* sites interwrap. The three subsites, each carrying two inversely repeated binding sites for Res, are indicated (sites I are the core sequences). Interwrapping of the two *res* sites around the Res filament constrains three crossings of negative supercoiling (indicated by the stars). (B) Synapse between two inversely repeated *hix* sites. The synapse is thought to be preferentially formed at a branch of negative supercoiling and involves an interaction between the Fis-bound enhancer sequence and the Hin-bound *hix* sites. Two monomers of Fis (the gray ellipses) bind at each edge of the enhancer.

inside a synaptic complex that contains a tet-ramer of invertase bound to the two *hix* sites plus the Fis-bound enhancer (Fig. 6B), requires negative supercoiling of the DNA, and is highly disfavored when recombination sites are in direct repetition or on independent DNA mol-ecules. This synaptic complex is another exam-ple of a topological filter since achieving the same complex between sites in direct repetition or on independent molecules is energetically disfavored on negatively supercoiled DNA.

FLEXIBILITY IN SITE-SPECIFIC RECOMBINATION: EXAMPLE OF THE Xer SYSTEM

The Xer system acts in maintenance of circu-lar chromosome integrity by ensuring the res-olution of chromosome dimers that may form by recombinational repair of DNA lesions. In *E. coli* recombination is accomplished by a pair of Y recombinases, XerC and XerD, which act in concert as a heterotetramer at the *dif* site located in the region of replication termina-tion. Recombination between two sites in a dimer chromosome allows separation of the sister chromosomes. XerC and XerD are found in many bacteria harboring circular chromo-somes and are thought to conform to the Cre model for synapse organization.

XerCD-mediated recombination (Xer recombination) at *dif* is highly controlled and integrated into the *E. coli* cell cycle. It requires FtsK, a component of the division septum, both in vivo and in vitro. It is thought that FtsK plays two distinct roles in XerCD/*dif* recombination. First, it positions and assists synapsis of the *dif* sites at the division septum by tracking chromo-somal DNA toward *dif*. Then it activates recombination by allowing XerD to catalyze the first pair of strand exchanges.

Xer sites homologous to *dif* are also found on plasmids and temperate phages. Plasmid-borne sites are present in numerous medium-to high-copy-number plasmids. They serve to resolve plasmid multimers that arise by recom-bination between sister copies and are delete-rious for plasmid maintenance in bacterial populations. Phage-borne sites have been found in phages infecting *Vibrio cholerae* and other bacteria and serve to integrate the phage genome at resident chromosomal Xer sites. Thus, different Xer sites may coexist in the same cell assuming different roles. Unnec-essary recombination between these sites is prevented mainly by the absence of homology in the central regions of their core sequences (see "Specificity and orientation of the core sequences: the role of homology," above).

Plasmid-borne sites have been studied in detail in two cases: the *cer* site of plasmid ColE1 (and homologous sites of its relatives) and the *psi* site of pSC101. These consist of a conserved 28- to 30-bp core sequence homol-ogous to *dif* and, in addition, adjacent nonho-mologous ~200-bp accessory sequences absent in *dif*. These accessory sequences are targeted by accessory proteins, which are also involved in other diverse cellular functions. This is consistent with the idea that, as for XerCD, they have been appropriated by plas-mids and diverted from their normal cellular roles. Both *psi* and *cer* recombination require PepA, an aminopeptidase. PepA plays a struc-tural role in Xer recombination and does not require its catalytic function for this activity. In addition to PepA, *cer* recombination requires binding of ArgR, the general repressor of the arginine biosynthesis regulon, to a specific sequence in the *cer* accessory region. ArcA plays a role equivalent to ArgR in *psi* recombi-nation. ArcA is the regulator element of the ArcAB two-component system that controls aerobic—anaerobic growth transitions. These accessory elements form higher-order com-plexes with the accessory sequences that impose a control analogous to the topological filter of the Tn*3* and γδ resolution systems. This ensures that Xer recombination resolves rather than creates plasmid multimers.

Section Summary

- There are two major types of site-specific recombinases, the S and Y recombinases.
- These recognize and recombine short spe-cific DNA sequences called core sequences.

- Recombination involves transient covalent linkage between the recombinase and its recombination substrate.
- Recombination is highly regulated and uses accessory DNA sequences that flank the core sites together with additional proteins.
- These create a highly specific DNA conformation that facilitates recombination.

TRANSPOSITION

Transposition occurs within DNA–protein complexes (called transpososomes) composed of two specific transposon sites located at the ends of the element and often representing short imperfect inverted repeats (IRs), bridged by a recombinase, the transposase (Tpase). The site into which transposition will occur, the target site, becomes engaged in the transpososome at one of the several steps involved in its assembly. Several types of Tpases are known. These include the DDE, Y2, Y, and S transposases. They exhibit very different reaction mechanisms.

DDE Transposases

STRUCTURE AND ORGANIZATION

Comparison of the primary sequence of different Tpases, including those from both prokaryote and eukaryote insertion sequences and transposons, has revealed that many carry a highly conserved triad of acidic amino acids known as the DDE motif together with several additional residues (in particular, a K or R located seven amino acids downstream; Fig. 7). These enzymes are also closely related to the integrases (IN) of retroviruses and retrotransposons (not to be confused with phage integrases described above which are Y recombinases) and, albeit less closely, to enzymes such as RnaseH and RuvC. Together with these the DDE Tpases are grouped in the super family of phosphotransferases. Although the spacing and conservation can be somewhat variable, the DDE signature of Tpases is embedded in three relatively well-conserved regions (called N2, N3, and C1) centered, respectively, on the D, D, and E residues (Fig. 7A). Moreover, the

structure of the catalytic domains of several of these enzymes has been determined. These show similar overall topologies and localization of the DDE triad. One essential role of this triad is to coordinate two divalent cations necessary for catalysis. Tpases must recognize and bind the ends of their cognate elements and assemble them into a synaptic complex (the transpososome) prior to chemical catalysis. As a general rule, the DDE Tpases carry DNA-binding functions toward their N-terminal ends and the catalytic domain toward the C-terminal end. They also include various regions necessary for protein multimerization within the transpososome (Fig. 7B).

Like the recombinases, transposases bind to specific DNA sites. These are located at the ends of the cognate element. In the insertion sequences (ISs), they are short imperfect terminal IRs (IRL [left] and IRR [right]), generally carrying a single binding site (Fig. 8A). In general, the sequences required for Tpase binding and those necessary for catalysis can be separated into two small domains of the IRs (Fig. 8B). The two ends are brought together in a synapse known as a transpososome, which includes a dimeric or higher-order transposase multimer. Other transposons may carry more complex patterns of binding sites at their ends (Fig. 8A). By accommodating different binding patterns at each end, such an arrangement can provide a functional distinction between the ends, which may be important either in the assembly or in the activity of the synaptic complex.

CATALYSIS BY DDE TRANSPOSASES AND THE IMPORTANCE OF SECOND-STRAND PROCESSING

DDE transposases catalyze two related reactions at the ends of their cognate transposons. The first is the hydrolysis of the terminal phosphodiester bond of the strand that will be transferred to the target DNA. This occurs by nucleophilic attack using H_2O as the nucleophile, and results in single-strand cleavage of the donor transposon, leaving a $3'$-OH group. In the second step, a transesterification reaction, this $3'$-OH group

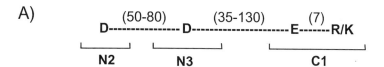

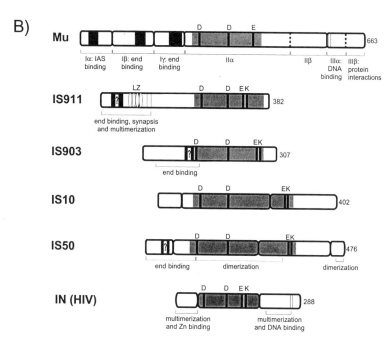

FIGURE 7 Transposase organization. The positions of protease-sensitive sites are delimited by the open boxes. This information is not yet available for IS*903* and IS*911*. Potential or real helix-turn-helix (HTH) motifs are shown in grey boxes. Potential HTH motifs are indicated by "?". The catalytic core is indicated by grey and carries the DDE motif. These residues, together with others referred to in the text, are indicated in upper-case letters above each transposase molecule. LZ indicates the leucine zipper motif with the four repeating heptads observed in the IS*911* transposase. A second region involved in multimerization is also shown slightly downstream. In those cases investigated, the catalytic core is also capable of promoting multimerization. Known functions of the different transposase regions are indicated below. Transposase alignments are centered on the second aspartate residue. The length of each protein in amino acids is indicated at the right. The function of each region, where known, is indicated under the respective proteins.

is used as a nucleophile in the attack of a target phosphodiester bond (Fig. 9). The chemical mechanism is understood in some detail and proceeds by what is called a one-step in-line attack by the nucleophile.

A key event in transposition reactions of this type is the way in which the second DNA strand is processed (Turlan and Chandler,

2000). Cleavage of a single strand followed by strand transfer gives rise to a branched intermediate, which must be processed further (Fig. 9). In principle, a second round of cleavage of the second donor strand would separate the transposon from the donor, leaving it attached to the target molecule. Alternatively, one of the branches can provide a focus for

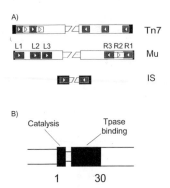

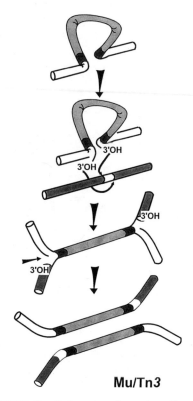

FIGURE 8 Organization of transposon ends. (A) General organization of the ends of different prokaryote transposable elements. Terminal inverted repeats (IRs) are indicated in black. Transposase-binding sites are indicated in grey or white, and their relative orientation is indicated by a small arrow within each box. (B) Functional organization of the ends of a typical insertion sequence. This figure shows an IR that can be divided into a domain required for catalysis and a domain required for transposase recognition. The 1 and 30 represent base pairs to indicate approximately the general length of IS terminal IRs.

Mu/Tn3

FIGURE 9 Cointegrate formation. From top to bottom, this figure shows synapsis of the transposon ends (dark gray); initial cleavage to generate a free 3'-OH at the ends of the transferred strand and a staggered attack on the target DNA; the formation of a branched molecule with a suitable 3'-OH capable of being used as a primer for replication of the element; and the result of replication and repair, which generated two copies of the element, leaving each attached both to donor DNA at one end and target DNA at the other.

assembly of a replication fork and replication of the element would give rise to a cointegrate molecule with a single copy of donor and target replicons separated at each junction by a single copy of the transposon (Fig. 9).

Classic examples of transposable elements that use this type of pathway are bacteriophage Mu and members of the Tn3 family of transposons. The Tn3 family transposons have recruited a second enzyme, a site-specific S recombinase, which catalyzes recombination between the two transposon copies generated by cointegration (see "S recombinases" and "The nature and role of accessory elements," above). This results in physical separation of the donor and target replicons, while it maintains a single transposon copy in each.

Different elements have adopted different strategies to deal with the second strand (Fig. 10). Insertion sequences IS10 and IS50 use the newly exposed 3'-OH to attack the opposite strand, separating the element from flanking DNA and generating a hairpin structure at the transposon tip. This then undergoes a second round of hydrolysis to regenerate the 3'-OH

prior to strand transfer (Fig. 10, left). Tn7 and its relatives have enlisted a second enzyme that resembles a restriction endonuclease rather than a DDE transposase (Fig. 10, center). An alternative solution has been adopted by members of the IS3 family, of which IS911 is a paradigm, and several other families of insertion sequences (Fig. 10, right). In these cases, cleavage at a single end is followed by a one-ended strand-transfer reaction in which the single 3'-OH is used to attack the opposite end of the transposon to generate a single-strand circle. This is resolved, presumably by replication, into a free double-strand transposon circle in

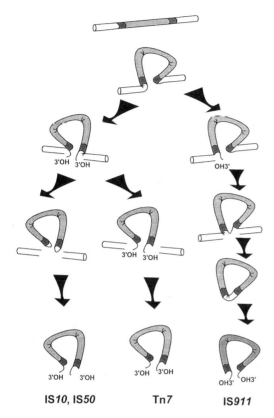

FIGURE 10 Variant pathways in DDE transposase-catalyzed transposition. This figure demonstrates alternative catalytic pathways observed for various transposons encoding DDE transposases. The basic initial enzyme-catalyzed hydrolysis to generate a 3′-OH end on the DNA strand that will be transferred into the target molecule and the final strand transfer step in which the 3′-OH are used as nucleophiles to attack the target phosphodiester bonds are chemically identical. The major differences arise from the manner in which the second nontransferred strand is processed. (Left panels) Initial hydrolysis of the strand to be transferred occurs at both ends to liberate a 3′-OH. This then attacks the opposite strand to generate an intermediate carrying a hairpin structure at each end and separates the transposon from the donor backbone DNA. A second round of hydrolysis, chemically identical to the first, generates an intermediate with two 3′-OH residues at its ends capable of inserting into the target DNA molecule. (Center panels) Initial hydrolysis of the strand to be transferred occurs at both ends to liberate a 3′-OH. The other strand is cleaved by a second non-DDE endonuclease (in the Tn7, this resembles a restriction endonuclease). (Right panels) Initial hydrolysis of the strand to be transferred occurs at only one end to liberate a single 3′-OH. This then attacks the opposite end to generate a single-strand

which the two ends are covalently joined. A subsequent round of single-strand cleavages at each end generates two 3′-OH groups, which can then undergo strand transfer into a suitable target to complete transposition.

TRANSPOSOSOME ASSEMBLY

Transpososome assembly is a central checkpoint in transposition. It has been investigated in detail for only a few systems. In the two best-characterized systems, phage Mu and transposon Tn5, fortuitous cleavage of single transposon ends is avoided by a simple reaction constraint: end cleavage occurs in *trans*. A Tpase molecule bound to one transposon end in the synaptic complex is programmed to cleave the partner end in that complex rather than the end to which it is bound. This, therefore, imposes the formation of a productive synaptic complex before catalysis can occur. In addition, certain transposons (e.g., Tn7) also normally require the presence of a target sequence before any catalysis can take place.

In certain cases, host factors, whose function in the cell is to modify DNA structure and topology, have been implicated at this level in influencing both assembly and the final architecture of the transpososome or of a complex that includes the target site (e.g., the DNA "chaperone" IHF in Tn10).

Assembly of the phage Mu transpososome is probably the best understood, although it is probably not the simplest. It occurs in a series of well-defined steps in which the overall structure exhibits an increasing stability as it evolves to a productive complex. This is presumably because of progressive interwrapping of the various protein and DNA components. The steps of transpososome assembly have been successfully reproduced in vitro. Although the ends of bacteriophage Mu each carry three Tpase-binding sites (Fig. 8), only three (one on the left end, L1, and two on the

bridge between the ends. In a further step, this form gives rise to a double-stranded covalently closed transposon circle, which undergoes subsequent cleavage at each end to regenerate a 3′-OH at each end.

A) Inversion systems

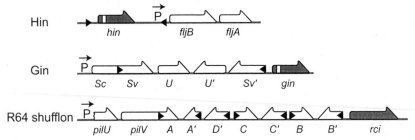

Hin
hin   fljB   fljA

Gin
Sc   Sv   U   U'   Sv'   gin

R64 shufflon
pilU   pilV   A   A'   D'   C   C'   B   B'   rci

B) Integrons

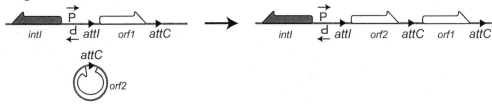

intl   attl   orf1   attC   →   intl   attl   orf2   attC   orf1   attC

attC
orf2

C) Resolution of plasmid multimers

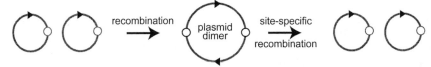

recombination   plasmid dimer   site-specific recombination

D) Resolution of cointegrates

replicative transposition   cointegrate   site-specific recombination

E) Integration/excision systems

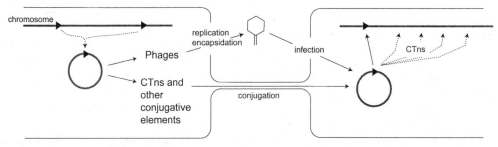

chromosome

Phages

replication encapsidation   infection   CTns

CTns and other conjugative elements   conjugation

FIGURE 11 Systems using site-specific recombination. Black arrowheads symbolize the recombination sites. See text for a description of the different systems. (A and B) Gene-shuffling systems. Arrows represent the various genes with their names indicated below. The genes coding for the respective recombinases are shown in gray and may contain the recombination enhancer sequence (the enclosed white bar in the *hin* and *gin* genes). P is the main promoters with the arrows above indicating the sense of transcription. (C and D) Resolution systems. The open circles symbolize replication origins. The open bars represent the ends of a replicative transposon. (C) Different copies of a circular replicon (in this case, two monomers of a plasmid, left drawing) may be fused by recombina-

right, R1 and R2) are essential for formation of a functional transpososome. The Tpase, MuA, binds as a tetramer to paired Mu ends. Different domains of the monomers are shared in the active transpososome. The additional MuA-binding sites (L2, L3, and R3) presumably increase the efficiency of directing Tpase molecules into the correct positions at the essential sites. Phage Mu has an additional level of control in vivo. Mu Tpase carries two different DNA-binding domains that recognize and bind two sets of specific DNA sequences: the Tpase-binding sites located at the phage ends and additional sites located at approximately 1 kb from the left end (the internal activating sequence; Fig. 11). This sequence increases the efficiency of transpososome formation, presumably, by presenting Tpase molecules in the correct configuration for assembly on the ends. There is some evidence that, as transpososome assembly progresses, DNA at the phage–host junction is slightly unwound, a process that probably renders the scissile phosphodiester bond more susceptible to hydrolysis. A similar requirement for strand opening has been suggested in processing the linear double-strand retrovirus genome prior to integration.

Synaptic complexes have been observed in vitro for several transposable elements such as Tn7, IS10, IS50, and IS911. The structure of one transpososome, that of IS50, has been solved at the atomic level. It is a dimeric symmetrical structure in which two Tpase monomers bridge two antiparallel transposon ends. As predicted, "*trans*" interactions occur between the ends and Tpase. The tip of the ends is unwound and the penultimate residue, a T, on the nontransferred strand was observed to have swung out from the double helix. The initial structure was obtained using precleaved transposon ends carrying a 3′-OH on the transferred strand and a 5′-OH on the nontransferred strand. The structure confirmed the projected interactions of the DDE motif with divalent metal ions, H_2O and the DNA substrate, although only a single metal ion was detected. The presence of a second divalent metal ion predicted to be involved in catalysis required the presence of a 5′-phosphate on the nontransferred strand of the precleaved extremity. This was observed in a second structure and highlighted the role of the second metal ion in coordinating the phosphate group in the uncleaved strand. Subsequent structures have addressed the interactions involved between Tpase and the hairpin transposition intermediate and its cleavage to form the donor end (see "Catalysis by DDE transposase and importance of second strand processing," above, and Fig. 10). Although the position of donor DNA and the target is not yet clear, a molecular modeling approach has provided some insights.

THE TARGET SITE

Transposable elements exhibit a wide range of specificity in choosing a target site. This can vary from highly sequence-specific insertion

---

tion (in this case, a dimer is formed, central drawing). Site-specific recombination resolves these multimeric forms to monomers (right). (D) Replicative transposition (symbolized by the dashed arrow; see also Fig. 9) fuses the DNA molecules carrying the transposon (the black line) and the transposition target site (the dotted circle), respectively (in this case, two plasmids, left drawing). This generates a cointegrate that consists of the fused DNA molecules separated by two directly repeated copies of the transposon (central drawing). Site-specific recombination resolves this cointegrate. This generates a copy of the donor molecule and a target molecule carrying the transposon at the target site. (E) Integration/excision systems. The elements are inserted into a DNA molecule of their host (here the chromosome, left drawing). Excision by site-specific recombination generates a DNA circle that carries one recombination site. The circular element is eventually transferred to another cell. Phage genomes are transferred by infection following encapsidation. These may be injected into the recipient cell as a linear molecule and are subsequently circularized. Conjugative transposons and other conjugative elements are transferred by conjugation (see text). These may be injected into the recipient cell as single-strand DNA molecules and are subsequently replicated. Integration may occur at a specific site (phages and site-specific integrative elements) or at a variety of different sequences whose specificity depends on the system considered (conjugative transposons).

of transposons such as Tn7 through regional specificity determined, for example, by AT and GC content (IS1, IS186) to relatively random insertion by elements such as Tn5. In certain cases there is also some evidence that local DNA conformation, replication, and transcription can also influence the attractiveness of a target molecule.

The exact step in the catalytic process at which the target DNA intervenes probably depends on the element. For Tn7, no reactions take place in the absence of target DNA (Peters and Craig, 2001), whereas for bacteriophage Mu, at least in vitro, target DNA can be introduced into the transpososome at various stages in its assembly. The majority of elements generate duplications of the target sequence at each junction. This is a consequence of the chemistry of the insertion event during which the two liberated 3'-OH ends of the transposon attack each of the DNA strands in the target in a staggered manner (see Fig. 9). This generates short, complementary single-strand sequences of the target at the target–transposon joints. These must be repaired to obtain a productive transposition. In cointegrate formation, this may be accomplished by the replication process, which duplicates the element (Fig. 9). However, in direct insertions (Fig. 10), host-repair enzymes presumably intervene. In both cases, the result is to generate short direct repetitions of target DNA flanking the insertion (see Fig. 9). The length of the duplication is relatively specific for a given element, presumably reflecting the particular architecture of the transpososome–target complex.

Y2 Transposases

Although most transposable elements appear to use DDE-catalyzed reactions for their displacement, some clearly do not fall into this category. Members of the IS91 family of ISs exhibit several unusual characteristics.

The Tpases of these ISs are members of a family of proteins that catalyze rolling circle replication (RCR) in certain bacteriophages and plasmids, and, indeed, the "left end" of IS91 has some resemblance to rolling circle origins of replication. The enzymes carry five conserved

blocks of amino acids, one of which includes a pair of tyrosine residues involved in catalysis. The best-characterized reactions using Y2 enzymes are those involved in phage replication (e.g., φX174; see Novick, 1998). Variations on this basic theme have been proposed to account for rolling-circle transposition (RCT). One difference between RCR and RCT is that, in RCT, strand transfer from donor to target molecule must occur. There are several possible ways in which strand transfer could be integrated into this type of process, although the exact mechanism has yet to be determined at the biochemical level.

The rolling-circle transposition mechanism confers several interesting features on IS91 family members. These elements do not carry terminal inverted repeats nor do they generate target-site duplications (see "The target site," above). IS91 itself inserts with a specific orientation at the 3' end of a conserved tetranucleotide sequence (5'-CTTG-3' or 5'-GTTC-3') probably involved in replication initiation and termination. This sequence is required for further efficient transposition. However, deletion of the downstream ("right") end results in so-called one-ended transposition, in which different lengths of vector DNA neighboring the deletion accompany the element to its new target site. These terminate with a 5'-CTTG-3' or 5'-GTTC-3' tetranucleotide located in the vector.

Y and S Transposases

Several transposons have been mobilized by S or Y enzyme-driven mechanisms, most often used in site-specific recombination. Examples of these are the Y recombinase-driven conjugative transposons, Tn916 and Tn1545, and the S recombinase-driven elements, Tn4451 and IS607. It has been proposed that these elements should be referred to as Y transposons and S transposons, respectively. Transposition occurs in two steps. The transposon first excises as a circle by recombination between its ends in a manner probably similar to phage excision. This generates the equivalent of an attP recombination site (see "The nature and role of accessory elements," above). The circular transposon is subsequently inserted into a new target in a way similar to phage integration (Fig. 12). These two steps presumably involve cleavage and strand-transfer reactions similar to those occurring in site-specific recombination. The main difference between transposition and site-specific recombination lies in the reduced sequence specificity both at the excision and at the integration steps in the transposons. These transposons can exhibit a wide range of tolerance to variation in the target sequence.

Most known Y transposases are associated with conjugative transposons and are generally related to the phage integrase subfamily of Y enzymes. For example, the recombinase of Tn916 is bivalent, like λInt, with two different DNA domains involved in binding to core and arm-like sequences of the appropriate att site. This element also encodes an equivalent of λXis required for excision. Tn916 integrates into different target sites; at present, the molecular basis of this relaxed specificity is poorly understood. (See also "Specificity and orientation of the core sequences: the role of homology," above.)

S transposases fall into two groups that differ from the S recombinase paradigm. The Tn4451 and Tn5397 transposases belong to the group of the "large" S recombinases together with the integrase of phage ΦC31. These are larger than classic S recombinases in that they contain an oversized C-terminal domain (Fig. 2A). The transposase of IS607 and related enzymes define a second group of S transposases. These have a "reversed" domain structure in which the DNA-binding domain is located at the N-terminal end of the protein. Both groups of S transposases display a catalytic domain typical of S enzymes. In contrast with Y transposons, the 2 bp immediately adjacent to the ends of integrated S transposons are invariant. These form part of the core att sequence on excision. This implies that homology between the central regions of the circular transposon att site and the target site is important. The recombination sites of S transposons are a notable exception among S enzyme sites (see "Specificity and orientation of the core sequences: the role of homology,"

above) in that their 2-bp central regions are not symmetrical. There is no evidence so far for the involvement of accessory elements in excision or integration.

Reaction Control in Transposition: Avoiding Collateral Damage

Transposable elements have developed numerous ways of controlling their activity since high transposition rates might be expected to be deleterious for the "host" cell. The more complex transposons, such as phage Mu, Tn7, and Tn552, have evolved their own set of regulatory pathways, which are too elaborate to consider in detail here (see Peters and Craig, 2001). On the other hand, some general novel regulatory mechanisms have emerged from the simple IS elements.

Regulation of IS activity can occur at several levels. Many of the classical mechanisms of controlling gene expression, such as the production of transcriptional repressors or translational inhibitors (e.g., antisense RNA) or Dam methylation, operate in Tpase expression. Moreover, endogenous transposase promoters are generally weak, and, in the IS elements, many are partially located in the terminal IRs. This would enable their autoregulation by Tpase binding.

AVOIDING ADVENTITIOUS ACTIVATION

Several elements have evolved mechanisms that attenuate their activation by impinging transcription following insertion into active host genes. One such mechanism is the sequestering of translation-initiation signals in an RNA secondary structure. Examples of elements that have adopted this mechanism are IS10 and IS50. In this case, inverted repeat sequences are located close to the left end. These include the ribosome-binding site or translation initiation codon for the Tpase gene. Transcripts from the resident Tpase promoter include only the distal repeat unit, which is unable to form the secondary structure, while transcripts from neighboring DNA include both repeats. In addition, simple transcription

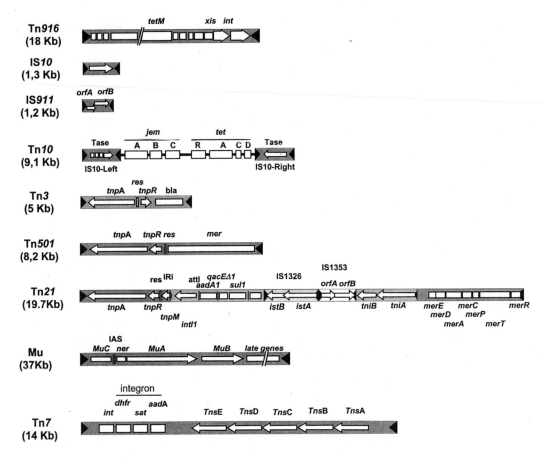

FIGURE 12 Diversity in transposable elements. A collection of transposable elements is presented to demonstrate their diversity. Reactive ends are indicated as dark-grey triangles, genes involved in recombinational mobility are shown in white with arrows indicating the direction of transcription/translation. Additional genes such as antibiotic resistance genes are shown as white bars. (Tn916) A conjugative transposon is an example of a transposon using a Y transposase (*int*) and excisase (*xis*). Also shown is the tetracycline resistance gene TetM and various genes involved in conjugal transfer. (IS10 and IS911) Two insertion sequences, with one and two transposase ORFs, respectively. (Tn10) A composite transposon based on the IS10 element showing the *tet* resistance and *jem* genes flanked by two IS10 copies. Note that the left IS10 copy has reduced activity and depends on the right copy for its activity. (Tn3) A member of the unit transposons showing the DDE Tpase gene (*tnpA*), the S recombinase involved in resolution of Tpase-generated cointegrates (*tnpR*), and the site on which it acts (*res*). The β-lactamase gene (*bla*) carried by Tn3 is also indicated. (Tn501) A second member of the "Tn3" unit transposons showing an alternative arrangement of *tnpA*, *tnpR*, and *res* together with the genes specifying resistance to mercury salts. (Tn21) Found as part of multiple antibiotic resistance plasmids, this member of the Tn3 family illustrates the high complexity of certain transposable elements. The basic module is one similar to Tn501 in organization. In addition, Tn21 also carries an integron landing platform with its own Y recombinase (integrase; *intI1*) and attachment site (*attI*) together with several integron cassettes (*aadA1*, *qacEΔ1*, *sul1*). This is inserted into a second structure bordered by short inverted repeats (IRi) and additional mobility functions (*tniB*, *tniA*). Finally, two insertion sequences, an IS21 family member (IS1326) with its transposase genes (*istA* and *istB*) and an IS3 family member, IS1353, are also present. Here, IS1353 is inserted into IS1326. (Bacteriophage Mu) This scheme shows the organization of the genes involved in transposition of this 37-kb bacteriophage. From left to right, it shows the repressor gene c, involved in regulating early gene expression; the internal activating sequence (IAS) involved in assembly of the transpososome; an accessory gene, *ner*, also involved in regulation of early gene expression; the transposase, MuA; and the accessory protein Mu. The late genes are not shown in detail. (Tn7) From left to right, the integrase gene and three integron resistance cassettes; the five genes involved in transposition, *TnsB*, the trans-

across the end of the element may reduce transposition activity, as has been shown for IS1, possibly by disrupting complexes between Tpase and cognate ends.

PROMOTER ASSEMBLY BY RECOMBINATION

An additional level of control at the transcriptional level for some elements has been observed. These elements transpose through an intermediate (e.g., transposon circles) in which the left and right ends must be joined to form a highly recombinogenic IRR–IRL junction. In these cases, the right end often carries a −35 promoter element orientated outward, while the left end may carry a −10 hexamer directed inward. When two ends of such an element are juxtaposed, by formation of head-to-tail dimers or of circular copies of the IS, the combination of the IRL −10 hexamer with a −35 hexamer resident in the neighboring IRR can generate relatively strong promoters. This arrangement can lead to high Tpase expression and consequent increases in transposition activity. Insertion into the target DNA results in separation of the two joined ends with consequent disassembly of the promoter and a return to the basal transposase levels specified by the indigenous weak IS promoter. Examples of elements that have adopted this type of mechanism are IS21, IS30, and IS911.

TRANSLATIONAL FRAMESHIFTING

Many ISs carry only a single open reading frame (ORF). However, members of several families carry two partially overlapping ORFs. The second frame, which generally includes the catalytic site of the enzyme, is positioned in a −1 reading phase compared with the first. The latter generally carries the determinants involved in sequence-specific binding to the terminal IRs. Here, the Tpase is expressed as a fusion protein by a programmed −1 translational frameshift, which combines the products of the two frames in a single protein with an N-terminal DNA-binding domain and a C-terminal catalytic domain. The product of the upstream frame modulates Tpase activity. The frameshifting frequency is thus critical in determining overall transposition activity. Examples of elements using this type of mechanism are IS1, and members of the IS3 family, such as IS3 itself, IS2, IS150, and IS911. Although it has yet to be explored in detail, frameshifting could be influenced by host physiology, thus coupling transposition activity to the state of the host cell.

TRANSPOSASE STABILITY

Transposase stability can also contribute to control of transposition activity. Some Tpases are sensitive to host proteases, such as the *E. coli* Lon protease. This sensitivity limits the activity of the Tpase both temporally and spatially and may provide an explanation for the observation that several Tpases function preferentially in *cis* (see below). Indeed, mutant IS903 Tpase derivatives have been isolated that exhibit an increased capacity to function in *trans*. These are more refractory to Lon degradation than the wild-type protein. An observation that might also reflect Tpase instability is the temperature-sensitive nature of transposition of several ISs.

DIFFERENTIAL *CIS-TRANS* ACTIVITY

Early studies with several transposable elements indicated that transposition activity was more efficient if the Tpase is provided by the element itself or by a Tpase gene located close by on the same DNA molecule. These elements included IS1, IS50, and IS903. Preferential activity in *cis* reduces the probability that Tpase expression from a given element will activate transposition of related copies elsewhere in the genome. The effect can be of several orders of magnitude. Its

posase, *TnsA*, a second endonuclease involved in processing the nontransferred strand, *TnsC* and *D* involved in the recognition of the specific chromosomal target site, and *TnsE*, which, in conjunction with *TnsC*, permits integration at many different target sites.

magnitude is characteristic for a given IS. This property presumably reflects a preference of the cognate Tpases to bind to transposon ends close to their point of synthesis and is likely to be the result of several phenomena. For some ISs, increased stability and expression increase the capacity for Tpase activity in *trans*.

An additional consideration that may promote preferential activity in *cis* is reflected in the N-terminal location of the DNA-binding domain in many Tpases. If the N-terminal domain is capable of folding independently of the catalytic domain, this arrangement would permit preferential binding of nascent Tpase polypeptides to neighboring binding sites. For several Tpases, the N-terminal portion of the protein exhibits a higher affinity for the ends than does the entire Tpase molecule, suggesting that the C-terminal end may in some way mask the DNA-binding activity of the N-terminal portion. It remains to be seen whether this is a general property of Tpases.

HOST FACTORS

Transposition activity is frequently modulated by host factors. In general, these effects are specific for each element. A nonexhaustive list of such factors includes the DNA chaperones (or histone-like proteins) IHF, HU, HNS, and FIS, the replication initiator DnaA, the protein chaperone/proteases ClpX, P, and A, the SOS control protein LexA, and the Dam DNA methylase. Proteins that govern DNA supercoiling in the cell can also influence transposition. These factors may intervene at a variety of levels modulating expression of the Tpase, its stability, or its intrinsic activity.

TRANSPOSITION IMMUNITY

Transposition immunity is a phenomenon that prevents insertion of an element into itself or in the neighborhood of a second copy. It has only been described for a limited number of transposable elements, which include phage Mu and transpososns Tn7 and Tn3. Most ISs do not exhibit this property. For both Mu and Tn7 the phenomenon depends on interactions between the Tpase and a second pro-

tein, which shows DNA-binding-dependent ATPase activity.

Section Summary
- There are several families of transposases of which the DDE family is by far the most widespread.
- DDE transposases recognize specific DNA sequences at the ends of the transposon and catalyze hydrolysis of the adjacent phosphodiester bond.
- No covalent DNA-transposase intermediate is formed.
- There are checks and balances that ensure the correct assembly of the transposition complex before cleavage can occur.
- Transposase expression and activity are highly regulated by a wide variety of mechanisms.

DIVERSITY OF NONHOMOLOGOUS RECOMBINATION SYSTEMS: SITE-SPECIFIC RECOMBINATION

An extremely large number of genetic elements have now been detected which use the nonhomologous recombination mechanisms described above. Some of these are relatively well characterized, while others have simply been detected as a result of genome-sequence analysis. Some of these are briefly described below.

Inversion Systems

DNA inversion systems are the basis of programmed switches that allow alternative expression of genes (Fig. 12). In general, these control expression of bacterial surface proteins often involved in pathogenicity and switches that determine the recognition of bacteria by certain bacteriophages. The simplest systems invert a promoter with respect to flanking genes or genes with respect to a promoter. The best-known members of this group are represented by the Hin and Gin switches, which use S recombinases, and the Fim switch, which uses two Y recombinases. Hin promotes phase variation in *Salmonella* spp. by inverting the promoter of the *fljBA* operon that encodes the H2 flagellin and a repressor of *fliC*, the gene for the H1 flagellin. Bacteriophage Mu Gin and a related system, Cin from phage P1 control phage, host specificity by inverting genes coding for phage tail fibers. The Fim system controls phase variation of type I fimbriae in *E. coli* via expression of *fimA*, which encodes type I pilin. This system is atypical in that it differentially uses two Y recombinases, FimB and FimE. FimB catalyzes inversions leading to both expression or silencing of *fimA*, whereas FimE catalyzes recombination in only one direction, that leading to silencing. This allows a control of phase variation by differential expression of the two recombinases. Other Y recombinase-driven switches use only one recombinase (e.g., the MR-P fimbriae variation system of *Proteus mirabilis*).

A third class of site-specific recombinases is now emerging from the study of DNA inversions of the pilin genes in *Moraxella* species. These systems use a recombinase, Piv, that does not display homology to S or Y recombinases. Instead, Piv is probably related to a subclass of DDE transposases thought to be a member of the large RnaseH-DDE family of enzymes.

Multiple inversion systems have been found in different bacteria and their plasmids. They are composed of various numbers of consecutive recombination sites and recombination between these sites can generate a multitude of different combinations. In these systems, recombination fuses variable segments with constant segments of a gene. The prototypical system, called a shufflon, is carried by plasmid R64, in which it generates diversity of the thin mating pilus. Seven different 3′ ends of the *pilV* gene may be fused to a single 5′ constant region. Recombination uses a Y recombinase, but analogous systems have been found that use S recombinases (e.g., the Min system of plasmid p15B).

Some inversion systems may also catalyze deletions (excisions). This is the case of the Flp/FRT system (Y recombinase) of the yeast 2μ plasmid and of the β-recombinase driven systems (S recombinase) of some gram-positive bacteria plasmids (e.g., the β-recombinase system of plasmid pSM19035 and the analogous Resβ-driven system of pAMβ1). In both cases,

it is thought that the biological role of the inversion event is to allow concatameric plasmid amplification. The recombinases have also been implicated in resolution of the concatamers leading to excision of monomers.

Temperate Phages

Temperate bacteriophages often integrate into their host chromosome to achieve the lysogenic state (see chapter 9). This renders the host cell immune to further infection. Phages may also carry other functions advantageous for the host, for instance toxin or hemolysin genes involved in pathogenesis (see chapter 16). Certain prophages may have evolved into genomic islands that have retained the ability to be mobilized (see below). Analysis of the sequence of numerous bacterial genomes has revealed a large population of cryptic and defective prophages that can represent a high percentage of genomic DNA and has revealed the crucial role of these phages and phage remnants in promoting genome diversity. For example, the chromosomes of *E. coli* K-12 and O157 contain 11 and 18 of these elements, respectively, which account for more than half of their strain-specific DNA. Defective prophages may remain mobilizable (e.g., P4 prophages can be transferred by functions of phage P2). They may even be implicated in bacterial differentiation. For instance, an excision of a phage-like element, the SKIN element, is required for completion of the sporulation process in *Bacillus subtilis*. Recent studies using comparative genomics of a large range of temperate phages have provided strong support for the idea that phage genomes are assembled in a modular manner from a common gene pool. They have underlined the mosaic nature of phage genomes and thrown into question the idea of a hierarchical evolution with a unique evolutionary tree. This has given rise to the idea of a reticulate or web-like phylogeny reflecting extensive horizontal gene exchange.

Integration of temperate phages usually occurs at a single site in the genome called the attachment site (*attB*) that may be embedded in a gene (most often coding for a tRNA), ensuring good conservation of this site. Since, in these cases, the phage *att* site generally includes the 3′ end of the target gene, insertion is not disruptive and the gene is reconstituted during the insertion event. Most known temperate phages encode their own Y recombinase (the integrase), which, in most cases, carries two distinct DNA-binding domains, one specific for core sequences and the other, for accessory sequences (called "arms sites" in the phages systems). The organization of the *attB* site usually conforms to that of phage λ with a core sequence only (Fig. 3). The organization of the phage recombination site, *attP*, can vary substantially in the nature and position of binding sites for various accessory proteins.

Although most phages encode their own recombinase, it has been reported recently that some filamentous phages use their host's XerCD recombinases to integrate into the *dif* site of their host. This has been shown for the CTX phage of *V. cholerae*, which encodes the cholera toxin. In this case, the core sequence of the phage recombination site is homologous to that of one of the *dif* sites of the two *V. cholerae* chromosomes. It is not known if accessory elements are required. Integration reconstitutes a *dif* site active for resolution of chromosome dimers at one end of the integrated phage. Sequence analysis indicates that other filamentous phages may integrate into the *dif* sites of their hosts (e.g., the ΦLf and Cf16-v1 phages of *Xanthomonas campestris*, the CUS phage of *Yersinia pestis*, and the XfΦ fl phage of *Xylella fastidiosa*).

Several temperate phages that infect gram-positive bacteria use S recombinases (e.g., phages ΦC31 and R4 of *Streptomyces* and TP901-1 of *Lactococcus*). These are "large S recombinases" that lack the C-terminal DNA-binding motifs of typical S recombinases and show instead a long C-terminal extension (Fig. 2A). The *attB* and *attP* sites that have been characterized are rather short (<50 bp), suggesting that control of these systems is simpler than in the case of λ. The ΦC31 integrase catalyzes recombination between its *attP* and *attB*

sites with no requirement for other factors. However, no recombination between the *attL* and *attR* sites of the integrated phage can be detected in the same conditions. This suggests a requirement for a functional equivalent of λXis (see "The nature and role of accessory elements," above). Such a product, specifically required for excision, has been reported for the related TP901-1 phage.

Genomic Islands

Genomic islands were first discovered as relatively large segments of DNA (from 10 to 200 kbp) carrying a variety of genes associated with pathogenicity (see chapter 16). They were found in the genomes of pathogenic bacteria but not in closely related nonpathogenic strains. These elements are called pathogenicity islands (PAIs) (Hentschel and Hacker, 2001). Analysis of the accumulating number of bacterial genomic sequences has revealed similarly large DNA segments that carry other types of functions, including genes involved in symbiosis, secretion, catabolism, and other metabolic functions (see also chapter 6). These genetic elements are now referred to more generally as "genomic islands" (Hentschel and Hacker, 2001). They appear to be mosaic structures that include other types of often degenerate mobile genetic elements, such as transposons and IS sequences, plasmid transfer and replication genes.

Many genomic islands show features related to temperate bacteriophages, and, from a mechanistic point of view, it is probably preferable to retain the term "genomic islands" exclusively for these phage-like types. Not only are many genomic islands associated with tRNA genes, as in many prophages, but they also carry phage-like Int genes. It is possible that these islands arrived as prophages and subsequently evolved by acquisition of additional genes and mobile element-driven rearrangements. However, the finding that certain islands occur at different tRNA loci in different strains implies that they might be mobilized as coherent units (e.g., the locus of enterocyte effacement, LEE, occurs at the *pheU* and *selC* loci in different *E. coli* strains or HPI in different asparagine tRNA genes in various *Yersinia* isolates).

Conjugative Elements That Use Site-Specific Recombinases

In contrast to the classical conjugative transposons such as Tn*916*, which integrate at several different sites (see "Y transposons," below), a variety of additional conjugative elements are known which integrate at single-attachment sites in their host genome (Burrus et al., 2002). Most use Y recombinases for integration and excision. These elements are quite heterogeneous and diverse in length and gene content and carry different types of conjugative functions. Some of the smaller elements, such as pSAM2 (10 kb) and SLP1 elements from *Streptomyces coelicolor* and *Streptomyces ambofaciens* carry genes involved in integration (*int*) and excision (*xis*) together with a replicase (*repSA*) related to rolling circle replicases. pSAM2 was originally thought be a plasmid since circular forms were often (but not always) observed. Transfer requires only a single gene (*traSA*), whose product is a member of the FtsK/SpoIIIE family of DNA translocases and which probably transfers an excised copy of the element as double-stranded DNA (see Burrus et al., 2002). Other, much larger, elements such as the IncJ R391 from *Providencia rettgeri* (88.5 kb) were also originally thought to be plasmids on the basis of their incompatibility functions. These are now known to be integrative elements related to SXT (100 kb) of *V. cholerae*. Most of these elements have been identified in pathogenic bacteria because they carry antibiotic resistance determinants. They also carry plasmid- and phage-related genes but appear to be devoid of replication functions. SXT and R391 are thought to be transferred by a mechanism similar to that of the F plasmid since they carry genes related to plasmid transfer operons such as those of F and R27. Transfer of these elements can occur between different bacterial species. Moreover, in addition to its own transfer, SXT can also promote transmission of parts of its host chromosome by Hfr-type transfer.

A variety of other less well characterized elements have been identified by virtue of their associated functions, such as antibiotic resistance, restriction and/or modification systems, catabolic or symbiotic functions, and by analysis of bacterial genome sequences (Burrus et al., 2002). Integration of many of these site-specific elements occurs in a tRNA gene and therefore resembles that of certain phages (see "Temperate phages," above) and genomic islands (see "Genomic islands," above). The 500-kb symbiosis island of *Mezorhizobium loti* might also be included among these elements since it appears capable of conjugative transfer and integrates into a tRNA gene (Burrus et al., 2002).

Finally, classification of conjugative elements into conjugative transposons and conjugative site-specific integrating elements is problematic since the specificity of integration of both kinds of elements may vary in different hosts (see Burrus et al., 2002).

Integrons

Integron cassettes are small diverse DNA sequences that carry single ORFs specifying a variety of functions. Initial examples were isolated as carriers of antibiotic resistance genes. Subsequent isolates have been defined from genome sequences by their organization and by their bordering sequences (Fig. 12B). These are more diverse and encode proteins that resemble restriction enzymes, toxins, and pathogenicity functions. Many integrons do not possess their own promoters, and expression of their associated genes is driven by a resident promoter carried by the DNA target into which they integrate (integration platform). Small arrays have been identified in various plasmids and transposons of gram-negative and gram-positive organisms (*Corynebacterium glutamicum*) and mycobacteria. Structures in which tens or hundreds of copies of such elements have been assembled in tandem arrays, known as superintegrons, have also been observed in various bacteria. They have been found largely in the gammaproteobacteria such as *Vibrio*, *Shewanella*, *Xanthomonas*, and *Pseudomonas* spp. but have also been observed in beta- and deltaproteobacteria and spirochetes.

Integron cassettes do not themselves encode recombination functions. They insert into a recombination platform that carries a Y recombinase gene (IntI) and a neighboring sequence resembling an attachment site known as *attI*. Individual cassettes carry unique *att* sites but these share a consensus. Integrons have been divided into four types (1, 2, 3, and the superintegron members) based on the sequence of their integrases and recombination sites. Integrons have been observed to circularize and, in so doing, create an integron *att* site, *attC*, equivalent to λ *attP*. Moreover, they can integrate consecutively into the integration platform to form tandem arrays. Integron cassette circles are presumably integration intermediates able to form an Int-type synaptic complex with the resident *att* site carried by the integration platform. Integration platforms have been observed in transposons of the Tn7 and Tn3 (in the Tn21 subgroup) families (Fig. 11) and are present in several plasmids. A typical *attI* recombination site has features common to all Y recombinase sites: a core sequence composed of a central region of 6 to 8 bp (depending on the integron type) flanked by two inverted recombinase-binding sites. The core sequence is bordered on one side by two accessory recombinase-binding sites in direct repeat. The integron *attC* site, originally called a "59 base element," has subsequently been shown to vary substantially both in length and sequence (from 57 to 141 bp) and is composed of a core site, like *attI*, with two additional inverted recombinase sites located to one side. Although these may represent a secondary core site, it seems more likely that they function as accessory IntI-binding sites. The structures of the recombination sites and the nature of the recombinase involved suggests that integron integration (and excision) occurs using a typical Y recombinase recombination pathway in which the additional recombinase-binding sites act as accessory elements in *attI* × *attC* synapsis. Each IntI appears to preferentially recognize its cognate *attI* site but all three Int types can accommodate a variety of *attC* sequences rendering the recombination system relatively promiscuous (Fig. 12B).

Multimer Cointegrate Resolution Systems

Cointegrates are produced by replicative transposition (see Fig. 9 and 11) and need to be resolved to separate the fused donor and target molecules. Multimers arise by recombination between sister copies of circular plasmids or chromosomes. These must be resolved if the replicons are to be maintained stably during cell growth. In both cointegrates and plasmid multimers, recombination occurs between two identical sites each carried by one of the partner molecules. Resolution systems generally use topological filters to sense the relative location of their recombination sites (see "The nature and role of accessory elements," above). Numerous plasmids carry transposons that are mobilized by replicative mechanisms, and the cointegrate resolution systems of these may also be coopted to resolve plasmid multimers. Cointegrate resolution may use S recombinases (e.g., Tn3, γδ, and the unrelated Mu-like transposon Tn552) or Y recombinases (the TnpI system of Tn4430). Some plasmids use S recombinases for resolution of multimers (e.g., pSM19035 and RP4) although Y recombinases seem to predominate. Of these, the Xer system, used both by chromosomes and plasmids, appears to be the system most generally adopted for resolution of multimers. The Xer/*dif* system is found to be conserved in many bacteria for resolving chromosome dimers. However, not all plasmids have coopted the Xer system. Several encode their own specific Y recombinases for multimer resolution (e.g., the ResD/*rsf* system of plasmid F and the Cre/*loxP* system of phage P1).

Telomere Resolution

This is another cellular process that has recently been shown to involve site-specific recombinases. Certain bacterial chromosomes, plasmids, and prophages are linear with hairpin ends (telomeres). Examples of these are the chromosome and some plasmids of *Borellia burgdorferi* and the bacteriophage N15 in its lysogenic state. Replication of these linear molecules proceeds bidirectionally from within the replicons. This leads to circular head-to-head dimers separated by two inverted telomere copies at each junction. These forms are resolved to linear monomers with hairpin ends by interstrand recombination at each telomere. These "telomerases" display homology with Y recombinases and are thought to use a similar mechanism for cleavage and strand transfer.

Section Summary

- Site-specific recombination has been adopted in a large variety of cellular processes.
- These range from the assembly of clusters of genes and programmed phenotypic changes to the maintenance of chromosome integrity.

DIVERSITY OF NONHOMOLOGOUS RECOMBINATION: TRANSPOSABLE ELEMENTS

Transposons and Insertion Sequences

Insertion sequences (Fig. 11) constitute one of the largest groups of mobile genetic elements. They have been observed in most bacterial genomes and plasmids, where they may be present in high numbers. They are defined as relatively short, genetically compact DNA segments of between 0.7 and 2.5 kb encoding no functions other than those involved in their mobility. Their transposase is generally encoded by a single or sometimes two ORFs and stretches nearly the entire length of the element. In the vast majority of cases ISs that use DDE transposases carry short terminal inverted repeat sequences (IR) of between 10 and 40 bp (see "Structure and organization," above). The IRs are recombinationally active and carry recognition signals for transposase binding and catalysis. The several elements that do not exhibit terminal IRs (e.g., IS91, IS110, and IS200/605) do not specify DDE transposases and use different catalytic strategies.

At present, an estimated 1,500 different ISs have been detected. These can be grouped into relatively distinct families based on their genetic organization, the similarities between their transposases, and the relationship of their terminal inverted repeats. These families should only be considered as an aid to classification and to managing the high number and variety, which are being identified in the various genome-sequencing projects. There are approximately 20 different families presently defined (www-is.biotoul.fr). While many of these are coherent and the elements appear quite closely related, several families are more heterogeneous. For example, all members of the larger IS3, IS21, IS30, IS256, and ISL3 families are clearly recognizable, while those of the IS4 and IS5 families clearly form very distinct subgroups. These differences can even extend to the number of ORFs they encode. As more "members" of these families are identified, they will undoubtedly give rise to more consistent groups.

Miniature Inverted Repeat Transposable Elements (MITEs)

Several bacterial genomes harbor large numbers of small repeated sequences often carrying terminal inverted repeats. These are called variously Rep (E. coli, Pseudomonas putida), Box (S. pneumoniae), Rup (S. pneumoniae) or Correia (Neisseria spp.). They also appear in the S. coelicolor and Photorhabdus luminescens genomes. Some have been implicated in control of gene expression by acting as transcription terminators or influencing the stability of mRNA when located within transcriptional units. They also often carry sites for binding of various host proteins, such as IHF and gyrase. Some resemble insertion sequences from which the transposase gene has been deleted. The ends are often closely related to those of members of the IS630 family of insertion sequences. They also resemble elements, generally called miniature inverted repeat transposable elements (MITEs), found in many eukaryote genomes. Many MITEs resemble the ends of members of the Tc/mariner family and are thought to be mobilized by complete copies of the parent element when present in the genome. The eukaryote Tc/mariner family is also related to the prokaryote IS630 family of elements. A second family of MITEs, related to the bacterial IS5 element, has been identified in some eukaryotes, although no prokaryote equivalents have yet been described. The length of many of these elements approaches the persistence length of native DNA. This means that synapsis of the ends cannot occur by random collision and would presumably require Tpase- and/or host factor-mediated DNA bending. The mobility of these elements in bacteria remains to be addressed.

Compound Transposons

In addition to simple displacement from one site to another, IS elements are capable of provoking genome rearrangements such as deletions and inversions. They are also capable of translocating extraneous DNA segments in the form of compound transposons, where two copies of an IS flank the DNA segment and act in concert to transport it to other sites (Fig. 11). Examples of classic transposons of this type are Tn5, Tn9, and Tn10. The interstitial genes carried by these early examples specify resistance to antibiotics, but other characteristics, such as factors involved in catabolic pathways, symbiosis, or virulence have subsequently been identified as part of compound transposons. Although the flanking ISs are expected to be able to transpose autonomously, compound transposons often have evolved characters that would tend to maintain the coherence of the entire structure. These include reduced activities of one of the flanking ISs (e.g., Tn5, or Tn10), differential activities of the "outside" ends of the flanking ISs compared with the "inside" ends (i.e., those located proximal to the DNA segment located between the two ISs) and driving expression of transposon genes from a promoter located in the upstream IS. There is also some evidence that the transposition of several of these types of elements decreases as the length of the interstitial DNA increases, thus tending to limit the effective size of the element.

Unit Transposons

Several other classes of transposable element have been characterized in the bacterial gene pool. Members of one large and extended family, the Tn3 family (Fig. 11), also use a DDE transposase in their displacement. Like ISs they carry characteristic terminal inverted repeats, which appear to include a single site for transposase binding. They generate cointegrates on transposition (see "Catalysis by DDE transposases and the importance of second-strand processing,"above) and have recruited a site-specific recombinase (see "Resolution of multimers and cointegrates," above) to assure recombination between the two directly repeated elements that separate donor and target replicons in these transposition intermediates. This process is known as resolution, and the recombinase is known as "resolvase." Members of this family also generally carry various genes specifying resistance to antimicrobials, and many of these occur as integron cassettes located at a resident integron-integration platform. Other members have been found that are devoid of both resistance and resolvase genes and therefore strongly resemble simple insertion sequences.

Transposable Phages

Transposable phages have been known for many years. The first described, Mu (Fig. 11), was named for its Mutator effect since its integration into the host genome to form stable lysogens occurs with little macrosequence specificity. Integration therefore leads to mutation by disrupting the target gene. Phage Mu was the first transposable element to undergo extensive analysis and has become the paradigm for transposition. In contrast to the relatively simple ends of the IS elements, Mu carries multiple and asymmetric transposase-binding sites at its ends (Fig. 8). The arrangement also reflects the complexity of assembling the transpososome. This occurs in several steps and provides a series of checkpoints preventing adventitious strand cleavages and transfers before the assembly of the productive complex. Assembly also involves two host proteins that assist in generating the correct transpososome architecture for recombination and a secondary transposase-binding region within the phage that aids correct delivery of the transposase to the phage ends. In addition to the transposase (MuA), a second protein (MuB), which is a DNA-dependent ATPase and whose expression is probably linked to that of MuA by translational coupling, modulates transposition activity. It does this in two ways. On the one hand, it prevents the Mu transpososome from inserting into a target that includes a neighboring MuA-binding site (see "Transposition immunity," above). MuB also stimulates the activity of MuA in the transpososome, facilitating its insertion into sites devoid of Mu end sequences.

Phages of this type are widespread and not limited to the enterobacteria. They have been isolated from pseudomonads and have appeared in the genomic sequences of several other bacterial species (e.g., *Deinococcus radiodurans, Haemophilus influenzae,* and *Neisseria meningitidis*).

Y Transposons

Y transposons form a broad and heterogeneous group of elements that use a Y transposase-driven mechanism for mobility (see also "Conjugative elements that use site-specific recombination," above). They also share the capacity to be transferred by a conjugative mechanism from cell to cell and are referred to as conjugative transposons (CTns) (see chapter 8). They were originally isolated from enterococci (Tn*1525*) and streptococci (Tn*916*) and are mostly found in gram-positive bacteria. They often carry antibiotic resistance determinants (usually the *tetM* gene) and are important vectors for disseminating this antibiotic resistance. Some are self-transferable (e.g., Tn*916*), whereas others are mobilizable by functions provided in *trans* by other elements such as conjugative plasmids or other CTns (e.g., the NBUs elements from *Bacteroides* spp.). CTns can be unusually long for transposons. For instance, Tn*5276* of *Lactococcus lactis* is 70 kb long and *Bacteroides* species carry several large CTns of more than 50 kb (e.g., the CTnDOT elements). They can have a very broad host range and appear to contribute significantly to gene transfer in complex bacterial populations.

S Transposons

S transposons use S transposases for mobility. They fall into two distinct classes. Some are

CTns found in *Clostridium* species and use "large S enzymes" for mobility. Of these, Tn*5397* and Tn*5398* are large self-transferable CTns, while Tn*4451* and Tn*4453* are shorter and may be mobilized by other elements. The second class consists of IS*607* of *Helicobacter pylori* and related ISs first identified as a result of analysis of bacterial genome sequences. These ISs encode atypical S enzymes with a reverse order of DNA binding and catalytic domains (see "Y and S transposases," above).

Section Summary
- There is an extremely wide range of transposable elements even within the major groups that use DDE transposases.
- They can vary in length from several hundred base pairs (the nonautonomous MITE sequences) to many kilobases.
- They also include elements that transpose by replicative mechanisms similar to that of certain plasmids or phage, elements that have appropriated site-specific recombination enzymes with degenerate sequence specificity.
- Most transposable elements require phage or plasmid vectors for their transmission from cell to cell, whereas some carry their own specific transfer functions.

SIGNIFICANCE OF SITE-SPECIFIC RECOMBINATION AND TRANSPOSITION

The importance of nonhomologous recombination in shaping bacterial genomes, in their evolution and their expression, is considerable. In view of the numerous genetic elements capable of undertaking these types of reaction and their wide distribution, it is perhaps surprising that the majority use one of only four basic mechanisms. These mechanisms and the enzymes involved (Y recombinases/transposases, S recombinases/transposases, DDE transposases, and Y2 transposases) were initially characterized genetically using well-defined model systems.

However, several important questions remain and they are often difficult to address experimentally. Among these are the following:

(i) the basis for the dramatic differences in the number of whole or partial IS elements frequently observed in closely related bacteria (e.g., *E. coli* and *Shigella*); (ii) the capacity of bacterial MITEs to be mobilized by complete indigenous IS copies and their effects on gene expression and genome structure; (iii) the assembly, plasticity, and transmission of genomic islands and their relationship to phages and conjugative elements; and (iv) the assembly of super integrons.

Genome-scale analysis is slowly providing clues to the evolution of the elements themselves. There is a general consensus that, at one level, they evolve by acquiring functional modules from a horizontal gene pool. This is clearly suggested from recent evaluations of phage genomes and appears similar for other genetic elements such as plasmids and conjugative transposons.

An additional area that merits attention is the "micro" variability of the recombination modules themselves. Thus, while the known S recombinases show a high degree of identity, at least in their catalytic domains, the Y recombinase family is significantly more heterogeneous and the DDE transposases are even more diverse. Presumably, in all these cases, the catalytic sites are well conserved since they carry out the same chemistry; the differences reflect constraints imposed by the specific system (e.g., the interactions required to unwind the penultimate base of the end during IS*50* transposition [see "Transpososome assembly," above] or those involved in formation of a composite catalytic site in the Flp Y recombinase [see "Control of cleavage: a requirement for recombinase multimers," above]).

QUESTIONS

1. What is the diversity of mobile elements in the bacterial world?

2. What are the details of the known mechanisms of mobility?

3. Are there additional, as yet undefined, mechanisms that contribute to mobility of the elements?

4. What are the global controls and driving forces of mobility?

5. How does all this relate to bacterial physiology and ecology?

REFERENCES

Burrus, V., G. Pavlovic, B. Decaris, and G. Guedon. 2002. Conjugative transposons: the tip of the iceberg. *Mol. Microbiol.* **46:**601–610.

Hentschel, U., and J. Hacker. 2001. Pathogenicity islands: the tip of the iceberg. *Microbes Infect.* **3:**545–548.

Morgan, G. J., G. F. Hatfull, S. Casjens, and R. W. Hendrix. 2002. Bacteriophage Mu genome sequence: analysis and comparison with Mu-like prophages in *Haemophilus*, *Neisseria* and *Deinococcus*. *J. Mol. Biol.* **317:**337–359.

Novick, R. P. 1998. Contrasting lifestyles of rolling-circle phages and plasmids. *Trends Biochem. Sci.* **23:**434–438.

Peters, J. E., and N. L. Craig. 2001. Tn7: smarter than we thought. *Nat. Rev. Mol. Cell Biol.* **2:**806–814.

Toussaint, A., and C. Merlin. 2002. Mobile elements as a combination of functional modules. *Plasmid* **47:**26–35.

Turlan, C., and M. Chandler. 2000. Playing second fiddle: second-strand processing and liberation of transposable elements from donor DNA. *Trends Microbiol.* **8:**268–274.

FURTHER READING

Craig, N. L. 1997. Target site selection in transposition. *Annu. Rev. Biochem.* **66:**437–474.

Craig, N. L., R. Craigie, M. Gellert, and A. M. Lambowitz (ed.). 2002. *Mobile DNA II*, p. 149–161. ASM Press, Washington, D.C.

Hallet, B. 2001. Playing Dr. Jekyll and Mr. Hyde: combined mechanisms of phase variation in bacteria. *Curr. Opin. Microbiol.* **4:**570–581.

Haren, L., B. Ton-Hoang, and M. Chandler. 1999. Integrating DNA: Transposases and Retroviral Integrases. *Annu. Rev. Microbiol.* **53:**245–281.

Higgins, N. P. (ed.). *The Bacterial Chromosome*, submitted for publication.

GENE DUPLICATION AND GENE LOADING

Renato Fani

4

The genes and genomes of the extant organisms are the result of 3.5 or perhaps 4 billion years of evolution. Studies at both the gene and genome levels have been greatly facilitated by the development of recombinant DNA technology and especially by the sequencing of entire genomes. The availability of the nucleotide sequences of complete genomes from several organisms belonging to the three cell domains, *Archaea*, *Bacteria*, and *Eucarya* is providing an enormous body of data concerning the structure and the organization of genes and genomes. As a result, it is now possible to

shed some light on the mechanisms involved in their evolution and responsible for the shaping of metabolic pathways. It has become possible to develop a clearer, more consistent picture of the evolution of genes and genomes that includes organisms both extant and extinct and goes back in time far beyond the fossil record.

What forces drove the evolution of the earliest genes and genomes? Very likely, those same forces drive the evolution of genomes today. Analysis of the current databases shows that large proportions of an organism's genes are related to each other and to genes in distantly related species. All known life forms share a common pool of highly conserved genetic information shaped to a considerable extent by the duplication and divergence of DNA sequences predating the prokaryote–eukaryote divide. Computer-aided whole-genome comparison demonstrates that a remarkable percentage (approximately 50%) of the gene set of different prokaryotes results from ancient gene duplications and that many of these genes group into numerous families of different sizes (Labedan and Riley, 1995; de Rosa and Labedan, 1998; Mushegian and Koonin, 1996). Thus, the duplication of DNA sequences appears to have played a very important role in the evolution of genes and genomes. It suggests that the earliest living organisms contained a few genes that gave rise to new genes by duplication followed by evolutionary divergence.

The evolutionary significance of gene duplication was first recognized in the 1930s by

Renato Fani, Dipartimento di Biologia Animale e Genetica, Università degli Studi di Firenze, Via Romana 17-19, I-50125 Florence, Italy.

Microbial Evolution: Gene Establishment, Survival, and Exchange
Edited by Robert V. Miller and Martin J. Day, ©2004 ASM Press, Washington, D.C.

Haldane and Muller, who suggested that a redundant duplicate of a gene might acquire divergent mutations and eventually emerge as a new gene. However, few examples of duplicate genes were discovered before the advent of biochemical and molecular biology techniques. In 1970 Ohno put forward the view that gene duplication is the only means by which a new gene can arise and, although other means of creating new genes or new functions are now known, Ohno's view remains fundamentally valid.

GENE DUPLICATION

During the early evolution of life, gene duplication, the production of two copies of a DNA sequence, allowed the rapid diversification of enzymatically catalyzed reactions and an increase in genome size, providing also material for the invention of new enzymatic properties and complex regulatory and developmental patterns. The possibility that functionally related genes arose from the duplication of a common ancestral gene and underwent a subsequent evolutionary divergence relies on the

fact that, after a gene duplicates, one of the two copies becomes dispensable and can undergo several types of mutational events, mainly substitutions (Fig. 1).

Orthologous and Paralogous Genes

Two structures or sequences that evolved from a single ancestral structure or sequence are referred to as "homologs." The terms "orthology" and "paralogy" were introduced to classify different types of homology. Orthologous structures or sequences in two organisms are homologs that evolved from the same feature in their last common ancestor, but they do not necessarily retain their ancestral function (Fig. 2). The evolution of orthologs reflects organismal evolution. In contrast, homologs whose evolution reflects gene-duplication events are called paralogs. In general, paralogous genes perform different, although similar, functions within the same microorganism, but gene duplication may generate many copies of genes with the same function, thereby enabling the production of a large quantity of RNAs or proteins. This is the case, for instance, in genes

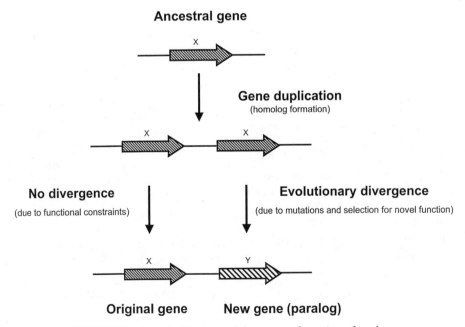

FIGURE 1 Gene duplication and divergence: formation of paralogs.

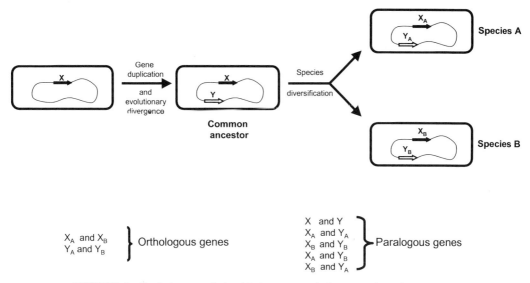

FIGURE 2   Evolutionary relationship between orthologous and paralogous genes.

encoding for rRNA molecules. In principle, two paralogous genes may also undergo successive and differential duplication events involving one or both of them giving rise to a group of paralogous genes, which is referred to as a paralogous gene family (Fig. 3).

Types of DNA Duplication

A duplication may involve (i) a part of a gene, (ii) a whole gene, (iii) DNA stretches including two or more genes involved in the same or in different metabolic pathways, (iv) entire operons, (v) a part of a chromosome, (vi) an entire chromosome, and finally (vii) the whole genome.

Therefore, any DNA sequence may undergo a duplication event(s), but the fate of the replicate depends on whether it provides an evolutionary advantage to the host cell. How is this achieved?

GENE ELONGATION

Gene elongation describes the increase in the size of a gene and represents an important step in the evolution of extant genes. A gene-elongation event can result from duplication of a DNA sequence that leads to two genes arranged in tandem, in that they are adjacent

on the same DNA molecule. If then a deletion of the sequence intervening between the two copies occurs coincidentally with a mutation to convert the stop codon of the first copy into a sense codon (Fig. 4), the elongation by fusion of the ancestral gene and its replicate results. Thus, the new gene is constructed from two paralogous moieties (hereinafter, modules). Potentially, each module or both of them might undergo further duplication events, leading to a gene constituted by more repetitions of amino acid sequences. Many proteins of present-day organisms show internal repeats of amino acid sequences, and the repeats often correspond to the functional or structural domains. A protein domain is a well-defined region within a protein that either performs a specific function (functional domain), such as substrate binding, or constitutes a stable, compact structural unit within the protein that can be distinguished from all the other parts (structural domain) of the protein.

What is the biological significance of gene elongation? Duplication has occurred in so many proteins that the process must have a considerable evolutionary advantage. It could allow for the improvement of the function of a protein by increasing the number of active

Ancestral gene

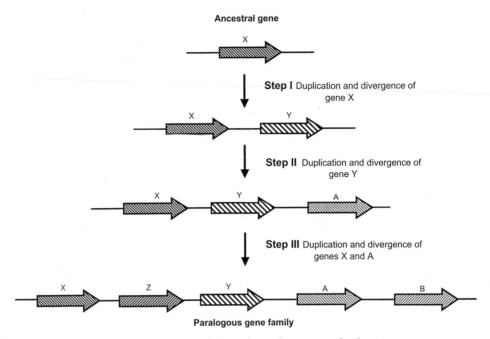

Paralogous gene family

FIGURE 3   Evolution of a paralogous gene family.

sites and/or the acquisition of an additional function by modifying a redundant segment. There are several examples of genes sharing internal sequence repetitions (i.e., duplications). For example, the *Escherichia coli thrA, thrB,* and *thrC* genes of the threonine biosynthetic operon each share a short module of about 35 amino acids. In another example, the *carB* gene of *E. coli,* which specifies a subunit of carbamoylphosphate synthetase, shows an internal duplication approximately half the size of the entire gene. Others have found that the two domains of gram-negative bacterial tetracycline efflux proteins are encoded by genes that evolved by duplication of an ancestral module having half the size of the present-day gene(s) (Rubin et al., 1990). Finally, the most extensively documented example comes from my research and pertains to the pair of genes, *hisA* and *hisF,* encoding two TIM barrel proteins.

OPERON DUPLICATION

Duplications of entire gene clusters, i.e., entire operons, are possible. For example, if operon A, responsible for the biosynthesis of amino acid

A, were to duplicate to give rise to a pair of paralogous operons, then one diverges and evolves so that the encoded enzymes catalyze reactions leading to a different amino acid, B. This event will both extend the metabolic abilities of the cell and increase its genome size. We should find vestiges of such duplications by comparing the amino acid sequences of the proteins encoded by operons A and B. An example has been recently reported for some genes involved in nitrogen fixation (*nif*). No other examples of paralogous operons have been described. How does and/or did operon duplication(s) occur during cellular evolution? Further sequence analysis is needed since molecular clocks and functional constraints are different for each protein. Evidence for the duplication event may now be blurred by evolution.

Fate of Duplicated Genes

STRUCTURAL FATE

Duplication events can generate genes arranged in tandem. In addition, duplication by recombination involving different DNA molecules or

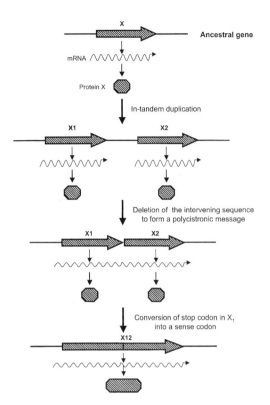

FIGURE 4 Gene elongation: the duplication of an ancestral gene and the subsequent fusion of the two homologs to produce a longer protein.

transposition can generate a copy of a DNA sequence at a different location within the genome. If an in-tandem duplication occurs, at least two different scenarios for the structural evolution of the two copies can be depicted: (i) the two genes undergo an evolutionary divergence and so become paralogous genes; or (ii) the two genes fuse, doubling their original size and forming an elongated gene constituted by two paralogous modules. If the two copies are not arranged in-tandem: (i) they may become paralogous genes or (ii) one copy may fuse to an adjacent gene, with a different function, giving rise to a mosaic or chimeric gene that potentially may evolve to perform another metabolic role(s). Tandem duplications of DNA stretches are often the result of an unequal crossing-over between two DNA molecules,

but other processes, such as replication slippage, may be invoked to explain the existence of tandemly arranged paralogous genes. The presence of paralogous genes at different sites within a microbial genome is probably the result of ancient activity of transposable elements.

FUNCTIONAL FATE

Given that DNA segments duplicate, then how can they evolve differentially? They may, for instance, both retain their original function; in this way, the organism produces a larger quantity of a unique RNA molecule and/or protein. Alternatively, one of the two copies may become inactivated by mutation and turn into a pseudogene, that is, a functionless gene. Finally, and mostly important for the evolution, duplication may result in the acquisition of novel metabolic abilities. This happens if one of the duplicated genes maintains its original function while the other accumulates base changes (a combination of point mutations), altering its structure/active site so it becomes able to perform a different metabolic task (new phenotype).

Section Summary
- One of the most important forces driving the evolution of genes and genomes is the duplication and divergence of DNA sequences of different sizes, which might include a whole gene, a part thereof, or a cluster of genes. This process may allow the formation of new genes from existing ones.
- Duplicated genes undergo different functional fates. One of the two copies might accumulate mutations, turning into a new, different gene, or could be inactivated, becoming a functionless gene. If the two copies do not diverge, they can both retain their original function, producing a larger quantity of the same product.
- Duplication events may generate elongated genes (in-tandem duplication) or chimeric genes (not in-tandem duplication) as a consequence of fusion events with the adjacent gene.

- What are the differences between orthologous and paralogous genes? Orthologs evolved in different organisms as a consequence of speciation; paralogs derived by duplication of a common ancestral gene and perform different, but often similar, functions in the same organism.

ROLE OF GENE DUPLICATION IN EARLY CELL EVOLUTION AND THE ACQUISITION OF NEW METABOLIC FUNCTIONS

It is proposed that life evolved from a primordial soup, containing different organic molecules (many of which are used by the extant life forms), probably formed spontaneously during Earth's first billion years. This soup of nutrient compounds was available for the early organisms, so they had to do a minimum of biosynthesis. This is frequently referred to as the Oparin-Haldane theory. Experimental support for this hypothesis was obtained in 1953 when S. L. Miller showed that amino acids and other organic molecules are formed under atmospheric conditions thought to represent those on the early Earth. If we assume that life arose in a prebiotic soup containing most, if not all, of the necessary small molecules, then there was a large potential supply of nutrients available on the primitive Earth. These nutrients were both the growth and energy supply for numerous ancestral organisms. From their growth there was a rapid depletion of these nutrients, and this depletion would have imposed a strong selective pressure on those primordial microorganisms able to synthesize these molecules.

Therefore, the emergence of basic biosynthetic pathways represents one of the major and crucial events during the early evolution of life. Their appearance and refinement allowed primitive organisms to become increasingly less dependent on exogenous sources of amino acids, purines, and other compounds. But how did these metabolic abilities originate? This is still an open question, but several different theories have been suggested to account for the establishment of anabolic routes. These

explanations include the following: the retrograde hypothesis; the "patchwork" theory; biosynthetic routes evolving forward, i.e., from simple precursors to complex end products; and metabolic pathways that appear as a result of the gradual accumulation of mutant enzymes with minimal structural changes. All these theories are based on gene duplication. Duplication rates were potentially high because of a lack of DNA repair mechanisms combined with high levels of ionizing and ultraviolet radiation in the early Precambrian environment. Since prokaryotes are haploid organisms, genetic changes would be expressed as soon as they arose, resulting in a rapid gene and metabolic evolution.

Important insights into the evolutionary development of microbial metabolic pathways can be obtained by sequence comparisons, comparative physiology, and laboratory studies in which new substrates are used as carbon, nitrogen, or energy sources. The latter are the so-called "directed-evolution experiments," in which a microbial (typically, prokaryotic) population is subjected to a strong selective pressure that leads to the establishment of new phenotypes capable of exploiting different substrates (Mortlock and Gallo, 1992). By assuming that the processes involved in acquiring new metabolic abilities are similar to those found in natural populations, "directed-evolution experiments" can provide useful insights in early cellular evolution.

The Horowitz Retrograde Hypothesis

The first attempt to explain in detail the origin of metabolic pathways was made by Horowitz in 1945. He based his explanation on two pieces of work. The first, in 1938, was Oparin's heterotrophic hypothesis on the origin of life, and the second was the one-to-one correspondence between genes and enzymes noticed by Beadle and Tatum. Horowitz suggested that biosynthetic enzymes had been acquired by way of gene duplication that took place in the reverse order found in current pathways. This idea, also known as the Retrograde Hypothesis, states that if the contemporary biosynthesis of compound "A" requires the sequential transformations of precursors "D," "C," and "B" via the corresponding enzymes, the final product "A" of a given metabolic route was the first compound used by the primordial heterotrophs (Fig. 5). When "A" became depleted from the primitive soup, the transformation of a chemically related compound "B" into "A" catalyzed by enzyme "a" would lead into a simple, one-step pathway. The selection of variants having a mutant "b" enzyme related to "a" via a duplication event and capable of mediating the transformation of a chemically related molecule "C" into "B," would lead into an increasingly complex route, a process that would continue until the entire pathway was established in a backward fashion, starting with the synthesis of the final product, then the penultimate pathway intermediate, and so on down the pathway to the initial precursor (Fig. 5). Twenty years later, the discovery of operons prompted Horowitz to restate his model, arguing that it was supported also by the clustering of genes, which could be explained by a series of tandem duplications of an ancestral gene; in other words, genes belonging to the same operon should have formed a paralogous gene family.

The retrograde hypothesis establishes a clear evolutionary connection between prebiotic chemistry and the development of metabolic pathways, and it may be invoked to explain some routes. However, the origin of many other anabolic routes cannot be understood in terms of their backward development because they involve many unstable intermediates. It has been also argued that the Horowitz hypothesis does not account for the origin of catabolic pathway-regulatory mechanisms and for the development of biosynthetic routes involving dissimilar reactions.

The Patchwork Hypothesis

Gene duplication has also been invoked by another hypothesis proposed to explain the origin and evolution of metabolic pathways, the so-called "patchwork" hypothesis, according to which metabolic pathways may have been assembled through the recruitment of primitive enzymes that could react with a wide range of chemically related substrates. Such relatively slow, nonspecific enzymes may have enabled primitive cells containing small genomes to overcome their limited coding capabilities.

Figure 6 presents a schematic three-step example of the patchwork hypothesis. In Fig. 6a, the ancestral enzyme E1 endowed with low substrate specificity is able to bind to three substrates (S1, S2, and S3) and catalyze three different, but similar reactions. In Fig. 6b, a paralogous duplication of the gene encoding enzyme E1 and the subsequent divergence of the new sequence lead to the appearance of enzyme E2 with an increased specificity. In Fig. 6c, a further duplication event occurred leading to E3 showing a diversification of function and narrowing of specificity. In this

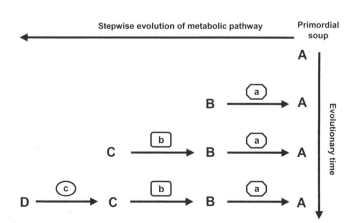

FIGURE 5 The Horowitz hypothesis. As each intermediate (A through D) becomes exhausted in the primordial soup, the successful organisms evolve the next step in the pathway. Thus, as A is exhausted, B (the next intermediate) becomes the prime material source, and so on.

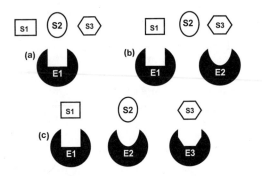

FIGURE 6 Patchwork hypothesis on the origin and evolution of metabolic pathways.

way, the ancestral enzyme E1 belonging to a given metabolic route is "recruited" to serve a single (Fig. 7a) or other novel pathways (T–Z, M–O, and F–I; Fig. 7b). In addition, the patchwork hypothesis may permit the evolution of regulatory mechanisms coincident with the development of new pathways.

The broad substrate specificity of some enzymes means that they can catalyze a class of chemical reactions, and this provides a support for the patchwork theory. Sequence comparisons of enzymes catalyzing different reactions in the biosynthesis of threonine, tryptophan, isoleucine, and methionine indicate that these proteins have evolved from a common ancestral molecule active in several metabolic pathways.

Case Study 1: a Cascade of Gene and Operon Duplications: Nitrogen Fixation

Nitrogen fixation, the biological conversion of atmospheric nitrogen to ammonia, is widespread in bacteria and archaea. The *nif* genes of the free-living nitrogen-fixer (diazotroph) *Klebsiella pneumoniae* are clustered in a single chromosomal region and are organized into several operons (Fig. 8). During the past few years, an increasing number of nitrogen fixation (*nif*) genes have been identified, cloned, sequenced, and analyzed. This has led to the conclusion that the basic features of nitrogen

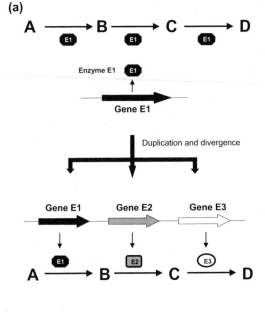

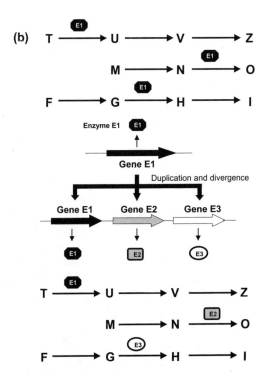

FIGURE 7 Evolutionary divergence of an ancestral enzyme (E1) with a low specificity catalyzing similar reactions in the same (a) or different (b) metabolic route(s).

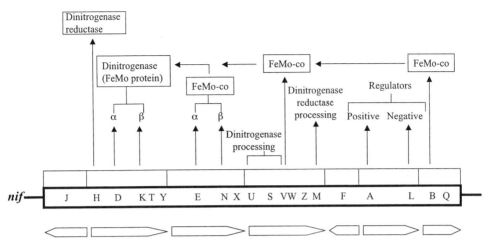

FIGURE 8 Organization of *nif* genes in *Klebsiella pneumoniae*.

fixation are strongly conserved. Nitrogenase, the enzyme responsible for nitrogen fixation, shows a high degree of conservation in structure, function, and amino acid sequence across phylogenetic groups. All known Mo-nitrogenases consist of two components, component I (also called dinitrogenase or Fe-Mo protein), an $\alpha_2\beta_2$ tetramer encoded by *nifD* and *nifK* genes, and component II (dinitrogenase reductase or Fe-protein), a homodimer encoded by the *nifH* gene. Nitrogenase contains two unusual rare metal clusters. One of them is the iron-molybdenum cofactor (FeMo-co), which is considered to be the site of dinitrogen reduction and whose synthesis requires the products of another pair of genes, *nifEN*. Despite the detailed information available, it is only recently that some light has been shed on the origin and evolution of *nif* genes and on the molecular mechanisms that might have been involved in shaping this pathway (Fani et al., 2000). A detailed analysis on the two pairs of genes mentioned above (*nifDK* and *nifEN*) shows that their products share some common features. First, both of them encode a tetrameric ($\alpha_2\beta_2$ and N_2E_2) enzymatic complex; second, the products of *nifE* and *nifN* genes are structurally homologous to the products of *nifD* and *nifK*, respectively (Brigle et al., 1987). Finally, those diazotrophs in

which *nifDK* and/or *nifEN* have been characterized share the same gene organization. The four genes are clustered in operons where the two genes of each pair are contiguous and arranged in the same order (*nifDK* and *nifEN*). The similarity between them shows that the four *nifDKEN* genes belong to a paralogous family, and we proposed a two-step model in the evolutionary process leading to these four genes. Starting with an ancestor gene, an intandem duplication event gave rise to a bicistronic operon. This then duplicated leading to the present-day *nifDK* and *nifEN* operons (Fig. 9).

If the ability to fix nitrogen was a primordial property, then the paralogous duplication events leading to the two operons predated the divergence of archaea and bacteria. Thus, the function(s) performed by the primordial enzyme (X) would have evolved because of the composition of the atmosphere. Therefore, the hypothesis of an ancient origin of nitrogen fixation raises the question of whether the composition of the early atmosphere was appropriate. There is no agreement on the composition of the primitive atmosphere. Theories vary from strongly reducing to neutral; but it is generally accepted that O_2 was absent, an essential prerequisite for the evolution of an ancestral nitrogenase, because free

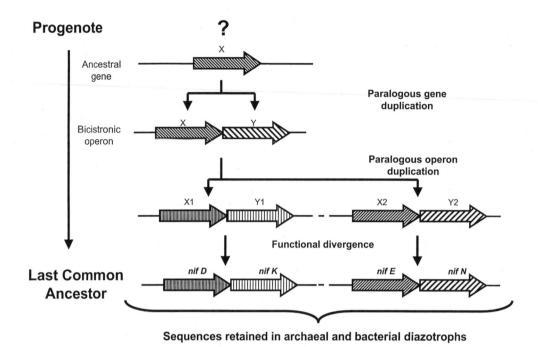

FIGURE 9 Origin and evolution of *nifD*, *nifE*, *nifK*, and *nifN* genes found in present-day bacterial and archaeal diazotrophs: a cascade of DNA duplications and divergence.

oxygen inactivates nitrogenase. The first living microorganisms were probably heterotrophic anaerobes and depended on abiotically produced organic matter for their metabolism. The evolution of nitrogenase represented a necessary event for the first cells on a planet whose atmosphere was neutral, containing dinitrogen, but not ammonia (Fig. 10, scenario 1). In fact, if ammonia was required by the primitive microorganisms for their macromolecular syntheses, then its absence must have imposed an important selective pressure. This would have strongly favored cells that had evolved a system to synthesize ammonia from an abundant source, such as atmospheric dinitrogen. Therefore, the ancestral enzyme might have been a "nitrogenase," albeit slow and inefficient. Yet with its low substrate specificity, it would be able to react with a wide range of compounds with a triple bond (see below). This primordial nitrogenase might too have been an iron-dependent, molybdenum-free enzyme. Alternatively, if the early atmosphere was a reducing one and contained free ammonia (Fig. 10, scenario 2), then the evolution of a nitrogen-fixation system was not a prerequisite because of the abundance of abiotically produced ammonia. Why evolve a nitrogenase in these conditions? The answer to this question might be found in the catalytic properties of nitrogenase. The enzymatic complex is known to reduce other molecules, such as acetylene, hydrogen azide, hydrogen cyanide, and nitrous oxide too and all contain a triple bond. The low substrate specificity of the enzyme has led some to speculate that nitrogenase was initially involved in detoxifying cyanides and other chemicals present in the primitive reducing atmosphere (Silver and Postgate, 1973). Thus, the primitive enzyme encoded by the ancestor gene would not have been a nitrogenase, but a detoxyase, a detoxifying enzyme that might have been molybdenum-free and iron-dependent. The first duplication event led to the ancestral bicistronic operon and was followed

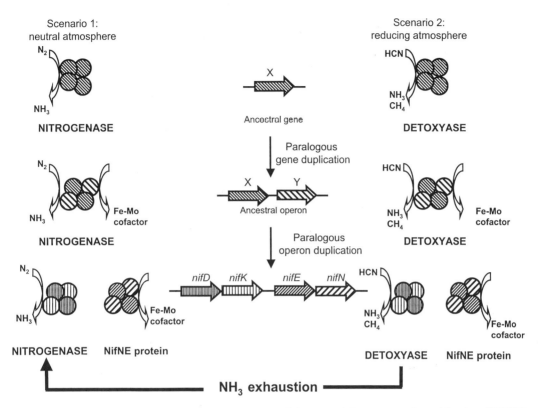

FIGURE 10 Two scenarios depicted for the evolution of the original functions performed by the *nifDKEN* genes and their ancestor genes.

by functional divergence, which refined the specificity of the primitive nitrogenase/detoxyase. Successive duplications of the ancestral operon and divergence led to the appearance of the present day *nifDK* and *nifEN* operons.

Scenario 2 implies that the progressive exhaustion or the limited availability of combined nitrogen would select for a refinement of the enzyme specificity to another triple-bond substrate, dinitrogen. This activity was retained by some bacterial and archaeal lineages to enable growth in nitrogen-deficient environments. The concomitant decrease of free ammonia and cyanides in the atmosphere provided the selective force for the evolution of the detoxyase to a nitrogenase.

Thus, both scenarios agree that the ability to fix nitrogen is an ancient phenotype and one of the pathways in the last common ancestor.

Case Study 2: Building Enzymes by Cassettes: Histidine Biosynthesis

Histidine biosynthesis is one of the best-characterized anabolic pathways. There is information on its operon structure, regulation, and genetic sequence. For more than 35 years, this pathway has been the subject of extensive studies, mainly in the enterobacterium *E. coli* and its close relative *Salmonella enterica* serovar Typhimurium, in both of which biosynthesis appears to be identical (Winkler, 1987). The pathway is unbranched, includes a number of complex and unusual biochemical reactions, and consists of nine intermediates. Three of the biosynthetic enzymes, encoded by the *hisB*, *hisD*, and *hisIE* genes, are bifunctional (Carlomagno et al., 1988).

Histidine biosynthesis plays an important role in cellular metabolism, because it is interconnected to both the de novo synthesis of purines and to nitrogen metabolism (Fig. 11). The connection with purine biosynthesis results from an enzymatic step catalyzed by imidazole glycerol phosphate synthase, a dimeric protein composed of one subunit each of the *hisH* and *hisF* gene products. This heterodimeric enzyme catalyzes the transformation of PRFAR (*N'*-[(5'-phosphoribulosyl)-formimino]-5 amino-imidazole-4-carboxamide-ribonucleotide) into AICAR (5'-phosphoribosyl-4-carboxamide-5-aminoimidazole), which is then recycled into the de novo purine biosynthetic pathway, and imidazole glycerol phosphate (IGP), which in turn is transformed into histidine. Histidine biosynthesis is connected to nitrogen metabolism by glutamine molecules, which are the source of the final nitrogen atom of the imidazole ring of IGP (Fig. 11). It is likely that histidine was just one of the many organic compounds formed during the long prebiotic period of chemical synthesis. There are several independent clues as to the antiquity of the histidine biosynthesis pathway. First, histidine is present in the active sites of enzymes because of the special properties of the imidazole group. Then the presence of conserved *his* genes in *Archaea*, *Bacteria*, and *Eucarya* suggests that it is an ancient metabolic route.

If primitive catalysts required histidine, then the eventual exhaustion of the prebiotic supply of histidine and histidine-containing peptides must have imposed an important pressure favoring those organisms capable of synthesizing histidine. Hence this metabolic pathway might have been assembled long before the divergence of the three cell lineages. How the *his* pathway originated remains an open question, but detailed analysis derived by the comparison of gene structure has shown that duplication, elongation, and fusion events have played a major role in its assembly (Fani et al., 1998).

Genetic and sequence data show that once the entire pathway was assembled it underwent major rearrangements during evolution. A wide variety of different clustering strategies of *his* genes have been documented, suggesting that many possible histidine gene arrays exist.

Structure and Evolution of Paralogous Histidine Genes

An important portion of the histidine biosynthetic pathway (*hisG*, *hisIE*, *hisH*, *hisC*, *hisA*, and *hisF* gene products) appears to have been assembled by recruitment of preexisting broad-specificity enzymes after gene duplications. The genes *hisA* and *hisF* are the most interesting ones. These two genes, whose products catalyze sequential reactions in histidine biosynthesis, are paralogs that share a similar internal organization. Comparison of these genes led to the suggestion that *hisA* and *hisF* are the result of two

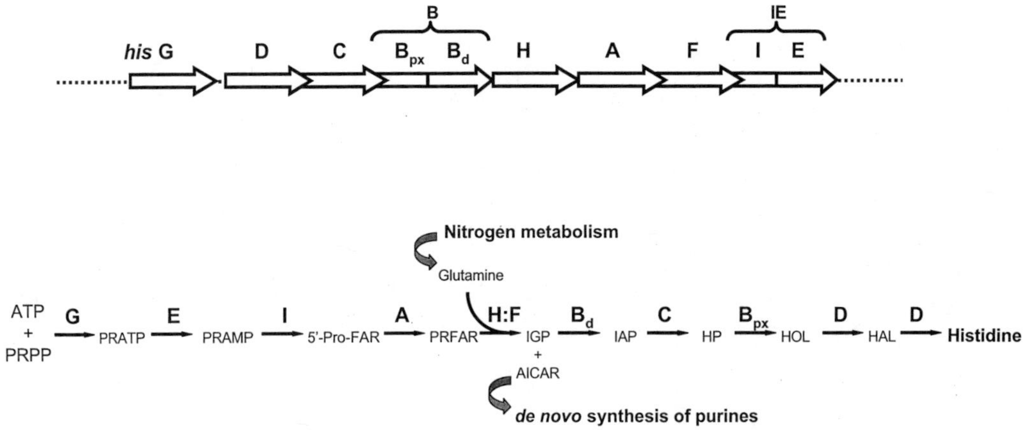

FIGURE 11 Schematic representation of the *E. coli* histidine biosynthetic operon (upper) and pathway (lower).

ancient successive duplications, the first one involving an ancestral module (half the size of the present-day *hisA* gene) and leading by a gene-elongation event to the ancestral *hisA* gene, which in turn underwent a duplication that gave rise to the *hisF* gene (Fig. 12). Since the overall structure of the *hisA* and *hisF* genes is the same in all known (micro)organisms, it is likely that they were part of the genome of the last common ancestor.

The biological significance of the subdivision of *hisA* and *hisF* into two paralogous modules half the size of the current genes resides in their physical structure. The proteins encoded by the two genes belong to the TIM-barrel superfamily, which includes several enzymes containing an α/β-barrel fold. The barrel structure is composed of eight catenated strand-loop-helix-turn units. The β-strands are located in the interior of the protein, forming the staves of a barrel, whereas the α-helices pack around the exterior. Figure 12 shows a plausible model concerning the biological significance of the *hisA* and *hisF* construction. According to this model the ancestral gene encoded for a half-barrel protein, consisting of four β/α-folds (Lang et al., 2000). The active form of this ancestral protein was very likely a homodimer. The elongation event leading to the ancestor of *hisA/hisF* genes enabled the covalent fusion of the two half-barrels to produce a protein (an entire β/α-barrel) capable of broad enzymatic activity. Its activity was refined and optimized by mutational changes occurring over time. Then, the whole-barrel gene underwent a paralogous duplication event, leading to the diversification in activities shown by *hisA* and *hisF*.

Section Summary

- Two hypotheses on the origin and evolution of metabolic pathways exist. The first one, the Horowitz retrograde hypothesis, predicts that an entire metabolic route was assembled by successive duplications of an ancestral gene in a backward fashion, starting with the synthesis of the final product, then the penultimate pathway intermediate, and so on down the pathway to the initial precursor. The patchwork hypothesis is based on the duplication(s) of ancestral gene(s) leading to the progressive increasing of specificity of low-specific

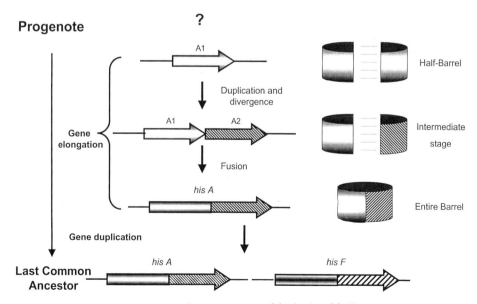

FIGURE 12 Evolutionary origin of the *hisA* and *hisF* genes.

enzymes, which then may be recruited to catalyze similar reactions in different metabolic pathways or sequential steps in the same route.

- The origin and evolution of some genes involved in nitrogen fixation (*nif* genes), which are the results of a cascade of in-tandem DNA duplications, first involved an ancestral gene encoding a nitrogenase or a detoxyase, depending on the composition of the early atmosphere. The resulting ancestral bicistronic operon in turn underwent an in-tandem duplication leading to the ancestors of the present day *nifD*, *nifK*, *nifE*, and *nifN* genes.

- A similar cascade of in-tandem duplication events was responsible for the origin and evolution of the paralogs *hisA* and *hisF*, involved in histidine biosynthesis. An ancestral gene, half the size of the extant *hisA* and *hisF*, and encoding a half-barrel protein was first elongated leading to the ancestor of the whole-barrel gene. This, in turn, duplicated leading to *hisA* and *hisF*.

- Duplication of DNA sequences may increase the genome size and the metabolic abilities of the ancestral organisms.

SUMMARY OF THE ROLE OF GENE DUPLICATION IN METABOLIC PATHWAY EVOLUTION

In the course of evolution different molecular mechanisms have acted cooperatively to provide for cellular demands for new metabolic abilities. It is clear that the duplication of DNA sequences is a major force in genome evolution. Sequences concerned with protein domains and motifs, entire genes, operons, parts of genomes, and entire chromosomes allowed the ancestral cellular genomes to increase their size and complexity.

Yet the recognition of the important role proposed for these ancient duplication events does not provide answers for the emergence of the "starter-types," i.e., enzymes that did not arise in this manner, and whose number has been estimated to range from 20 to 100 (Lazcano and Miller, 1996). It is plausible that the starter types encoded enzymes of low substrate specificity as suggested by the patchwork hypothesis. Even if primitive enzymes were less specific, ancient cells would have been endowed with significant metabolic abilities, this despite having smaller genomes. The evidence for gene-elongation, gene-duplication, and operon-duplication events suggests that the ancestral forms of life might have expanded their coding abilities and their genomes by "simply" duplicating a small number of mini-genes via a cascade of duplication events, involving DNA sequences of different size.

QUESTIONS

1. What could be the evolutionary significance of duplication of DNA sequences?

2. If we agree with Oparin's idea on the origin of life, how and why did the present-day metabolic pathways originate and evolve?

3. Genes encoding the extant nitrogenase very likely existed long before the appearance of the last common ancestor; is this in accordance with the current theories on the composition of the early atmosphere?

4. Can you find similarities between the origin and evolution of *nifDKEN* and the pair *hisA*–*hisF*? And, if yes, is it possible to depict a "common rule" which might in principle be applied to other genes or gene clusters?

5. How can the starter types originate?

REFERENCES

Brigle, K. E., M. C. Weiss, W. E. Newton, and D. R. Dean. 1987. Products of the iron-molybdenum cofactor-specific biosynthetic genes, *nifE* and *nifN*, are structurally homologous to the products of the nitrogenase molybdenum-iron genes, *nifD* and *nifK*. *J. Bacteriol.* **169:**1547–1553.

Carlomagno, M. S., L. Chiarotti, P. Alifano, A. G. Nappo, and C. B. Bruni. 1988. Structure of the *Salmonella typhimurium* and *Escherichia coli* K-12 histidine operons. *J. Mol. Biol.* **203:**585–606.

de Rosa, R., and B. Labedan. 1998. The evolutionary relationships between the two bacteria *Escherichia coli* and *Haemophilus influenzae* and their putative last common ancestor. *Mol. Biol. Evol.* **15:**17–27.

Fani, R., R. Gallo, and P. Liò. 2000. Molecular evolution of nitrogen fixation: the evolutionary

history of *nifD*, *nifK*, *nifE*, and *nifN* genes. *J. Mol. Evol.* **51**:1–11.

Fani, R., E. Mori, E. Tamburini, and A. Lazcano. 1998. Evolution of the structure and chromosomal distribution of histidine biosynthetic genes. *Origins Life Evol. Biosph.* **28**:555–570.

Horowitz, N. J. 1945. On the evolution of biochemical synthesis. *Proc. Natl. Acad. Sci. USA* **31**:153 157.

Labedan, B., and M. Riley. 1995. Widespread protein sequence similarities: origin of *Escherichia coli* genes. *J. Bacteriol.* **177**:1585–1588.

Lang, D., R. Thoma, M. Henn-Sax, R. Sterner, and M. Wilmanns. 2000. Structural evidence for evolution of the b/a barrel scaffold by gene duplication and fusion. *Science* **289**:1546–1550.

Lazcano, A., and S. L. Miller. 1996. The origin and early evolution of life: prebiotic chemistry, the pre-RNA world, and time. *Cell* **85**:793–798.

Mortlock, R. P., and M. A. Gallo. 1992. Experiments in the evolution of catabolic pathways using modern bacteria, p. 1–13. *In* R. P. Mortlock (ed.), *The Evolution of Metabolic Functions*. CRC Press, Boca Raton, Fla.

Mushegian, A. R., and E. V. Koonin. 1996. Gene order is not conserved in bacterial evolution. *Trends Genet.* **12**:289–290.

Rubin, R. A., S. B. Levy, R. L. Heirinkson, and F. J. Kezdy. 1990. Gene duplication in the evolution of the two complementing domains of Gram-negative bacterial tetracycline efflux proteins. *Gene* **87**:7–13.

Silver, V. S., and J. R. Postgate. 1973. Evolution of asymbiotic nitrogen fixation. *J. Theor. Biol.* **40**:1–10.

Winkler, M. E. 1987. Biosynthesis of histidine, p. 395–411. *In* F. C. Neidhardt, J. L. Ingraham, K. B. Low, B. Magasanik, M. Schaechter, and H. E. Umbarger (ed.), *Escherichia coli and Salmonella typhimurium: Cellular and Molecular Biology*, vol. 1. American Society for Microbiology, Washington, D.C.

FURTHER READING

Alifano, P., R. Fani, P. Li, A. Lazcano, M. Bazzicalupo, M. S. Carlomagno, and C. B. Bruni. 1996. Histidine biosynthetic pathway and genes: structure, regulation and evolution. *Microbiol. Rev.* **60**:44–69.

Bryson, V., and H. J. Fogel (ed.). 1965. *Evolving Genes and Proteins*. Academic Press, New York.

Carlile, M. J., and J. J. Skehel (ed.). 1974. *Evolution in the Microbial World*. Cambridge University Press, Cambridge, England.

Go, M., and M. Nosaka. 1987. Protein architecture and the origin of introns. *Cold Spring Harbor Symp. Quant. Biol.* **52**:915–924.

Gogarten, J. P., and L. Olendzenski. 1999. Orthologs, paralogs and genome comparisons. *Curr. Opin. Genet. Dev.* **9**:630–636.

Haldane, J. B. S. 1932. *The Causes of Evolution*. Longman and Green, London, England.

Hartman, H., and K. Matsuno (ed.). 1992. *The Origin and Evolution of the Cell*, p. 163–182. World Scientific, River Edge, N.J.

Jensen, R. A. 1976. Enzyme recruitment in evolution of new function. *Annu. Rev. Microbiol.* **30**:409–425.

Li, W. H., and D. Graur. 1991. *Fundamentals of Molecular Evolution*. Sinauer Associates, Sunderland, Mass.

McLachlan, A. D. 1987. Gene duplication and the origin of repetitive protein structures. *Cold Spring Harbor Symp. Quant. Biol.* **52**:411–420.

Miller, S. L. 1953. A production of amino acids under possible primitive earth conditions. *Science* **117**:528–529.

Muller, H. J. 1935. The origination of chromatine deficiencies as minute deletions subject to insertion elsewhere. *Genetics* **17**:237–252.

Ohno, S. 1970. *Evolution by Gene Duplication*. Springer-Verlag, New York, N.Y.

Oparin, A. I. 1938. *The Origin of Life*. MacMillan Publishing, New York, N.Y.

MULTIPLE CHROMOSOMES

Chris Mackenzie, Samuel Kaplan, and Madhusudan Choudhary

5

Chris Mackenzie, Samuel Kaplan, and Madhusudan Choudhary,
Department of Microbiology and Molecular Genetics,
University of Texas Medical School-Houston, 6431 Fannin
St., Houston, TX 77030.

A mind is not a vessel to be filled, rather a fire to be kindled. . . .

—Plutarch

We begin in this fashion because this chapter has few facts to provide to you, but we hope it will set you thinking. Many microbiologists had been slow to accept or even skeptical to consider the idea of multiple different chromosomes in prokaryotes. Consequently, although we may know something, or even a great deal about the genes or their interactions that reside on these different chromosomes, we *still* know little about the interplay or status of the chromosomes themselves. We know nothing as to how or why or when multiple chromosomes evolved, why some bacteria have them, and why others do not and so on. Therefore, much of what you read in this chapter is not based on hard experimental evidence; instead, it is based on our musings and speculations. Hopefully this will provide kindling for your fire.

If you were reading this book just over 10 years ago (1989), this chapter would not have been present. Instead there might have been one or more chapters detailing relationships between chromosomes and plasmids, the origins of plasmids, their replication, etc. Until that date, the one immutable fact that was branded into the minds of all microbiology students, including the authors' students, was that bacteria had one single circular chromosome, end of story! This chapter will provide you with a brief time-line from that date to the present, providing both an historical and a

Microbial Evolution: Gene Establishment, Survival, and Exchange,
Edited by Robert V. Miller and Martin J. Day, ©2004 ASM Press, Washington, D.C.

scientific perspective to the dissolution of the earlier dogma. You will also see that scientific debate pertaining to this issue is alive and well. The idea of multiple chromosomes in bacteria, though no longer novel, is still subject to controversy. The controversy stems from questions that require answering and that are crucial to understanding prokaryote/eukaryote cellular evolution: what is a chromosome and how does it differ from a plasmid? This question, perhaps more than any other, is the central thread that is woven into this chapter. You will see that this is not as easy a question to answer as you might think. Even if we don't answer it to your satisfaction, if this chapter serves to address your imagination, to make you think critically, or perhaps experimentally, or even be willing to take sides in a solid intellectual discussion, then we have succeeded.

THE WIND OF CHANGE

When it became appreciated that bacteria possessed, in the eukaryotic sense, genes and genetic systems, it was logical to suggest this genetic information must reside within a "chromosome." It was then shown that *Escherichia coli* had a single chromosome that was "circular" both genetically and physically. This was a surprise, because eukaryotic chromosomes were known to exist in greater multiplicities and were linear. Even though the *E. coli* genome differed markedly from those of eukaryotes, by being the first bacterial chromosome to be analyzed, it became the model for all subsequent bacterial genomes. Later, similar single circular genomes were found in other bacteria, e.g., *Bacillus subtilis*, eventually leading to the dogma that only single, circular chromosomes exist in bacteria.

By the mid to late 1980s pulsed-field gel electrophoresis had become a valuable technique, and many laboratories began the process of making physical restriction maps of the genome of their favorite organism. In our laboratory the favorite topic of discussion was the facultative photoheterotroph, *Rhodobacter sphaeroides* 2.4.1. This bacterium, which is used as a model system for studies into the biophysical, structure–function, and regulatory

elements of bacterial photosynthesis and redox flow, is a member of the α-subgroup of the purple, nonsulfur Proteobacteria. In 1987, Antonius Suwanto, a young graduate student, was assigned the task of constructing a physical map of the *R. sphaeroides* genome. This assignment was not a trivial one, but was essential to the placement of many genes involved in photosynthetic membrane function and regulation relative to one another within the genome of this organism. Antonius completed the physical map in less than two years. It was the second physical map to be completed, *E. coli* being the first. What he found on completion was that, rather than forming a single closed circular structure, the genome formed two relatively large (2.9 and 0.9 Mb) closed circles, which he designated chromosome I (CI) and chromosome II (CII), respectively (Suwanto and Kaplan, 1989a, 1989b). There were also five additional, much smaller replicons (pRS241a–e that ranged from ~100 to 42 kb) that he designated plasmids. However, it was the designation of CII as a second, different "chromosome" that was considered controversial. Here, we must explore the rationale as to why Antonius designated CII as a chromosome and, more generally, what really constitutes a prokaryotic chromosome or plasmid.

Section Summary

- Some species of bacteria have more than one chromosome.
- Scientific dogma can be challenged at any time by the acquisition of new data or the reinterpretation of old data, but dogma is not easy to change.

THE METAMORPHOSIS OF A NAME: PLASMID

Before we address what a chromosome might be and also address what a plasmid is, let us turn back the clock. The term "plasmid" was coined in 1952 by Joshua Lederberg. It was an all-encompassing term for any stably maintained extrachromosomal genetic element. The term made no distinction between prokaryotic or eukaryotic origin; therefore, under the plasmid umbrella lay a diverse mixture that included mitochondria, chloroplasts, viruses, phage particles, and "our modern concept of plasmids." The term lay almost unnoticed for many years but was revitalized to the fullest after the advent of gene-splicing technology (cloning) in the mid-1970s. The excitement generated by this new technology resulted in a deluge of publications describing the use of small circular DNA plasmids in cloning experiments. Before long, the word plasmid came to suggest small to relatively small circular, extrachromosomal bacterial DNAs with or without a defined function. By the early eighties, the original meaning of the word had been fully eclipsed, and the new definition had widespread usage.

The problem with definitions in biology is that, sooner or later, someone discovers something that is a legitimate exception to the definition, i.e., definitions are often short lived. So it was with plasmids, where an assortment of different topologies and sizes were soon discovered. For example, it was shown that some plasmids, such as those found in some *Streptomyces* species, were linear rather than circular. In terms of size, it was already known that some plasmids were not small (2 to 25 kb) but could be relatively large (25 to 100 kb or more). This upper size limit was extended with the discovery of the very large circular, "nonchromosomal" DNA in *Rhizobium meliloti*, *Pseudomonas solanacearum*, and others that led, perhaps not surprisingly, to the term megaplasmid. Some of these so-called megaplasmids have been shown to be larger than some bona fide bacterial chromosomes. For example, the pSymA and pSymB megaplasmids of *Sinorhizobium meliloti* are 1.35 and 1.68 Mb, respectively. This makes them slightly larger than the chromosome of the syphilis spirochete *Treponema pallidum* (~1.1 Mb), and two to three times larger than the chromosome of *Mycoplasma genitalium* (0.58 Mb). So why did the discoverers of these megaplasmids not just call them chromosomes, and why would it take nearly 8 more years for Suwanto to further alter the concept by defining the existence of a second chromosome?

Section Summary
- The word "plasmid" has drifted considerably from its original meaning. Chromosomes and plasmids can be circular or linear and both vary widely in size.
- With hindsight and the acquisition of new data, initial definitions and hypotheses often appear naïve.

PLASMIDS: A FUNCTIONAL DEFINITION

A generality that unites most conceptualizations of what constitutes a plasmid is that their genes often encode optional or dispensable characteristics, i.e., ones not necessary for growth under all conditions. That is, many plasmids are known to confer traits, such as heavy-metal tolerance, antibiotic resistance, or substrate utilization abilities, on their bacterial hosts. If the selection for their maintenance is removed, the plasmids can be lost over time with some finite possibility. Such a loss does not make the cell inviable under all possible conditions; therefore, the plasmid does not encode an essential function. Like their smaller counterparts, some megaplasmids can also be lost and therefore can be considered nonessential. Certainly, the rhizobial megaplasmids pSymA and pSymB are absolutely required for the infection of root hairs, nodulation, and nitrogen fixation, but they are not required for host-cell survival. At least, that was what was originally believed. It is now known that pSymB is essential because it encodes the only tRNA that recognizes the CCG codon (Galibert et al., 2001). That is, not all isolates of *Rhizobium* possess both megaplasmids, and pSymA can be cured from cells by using a variety of chemical agents. Therefore, in this context, megaplasmids were plasmid-like rather than chromosome-like, and probably for this reason they were never given the designation of chromosome. Another characteristic associated with some plasmids is their ability to move from one cell to another by virtue of plasmid-specific encoded information. How then does CII of *R. sphaeroides* differ from the megaplasmids?

Section Summary
- Plasmids encode nonessential functions and can be considered dispensable.
- The word "plasmid" is a bit like the word "dog," in that it gives a general concept but fails to describe size, shape, or purpose of possession, and, like plasmids, dogs are not essential but may be highly advantageous under certain circumstances.

CHROMOSOMES: A FUNCTIONAL DEFINITION

The Encyclopedia Britannica (2003) defines a chromosome as "the microscopic, threadlike part of the cell that carries hereditary information in the form of genes." Unfortunately this does not really help us to distinguish plasmids or megaplasmids from chromosomes. Perhaps we can extend the functional distinction between the two, i.e., if we think of plasmids as being dispensable under some conditions then chromosomes can be thought of as being indispensable under any condition. The logic then followed, that if we could cure or remove CII from *R. sphaeroides* and if the cell survived then CII is a megaplasmid and not a chromosome.

Numerous attempts using a variety of methodologies that had been used to successfully cure plasmids in other bacteria were applied in order to cure *R. sphaeroides* of CII. These efforts resulted in the loss of three of the five plasmids, but not CII itself. This is an ambiguous result. Did CII remain because it carried truly essential functions or due to the lack of efficacy in the curing method? This result is even less convincing since we could only cure three of the five plasmids. As might be anticipated from such an approach, it was destined to fail, due to the possibility of an inherent ambiguity from the outset. However, during the course of these early studies it was determined that CI and CII were in single copy in a 1:1 ratio, whereas the five plasmids appeared to have greater copy numbers.

Another approach was to look for and catalog the essential genes on CII. One set of genes that we considered essential for cell survival was the ribosomal RNA genes (*rrn*). These are usually found as *rrn* operons, and hybridization indicated that *R. sphaeroides* possesses three such operons; one (*rrnA*) is located on CI, and two (*rrnB* and *rrnC*) are located on CII. Sequencing showed that within each of these operons (between the 16S and 23S genes) lay an isoleucine and an alanine tRNA. It was further revealed that each possessed an fMet-tRNA gene following the 5S rDNA region. It was also shown that the *rrnA* operon on CI could be disrupted, and all the rRNA in the cell could be supplied by CII. Further, it was revealed that the most actively transcribed *rrn* operon was *rrnB* on CII (Dryden and Kaplan, 1993). These features were thought sufficient to suggest that CII was not plasmid-like but more chromosome-like in nature, hence the designation, chromosome II (CII).

Of course, this finding by itself was not sufficient to convince everyone and many thought that naming CII a chromosome was grandstanding. Spurred on in part by this criticism, we looked for other genes that mapped to CII and might be considered truly "essential." We found many, such as the chaperone *groEL*, alternative sigma factor *rpoN*, and enzymes of the reductive Calvin cycle *cbbA* and *cbbP*, and *hemT* encoding 5-aminolevulinic acid synthase to name but a few. However, all these were found to exist as duplicate copies or functions on CI.

By this juncture, we had discovered that we could disrupt any pair of the three rRNA operons and the cell would survive, albeit at a slower growth rate. We therefore required some unimpeachable way to determine that CII was truly essential. We set upon another tack; we made many random transposon (Tn5) insertions into the genome and looked for those insertions that mapped to CII and that also caused auxotrophy. A number of these were found, including those for histidine, thymine, serine, uracil, and tryptophan (Choudhary, 1994). This result suggested that the transposon insertions into CII had disrupted genes that could not be complemented by functions on CI. Therefore, CII did indeed encode unique and essential functions, i.e., essential under all conditions of growth using the wild type as the standard.

We analyzed the genes of the tryptophan biosynthetic pathway in considerable detail and found that they were localized to three distinct regions of the genome: two locations on CI and one location on CII. Of particular interest, the final step of the pathway, which is carried out by the heterodimeric enzyme tryptophan synthase, was encoded by the gene *trpA* on CI and *trpB* on CII. A division of labor clearly exists between the two chromosomes. This was further reinforced by two additional findings: *rpsA1*, the gene that encodes the S1 subunit (the largest subunit of the small ribosome, which is essential for translation), and *cmk* (a gene that encodes the essential enzyme cytidylate monophosphate kinase) existed solely in single copy on CII (Mackenzie et al., 1999). We therefore felt fully justified in our earlier conclusion, i.e., designating CII as a chromosome. Indeed, we believe this idea of essentiality is the keystone for granting a replicon chromosomal status. The advantage of the essentiality test is that it removes topology, size, replication origin type, etc., as criteria for judging the status of a replicon. In essence everything resolves around the concept of "essential" or "nonessential," which are basic plasmid and chromosome definitions. These definitions themselves are not without their own problems, but let us wait until later to address them.

Section Summary

- Chromosomes encode essential functions. The functions encoded by a replicon may be the most useful method in defining it as a plasmid or a chromosome.
- "Form follows function"—Frank Lloyd Wright

MULTIPLE CHROMOSOMES AND MEGAPLASMIDS: WHO HAS THEM?

In carrying out this survey let us not differentiate, for the moment, between bacteria with multiple chromosomes and those with essential megaplasmids. Let us place them in a single contrived group; organisms with multiple chromosome-like replicons (MCLRs). The existence of MCLRs was initially demonstrated in *R. meliloti* and *P. solanacearum* with standard agarose gel electrophoresis. The advent of pulsed-field gel electrophoresis gradually revealed that MCLRs were widespread in bacteria, especially among members of the α-subgroup of the Proteobacteria (of which *R. sphaeroides* is a member). For some organisms, specifically the vibrios and deinococci, the revelation of their true genome architecture and the presence of MCLRs awaited complete genome sequencing.

At least 44 species of bacteria, as of this writing, are known to have MCLRs (see Table 1). Bacteria containing MCLRs are dispersed throughout many bacterial groups including the spirochetes, deinococci, and Proteobacteria. To date most of the species containing MCLRs have been found within the Proteobacteria, especially within members of the α-subgroup. More species within this subgroup, which includes the genera *Agrobacterium*, *Rhizobium*, *Sinorhizobium*, *Brucella*, and *Rhodobacter*, have been shown to contain more MCLRs than all the other groups combined (Jumas-Bilak et al., 1998). It has been observed that many members of this subgroup have unusual associations with plants or animals (such as the formation of nodules, tumors, or galls), and some researchers have speculated that the genome architecture may in some way facilitate these relationships. We can also speculate that this unusual association has been manifest over evolutionary time and perhaps has played a role in the origins of intracellular symbionts. We still lack scientific evidence to substantiate this hypothesis, and it will be interesting to see if over time this is supported or whether the possession of such architecture is purely coincidental.

Not all members of the α-subgroup have MCLRs, indeed very closely related species differ widely in their genome architecture, e.g., *Bradyrhizobium japonicum* strain 110 has a single 8.7-Mb chromosome, whereas many other members of the family *Rhizobiaceae* have been shown to possess MCLRs (Jumas-Bilak

TABLE 1 Bacterial genomes containing multiple chromosome-like replicons

Group or subdivision	Bacterium	Strain/origin (T = type strain)	No. of replicons[a]	Sizes of replicons (Mb)[b]	Genome size (Mb)	Method of analysis[c]
Alpha-proteobacteria	Agrobacterium radiobacter bv. 1	CFBP 2414[T]	3(2)	3.0*, 2.1(L)*, 0.68	5.78	PFGE
	A. tumefaciens bv. 1	ATCC 23308[T]	4(2)	3.0*, 2.1(L)*, 0.55, 0.25	5.90	PFGE
	A. tumefaciens	C58	4(2)	3.0*[2], 2.0*[2](L), 0.55, 0.21	5.70	Genome
	A. tumefaciens bv. 3 (vitis)	CFBP 2721	4(2)	3.5*, 1.1*, 0.5, 0.25, 0.2	5.55	PFGE
	A. tumefaciens bv. 3 (vitis)	CFBP 2602	4(2)	3.5*, 1.05*, 0.25, 0.18	4.98	PFGE
	A. rhizogenes bv. 2	ATCC 11325[T]	4(1)	4.0*, 2.7, 0.285(NS), 0.25(NS)	7.23	PFGE
	A. rhizogenes bv. 2	K84	4(1)	4.0*, 2.7, 0.365, 0.2	7.26	PFGE
	A. rubi	ATCC 13335[T]	4(2)	3.1*, 1.8(L), 0.55, 0.285	5.73	PFGE
	Azospirillum brazilense	FP2	7(4)	2.5*, 1.72*, 0.81(L)*, 0.7(L), 0.63(L)*, 0.17, 0.15	6.70	PFGE
	A. lipoferum	Sp59b	7(6)	2.6*, 1.8*, 1.38*, 1.18(L)*, 0.97(L)*, 0.71(L), 0.65(L)*	9.70	PFGE
	A. amazonense	Y2	4(2)	2.7*, 1.2, 1.7*, 0.75	7.30	PFGE
	A. irakense		4(3)	2.4*, 1.2*, 0.95*, 0.22	4.80	PFGE
	A. halopraeferens		5(4)	2.6*, 1.2*, 0.98*, 0.92*, 0.22	5.90	PFGE
	Brucella melitensis	16M	2(2)	2.1*[2], 1.2*	3.30	Genome
	B. abortus	544	2(2)	2.1*[2], 1.2*	3.30	PFGE
	B. suis	1330	2(2)	2.1*[2], 1.2*	3.30	PFGE
	B. canis	RM/666	2(2)	2.1*[2], 1.2*	3.30	PFGE
	B. ovis	63/290	2(2)	2.1*[2], 1.2*	3.30	PFGE
	B. neotomae	5K33	2(2)	2.1*[2], 1.2*	3.30	PFGE
	Mesorhizobium loti	MAFF303099	3(1)	7.0*[2], 0.35, 0.2	7.60	Genome
	Mycoplana dimorpha	ATCC 4279[T]	3(NE)	3.2, 0.3, 0.15	3.15	PFGE PFGE
	Ochrobactrum anthropi	ATCC 49188[T]	4(2)	2.7*, 1.9*, 0.15, 0.1	4.85	PFGE PFGE
	O. anthropi	LMG 3301	4(NE)	2.7, 1.9, 0.05, <0.05	4.70	PFGE
	Paracoccus denitrificans	Pd1222	3(NE)	2.0, 1.1(L), 0.64(L)	3.74	PFGE PFGE
	Phyllobacterium myrsinacearum	ATCC	5(1)	3.5*, 0.68, 0.5, 0.365, 0.285	5.33	PFGE PFGE
	Rhizobium leguminosarum bv. trifolii	ATCC 14480[T]	4(1)	4.6*, 1.1, 0.65, 0.45	6.80	PFGE PFGE PFGE
	Rhizobium leguminosarum bv. phaseoli	ATCC 14482[T]	4(NE)	4.6, 1.1, 0.45, 0.285	6.435	PFGE PFGE PFGE
	Rhodobacter capsulatus	ATCC 11166	2(1)	3.8*, 0.15	3.95	PFGE
	R. capsulatus	SB1003	2(1)	3.8*, 0.134	3.93	Genome
	R. sphaeroides[d]	2.4.1	7(2)	3.110*, 0.9*[2], 0.114, 0.113, 0.104, 0.101, 0.037	4.48	Genome
	Sinorhizobium fredii	ATCC 35423[T]	3(1)	4.0*, 2.2, 0.45	6.65	PFGE
	S. meliloti	1021	3(1)	3.65*[3], 1.68, 1.35	6.70	Genome
	S. meliloti	2011	3(1)	3.5*, 1.7, 1.4	6.60	PFGE

(continues)

TABLE 1 *(continued)*

Group or subdivision	Bacterium	Strain/origin (T = type strain)	No. of replicons[a]	Sizes of replicons (Mb)[b]	Genome size (Mb)	Method of analysis[c]
Betaproteo-bacteria	*Burkholderia cepacia*[e]	ATCC 25416	4(3)	3.7*[4], 3.2*, 1.1*, 0.2	8.20	PFGE
	B. cepacia[e]	C1 (CF patient, Wales)	3(NE)	3.1, 1.9, 0.1	5.10	PFGE
	B. cepacia[e]	NCPPB 2993 (rotting onion, USA)	3(NE)	3.4, 3.1, 1.1, 0.2	7.80	PFGE
	B. cepacia[e]	NCIMB 9087 (rotting tree trunk, Trinidad)	3(NE)	3.2, 3.0	6.20	PFGE
	B. gladioli[e]	NCPPB 2478	2(NE)	3.2, 3.0	5.20	PFGE
	B. glumae[e]	PG1	2(NE)	3.8, 3.0	6.80	PFGE
	Ralstonia (Alcaligenes) eutropha	H16	3(2)	4.1*, 2.9*, 0.44	7.10	PFGE
	Ralstonia solanacearum	GMI1000	2(2)	3.7*[3], 2.1*	5.80	Genome
Gammaproteo-bacteria	*Buchnera aphidicola* sp. APS		3(1)	0.64*, 0.078, 0.073	0.80	Genome
	Vibrio cholerae	El Tor N16961	2(1)	3.0*[8], 1.0	4.00	Genome
	V. parahaemolyticus	AQ4673	2(2)	3.2*[4], 1.9*	5.10	PFGE
	V. mimicus	2031	2(NE)	2.4, 1.6	4.00	PFGE
Thermus/deino-coccus group	*Deinococcus radiodurans*	R1	4(1)	2.6*[6], 0.4, 0.2, 0.045	3.30	Genome
Spirochetes	*Leptospira interrogans* serovar ictero-haemorrhagiae	Verdun	2(1)	4.6*, 0.35	5.00	PFGE

[a] This column gives the number of replicons that make up the genome of each organism. In parentheses are the numbers of replicons that hybridize or are shown to have 16S ribosomal RNA (*rrn*) genes through sequencing. If 16S ribosomal RNA hybridization was not carried out, this is indicated by NE (not examined).

[b] The sizes of the replicons are given in megabase pairs. (L) indicates that a replicon is linear rather than circular. An asterisk (*) indicates that a 16S *rrn* gene or a *rrn* operon has been detected by hybridization or sequencing. A number after an asterisk indicates the number of *rrn* hybridization signals or operons that were detected on the replicons by hybridization or sequencing, e.g., *A. tumefaciens* C58 has the following entry; 3.0*[2], 2.0*[2](L), 0.55, 0.21, meaning that there are two *rrn* operons on the 3.0-Mb replicon, two *rrn* operons on the 2.0-Mb linear replicon, and no *rrn* operons on the smaller replicons.

[c] Indicates whether pulsed-field gel electrophoresis (PFGE) or genome sequencing was the method used to determine the size and number of the replicons and the locations of their ribosomal RNAs.

[d] In addition to *R. sphaeroides* 2.4.1 at least 23 other strains are known to contain multiple chromosomes.

[e] In addition to the four strains of *B. cepacia* given as examples in the table, 25 other strains are known to contain multiple large replicons. Two other *Burkholderia* species, *B. glathei* and *B. pickettii,* have also been shown to contain multiple replicons greater than 1.0 Mb in size.

et al., 1998). It has been suggested that such differences reflect differences of lifestyle. However, these differences may also be a consequence of the criteria that we use to classify bacteria. Perhaps through 16S RNA analysis *Bradyrhizobium* spp. are members of the *Rhizobiaceae*; however, if a different method for classification were used they might be more distantly related. It is also possible that

Bradyrhizobium at one time had a second chromosome that formed a stable cointegrate to form a single large chromosome or simply that such events never arose in this genus.

Section Summary

- MCLRs are widespread among bacteria. Genome architecture varies widely even between closely related bacterial species and may influence lifestyle.

- It has been hypothesized that genome architecture may reflect lifestyle; however, there is still no clear correlation between genome architecture and lifestyle.

ESSENTIALITY: HOW GOOD A TEST IS IT?

The commendable thing about using the "essentiality" criterion is that, as stated previously, it removes all the bias pertaining to replicon size, topology, composition (DNA/RNA), or origin of replication type from the judgment process. However, this single criterion does have a problem of its own, i.e., deciding what "essential" really means. We should also be aware that an essential gene in one organism is not necessarily an essential gene in another, e.g., the ability to make nucleosides de novo may be considered essential in *E. coli* but not in *Mycoplasma genitalium*, where these compounds are supplied by the host. Therefore, we are looking at essentiality within a genome (intragenomic essentiality) rather than between different genomes.

For a replicon to be essential, it must encode at least one gene essential to the survival of the organism under all possible conditions of growth. This moves us forward a bit but still leaves us asking "what is an essential gene" and further, is the occurrence of "one" sufficient to characterize the genome? There are several schools of thought on this. The narrowest view is that an essential gene is one that cannot be disrupted (assuming no polarity effects) because disruption would be lethal (no growth) under ALL growth conditions (let us call these lethal disruption genes [LDGs]). If we assume that these genes exist only in single copy within the genome, this class would include, but not be limited to, genes such as those that encode the core subunits of RNA polymerase (where disruption would prevent transcription) or a protein-encoding gene (that is required for ribosome assembly and ultimately translation). What about a gene or genes encoding steps in the biosynthetic pathway for an amino acid or a nucleic acid? Granted there are many organisms whose wild-type pheno-

type is such that they have an absolute requirement for an essential amino acid. If we subscribe to the unity of protein structure, there are 20 essential amino acids for all living systems. Thus, if an organism in its wild-type form contains the genetic information for an amino acid biosynthetic pathway but loses that ability by mutation, we consider this the loss of an "essential" activity even if the amino acid can be provided exogenously.

A broader class of essential genes would be those where disruption does not render the cell inviable, but makes it "very sick." Such genes would include *recA* or *mutL*. Let us call these genes sick disruption genes, or SDGs. Cells having mutations in these genes would be able to live on "life support" in laboratory conditions, but placing them under natural conditions or under a severe competitive disadvantage in a chemostat would be the equivalent of effectively sentencing them to elimination from the gene pool. (We assume that reversion of the gene to the wild type does not occur.) A more complete list of LDGs and SDGs that reside on CII of *R. sphaeroides* is provided in Table 2.

Section Summary

- Defining a gene as essential is not a simple task. Essential genes fall into two broad classes, those where disruption of the gene results in lethality and those where disruption severely compromises the survival of the cell.
- The disruption of an essential gene ultimately results in the removal of the cell from the gene pool.

PROBLEMS ASSOCIATED WITH MAKING A DECISION

Let us invent for the purposes of classification our first hypothetical bacterium. We isolate this one from a Houston sewer, sequence its genome, and find the following. The genome comprises two replicons: one is 100 kb and encodes all the cell's LDGs, and the other is 5 Mb and encodes all the cell's SDGs. If we use our narrow definition (namely, the LDGs are those that define a chromosome), then we have an organism with a 100-kb chromosome and a

TABLE 2 Examples of the distribution of lethal and sick disruption genes in selected organisms with multiple chromosome-like replicons (MCLRs)

Protein[a]	Function	Species replicon[b]
Translation		
Ribosomal protein S1	Translation of highly structured mRNAs	Bm CI, Dr CI, Rs CII, Vc CI
Ribosomal protein L21	Ribosome assembly	Bm CI, Dr CI, Rs CII, Vc CI
Ribosomal protein L27	Ribosome assembly	Bm CI, Dr CI, Rs CII, Vc CI
Histidyl tRNA synthetase	Aminoacyl-tRNA synthesis	Bm CII, Dr CI, Rs CII, Vc CI
Glutamyl-tRNA amidotransferase A	Transamidation of glutamyl-tRNA	Bm CI & CII, Dr CI, Rs CI & CII
DNA replication and repair		
DNA polymerase III, α-subunit	Chromosome replication, elongation	Bm CI & CII, Dr CI, Rs CI & CII, Vc CI
Host integration factor β-subunit	Recombination, transcription, DNA replication	Bm CI, Dr CI, Rs CII, Vc CI
DNA topoisomerase I	RecA-mediated recombination and resolution	Bm CII, Dr CI, Rs CI, Vc CI
Cell division		
FtsK	Resolution of chromosome dimers	Bm CI & CII, Dr CI, Rs CI, Vc CI
FtsH	Protease maintains balance of cellular components	Bm CI, Dr CI & CII, Rs CI, Vc CI
MinE	Correct positioning of division septum	Bm CII, Sm pSymB, Vc CI
MinD	Correct positioning of division septum	Bm CII, Sm pSymB, Vc CI
MinC	Correct positioning of division septum	Bm CII, Sm pSymB, Vc CI
Chaperones		
GroEL Hsp60	Folding cytosolic proteins	Bm CII, Dr CI, Sm pSymA & B, RsvCI & CII
GroEL Hsp60	Folding cytosolic proteins	Bm CII, Dr CI, Sm pSymA & B, Rs CI & CII, Vc CI & CII
DnaJ Hsp40	Nascent protein stabilization, folding	Bm CI, Dr CI, Rs CI, Vc CI & CII
DnaK Hsp70	Nascent protein stabilization, folding	Bm CI, Dr CI, Rs CI & CII, Vc CI
Amino acid biosynthesis		
Tryptophan synthase, β-chain	Tryptophan biosynthesis	Bm CI, Dr CI, Rs CI & CII, Vc CI
Phosphoribosylanthranilate isomerase	Tryptophan biosynthesis	Bm CI, Dr CI, Rs CII
Anthranilate synthase component II	Tryptophan biosynthesis	Bm CII, Dr CI, Rs CI & CII, Vc CI
Gamma glutamyl phosphate reductase	Proline biosynthesis	Bm CI, Dr CI, Rs CII, Vc CI
Gamma glutamyl kinase	Proline biosynthesis	Bm CI, Dr CI, Rs CII, Vc CI
Pyrroline carboxylate reductase	Proline biosynthesis	Dr CI, Rs CII, Vc CI
Phosphoshikimate carboxy-vinyl transferase	Aromatic amino acid biosynthesis	Dr CI, Rs CII, Vc CI
Shikimate 5-dehydrogenase	Aromatic amino acid biosynthesis	Bm CI & CII, Dr CI, Rs CI, Vc CI
threonine synthase	Threonine biosynthesis	Bm CI, Rs CI, Vc CI & CII
ATP phosphoribosyltransferase	Histidine biosynthesis	Bm CII, Rs CII
S-Adenosylmethionine synthetase	S-Adenosylmethionine biosynthesis	Bm CI, Dr CI, Rs CII, Vc CI
Miscellaneous functions		
Cytidylate monophosphate kinase 1	Nucleotide metabolism	Bm CI, Rs CII, Vc CI
Thymidylate synthase	Nucleotide metabolism	Bm CI & CII, Dr CI, Rs CI, Vc CI
Polypeptide deformylase	N-terminal formylation of polypeptides	Bm CII, Rs CI
Biotin synthase	Biotin synthesis	Bm CII, Rs CI, Vc CI
Dethiobiotin synthetase	Biotin synthesis	Bm CII
Uroporphyrin III C-methyltransferase	Siroheme biosynthesis	Bm CI & CII, Rs CI

[a] The names of the gene products encoded by the different replicons are given.

[b] The following abbreviations were used for the organisms: Bm, *Brucella melitensis*; Dr, *Deinococcus radiodurans*; Rs, *Rhodobacter sphaeroides*; Vc, *Vibrio cholerae;* Sm, *Sinorhizobium meliloti*. After the abbreviation is the MCLR that encodes the gene, e.g., in *Brucella melitensis*, *Deinococcus radiodurans*, and *Vibrio cholerae* the gene encoding ribosomal protein S1 is on chromosome I (CI) whereas in *Rhodobacter sphaeroides* it is on CII.

5-Mb plasmid. Although this follows the rules, we believe it would leave most researchers feeling uneasy and we would have great difficulty getting a manuscript passed by reviewers. This may not appear to be a very scientific justification for the designation, but, as you shall see later, even when there are clear rules, we as scientists are often reluctant to change our mindset.

Consider a second hypothetical organism. This bacterium comes from the bottom of the ocean, we cannot grow it easily under laboratory conditions and there are no known genetic tools for its analysis; undeterred, we sequence its genome. We discover it has three MCLRs. The largest of these (MCLR1) encodes 80% of the genome but contains only computer-predicted genes, i.e., "open reading frames of unknown function." MCLR2, a smaller replicon, contains numerous genes that are essential for growth under aerobic conditions. The smallest replicon, MCLR3, contains only LDGs and SDGs.

Where do the essentiality rules take us in defining the type of replicon within this organism? MCLR3 is clearly a chromosome. It encodes functions that are essential under all growth conditions. MCLR2 is more problematic; it looks like it might be essential, and yet we only know this organism as an obligate anaerobe. However, is there some set of conditions that allow it to grow aerobically? If so, do we consider this a dispensable replicon because some of the time it is an aerobe? As for MCLR3, even though it encodes by far the greatest number of genes, we end up with no essentiality information and are unable to classify it. Some of the genes may be LDGs, but we will never know until we are able to carry out an immense amount of experimental work to define the role of each open reading frame (ORF), a process that is almost impossible, because it appears to lack any genetic system to facilitate its analysis. We now end up with an organism with two very small chromosomes and one other replicon of unclear classification. Thus, the question of "either/or" requires some knowledge to assign a designation to the large replicon. We might consider it a "virtual" chromosome or a megaplasmid until such time as more information becomes available.

The hypothetical case above may sound fanciful, and at the moment it is. However, let's take a brief look at the Lyme disease spirochete *Borrelia burgdorferi* B31, which lives in both arthropod (ticks) and mammalian hosts. The genome of this organism is sequenced, and it is shown to be made up of a single linear chromosome (~0.9 Mb) and 17 plasmids (total size, ~0.5 Mb), two of which are circular, the remainder linear. One of the circular plasmids called cp26 (26.5 kb) encodes two genes, *guaA* and *guaB*. *guaA* encodes GMP synthetase, *guaB* encodes IMP dehydrogenase. Both enzymes are responsible for the de novo synthesis of purines, which are thought to allow the organism to survive in mammalian blood where the levels of purines are low. Recent work has shown that all infective strains (for mammals) of *B. burgdorferi* contain cp26 (Purser and Norris, 2000). Until a few years ago, it was not known that *B. burgdorferi* was the causative agent of Lyme disease. Had we been working with this organism back then and looking at it only in the environment of the tick and knowing nothing of its life in mammals, we may have concluded that cp26 could only be a dispensable plasmid. But even today when we know cp26 is essential for survival in mammalian blood we would be hard pressed to get anyone to call cp26 a chromosome.

Before we continue to examine the problems biologists face, it might be both comforting and worthwhile to examine a recent astronomical parallel involving the planet Pluto and whether it should be redesignated as an asteroid or minor planet. Even though the International Astronomical Union had come up with a set of rules for the definition of planets (based on a number of criteria of which size was one), there was still vigorous debate over whether these rules should apply to Pluto, e.g., "I oppose any action by the International Astronomical Union which would result in the loss of Pluto's official designation as a planet . . . there is little scientific or historical justification for such an action." Even where there was an established protocol, the wider community did not want to take it up, because they, the astronomers, already had the "Pluto is a planet" mindset.

If cp26 were to fit the rigid functionality criterion of a chromosome, then we would be faced with the dilemma that would naturally arise by calling it a minichromosome. This would be wrong, because size should not become a parameter in defining a chromosome. Thus, on the basis of functional precedence, we could refer to cp26 as a supernumerary chromosome (supernumerary chromosome: a chromosome present, often in varying numbers, in addition to the characteristic "invariable" complement of chromosomes). This way, we avoid the natural tendency of wondering how large a minichromosome would have to be for it to become a full-grown chromosome. This designation of supernumerary also takes into account ticks/mammals in the life-cycle requirements of the organism. It also reminds us of the New Mexico astronomer, Alan Hale, codiscoverer of Comet Hale-Bopp, who suggested, ". . . the Pluto debate is somewhat silly since there's really no clear definition of what a planet is. And, besides, a hypothetical resident of Jupiter would probably laugh at our calling Earth a 'major planet.'"

As more genomes are sequenced, more anomalies are likely to arise. Consequently, it would seem foolish to impose hard guidelines that will eventually be replaced. Indeed, to paraphrase Lederberg, "As history has taught us with regard to the gene, the concept of plasmid or chromosome is more important than a probably futile effort to police their usage." There is nothing wrong with the effort to seek order and, assuming that these guidelines yield over time, they help to focus our thinking to construct new paradigms and to provide benchmarks in our quest for the ultimate reality.

With this in mind, we can perhaps think of replicons within bacteria as forming two discrete camps. On the one hand, we have the clearly dispensable replicons or parasitic plasmids and, on the other hand, we have the clearly indispensable "chromosome" harboring "essential" genes. In between, there are clearly examples where the call is presently difficult. But there are not many and some rigid terminology will restore perspective, e.g., supernumerary chromosome, virtual chromosome, or protochromosome.

We can provide an example by examining the lifestyle of *Buchnera aphidicola*, an organism that lives an obligatory mutualistic lifestyle within the pea aphid (a small insect), *Acyrthosiphon pisum* (Shigenobu, 2000). These two organisms have had this symbiotic relationship for 200 to 250 million years. The bacterium lives within specialized cells within the aphid called bacteriocytes, and there it provides essential amino acids for the aphid (these are essential amino acids for animals, i.e., the animal cannot synthesize them de novo and obtains them from its food source; they include arginine, valine, leucine, phenyl-alanine, and tryptophan; however, the aphid's food source is plant sap and it is deficient in these nutrients); in return, the aphid supplies nonessential amino acids (tyrosine, proline, serine, etc.) and vitamins and other cellular components to the bacterium. The genome of *Buchnera* consists of a large chromosome and two small "plasmids," which are ~70 to 80 kb. These smaller replicons are of plasmid origin, because they have the *repA1* and *repA2* origins of replication (Roeland, 2000), and are evolutionarily related to the IncFII replicon (which occurs on many antibiotic resistance and virulence plasmids in enterobacterial species). However, these replicons carry the only copies of those genes encoding for critical steps of the tryptophan (*trpE* and *trpG*) and leucine (*leuA,B,C,D*) biosynthetic pathways. Therefore, if we apply the functionality guidelines, these are chromosomes not plasmids. But it is also clear that they possess plasmid-like characteristics in that they possess limited capacity for essential functions, and they have plasmid origins of replication. The bacteria cannot survive without these replicons under any conditions and consequently neither can the host. The host appears essential for the survival of the bacterium and, therefore, these replicons are fully integrated players in the life of both organisms. Therefore, rather than cast aside the guidelines of essentiality, we need to expand them. Unlike a supernumerary chromosome that possesses a chromosomal origin of replication, why not refer to these replicons as protochromosomes?

Section Summary

- To define a gene as essential or nonessential we need to have detailed knowledge of the lifestyle of the organism. Definitions, if used wisely, provide a useful structure and order for understanding the world around us. However, the definitions should not be allowed to imprison our thought processes.
- Scientists often become entrenched in their thinking and resist change even when faced with mounting evidence contrary to their favored viewpoint.

MULTIPLE CHROMOSOMES: WHERE DID THEY COME FROM?

There are at least three main hypotheses regarding the evolution of multiple chromosomes or MCLRs: they are the resulting fractions of an original single chromosome that was disrupted by a genetic event, usually recombination, i.e., our equivalent to the "Big Bang"; they have evolved by the addition of essential chromosomal genes into an originally dispensable plasmid; or, finally, they acquired a single chromosome, "captured" in some epigenetic event, a second chromosome.

In support of the first hypothesis is *Brucella melitensis* 16M. It has a genome that has been fully sequenced and shown to be composed of two circular chromosomes, ~2.1 Mb (CI) and ~1.1 Mb (CII) (DelVecchio et al., 2002). Plasmids have not been observed in the genome, nor were plasmid-like origins of replication. Both chromosomes are found to have bacterial chromosomes like *ori*'s with a characteristic AT-rich signature region (low GC skew). The *B. melitensis* genome contains three

rrn operons, two on CI and one on CII. The *rrn* structure is identical with that found in some *Brucella suis* serovars and consistent with the idea that the two chromosomes, at some time in the past, arose from a single chromosome, possibly through recombination between two of the *rrn* operons. That such an insult could result in a viable organism was demonstrated by Itaya and Tanaka (1997) when they disrupted the single chromosome of *B. subtilis* into two smaller chromosomes of roughly equal size. The resulting bacteria are viable as revealed by survival under laboratory conditions. A critical yet unanswered question is their survival in situ.

The second approach to the evolution of multiple chromosomes, i.e., integration of chromosomal genes into a plasmid, with the evolution of the replication system, has become especially clear with the completion of the genome sequence of *S. meliloti* (Galibert et al., 2001) and *Agrobacterium tumefaciens* (Goodner et al., 2001; Wood et al., 2001). The former contains a single circular chromosome and two megaplasmids, the latter contains a single circular chromosome, a single linear chromosome, and two megaplasmids. Even though their overall genome architecture is very different, proteome analysis suggests that the two species are very closely related, having diverged relatively recently from a common ancestor. Furthermore, the gene order of the two circular chromosomes shows extensive gene-order conservation. The linear chromosome of *A. tumefaciens* has an origin of replication and segregation proteins (*repABC*) similar to those found on its two megaplasmids and therefore it is hypothesized that the linear chromosome has been derived from an ancestral plasmid. It has also been noted that the gene density between the linear and circular chromosomes is very similar; however, genes representing essential functions, although pres-ent, are underrepresented on the linear chromosome. This asymmetry is consistent with the movement of essential genes from a chromosome to a plasmid with the eventual evolution of the chromosome in its own right by our functional definition. Examination of the dinucleotide frequencies between the linear and circular chromosomes shows that they are virtually identical with each other but differ significantly from the plasmids. This is consistent with the idea that the megaplasmid may be a relatively recent acquisition and their dinucleotide frequencies have not yet evolved to the point where they are normalized with the two chromosomes, whereas earlier acquisitions, in this case the "backbone" of the megaplasmid that became the linear chromosome, have had time for its dinucleotide frequencies to drift toward the norm of the circular chromosome. If we assume that the evolution is from plasmid to chromosome then we could describe the linear chromosome as a protochromosome. We assume that the reverse evolution can either not occur or is only compatible with an obligate commensal lifestyle.

The finding that so many members of the alphaproteobacteria possess MCLRs may be explained, in part, by their plasmid origins. It could be imagined that a transmissible plasmid that captured essential genes would be rapidly disseminated among closely related species or genera by horizontal transfer, whereas, if the "Big Bang" of a single chromosome occurred, it is likely to be transferred only vertically from mother to daughter. In the second case, it might be expected that different species or genera with MCLRs arose after the fractious event as a consequence of speciation. It seems likely that this would be a slower process and might result in fewer species having MCLRs derived from chromosomes than plasmids. If the former had occurred, we imagine that there would be molecular footprints showing vestigial relationships between the "second" chromosomes of different genera. However it would appear that in the alphaproteobacteria both mechanisms may have occurred to generate MCLRs. In the third case, i.e., capture of an exogenous chromosome, we might predict substantial gene duplication and a significant difference in dinucleotide frequencies.

Section Summary

- A bacterium can acquire a second chromosome by chromosome disruption, gene insertions into preexisting plasmids, and chromosome capture from another bacterium.
- To date, evidence suggests that multiple chromosomes have arisen during the course of evolution by two of the three possible mechanisms.

DO MULTIPLE CHROMOSOMES ACT AS A MECHANISM FOR GENERATING SEQUENCE DIVERSITY?

It has been suggested that recombination between multiple chromosomes may have played a major role in the evolution of gene diversity. We can imagine a duplication event followed by segregation that results in a copy of a gene residing on each of the two chromosomes. Both genes are now subject to evolutionary forces, and one gene may retain its original role, whereas the other may evolve into a gene whose product has completely novel functions. Such a mechanism would allow bacteria possessing multiple chromosomes to acquire a wider repertoire of functions and lifestyles by both recombination and drift, whereas crossing over within a single chromosome could have the deleterious effect of generating large deletions. However, we also know that very closely related organisms, e.g., the photosynthetic bacteria R. sphaeroides and Rhodobacter capsulatus, although sharing very similar lifestyles and genetic complement, show a major difference in that the former possesses two chromosomes and the latter only possesses one. This would suggest that having one chromosome has not left Rhodobacter capsulatus at an evolutionary disadvantage in terms of its abilities to exploit their common ecological niche. It is also not clear why recombination and duplication within a chromosome (intrachromosomal) could not be equally efficient and productive as recombination between two or more chromosomes (interchromosomal). We may speculate that having multiple chromosomes allows for such events to occur more readily; however, to our knowledge there has never been an attempt to measure the difference in recombination rates or gene evolution between these different forms of genomic architectures.

Where multiple chromosomes may assist in generating gene diversity is if the chromosome in question is a plasmid derivative with intact transfer functions. Recall that during transfer the chromosome would be transferred as a single strand of DNA to the host (the donor is not giving up its chromosome). We can imagine that such a chromosome may be transferred to a range of species and confer new properties on them, permitting their exploitation of new ecological niches. The genes on such transferred chromosomes will be under different selective pressures for two main sources, the external environment and the environment within the host. These pressures are likely to result in changes of the original genes and the formation of new genes, and they in turn may be redistributed to other bacteria. Such a cycle of transfer and selection could rapidly increase the genetic diversity of an ecosystem.

Section Summary
- Multiple chromosomes may increase sequence diversity by facilitating intrachromosomal recombination and/or horizontal gene transfer.
- Gene duplication without deletion may occur between multiple chromosomes.
- In theory, multiple chromosomes may act as a mechanism for increasing sequence diversity; however, there currently is no experimental evidence to validate this hypothesis.

WHAT IS THE DISTRIBUTION OF GENE FUNCTIONS ON MULTIPLE CHROMOSOMES?

There does not appear to be any hard evolutionary drive that has directed what individual or classes of genes have ended up on second chromosomes (see Table 2). The type of genes reflect how plasmid-like or how chromosome-like the second chromosome is, which in turn we can think of as a way of judging how recently the second chromosome came into being. Strikingly, to date, all organisms with multiple chromosomes have an underrepresentation of LDGs and an overrepresentation of ORFs and computer-predicted genes

on their second chromosomes. It may be suggested that the large numbers of ORFs and computer-predicted genes on second chromosomes are indicative of their having originated from plasmids. However, we should ask why these ORFs and computer-predicted genes do not match genes in the database? It should be remembered that the genes in databases reflect the organisms that researchers study. Therefore, databases are biased in that they do not contain a random sampling of genes from the biosphere. Generations of researchers have examined *E. coli* and a handful of other bacteria and have shown experimentally the functions of many genes in these organisms. If instead they had been working on *S. meliloti* or other environmental organisms, we would have had a totally different perspective of bacterial genes and genomes. We may today have been wondering how *E. coli* could survive with "just one chromosome." Indeed, it has been interesting to note that as the databases have become more extensive many ORFs that are not found in *E. coli* have been found to be widespread and highly conserved among many other species. It should always be remembered that, at least in part, our "*E. coli*-centric" view of the world has molded our perceptions of what constitutes chromosomal genes and bacterial chromosomes.

We will examine briefly the genomes of three bacteria with MCLRs, the human pathogen *Vibrio cholerae* (Heidelberg et al., 2000), the environmental organism *R. sphaeroides* (Mackenzie et al., 2001), and the plant pathogen *Ralstonia solanacearum* (Salanoubat et al., 2002). *V. cholerae* and *R. sphaeroides* have both large and small chromosomes. The small chromosomes are truly chromosome-like in that they carry a substantial number of essential genes. However, not all small chromosomes are made equal. As we shall see, the "megaplasmid" (which we think should be called a protochromosome) of *R. solanacearum* appears to encode fewer essential housekeeping genes and a greater number of specialized genes than the other two examples.

In *V. cholerae* there is significant imbalance in the types of genes distributed between the two chromosomes. Genes essential for DNA replication and repair, transcription, translation, cell wall biosynthesis, and a variety of catabolic and biosynthetic pathways are mainly encoded on CI, as are genes known to be essential in pathogenicity, i.e., those involving the toxin coregulated pilus, the cholera toxin, lipopolysaccharide, and extracellular protein-secretion machinery. CII encodes a gene-capture system or integron island, which incorporates genes from outside sources (horizontal gene transfer) into the genome. Probably as a consequence of this feature, CII contains a larger fraction of hypothetical genes, genes of unknown function, and 3-hydroxy-3-methylglutaryl-CoA reductase, which was thought to be acquired from the archaea. There are, however, essential housekeeping genes, e.g., D-serine deaminase (*dsdA*), threonine tRNA ligase (*thrS*), and ribosomal proteins L20 and L35.

Several ORFs with apparently identical functions exist on both chromosomes and are thought to have been acquired by horizontal transfer. Two copies of *glyA* encoding serine hydroxymethyltransferase are found on each chromosome. Phylogenetic analysis suggests the *glyA* genes on CI and CII branch with the alphaproteobacteria and gammaproteobacteria, respectively. The *glyA* on CII is flanked by genes encoding transposases, suggesting that this gene may have been acquired by a transposition event.

In *R. sphaeroides* the genome appears to have undergone a high level of gene duplication with nearly one-fifth of the genes in the genome appearing to be duplicated. Both duplicate copies often reside on CI, but, in many instances, one copy resides on CI and the other on CII. It has been noted that some of the amino acid biosynthesis pathways, such as for tryptophan biosynthesis, are divided between the two chromosomes. The evolutionary drive toward this kind of arrangement is still unclear, because CII does not appear to be specialized for any particular growth mode. In this respect CII appears to have arisen as a segment of what may have once been a larger chromosome, "The Big Bang." If that were

the case, the gene classes might be expected to be evenly distributed between the two chromosomes. However, on CII there is a slight overrepresentation of genes involved in amino acid transport and metabolism and a clear underrepresentation of the genes involved in cell division, translation, cell envelope biogenesis, and cell motility and secretion. It is clear, however, that CII does not appear to code for any specialized function or lifestyle, and in all respects it looks like a piece of any other typical bacterial chromosome. The origin of replication of CII has yet to be determined; therefore, the mechanism of establishment of this replicon remains unknown. Because of the high level of gene duplication in *R. sphaeroides*, it would be of interest to ascertain if they, the gene duplications, occurred before or after the "Big Bang."

Unlike *V. cholerae* or *R. sphaeroides*, the megaplasmid of *R. solanacearum* GMI1000 appears to be much more specialized. The genome of this organism consists of a 3.7-Mb chromosome that carries all the essential housekeeping genes, including all the genes for DNA replication and repair, transcription, and translation (which includes all the ribosomal proteins, three complete ribosomal RNA loci, and 55 tRNAs that allow recognition of all possible codons). The 2.1-Mb "megaplasmid" carries several metabolically essential genes, e.g., an rRNA operon with two tRNA genes, a gene encoding the α-subunit of DNA polymerase III and a gene for the protein elongation factor G. However, this replicon may be dispensable as additional copies of these reside on the chromosome and we still do not know if these genes are actually functional. There are several genes on the megaplasmid for amino acid and cofactor biosynthesis, however, that have no counterpart on the chromosome; therefore, it is predicted that a megaplasmid-deleted derivative would be auxotrophic for several metabolites. Thus the second replicon might be designated a protochromosome, i.e., an early-stage chromosome.

Analysis of the genes present on the protochromosome suggests that it may have a significant function in the overall fitness and adaptation of the organism to various environmental conditions. The protochromosome carries all the *hrp* genes encoding a type III secretion system that are involved in causing plant pathogenesis, a trait that allows this bacterium to colonize a rather exclusive ecological niche. The protochromosome also encodes elements of the flagellum and most of the genes governing exopolysaccharide synthesis. Although the formal possibility exists that the protochromosome is dispensable, no derivative of strain GMI1000 in which the protochromosome has been deleted has ever been isolated. Perhaps the major reason for referring to the 2.1-Mb replicon as a protochromosome is because 315 of the 748 genes are of unknown function and the origin of replication is plasmid-like. The origin is flanked by the *repA* gene and has at least 14 repetitions of a conserved motif that may be RepA-binding boxes. Given this description we feel comfortable with describing the "megaplasmid" as a protochromosome.

Section Summary

- There does not appear to be an evolutionary rule that mandates which genes reside on which chromosome.
- There appears to be an underrepresentation of LDGs and an overrepresentation of computer predicted hypothetical genes on second chromosomes.
- The extensive study of *E. coli* as a model system may have colored our perspective of microbial genetics and genomes.
- Through the acquisition of new functions, plasmids can evolve into chromosomes.
- Bacterial genomes are diverse in topology, architecture, and function, but all have at minimum one chromosome and in some species additional replicons that have varying degrees of plasmid-like or chromosomal qualities.

WHY DO ORGANISMS HAVE MULTIPLE CHROMOSOMES?

The answer to this question could be called the finish line in the quest for the Holy Grail.

We simply do not know, and research conducted to date has not given us many clues as to the avenues of approach. It may be expected that an organism with two chromosomes would eventually form a cointegrate and form a single stable chromosome. However, there is little evidence that this occurs in nature. In the laboratory we have estimated that if this occurs it is in <1% of cells. The advantage of integration could be suggested to derive from the process of cell division and chromosome segregation. On the other hand, this presumed difficulty might exist only in our minds, and not in reality. Therefore, we are left to conclude that there may be some inherent benefit of the multichromosomal genome architecture that offsets the added complexity of division, or that there is simply no disadvantage that would provide the drive for their removal.

It has been suggested that *Vibrio* responds to environmental cues by carrying out aberrant segregation, i.e., one daughter chromosome may partition to daughter cells in the absence of the other chromosome. Such daughter cells would be replication defective, but still retain partial metabolic activity ("drone" cells) and therefore would contribute to cell numbers but not viable count observed to occur in *V. cholerae*. It could be envisaged that such drones may serve as feeder cells supplying the viable daughter cells with nutrients. Thus, aberrant segregation would not be deleterious to survival but a selective advantage. To carry out such segregation, the possession of multiple chromosomes would be a requirement.

Section Summary

- Multiple chromosomes may provide a selective advantage, but what that advantage is remains unclear.
- Faced with a lack of data, it is wise to keep an open mind!

CHROMOSOME COPY NUMBER

During exponential growth, fast growing bacteria with a single chromosome, such as *E. coli*, may have four or more times as many copies of the DNA near the origin of replication than near the terminus. We know that in at least one multichromosomal bacterium, *R. sphaeroides*, the chromosome copy number ratios are 1:1, even during exponential growth. However, we can imagine a hypothetical organism where CI is three times the size of CII. If the rates of DNA replication per base pair of each chromosome are the same and the two chromosomes begin replication at the same time, then CII will have finished being replicated when CI is only a third of the way through. That is, there will be two copies of CII but only one and a third copies of CI. If we allowed replication to continue (i.e., CII replication doesn't stop when the first CII copy is complete), we would end up with a 2:6 ratio of CI to CII. Perhaps in such an organism CII could encode genes that are required at a higher copy number during growth than genes on CI. Alternatively, perhaps during lag phase or stationary phase, the numbers of copies of the two chromosomes are not the same; maybe there is just one copy of CI and perhaps multiple copies of CII. Maybe the additional copies of CII could in some way help jump-start logarithmic growth and may be considered analogous to gene amplification of ribosomal genes in *Xenopus* oocytes. These are interesting questions to which we have no answers.

In both *R. sphaeroides* and *R. solanacearum* there is a copy of *dnaE*, the gene that encodes the α-subunit of DNA polymerase III (the polymerization subunit), on each replicon. Perhaps the polymerization activities of the copies are different and allow the large and small replicons to replicate at the same rate so that the mole ratio of the replicons is always 1:1.

Section Summary
- Multiple chromosomes may be a way of varying gene dosage during replication.

THE NO-DISADVANTAGE HYPOTHESIS

Perhaps multiple chromosomes don't offer an advantage to the cell at all, but, rather, they are not disadvantageous enough. It is not known what factor decides chromosome number, after all why do you have 23 pairs of chromosomes compared to a goldfish with 94 pairs? Perhaps chromosome number and partitioning is not the burden that we might believe it is.

Imagine *R. sphaeroides* living in its ecological niche, a swine runoff lagoon. It receives relatively little light and the light that it does receive is of poor quality in that much of the better (more energetic wavelengths) have been used for photosynthesis by algae in the upper-liquid layers. The water temperature is low, and it is probably dividing fairly slowly (hours, days, or weeks rather than minutes). For *R. sphaeroides* its rate-limiting step for growth and division is going to be how much light it can capture, not how fast it can partition its chromosomes. Therefore, having two chromosomes rather than one may not be disadvantageous enough for evolution to have eliminated them. It has plenty of time to replicate and partition them evenly (or aberrantly) between the daughter cells. Conversely, a hypothetical *E. coli* having two chromosomes and a doubling time of tens of minutes may have real problems with replication and segregation of multiple chromosomes. Perhaps we can think of the single chromosome as a way of evolution having streamlined replication and division in a fast-growing organism (though it should be remembered that not all single chromosome organisms are fast growers). In slower growing organisms the focus of selective pressure may not lie in the number of chromosomes but in the methods of energy acquisition or an efficient chemotaxis mechanism. In this respect, multiple chromosomes may be a more ancient or less refined way of maintaining and storing information in some slower growing organisms.

Section Summary
- Some bacteria may possess multiple chromosomes because their existence does not leave the cell at a competitive disadvantage.
- The evolution of bacteria with single chromosomes may have been a result of having to carry out rapid, streamlined cell division.

- Multiple chromosomes may be a less evolved way of storing genetic information. The evolutionary pressures that resulted in the formation and maintenance of multiple chromosomes remain obscure.

ORGANELLE DEVELOPMENT

Ribosomal RNA analysis together with other data suggests that eukaryotic mitochondria evolved from a member of the α-subgroup of Proteobacteria. It is possible that the existence of multiple chromosomes in bacteria may have had a pivotal role in the evolution of the eukaryotic cell. We can imagine the ancestral "eukaryotic" cell as consisting of two or more essentially independent organisms, the host cell and one or more "associated" cells, perhaps one of which is an engulfed cell which would be our ancestral member of the alphaproteobacteria (symbiont). In *R. sphaeroides* we know that many genes exist as duplicates on both CI and CII; therefore, we could envision that with time one of these copies could migrate and become integrated into the genome of the dominant host while at the same time maintaining a viable genome in the symbiont. Having multiple chromosomes, one within the host and the other in the symbiont, would make for a much easier transition for the functional integration of the original genomes. If there were but a single chromosome, we would be asking for two things to occur simultaneously: (i) the successful integration of the symbiont genes into the host and (ii) a plan "decided" by the host to marshal supplies to the symbiont for gene products of that chromosome. In reality, the truly subordinate contributor has nothing more to give and could be readily lost. However, if we have two chromosomes of unequal content and one is relinquished to the host, the symbiont can continue making its own unique supply of its gene product, thereby sustaining its own viability but yet not disturbing the developing relationship, since it would still be dependent on what it has donated. That is, integration becomes a two-step process, integration followed by an evolutionary process where genes are swapped between the two genomes. We can imagine that with time more and more genes may have migrated to the dominant host and have been selected in this way. If the host provides the symbiont with sugars, nucleotides, essential amino acids, and other necessary compounds, greater and greater numbers of genes in the symbiont would have either migrated to the host or would have become nonfunctional and have been lost from its genome. Eventually the only remaining genes in the symbiont would be those whose products could not be exported from the host. The retention of the symbiont is likely to have been dependent on its unique structure, which could not be duplicated de novo by the host, e.g., thylakoid- or cristae-like membrane systems. Therefore, we can think of modern eukaryotic nuclear DNA as consisting of a mosaic of genes that were acquired from these (and there may have been more than two or three) separable sources with those of the mitochondrial genome encoding the few genes where function could not be supplied by the host to the symbiont, the mitochondrion. However, control over the mitochondria is fully exercised by the host. Another possible legacy that multichromosomal alphaproteobacteria may have ceded to eukaryotic organisms is ploidy and ultimately diploidy. We know that organisms like *R. sphaeroides* have multiple copies of genes with the different copies residing on the different chromosomes. Could this have been the forerunner of full diploidy and multiple chromosomes in eukaryotes?

Section Summary

- A member of the alphaproteobacteria was probably the ancestor of the mitochondria.
- The possession of multiple chromosomes would permit the evolution of the mitochondrion by a two-step process.
- The ploidy of eukaryotes may have had origins in bacteria.
- Bacteria with multiple chromosomes may have been intimately involved in the evolution of the modern eukaryotic cell.

SUMMARY OF MECHANISMS FOR GENERATING SEQUENCE DIVERSITY

It is clear that technological advances and the resulting flood of new genome information that has become available in the past 10 years have changed our view of biology forever. We have gone from making restriction maps of plasmids to sequencing entire bacterial genomes with almost the same ease in almost the same time. These advances in our knowledge have also highlighted the anomalies in the old rules of plasmids being circular and small, and chromosomes being single, circular, and large. There is now a diverse range of replicons that lie in the "no man's land" between these two genetic extremes, and we are certain that as more genomes are mapped and sequenced, there are going to be more replicons found that fall into the suspension between plasmid and chromosome. Therefore, we think it likely that the confusion over the nomenclature is going to continue. We have provided here several possible ways to address this problem. On the one hand, we could throw up our hands and declare it is too difficult to sort out or not worthwhile to bother with. On the other hand, we can take a systematic approach with but one assumption, plasmids can become chromosomes, but not the other way around. At the beginning of the chapter we said how little we knew about the interplay between multiple chromosomes in bacteria, why they evolved and have been maintained. If nothing else, you should now be convinced of our unfortunate ignorance. You will have seen that we have many theories but few data to substantiate them. For a student or any researcher looking for a field that is wide open and rich in pickings, this could be the place to dine. After all, is ignorance not the staple diet of science?

QUESTIONS

1. In the text we have given you a few examples of genes that could be considered essential for defining chromosomes (LDGs and SDGs). Using your knowledge of the cell, think of three additional unrelated genes that fall into each of these two classes and, of course, explain your answers.

2. Can you think of experimental strategies that would test the hypothesis that the genes listed in question 1 are essential? Negative results (e.g., inability to disrupt the genes) are not considered good strategies unless they are combined with other positive data.

3. Draw (pencil and paper is fine, a computer is not required) how gene duplications may occur between two chromosomes and within a single chromosome.

4. When and how does a replicon transition from a protochromosome to a supernumerary chromosome and ultimately to a chromosome? Should there be a prescribed order for these events?

5. The genetic information contained in a bacterial cell has to be conserved, but it also has to be adaptable to environmental changes. How might the possession of multiple chromosomes enhance this adaptability while sustaining conservation?

REFERENCES

Choudhary, M., C. Mackenzie, K. S. Nereng, E. Sodergren, G. M. Weinstock, and S. Kaplan. 1994. Multiple chromosomes in bacteria: structure and function of chromosome II of *Rhodobacter sphaeroides* 2.4.1[T]. *J. Bacteriol.* **176:** 7694–7702.

DelVecchio, V. G., V. Kapatral, R. J. Redkar, G. Patra, C. Mujer, T. Los, N. Ivanova, I. Anderson, A. Bhattacharyya, A. Lykidis, G. Reznik, L. Jablonski, N. Larsen, M. D'Souza, A. Bernal, M. Mazur, E. Goltsman, E. Selkov, P. H. Elzer, S. Hagius, D. O'Callaghan, J. J. Letesson, R. Haselkorn, N. Kyrpides, and R. Overbeek. 2002. The genome sequence of the facultative intracellular pathogen *Brucella melitensis*. *Proc. Natl. Acad. Sci. USA* **99:**443–448.

Dryden, S. C., and S. Kaplan. 1993. Identification of *cis*-acting regulatory regions upstream of the rRNA operons of *Rhodobacter sphaeroides*. *J. Bacteriol.* **175:**6392–6402.

Galibert, F., et al. 2001. The composite genome of the legume symbiont *Sinorhizobium meliloti*. *Science* **293:**668–672.

Goodner, B., G. Hinkle, S. Gattung, N. Miller, M. Blanchard, B. Qurollo, B. S. Goldman, Y. Cao, M. Askenazi, C. Halling, L. Mullin, K. Houmiel, J. Gordon, M. Vaudin, O. Iartchouk, A. Epp, F. Liu, C. Wollam, M.

Allinger, D. Doughty, C. Scott, C. Lappas, B. Markelz, C. Flanagan, C. Crowell, J. Gurson, C. Lomo, C. Sear, G. Strub, C. Cielo, and S. Slater. 2001. Genome sequence of the plant pathogen and biotechnology agent *Agrobacterium tumefaciens* C58. *Science* **294:**2323–2328.

Heidelberg, J. F., J. A. Eisen, W. C. Nelson, R. A. Clayton, M. L. Gwinn, R. J. Dodson, D. H. Haft, E. K. Hickey, J. D. Peterson, L. Umayam, S. R. Gill, K. E. Nelson, T. D. Read, H. Tettelin, D. Richardson, M. D. Ermolaeva, J. Vamathevan, S. Bass, H. Qin, I. Dragoi, P. Sellers, L. McDonald, T. Utterback, R. D. Fleishmann, W. C. Nierman, and O. White. 2000. DNA sequence of both chromosomes of the cholera pathogen *Vibrio cholerae*. *Nature* **406:**477–483.

Itaya, M., and T. Tanaka. 1997. Experimental surgery to create subgenomes of *Bacillus subtilis* 168. *Proc. Natl. Acad. Sci. USA* **94:**5378–5382.

Jumas-Bilak, E., S. Michaux-Charachon, G. Bourg, M. Ramuz, and A. Allardet-Servent. 1998. Unconventional genomic organization in the alpha subgroup of the proteobacteria. *J. Bacteriol.* **180:**2749–2755.

Mackenzie, C., M. Choudhary, F. W. Larimer, P. F. Predki, S. Stilwagen, J. P. Armitage, R. D. Barber, T. J. Donohue, J. P. Hosler, J. E. Newman, J. P. Shapleigh, R. E. Sockett, J. Zeilstra-Ryalls, and S. Kaplan. 2001. The home stretch, a first analysis of the nearly completed genome of *Rhodobacter sphaeroides* 2.4.1. *Photosynth. Res.* **70:**19–41.

Mackenzie, C., A. E. Simmons, and S. Kaplan. 1999. Multiple chromosomes in bacteria. The yin and yang of *trp* gene localization in *Rhodobacter sphaeroides* 2.4.1. *Genetics* **153:**525–538.

Purser, J. E., and S. J. Norris. 2000. Correlation between plasmid content and infectivity in *Borrelia burgdorferi*. *Proc. Natl. Acad. Sci. USA* **97:**13865–13870.

Roeland, C. H., R. C. V. Ham, F. Gonzalez-Candelas, F. J. Silva, B. Sabater, A. Moya, and A. Latorre. 2000. Postsymbiotic plasmid acquisition and evolution of the *repA1*-replicon in *Buchnera aphidicola*. *Proc. Natl. Acad. Sci. USA* **97:**10855–10860.

Salanoubat, M., S. Genin, F. Artiguenave, J. Gouzy, S. Mangenot, M. Arlat, A. Billault, P. Brottier, J. C. Camus, L. Cattolico, M. Chandler, N. Choisne, C. Claudel-Renard, S. Cunnac, N. Demange, C. Gaspin, M. Lavie, A. Moisan, C. Robert, W. Saurin, T. Schiex, P. Siguier, P. Thebault, M. Whalen, P. Wincker, M. Levy, J. Weissenbach, and C. A. Boucher. 2002. Genome sequence of the plant pathogen *Ralstonia solanacearum*. *Nature* **415:** 497–502.

Shigenobu, S., H. Watanabe, M. Hattori, Y. Sakaki, and H. Ishikawa. 2000. Genome sequence of the endocellular bacterial symbiont of aphids *Buchnera* sp. APS. *Nature* **407:**81–86.

Suwanto, A., and S. Kaplan. 1989a. Physical and genetic mapping of the *Rhodobacter sphaeroides* 2.4.1 genome: genome size, fragment identification, and gene localization. *J. Bacteriol.* **171:**5840–5849.

Suwanto, A., and S. Kaplan. 1989b. Physical and genetic mapping of the *Rhodobacter sphaeroides* 2.4.1 genome: presence of two unique circular chromosomes. *J. Bacteriol.* **171:**5850–5859.

Wood, D. W., et al. 2001. The genome of the natural genetic engineer *Agrobacterium tumefaciens* C58. *Science* **294:**2317–2323.

Yang, D., Y. Oyaizu, H. Oyaizu, G. J. Olsen, and C. R. Woese. 1985. Mitochondrial origins. *Proc. Natl. Acad. Sci. USA* **82:**4443–4447.

FURTHER READING

Casjens, S. 1998. The diverse and dynamic structures of bacterial genomes. *Annu. Rev. Genet.* **32:**339–377.

Krawiec, S., and M. Riley. 1990. Organization of the bacterial chromosome. *Microbiol. Rev.* **54:**502–539.

Lederberg, J. 1998. Personal perspective, plasmid (1952–1997). *Plasmid* **39:**1–9.

Moreno, E. 1998. Genome evolution within the alpha *Proteobacteria*: why do some bacteria not possess plasmids and others exhibit more than one different chromosome? *FEMS Microbiol. Rev.* **22:**255–275.

GENERATING INTRACELLULAR PROCESSES

Martin J. Day and Robert V. Miller

6

WHY WOULD CELLS HAVE A STRATEGY FOR GENOMIC CHANGE?

HOW IS CHANGE ACHIEVED?

WHEN IS CHANGE DONE?

WHAT ARE THE CONSEQUENCES OF CHANGE?

WHY WOULD CELLS HAVE A STRATEGY FOR GENOMIC CHANGE?

The old adage "If it ain't broke, don't fix it!" applies to genomes—well, almost. Intellectually, it is clear that for a bacterium short-term survival depends on the high fidelity of the replication machinery to prevent genomic alteration, but in the long term, evolution (i.e., mutational change) needs to occur. There are two basic themes, one providing for genome modification and the other providing for an increase in genomic content. These themes operate at intracellular and intercellular levels. The latter is mediated by gene exchange between bacteria and will be considered later in chapter 11.

The processes considered in this section are those of replication, amplification, and deletion. In fact, if gene acquisition, through transfer processes, is as high as is considered likely, then

processes driving deletion reactions will be strategically important to conserving genome size. The maximum genome size of a species is governed by the population size, mutation rate, and frequency of gene replacement through exchange. Thus, genes that do not provide selective fitness will be lost through a combination of genetic drift and deletion. For example, the genome sizes of *Mycoplasma genitalium* and *Mycobacterium lepra* have been reduced over time as these intracellular parasites have learned to rely on the host for the synthesis of a whole range of nutrients. Bacteria also acquire transposable elements, phage, etc., and some are deleterious. Having a mechanism for their removal is of survival value. Thus, there is a selective advantage to having structured mechanisms for manipulating DNA to conserve the size of the genome. As John Drake (1991) has pointed out, for a mutation rate to occur, there must be a threshold above which the repair and gene-exchange processes cannot compensate, and, at this point, the information in the genome becomes irretrievably lost, and the species becomes extinct. However, this does not mean that mutation rates above this value never occur and cannot be selected for under certain conditions. For example, what if the cell is going to die because it lacks a metabolic capacity? By switching into a "mutation mode" the population is provided a spectrum of mutants; some of the individuals may be able to survive while the rest of the clone is sacrificed. This mechanism, which will be discussed in

Martin J. Day, Cardiff School of Biosciences, Cardiff University, Cardiff CF10 3TL, Wales, United Kingdom, and *Robert V. Miller,* Department of Microbiology and Molecular Genetics, Oklahoma State University, Stillwater, OK 74074.

Microbial Evolution: Gene Establishment, Survival, and Exchange
Edited by Robert V. Miller and Martin J. Day, ©2004 ASM Press, Washington, D.C.

more detail in chapter 12, provides for immediate survival and the potential, at a later date, to recover any information lost by the transfer of DNA from species migrating in from other areas. Thus, for individual cells to err is clearly beneficial, but when is it good to err and how many errors should be allowed? Mutation rates are low, roughly one base in 10^9 replicated, whether the organism is a bacterium or a human. At this mutation rate it is calculated that an organism could effectively encode 60,000 essential proteins. Raise the mutation rate by a factor of 10 and the number of essential proteins that can be effectively produced falls to 6,000! *Escherichia coli* has almost 4,300 proteins and so could accommodate a short-term elevated mutation rate to survive the challenge of a lethal selection pressure.

HOW IS CHANGE ACHIEVED?

DNA sequences are subject to change for better or for worse, so selection would appear to favor individuals with a mutation rate approaching zero *and* an accurate replication machinery. Replication consists of three phases: polymerization (10^{-5}), proof reading (10^{-2}), and mismatch repair (10^{-2}), providing a basal level of accuracy of 10^{-9}. Additionally, it is estimated that during chromosome replication *E. coli* repairs some 3,000 to 5,000 single-stranded breaks as they occur. This provides some idea of the problems overcome by the repair and replication machinery to continually produce clonal progeny. Overlay this with all the other factors that have an impact on the genome sequence (tautomerism, thymine dimer formation, etc.), and you will begin to see the level of work that a bacterium must do to maintain the sense of its genomic sequence (Fig. 1).

Let us reflect on the evolution of prokaryotes over the past few billion years. In the dim and distant past, when all cells were simpler and haploid, imagine that two types of populations arise, one with a low (relatively) and

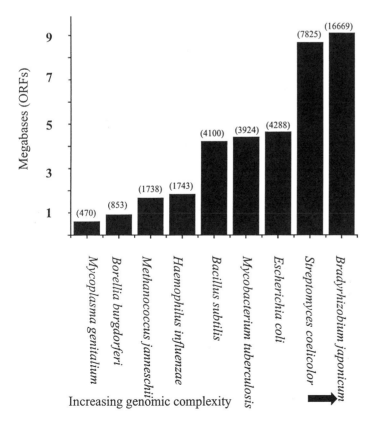

FIGURE 1 Data to illustrate that the increase in genome size parallels an increase in the complexity of the organism's life cycle.

the other with a higher mutation rate. What might the consequences be? Well, the higher mutation rate would deliver much genetic diversity but would "find" enlarging and keeping genomic complexity difficult. So it would retain the characteristics required for rapid reproduction and produce lots of progeny. What happens to the second imaginary group in which the mutation rate was lower? The genome would be more stable and thus capable of increasing in size. The cell could steadily evolve to become more complicated, maintain more plasmids, even additional chromosomes! These organisms could acquire more complex characteristics, become diploid and develop a sexual cycle.

However, no organism lives in a constant environment and, as environments fluctuate, we would expect to find mutation rates also respond to the changing selection pressures imposed. Thus, within bacterial populations there are atypical cells, displaying a spectrum of mutational rates due to physiological stress and mutations. As pointed out by Lenski and Travisano (1994), these constantly provide for a genetic diversity within the clone. Thus, evolution proceeds by integration of the two processes of mutation and gene exchange. Overlying the genomic changes produced by mutation with gene acquisitions from other cells ensures that there is another layer of diversity-generating processes available to the cell.

The mutational changes that occur to a genomic sequence can be directed (as occurs in site-specific recombination) or, at the other extreme, they can be random. They can have a biological, physical, or chemical basis. An example of a random event is a base-pair change promoted by misincorporation driven by a polymerase. If the sequence defines a gene, then the characteristics of that gene may be altered, for better or worse. Sequences not classified as a gene can also have functional roles (e.g., regulatory sites and protein-binding sites, like operators and promoters), and thus sites other than those participating in translation are also subject to alteration in function through change. Consequently, no base in the genome is completely immutable. Perhaps surprisingly, there is a high probability that in many gene sequences most nucleotide changes have little-to-no phenotypic effect on the cell. Thus, as a mechanism for generating diversity, point mutations are a potent force, but one that does not have the degree of impact that one might predict.

However, in principle, mutation is an event that is more likely to adversely effect the activity of a gene product. This is because most genes are the result of evolutionary selection on their activity in their host and over time they have become "best adapted" to suit the physiology of the organism. If the organism succeeds, then so does the gene sequence. It is also worth mentioning Muller's ratchet at this point (chapter 10). By 1964, Muller had already thought long and hard for at least 30 years about a conundrum that still remains unanswered today. His thesis was that asexually reproducing organisms were proceeding to an evolutionary dead end. Why? Simply, an organism, haploid or diploid, reproducing asexually has no other genetic resource than its own genome; the population resulting is in every sense clonal. As mutations occur there is no mechanism for identifying them or replacing them by recombination. Since most mutations are neutral or unbeneficial, the organism is likely to accumulate mutations until a lethal one or a combination of sublethal ones becomes lethal. Logically, asexual populations should cease to exist. Sex and the resulting diversity in the offspring reduces the mutational load and so enables an escape from the adverse effects of mutations. There should be no asexual organisms! As intellectually illogical as it is in the face of Muller's ratchet, there are several organisms that do not (as far as we know) have a sexual cycle.

Sexual reproduction is a much slower process than the asexual strategy, since males cannot themselves produce offspring. Every member of an asexual population can clone itself. This results in asexual populations growing twice as fast as sexual populations. Because the asexual species breeds more quickly than the sexual species, it corners more of the resources. Clonal species are strongly driven by competition because one clone is "much like" another. Also, individuals in the asexual species tend to be more specialized. Individuals from sexual species are more varied; they are generalists. However, bacteria, held as the classical asexual organism, do have a nonobligatory (para)sexual cycle. This allows them to acquire by recombination large amounts of genetic sequence. Thus, intraspecific recombination is possible. Do they thus have the best of both strategies?

The integration of DNA acquired in these processes is an important event in the life cycle of haploid asexual reproducing organisms, and RecA, the enzyme mediating this integration, is universally present in bacteria. Because recombination via RecA is done through homology, it signifies that there is pressure on those interacting genomes to maintain sequence similarity and a tendency for DNA from closely related organisms to be integrated more successfully. Thus, there will be a tendency to replace sequences that perform the same function or a slightly different one. RecA is key to the exchange of homologous DNA sequences within and between compartments of the genome. In a way, recombination can be seen as a mutational process also. It takes two different sequences and constructs a hybrid that, when expressed, ensures the organism is phenotypically tested for fitness. This potentially selects out the "wheat from the chaff"!

Homology-independent (site-specific) pathways of DNA incorporation (i.e., not RecA-mediated) are likely to lead to increases in genome complexity. RecA also has another key role in the cell, that of repair, and so it is not unexpected to find its activity and expression is regulated. But nonhomologous (site-specific) recombination is also important in shaping bacterial genomes. There are a large number of diverse genetic elements (IS, Tn, etc.) widely distributed in bacteria that use one of only four basic recombinase-transposase types. Why then is there a dramatic difference in the numbers of elements present in closely related bacteria (e.g., *E. coli* and *Shigella*)? How do these elements influence gene expression and genome structure? How are they involved in the assembly, plasticity, and transmission of genomic islands? What are their evolutionary relationships to phages, plasmids, and conjugative elements? How do they act in the assembly of super integrons? Genome-scale analysis is slowly providing clues to their evolution. There is a consensus that they evolved by acquiring functional modules from a horizontal gene pool. This is clearly suggested from analysis of phage and plasmid genomes, the aggregation of antibiotic resistance genes, and the evolution of conjugative transposons. Thus, homologous recombination is a process that can operate to repair damaged DNA, promote DNA loss by deletion, and enable the exchange of homologous DNA sequences. Site-specific or nonhomologous recombination complements this by enabling novel sequences to be added to a genome. The two processes clearly perform different roles in the evolution of a cell.

WHEN IS CHANGE DONE?

Let us now reflect on the event of duplication and envisage a time when cells were less evolved. At this time, these individual ancestral cells could not acquire a novel phenotype by gene exchange because the gene had not yet evolved. Duplication copies a sequence as small as a domain and as long as an entire chromosome. It allowed ancestral genomes to increase in size and, subsequently, the organism to evolve in complexity. Evidence suggests that gene elongation and duplication was the route taken by ancestral life forms to enlarge their genomes and increase their biochemical capacity. Thus microbial genomes are perhaps 4 billion years old. Genomic analysis of the three cell domains, *Archaea*, *Bacteria*, and *Eukarya* is allowing an analysis of their relationships. Is it too simplistic to state that the forces that drove the evolution of the earliest genes and genomes are those that drive their evolution today? A large proportion of any organism's genes are related to those in many other organisms. Thus, all life

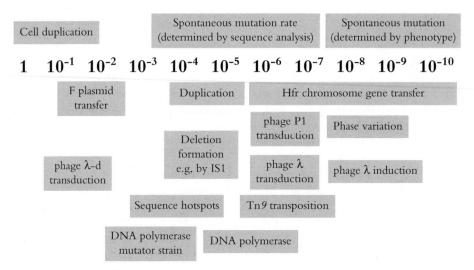

FIGURE 2 The spectrum of factors contributing to genomic change. The frequencies shown are approximate and indicate a general relationship.

shares in a pool of genetic information. The genes of the earliest organisms are present in today's organisms because of the processes of duplication and evolutionary divergence. The evolutionary significance of gene duplication was first recognized by Haldane (1932), who suggested that a redundant duplicate of a gene might acquire divergent mutations and eventually emerge as a new gene. However, few examples of duplicate genes were discovered before the advent of biochemical and molecular biology techniques. In 1970, Ohno put forward the view that gene duplication was the only means by which a new gene could arise and, although other means of creating new genes or new functions are now known, Ohno's view remains largely valid.

WHAT ARE THE CONSEQUENCES OF CHANGE?

So why did prokaryote genomes not continue to grow and grow larger? The answer to that question would appear to be—some did! Those that did not grow, have not done so because there was a growth advantage (for all the reasons stated above) for them to remain small. Figure 2 shows examples of species that

retain a small genome size and compares these with those genomes that have opted to increase or decrease their coding capacity and hence affect their life cycle complexity. This expansion leads neatly into the development of multiple chromosomes seen in some prokaryotes. It is now possible to see that genome size does reflect an increasingly differentiated lifestyle. Exactly where this is continuing to lead the evolution of these organisms remains a matter for conjecture, but, for certain, what we are seeing at present and what we are finding out about how these cells got here shows us that they will continue to evolve! You and I will have to wait a while to find out what the next episode in this exciting story will be.

REFERENCES

Cox, E. C., and T. C. Gibson. 1974. Selection for high mutation rates in chemostats. *Genetics* **77:** 169–184.

Lenski, R. E., and M. Travisano. 1994. Dynamics of adaptation and diversification: A 10,000-generation experiment with bacterial populations. *Proc. Natl. Acad. Sci. USA* **91:**6808–6814.

Ohno, S. 1970. *Evolution by Gene Duplication.* Springer-Verlag, New York, N.Y.

FURTHER READING

Drake, J. W. 1991. Spontaneous mutation. *Annu. Rev. Genet.* **25:**125–146.

Haldane, J. B. S. 1932. *The Causes of Evolution.* Longman and Green, London, England.

Muller, H. J. 1964. The relation of recombination to mutational advance. *Mutat. Res.* **1:**2–9.

Ochman, H. 1997. Miles of isles. *Trends Microbiol.* **5:** 222.

INTERCELLULAR MECHANISMS FOR GENE MOVEMENT

BACTERIAL CONJUGATION: CELL–CELL CONTACT-DEPENDENT HORIZONTAL GENE SPREAD†

Günther Koraimann

7

Günther Koraimann, Institut für Molekularbiologie, Biochemie und Mikrobiologie (IMBM), Karl-Franzens-Universität Graz, Universitätsplatz 2, A-8010 Graz, Austria.
†I dedicate this chapter to the memory of Brian Wilkins.

PLASMID CONJUGATION BASICS

Plasmid-Mediated Lateral Gene Transfer in Bacteria: a Short History

In 1946, some years before Watson and Crick unraveled the structure of DNA, Lederberg and Tatum discovered that phenotypic traits carried by certain *Escherichia coli* strains can be transmitted horizontally. Both researchers were awarded the Nobel Prize in Physiology or Medicine in the year 1958. At that time it was proposed that this transmission of genetic material from "male" bacterial donors to "female" bacterial recipients was a process related to sexual reproduction in eukaryotes. This view created a persistent paradigm, which, however, is barely more than an analogy. It became evident in those early days that the capacity of *E. coli* to function as a "male" donor is related to a genetic element that was found integrated into the chromosome or as a separate extrachromosomal entity in the cytoplasm—a plasmid. "For a lack of better inspiration . . ." this element was called F (for fertility), as Joshua Lederberg pointed out in his Nobel Lecture in 1959. A key feature of the transmission of genetic material promoted by this sex factor was that, in contrast to transduction (see chapter 9) mediated by bacterial viruses, cell–cell contact (the conjugation of cells, also called mating of cells) was required. For many years, the F factor was the only conjugative plasmid that had been identified and conjugation was thought to be an unusual biological phenomenon. It was recog-

Microbial Evolution: Gene Establishment, Survival, and Exchange
Edited by Robert V. Miller and Martin J. Day, ©2004 ASM Press, Washington, D.C.

nized during the seventies and eighties of the past century that bacterial plasmids are ubiquitous and that many of them are conjugative. They were grouped into incompatibility groups according to their inability to stably coexist in a single host cell. All conjugative plasmids, among them conjugative resistance plasmids or R factors have been found to encode a filamentous surface appendage, called the sex-pilus (or conjugative pilus), which is essential for recognition and mating-pair formation with the potential recipient cell. During the following decades the identification and characterization of genes and gene products required for conjugative DNA transport were central to investigations in many laboratories. Work on F-like plasmids dominated the field until the late eighties. With dramatic advancements in DNA-sequencing techniques and the emergence of whole-genome-sequencing projects, conjugative plasmids from different incompatibility groups with a wide host range entered the scene; one example is the IncP plasmid RP4. Regulation of transfer genes, DNA-processing reactions, and biochemical characterization of proteins involved in these reactions were major research areas during these years. Another level in the field, toward a structural biology of bacterial conjugation, has been entered through the work of different laboratories solving the three-dimensional structure of selected transfer proteins. A very exciting development in the field of bacterial conjugation that introduced a new view was the recognition that a distinct bacterial secretion system, the type IV secretion system, exemplified by the *Agrobacterium tumefaciens* T-DNA transfer system, is evolutionarily related to bacterial conjugation systems. The structure, function, and assembly of the type IV secretion apparatus are now central to investigations in many laboratories.

Bacterial Conjugation Systems
Although sDNA-transfer systems and conjugative plasmids also exist in the archaeal world and in gram-positive bacteria, the focus here will be on conjugation in gram-negative bacteria. Conjugative plasmids in archaea have

so far been found only in closely related strains of the extremely thermophilic crenarchaeon *Sulfolobus*, where they occur frequently. Only two open reading frames (ORFs) are present on these plasmids, one with some similarity to the coupling protein TraG of RP4 and the second with a higher similarity to the TrbE/VirB4 family of proteins (Stedman et al., 2000).

In gram-positive organisms, well-known and extensively studied are the transfer systems of enterococci, which allow mating in liquid media. Enterococci carrying plasmids induced for transfer secrete an aggregation substance that has been found to be a potent virulence factor (for more information, see "Regulators and triggers"). In contrast to enterococci, conjugative transfer in *Streptomyces* only takes place on a solid medium. A single protein, Tra, is required for transfer of the rolling-circle-replication (RCR) plasmids of *Streptomyces*. Tra is an inner-membrane protein and has extensive sequence homology to the SpoIIIE protein of *Bacillus subtilis* and to FtsK of *E. coli*. Possibly, Tra functions similarly to proteins that promote chromosome partitioning during cell division and sporulation. In both cases, the DNA needs to be positioned in the right location, which probably involves active transport by these proteins. A similar function could be fulfilled by the coupling proteins (e.g., TraG of RP4) from the conjugation systems of gram-negative bacteria, which are described in "Coupling DNA replication to DNA transport" in more detail.

Steps in Plasmid Conjugation
Transfer of DNA via bacterial conjugation follows a time-line with distinct steps. First, the bacterium that will transfer DNA must harbor a plasmid carrying the conjugation genes, called transfer or *tra* genes. Another prerequisite for successful mating is that the *tra* genes are expressed in donor cells carrying a conjugative plasmid. This is not always the case and can be regulated in many ways as will be detailed in "Regulators and triggers." Therefore, we start with a potential donor cell har-

boring the conjugative plasmid, expressing the *tra* genes. Furthermore, a *tra*-gene-encoded surface appendage, the pilus, needs to be synthesized and assembled, which is probably a function of the type IV secretion apparatus (the supramolecular protein complex spanning the cell envelope). The relaxosome (the pre-replication complex at the transfer origin of the plasmid DNA) needs to be assembled as well. *tra* gene expression and the subsequent assembly of multiprotein and nucleoprotein complexes makes a cell ready to transfer DNA, in other words, DNA-transfer competent. Proteins encoded by the host cell are also required to develop DNA-transfer competence. In the gram-negative conjugation systems described here, DNA-transfer competence can be reached in the absence of recipients. Second, potential recipients must be present and these recipients must be recognized by the donor cell and distinguished from other donors. Recipient cells are very likely identified by pilus contacts involving the pilus tip and surface components on the recipient cell. Self-mating of donors is prevented by plasmid-encoded surface exclusion or entry exclusion systems. At this stage, the pilus both connects and separates the mating cells. In contrast to what is commonly believed, it is very unlikely that the DNA is transported through the entire length of the pilus. Therefore, in the next step, the pilus needs to be retracted (disassembled in the donor) to allow an extended physical contact between the donor and the recipient. What actually happens at this stage, and how the contacts occur at the molecular level to allow the formation of stable mating pairs, is completely unknown. Distinct contact zones that could be visualized by electron microscopy are formed and it is assumed that the transfer of DNA occurs within these zones by virtue of the type IV secretion apparatus. After successful transfer of single-stranded DNA (ssDNA) from the donor to the recipient, DNA replication in both the donor and recipient cell, and concomitant expression of the leading-region genes in the recipient, the cells actively separate, and the recipient cell has become a potential donor (also termed transconjugand), ready for further rounds of DNA transfer.

Surface Appendages Called Sex Pili

The sex pilus is a surface structure that is present on the surface of cells that harbor a conjugative plasmid expressing the *tra* genes. Many different types of sex pili have been described. Long, thin, and flexible pili are ideal for mating in liquids (F-like plasmids); short, rigid pili are thought to be ideal for mating on surfaces like plant roots (e.g., plasmid RP4). In every case, there is one protein that constitutes the major pilus subunit, pilin. Pilus synthesis and assembly includes processing and in many cases posttranslational modification of the pilin precursor. The pilus precursor protein is targeted to the cell membrane, a signal sequence is cleaved off the precursor, and subsequent modifications include *N*-acetylation or further proteolytic processing and cyclization of the major pilus subunit (Eisenbrandt et al., 1999). The assembly of the pilus into a filamentous surface appendage occurs in the cell envelope, and numerous genes have been identified in different systems that are required for this process. Sex pili encoded by F-like plasmids are long (1 to 2 μm) but thin structures, comparable in diameter (8 nm for F-pili) to filamentous phage particles. Since they are surface-exposed, sex pili are receptors for bacterial viruses. For example, the RNA phage R17 attaches laterally to F-like sex pili. The ssDNA phage M13 recognizes the pilus tip via the terminally located adsorption protein pIII. The membrane containing phage PRD1 only infects cells that carry the broad-host-range conjugative plasmid RP4-expressing mating functions. The expression of sex pili encoded by conjugative plasmids also enhances the capability of bacteria to form biofilms (Ghigo, 2001; Reisner et al., 2003).

What is the function of the sex pilus in bacterial conjugation? Certainly, the pilus and the pilin protein are essential for DNA transfer, not only in conjugation systems but also in other type IV secretion systems, like the T-DNA transfer from *A. tumefaciens* to plant cells. Yet it is still not clear whether pilin is directly involved in DNA transport. The F pilus consists of helically arranged pilin subunits and is a hollow fiber with an inner diameter of 2 nm. However, it has not been convincingly shown that the ssDNA is transported through the pilus lumen. On the other hand, extensive mutational analysis of F pilin strongly suggests a role for the pilus beyond recipient recognition, i.e., DNA transport (Manchak et al., 2002). Pilus retraction brings the mating cells in close contact and the infecting phage particle to the cell surface. Attachment of phage to the pilus or interaction with a recipient cell could induce a conformational change in pilin subunits and the information could be transduced into the cytoplasm of the donor cell, triggering pilus disassembly and subsequent DNA transfer. Pilus subunits present in the cell envelope and not part of a visible pilus extending into the surroundings of a donor cell could function as a part of the envelope spanning the type IV secretion apparatus. The secretion apparatus could act in such a way that it first pushes the pilus or a pilot protein via the pilus across the membranes that are to be traversed and then pumps DNA into the recipient cell in a two-step scenario.

Recipients Become Donors: the Horizontal Gene Spread

Via conjugation, the original recipient cells are transformed in a way that, after a successful mating event, they also harbor a conjugative or a mobilizable plasmid. Thus, conjugative or mobilizable plasmids are infectious genetic elements that can spread throughout entire populations and subspecies and even to different species and kingdoms. Conjugative and mobilizable plasmids can carry genes that confer specific phenotypic traits (antibiotic resistance, production of toxic compounds, degradation of pollutants, metabolic functions, nitrogen fixation, plant tumor formation, virulence, etc.) to their hosts. Therefore, they constitute a particularly fluid part of the horizontal gene pool of bacteria and contribute to

the medically important spread of virulence and antibiotic resistance genes. Via transposition and integration events (see chapter 3), the genes carried by a plasmid can also integrate into the chromosome of the new host. On the other hand, complete conjugative plasmids can integrate into the chromosome and then the DNA that is transferred from the donor to the recipient is not only plasmid DNA but also chromosomal DNA. This is what was seen in the early experiments by Lederberg and his colleagues, when they found that a strain can acquire a novel phenotype after mixing with a donor that was an *E. coli* hfr (high frequency of recombination) strain, which carried the F plasmid integrated into the circular chromosome. In that case, after transfer, the chromosomal DNA of the donor needs to recombine with the chromosomal DNA of the recipient (see chapter 2).

An interesting question here is where and when conjugational DNA transfer takes place in the environment under natural conditions (in soil, on the surface of plant leaves, in the gut of mammals, in sewage plants, in biofilms). The answer is, we know that it happened frequently and it still happens in all the mentioned environments. Strikingly, plasmids even transfer inside cultured human epithelial cells (Ferguson et al., 2002). For many years plasmid conjugation was monitored in laboratory strains by using standard cultures and conditions. Although some research groups are trying to create more natural conditions for testing plasmid transfer, we are still far away from being able to mimic conditions that exist in complex environments. Nevertheless, steps toward understanding plasmid transfer using specially designed bioreactors or flow chambers are currently being made. Studying plasmid transfer under more natural conditions will supply an answer to the question when transfer genes are expressed and when they are not. Important areas of investigation here are the influence of cell densities, growth on surfaces and in biofilms, formation of microcolonies, and the composition of bacterial communities. In general, all conditions that influence the host, including the genetic background of the host cells, might also influence the frequency of plasmid transfer by conjugation.

Self-Transmissible or Mobilizable?

Conjugative plasmids are called self-transmissible because they carry all genes that are necessary for a successful conjugative plasmid transfer. Mobilizable plasmids can be transferred to recipient cells only if a second self-transmissible plasmid resides within the same donor cell. Mobilizable plasmids exploit the existing transporter (the type IV secretion complex) in the cell envelope provided by the conjugative plasmid. They carry their own *mob* (for mobilization) genes, including an *oriT* for relaxosome formation, and the genes required for linking the DNA to be transported to the "foreign" type IV secretion machinery. A well-known mobilizable, broad-host-range plasmid is RSF1010. Remarkably, RSF1010 can be mobilized by dedicated DNA transfer systems of many conjugative plasmids, the VirB system of *A. tumefaciens* and by type IV secretion systems of pathogens like *Legionella pneumophila* (see "Envelope spanning DNA and protein transporters").

Trans-Kingdom DNA Transfer

DNA transfer via conjugation and conjugation-like mechanisms has been shown from the bacterium *E. coli* to the yeast *Saccharomyces cerevisiae* to cultured mammalian cells (Waters, 2001), and from *Agrobacterium* to plants to yeast and to HeLa cells. *Agrobacterium*-to-plant-DNA transfer is a natural conjugation-like DNA-transfer system in which one part of the Ti plasmid, the T-DNA, is transported via the VirB proteins into plant cells, where it induces tumor formation. The VirB proteins form the envelope-spanning type IV secretion apparatus. This system has been extensively used to create transgenic plants. The other mentioned DNA transfer events have only been observed in the laboratory by using dedicated experimental settings.

Section Summary

- Bacterial conjugation is a form of horizontal DNA transfer that is cell-cell contact dependent.
- Conjugative plasmids harbor genes that are required for conjugative self-transfer.
- Conjugative plasmids are abundant and can be found in archaea and eubacteria.
- Conjugation occurs in temporally ordered steps.
- Transfer genes need to be expressed before DNA transfer can occur.
- Donors harboring an F-like plasmid contact the recipient via a plasmid-encoded surface appendage, the sex pilus.
- DNA is not transported through the sex pilus.
- After transfer, the recipient cell becomes a new donor.
- In conjugation, DNA replication is coupled to DNA transfer.
- The DNA is replicated both in the donor and the recipient.
- Conjugation in nature is frequent and can even occur inside eukaryotic cells.
- For transfer, mobilizable plasmids require the presence of a self-transmissible (or conjugative) plasmid in the same cell.
- T-DNA transport from bacteria to plants is similar to conjugation.
- DNA transfer via conjugation is possible across species and even kingdoms.

THE TRANSFER GENES: MODULES FOR CELL-CELL CONTACT-DEPENDENT GENE TRANSFER

Conjugative Plasmids from Gram-Negative Bacteria

According to the specificity of their replication machinery, conjugative antibiotic resistance plasmids from gram-negative bacteria were classified into about 20 incompatibility (Inc) groups. From these, several classes of self-transmissible plasmids have been analyzed in more detail: IncF plasmids like plasmids F and plasmid R1; IncH plasmids like R27; IncI plasmids like R64 or ColIb-P9; IncW and

IncN plasmids like R388 and R46, respectively; IncP plasmids like RP4/RK2 and R751; and IncX plasmids like R6K. The transfer genes of IncW and IncN plasmids are closely related to each other and to the Vir (for virulence) system of the Ti plasmids of *Agrobacterium* reflecting the notion that T-DNA transfer from *Agrobacterium* to plant cells is a specialized form of conjugation.

Transfer (*tra*) Genes

The genes required for plasmid DNA transfer from a particular host cell to a recipient are collectively called *tra* genes, although, unfortunately, no uniform nomenclature exists (see Table 1). Depending on the transfer system, 15 to 35 genes are necessary for conjugation. Based on the functional group to which they belong, the *tra* genes have been grouped into *mpf* (mating pair formation) genes responsible for donor-recipient contact and *dtr* (DNA transfer and replication) genes required for DNA processing and transport. However, it is still not known what proteins transport the DNA through the cellular envelope. The *mpf* genes encode the proteins necessary for the synthesis and assembly of the type IV secretion machinery, including the sex pilus, whereas the *dtr* genes encode proteins involved in relaxosome formation and DNA replication, including the coupling protein that connects the transporter and the relaxosome and presumably transports ssDNA across the inner membrane (Fig. 1). In the F-plasmid group

another subclassification of *tra* genes has been made. Besides the Mpf and Dtr functions, *sfx* (surface exclusion) or *eex* (entry exclusion) genes prevent self-mating of donors. Several plasmid-encoded genes are involved in regulation of transfer genes, and important functions are encoded by genes present in the so-called "leading region." The leading region encompasses the DNA that is transferred first during conjugation. Among the leading-region genes are antirestriction genes and genes inhibiting an SOS response in the recipient cell. Also present is a *ssb* gene encoding a ssDNA-binding protein. The expression of the leading-region genes is induced after transfer, and the encoded proteins facilitate the successful establishment of the plasmid DNA in the new cell.

The *oriT* and the Relaxosomal Genes

The *oriT* (origin of transfer) is an essential *cis*-acting element that must be present on the DNA to be transferred by conjugation. The *oriT* locus comprises approximately 300 bp of DNA and it includes the specific site where one DNA strand is cleaved, termed *nic* (for nicking). The cleavage of the phosphodiester bond at the *nic* site is accomplished by the action of a "nicking" enzyme, a transesterase that covalently attaches to the 5' end of the DNA strand that will be transported to the recipient cell. Plasmid DNA isolated from cells after proteins are removed by protease treatment appears to be relaxed and not supercoiled; thus, the enzyme accomplishing this reaction is commonly

TABLE 1 Transfer proteins and their homologues[a]

Proteins	F (IncF)	RP4 (IncP)	R388 (IncW)	Ti (Vir)
Pilin	TraA	TrbC	TrwM	VirB2
OM lipoprotein	TraV		TrwH	VirB7
Periplasmic protein	TraK		TrwF	VirB9
Lytic transglycosylase	P19 (Orf169)	TrbN	TrwN	VirB1
IM protein	TraB	TrbI	TrwE	VirB10
Coupling protein	TraD	TraG	TrwB	VirD4
Substrate selector	TraM (only in IncF plasmids)			
Relaxase	TraI	TraI	TrwC	VirD2

[a]Only selected Tra proteins corresponding to those shown in Fig. 1 are listed. OM, outer membrane; IM, inner membrane.

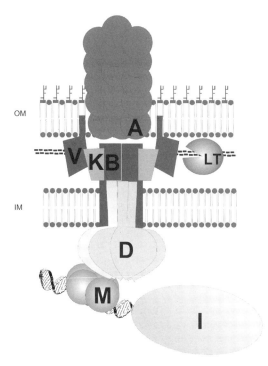

FIGURE 1 Model of the assembly of Tra proteins in the cell envelope. From the outer membrane a filamentous appendage (the F or sex pilus) extends into the exterior of the donor cell. A complex of proteins bridges the outer and inner membrane, forming a conduit for DNA transfer. A cell-wall-degrading enzyme (LT) enables the penetration of the peptidoglycan by the complex. The coupling protein (D) mediates contact to relaxosomal components via protein-protein interactions and possibly transports DNA through the inner membrane. Characters indicate Tra proteins according to the F-system nomenclature (see Table 2). Both the assembly of the secretion machinery and the transfer of the substrate DNA require energy in the form of ATP. ATPases, which are present in all type IV secretion systems, are not shown in this simplified model.

termed relaxase. Nicking, and closing by the same enzyme, occurs when *tra* genes are expressed even in the absence of a recipient. This reaction is similar to the replication-initiation events of ssDNA phages and of rolling-circle replicating plasmids from gram-positive bacteria. Besides the nicking enzyme, other accessory proteins (like TraM encoded by IncF plasmids) bind to the *oriT* region and together set up a nucleoprotein complex called the relax-

osome (Fig. 1). The relaxosome can be viewed as a prereplication complex. Only after the formation of stable mating pairs and initiation by an unknown signal does replication commence. The replication start also activates a 5′ to 3′ DNA helicase activity to unwind the double-stranded DNA (dsDNA) and to liberate the ssDNA that is subsequently transported into the recipient cell. In some plasmid systems (in the F family, the IncW plasmid R388, and the IncN plasmid pKM101), the helicase is physically coupled to the relaxase, forming one large polypeptide. Both domains of the polypeptide are absolutely required for DNA transfer (Matson et al., 2001). We will see later in this chapter how the replication machinery is coupled to the transporter complex.

Mating-Pair Formation Genes

Ten to approximately twenty of the *tra* genes, depending on the type of the conjugative plasmid, are required for the synthesis and assembly of the DNA-secretion machinery, including the sex pilus. Proteins from the Mpf complex display significant sequence similarity to the genes identified in type IV secretion systems. Hence, conjugation systems and type IV secretion systems are ancestrally related. Little is known about how these proteins interact and how this supramolecular envelope-spanning complex is assembled. In contrast to type III secretion systems, where the secretion machinery could be isolated in the form of stable needle complexes, the type IV secretion complex seems to be less stable and more transient. Whereas studies on the interactions of components of the Mpf complex revealed a picture of how the proteins might work together (Fig. 1 and Table 1), little is known about the function of the individual proteins present in the complex. One exception is the proteins that have been found to belong to the family of lytic transglycosylases (Bayer et al., 2001). These small periplasmic enzymes can degrade peptidoglycan and are believed to facilitate the assembly of the type IV secretion machinery. Well-studied representatives of these specialized lytic transglycosylases are VirB1 (Ti plasmid) and P19 (IncF plasmid R1).

Section Summary

- The conjugative plasmids of gram-negative bacteria are grouped into incompatibility classes.
- The *tra* (transfer) genes are required for conjugation.
- The *tra* genes can be subdivided into functional classes.
- Depending on the transfer system, 15 to 35 genes are necessary for conjugation.
- The *oriT* is a *cis*-acting DNA element required for DNA transfer.
- The *oriT* contains the *nic* site in which one DNA strand is cleaved by the nicking enzyme, termed the relaxase.
- The nucleoprotein complex formed at the *oriT* is called a relaxosome.
- The DNA in the donor is replicated via a rolling-circle model.
- The envelope-spanning DNA transport complex is made up of the majority of the *tra*-encoded proteins.
- These proteins include proteins necessary for pilus synthesis and assembly and represent the core components of all type IV secretion systems.
- The functions of individual proteins in the complex are unknown except for the specialized lytic transglycosylases (VirB1/P19 family of proteins).

COUPLING DNA REPLICATION TO DNA TRANSPORT

Coupling Proteins

Coupling proteins are essential proteins that physically link the relaxosome to the transporter. Coupling proteins have been identified in different systems and are thought to interact both with one of the proteins that is a part of the relaxosome and with a component of the transporter complex. Interactions of the coupling proteins TraG (of plasmid RP4) and TraD (of plasmid F) with the relaxosomal proteins, and with the transport complex partner, have been shown (Fig. 1). These proteins also bind to ssDNA in a sequence-independent way (Schröder et al., 2002). Due to localization stud-

ies and the known structure of one member of the coupling protein family, it can be inferred that these proteins reside in the inner membrane and function as gatekeepers for the transport of the ssDNA through the inner membrane. The structural prototype of the coupling proteins is TrwB from R388. TrwB forms a hexameric ring structure. Its monomeric molecular structure is reminiscent of ring helicases and AAA ATPases. The quaternary structure is made up by six equivalent protomers featuring a flattened sphere resembling F1-ATPase, with a central channel traversing the particle, thus connecting cytoplasm and periplasm (Gomis-Rüth et al., 2001). It is an attractive hypothesis that the DNA to be transported travels through the central cavity of the coupling protein and traverses the inner membrane in this way. It has been mentioned earlier that the Tra protein of the *Streptomyces* conjugative plasmids displays similarities to the coupling proteins.

Substrate Selection

In bacterial conjugation the substrate that is transported is single-stranded plasmid DNA. How is the correct DNA molecule recognized and selected? First of all, the DNA to be transported is recognized by proteins binding specifically to sequences in the *oriT* region of their cognate plasmid. Second, relaxosomal proteins are known to specifically interact with the coupling protein, thereby providing a physical link between the relaxosome and the secretion apparatus. Depending on the specificities of these interactions, the DNA-transport system is either very specific and exclusively transports the cognate plasmid, or more promiscuous and recognizes and transports different plasmids. In the F-like plasmids, a very specific recognition system is defined by the interaction of TraD (the coupling protein) with TraM (the *oriT*-binding protein). Mutations or removal of the specificity domain can increase the capacity of a coupling protein to recognize diverse substrates. This interaction excludes non-F-like plasmids from being selected as a substrate and transported. Within the F group, the exclusion of closely related plasmids is defined by the sequence-specific binding of the TraM protein to the cognate DNA in the *oriT* region. For instance, TraM of plasmid F only recognizes *oriT* DNA of the F plasmid; TraM of the closely related F-like plasmid R1 only recognizes R1 *oriT* DNA. Therefore, within the F group, substrate recognition and correct DNA selection involves TraD-TraM interaction and sequence-specific DNA binding by the relaxosomal protein TraM (Fig. 1).

Section Summary

- Coupling proteins are multimeric inner-membrane proteins that physically link the relaxosome to the type IV secretion apparatus.
- The structure of a coupling protein, TrwB from R388, resembles F1-ATPase and has a central channel connecting cytoplasm and periplasm.
- Coupling proteins interact with relaxosomal proteins.
- Substrate recognition, i.e., selection of the plasmid to be transported, is achieved through sequence-specific DNA binding by a relaxosomal protein and by a specific interaction between a relaxosomal protein and the coupling protein.

ENVELOPE-SPANNING DNA AND PROTEIN TRANSPORTERS

Transport of DNA and Protein

The main substrate that is transported via conjugation is DNA. However, it has been shown in several conjugation systems that not only

DNA but also protein is transported (Table 2). Provided selectable markers are available, showing that plasmid DNA is transferred from donor cells to recipients is straightforward, yet it is quite difficult to show a concurrent protein transport. A plasmid-encoded DNA primase, which is required to synthesize primers for complementary DNA synthesis on the transferred ssDNA, has been shown to be transported into the cytoplasm of recipient cells in the plasmids ColIb-P9 and RP4. The ColIb-P9 conjugation system also supports intercellular transport of primase without DNA transfer (Wilkins and Thomas, 2000). In contrast, none of the Tra proteins encoded by the F plasmid was found in recipient cells after mating. Possible exchanges of membrane proteins in these systems have not been investigated. In more recent studies, it has been elegantly shown that the T-DNA-transfer system encoded on Ti plasmids of *Agrobacterium* also selectively transports Vir proteins that have essential roles in plant cells (Vergunst et al., 2000). In this system, transport of VirE2, VirF, and VirE3 is independent of DNA transport but requires the type IV secretion machinery specified by VirB proteins and VirD4 (the coupling protein in this system). These examples show that the conjugation systems of IncI and IncP plasmids and related T-DNA-transport systems are able to transport both DNA and protein. What about dedicated protein-secretion systems? Is there evidence for DNA transport in these systems? As will be detailed below, conjugation systems and certain protein-secretion systems found in gram-negative bacteria are evolutionarily related and

TABLE 2 DNA and protein transport in type IV secretion systems

System	Genetic element	DNA	DNA + protein	Protein without DNA
Plasmid transfer	F (IncF)	Yes	?	?
	ColIb-P9 (IncI)	Yes	Yes	Yes (primase)
	RP4 (IncP)	Yes	Yes	?
T-DNA transport	Ti	Yes	Yes	Yes (VirE2, VirF)
Protein secretion	Dot/Icm (*L. pneumophila*)	Yes	?	Yes (DotA, RalF)
	Cag (*H. pylori*)	?	?	Yes (CagA)
	VirB (*B. suis*)	?	?	Yes (?)
	Ptl (*B. pertussis*)	?	?	Yes (pertussis toxin)

derived from a common ancestor that probably was a DNA-transport system. The *L. pneumophila dot/icm* genes (*dot* for defect in organelle trafficking and *icm* for intracellular multiplication) mediate key steps in intracellular survival and replication of this gram-negative pathogen. It has been demonstrated that this system, which is absolutely essential for virulence, can mediate conjugative transfer of the mobilizable plasmid RSF1010. The interpretation of these findings is that the Dot/Icm complex transfers not only one or more effector proteins to human host cells but also is able to transport DNA using a type IVB secretion system that is closely related to the conjugative transfer machinery of the IncI plasmids R64 and ColIb-P9 (Table 2).

Type IV Secretion Systems

An increasing number of bacterial pathogens are being found possessing type IV secretion systems that have been described as "adapted conjugation systems" because of their homology to plasmid transfer systems. Whereas the conjugation systems described in this chapter are dedicated to DNA transfer, the type IV secretion systems of bacterial pathogens transport proteins. These proteins can act as toxins (an example is the pertussis toxin secreted into the extracellular milieu by the type IV secretion machinery of *Bordetella pertussis*) or as effector molecules delivered directly into the cytoplasm of the eukaryotic host cell (an example is the pathogen *Helicobacter pylori*). *H. pylori* translocates a protein called CagA into the target host cells, which is then phosphorylated and elicits a number of changes enabling the subversion of the host and the establishment of the bacterial infection. There are two classes of type IV secretion systems. One, termed type IVA, including the *B. pertussis* and *H. pylori* secretion machineries, is related to the VirB system of *A. tumefaciens*; the second one, termed type IVB, is related to the *tra* genes of IncI plasmids R64 and ColIb-P9. The intracellular pathogen *L. pneumophila* harbors both types of type IV secretion systems. Only the type IVB system (the Dot/Icm sys-

tem) is required for virulence and intracellular multiplication; however, both can also transport DNA.

Section Summary

- Some DNA-transfer systems encoded by conjugative plasmids can transport DNA and protein.
- A plasmid-encoded primase is transported into recipients in the case of ColIb-P9 and RP4.
- The ColIb-P9-encoded primase can be transported independently of DNA.
- In the *Agrobacterium* T-DNA-transport system, T-DNA and effector proteins are translocated into the plant cell.
- Effector protein translocation occurs independently of DNA.
- The Dot/Icm type IV secretion system, which is an essential virulence determinant of *L. pneumophila*, can mediate DNA transfer of the mobilizable plasmid RSF1010.

REGULATORS AND TRIGGERS

Regulators of *tra* Gene Expression

Some conjugative plasmids, like most F-like plasmids, are repressed for transfer, because of a system called fertility inhibition or *fin*. That means that, in a bacterial culture in which all cells harbor the conjugative plasmids, only 1 of 100 to 1,000 cells is transfer competent. The repression system in F-like plasmids (R1 and R100 are naturally repressed) involves a negative element, the FinP antisense RNA, which represses the expression of a plasmid-encoded transcription activator, TraJ. FinP indirectly represses *tra* gene expression only in conjunction with the FinO protein. A complex regulatory circuitry involving positive regulatory elements enables the escape from FinOP-dominated negative control so that environmental cues can lead to expression of *tra* genes and transfer competence (Pölzleitner et al., 1997). The repressor system in the F plasmid is inactivated due to the insertion of an IS element into the *finO* gene. It is believed that the repression system, while ensuring a minimization of the metabolic burden on the host population carrying the conjugative plasmid, does not interfere with the ability to transfer the plasmid efficiently to recipients.

Regulation of *tra* genes is not only seen in the IncF group of conjugative plasmids. In IncP plasmids, which transfer at high frequencies on solid surfaces, the *tra* genes are not expressed constitutively but are under the control of a complex regulatory system. The purpose of the IncP regulation system seems to be, after an initial expression burst, the maintenance of a low level of *tra* gene expression to ensure transfer competence of each cell carrying an IncP plasmid like RP4. This control mechanism is thought to significantly lower the burden to the metabolic load of the bacterial host cell.

The Ti plasmid of *A. tumefaciens* not only harbors the VirB system but also carries *tra* genes, which are closely related to the *tra* genes of IncP plasmids. However, they are subject to a completely different control system. Conjugative transfer of the Ti plasmid is cell density dependent and depends on the presence of amino sugar compounds that are produced by plant cells transformed by T-DNA. Only when the amino sugar nutrient octopine is present and taken up by agrobacteria carrying plasmid pTiC58 is the regulator protein TraR expressed. TraR in turn is activated by *N*-acyl homoserine lactone, which is called the *Agrobacterium* autoinducer (AAI). TraR and AAI constitute a quorum sensor that couples Ti plasmid conjugation to high cell densities.

Another very interesting case of regulated conjugation is encountered in several enterococcal plasmids. In the absence of suitable recipients that do not carry the same plasmid that is present in the donor population, transfer and transfer gene expression is repressed. Only when plasmid-free bacteria are around, or at least bacteria that do not harbor the same conjugative plasmid, is transfer gene expression induced. How do the donors know? Plasmid-free bacteria secrete sex pheromones, peptides of seven or eight amino acids that are able to activate conjugative transfer in the donors. The donor response to a sex pheromone is the production of an aggregation substance facilitating clumping of cells and subsequent conjugative DNA transfer.

Host Factors That Modulate *tra* Gene Expression

Systems for regulating *tra* gene expression and gene transfer by conjugation have evolved to transduce signals to the conjugative plasmid residing in the cytoplasm. These signals may come from the exterior of the cell, as in some of the cases described above (the presence of recipients signalled by sex pheromones, the successful invasion of plants and good nutrient conditions signalled by opines). Signals transduced to transcription factors controlling *tra* genes can also arise within the cell and provide information about the metabolic state and the physiology of the cell. It is known that several plasmids respond to temperature and the growth phase, yet the molecular mechanisms governing these effects are largely unknown. F-like plasmids are down-regulated for conjugation in stationary phase and no conjugation occurs below 23°C. Other plasmids, like the IncH plasmid R27 from *Salmonella enterica* serotype Typhimurium, respond to temperature in a different way; mating is optimal between 22 and 30°C, but is inhibited at 37°C (Gilmour et al., 2001).

The ability of a plasmid to respond to changes in its environment and the ways in which signals are transduced to elicit changes in transcriptional activities reflect the adaptation of a plasmid to its host. Plasmid regulators are abundant and present in all the systems investigated so far (see above), yet only limited information is available on host factors that directly influence the transcription of *tra* genes by acting as transcription factors or indirectly by acting on plasmid regulators. The extent of the influence of host factors on *tra* gene expression might be indicative for a narrow or broad host range. For instance, in F-like plasmids known to replicate and transfer only in the group of enterobacteria, several host-encoded regulators have been identified that can modulate *tra* gene expression. The integration host factor (IHF) is required for *tra* gene expression. ArcA, the response regulator of a two–component regulatory system that senses changes in the redox state of the cell, is required for transcribing the genes from the major *tra* promoter together with the plasmid-encoded transcriptional activator TraJ. Yet another two-component signal transduction system, the *cpxA/cpxR* system, which senses cell envelope stress, when activated, is known to reduce the level of TraJ protein in the cells, thereby reducing *tra* gene expression.

From Ready to Go: Triggering DNA Transfer

To make a cell ready for DNA transfer *tra* genes must be expressed, the transport complex in the cell envelope must be assembled, and the relaxosome must be formed. Why is the DNA not transported into the medium? When does a bacterium know that stable mating pairs have been formed? What is the signal that initiates the transition from the prereplication complex termed relaxosome to an actively replicating machinery? The DNA strands at the *oriT* must be made single-stranded and a protein, the relaxase, binds to that region and it specifically nicks the DNA at the *nic* site. This happens without recipients being present. In some conjugative plasmids, a second domain in the relaxase is required for DNA unwinding, which liberates the ssDNA that is transported. It is this DNA helicase activity that is inhibited in the prereplication complex. When this activity is unleashed, the DNA-transfer event begins. Yet, we do not have the answers to the questions raised above. What exactly triggers DNA replication and transfer is unknown. There might be conformational changes in the envelope-spanning transport complex that are transmitted via protein-protein interactions to the prereplication complex (Fig. 1). These interactions are extensive, and it might well be that subtle conformational changes lead to the initiation of DNA replication and transfer.

Section Summary

- Some plasmids are repressed for DNA transfer.
- Plasmid and host encoded regulators control *tra* gene expression.
- In F-like plasmids, *tra* gene expression is repressed by the FinOP system.
- Ti plasmids of agrobacteria respond to the availability of nutrients and to cell densities.
- Plasmid transfer of enterococci is induced by plasmid-free recipients.
- Host factors transmit changes in the physiology of the cell to the level of *tra* gene expression.
- The signal that triggers DNA replication and transfer in donors is unknown.

WHERE DO WE GO FROM HERE?

Bacterial conjugation mediated by conjugative plasmids is now in a transition phase from the classical view on this important variant of horizontal gene transfer to a modern view. The classical view is that conjugative DNA-transfer systems were only recognized in the context of an additional function that plasmids can have to spread from cell to cell. The modern view is that cell-cell contact-dependent gene-transfer systems are related to type IV secretion systems of gram-negative bacteria. These systems are evolutionarily related as is evidenced by sequence conservation of the major components of the envelope-spanning secretion apparatus. Furthermore, they are similar in their capacity to transport protein and DNA. Therefore, conjugation systems can serve as model systems for type IV secretion systems in general. A major area of investigation is and will be the topology, the function, and the structure of the secretion complexes found in different type IV secretion systems. It will be of utmost importance to unravel the functions of the single components of the secretion machinery, and, finally, it will be a goal to reconstitute the system in vitro. Other important questions that will need to be addressed are the regulation and induction of transfer gene expression in natural situations, in complex environments, and mixed bacterial populations. The phenomenon of bacterial conjugation, despite its discovery more than fifty years ago, still carries its mysteries. One big question mark is the trigger that initiates DNA transfer, another is how substrates to be transported are selected and differentiated at the molecular level.

Clearly, conjugation systems have contributed and still contribute to the evolution of bacteria by providing access to the horizontal gene pool. Bacterial plasmids carrying transfer genes can be viewed as selfish and parasitic DNA elements, but they may enhance the fitness of bacteria in a way that has been discovered only recently. The expression of transfer functions, including surface appendages, greatly increases the ability to form biofilms and enhances virulence. This rather unexpected finding may explain why conjugative plasmids have been kept in the gene pool of bacteria and developed a fascinating diversity.

QUESTIONS

1. Why do you think conjugative plasmids have been kept within bacterial populations?

2. What are the advantages of plasmids that can transfer via conjugation versus DNA elements that are integrated within the bacterial chromosome?

3. Are conjugative plasmids selfish elements ("selfish DNA") or do they contribute to the fitness of their bacterial host?

4. Are conjugative plasmids important for the reemergence of antibiotic resistance during past years?

5. What are the hallmarks of bacterial conjugation in comparison with other DNA-transfer systems like transformation and transduction?

REFERENCES

Bayer, M., R. Iberer, K. Bischof, E. Rassi, E. Stabentheiner, G. Zellnig, and G. Koraimann. 2001. Functional and mutational analysis of P19, a DNA transfer protein with muramidase activity. *J. Bacteriol.* **183:**3176–3183.

Eisenbrandt, R., M. Kalkum, E. M. Lai, R. Lurz, C. I. Kado, and E. Lanka. 1999. Conjugative pili of IncP plasmids, and the Ti

plasmid T pilus are composed of cyclic subunits. *J. Biol. Chem.* **274:**22548–22555.

Ferguson, G. C., J. A. Heinemann, and M. A. Kennedy. 2002. Gene transfer between *Salmonella enterica* serovar Typhimurium inside epithelial cells. *J. Bacteriol.* **184:**2235–2242.

Ghigo, J. M. 2001. Natural conjugative plasmids induce bacterial biofilm development. *Nature* **412:**442–445.

Gilmour, M. W., T. D. Lawley, M. M. Rooker, P. J. Newnham, and D. E. Taylor. 2001. Cellular location and temperature-dependent assembly of IncHI1 plasmid R27-encoded TrhC-associated conjugative transfer protein complexes. *Mol. Microbiol.* **42:**705–715.

Gomis-Rüth, F. X., G. Moncalian, R. Perez-Luque, A. Gonzalez, E. Cabezon, F. de la Cruz, and M. Coll. 2001. The bacterial conjugation protein TrwB resembles ring helicases and F1-ATPase. *Nature* **409:**637–641.

Manchak, J., K. G. Anthony, and L. S. Frost. 2002. Mutational analysis of F-pilin reveals domains for pilus assembly, phage infection and DNA transfer. *Mol. Microbiol.* **43:**195–205.

Matson, S. W., J. K. Sampson, and D. R. Byrd. 2001. F plasmid conjugative DNA transfer: the TraI helicase activity is essential for DNA strand transfer. *J. Biol. Chem.* **276:**2372–2379.

Pölzleitner, E., E. L. Zechner, W. Renner, R. Fratte, B. Jauk, G. Högenauer, and G. Koraimann. 1997. TraM of plasmid R1 controls transfer gene expression as an integrated control element in a complex regulatory network. *Mol. Microbiol.* **25:**495–507.

Reisner, A., J. A. Haagensen, M. A. Schembri, E. L. Zechner, and S. Molin. 2003. Development and maturation of *Escherichia coli* K-12 biofilms. *Mol. Microbiol.* **48:**933–946.

Schröder, G., S. Krause, E. L. Zechner, B. Traxler, H. J. Yeo, R. Lurz, G. Waksman, and E. Lanka. 2002. TraG-Like Proteins of DNA transfer systems and of the *Helicobacter pylori* type IV secretion system: Inner membrane gate for exported substrates? *J. Bacteriol.* **184:**2767–2779.

Stedman, K. M., Q. She, H. Phan, I. Holz, H. Singh, D. Prangishvili, R. Garrett, and W. Zillig. 2000. pING family of conjugative plasmids from the extremely thermophilic archaeon Sulfolobus islandicus: insights into recombination and conjugation in Crenarchaeota. *J. Bacteriol.* **182:**7014–7020.

Vergunst, A. C., B. Schrammeijer, A. den Dulk-Ras, C. M. de Vlaam, T. J. Regensburg-Tuink, and P. J. Hooykaas. 2000. VirB/D4-dependent protein translocation from *Agrobacterium* into plant cells. *Science* **290:**979–982.

Waters, V. L. 2001. Conjugation between bacterial and mammalian cells. *Nat. Genet.* **29:**375–376.

Wilkins, B. M., and A. T. Thomas. 2000. DNA-independent transport of plasmid primase protein between bacteria by the I1 conjugation system. *Mol. Microbiol.* **38:**650–657.

FURTHER READING

Bradley, D. E. 1980. Determination of pili by conjugative bacterial drug resistance plasmids of incompatibility groups B, C, H, J, K, M, V, and X. *J. Bacteriol.* **141:**828–837.

Cavalier-Smith, T. 2002. Origins of the machinery of recombination and sex. *Heredity* **88:**125–141.

Christie, P. J. 2001. Type IV secretion: intercellular transfer of macromolecules by systems ancestrally related to conjugation machines. *Mol. Microbiol.* **40:**294–305.

Dürrenberger, M. B., W. Villiger, and T. Bachi. 1991. Conjugational junctions: morphology of specific contacts in conjugating *Escherichia coli* bacteria. *J. Struct. Biol.* **107:**146–156.

Heinemann, J. A., and G. F. Sprague, Jr. 1989. Bacterial conjugative plasmids mobilize DNA transfer between bacteria and yeast. *Nature* **340:**205–209.

Lederberg, J., and E. L. Tatum. 1946. Gene recombination in *Escherichia coli*. *Nature* **158:**558.

Lessl, M., and E. Lanka. 1994. Common mechanisms in bacterial conjugation and Ti-mediated T-DNA transfer to plant cells. *Cell* **77:**321–324.

Llosa, M., F. X. Gomis-Ruth, M. Coll, and F. de la Cruz. 2002. Bacterial conjugation: a two-step mechanism for DNA transport. *Mol. Microbiol.* **45:**1–8.

Rees, C. E. D., and B. M. Wilkins. 1990. Protein transfer into the recipient cell during bacterial conjugation: studies with F and RP4. *Mol. Microbiol.* **4:**1199–1205.

Thomas, C. M. 2000. *The Horizontal Gene Pool: Bacterial Plasmids and Gene Spread*. Harwood Academic Publishers, Amsterdam, The Netherlands.

Zhu, J., P. M. Oger, B. Schrammeijer, P. J. Hooykaas, S. K. Farrand, and S. C. Winans. 2000. The bases of crown gall tumorigenesis. *J. Bacteriol.* **182:**3885–3895.

CONJUGATIVE AND MOBILIZABLE TRANSPOSONS

Abigail A. Salyers, Gabrielle Whittle, and Nadja B. Shoemaker

8

Much of the literature on bacterial evolution has focused on the accumulation of point mutations and the subsequent selection for or against the changes these mutations produced. The collective effect of such evolutionary "baby steps" can be impressive, especially if the process occurs over millions of years. But bacteria can also experience evolutionary "giant steps" in the form of one-step acquisition of large segments of DNA, such as plasmids, bacteriophages, or conjugative trans- posons. Such giant steps can take place in a matter of hours and can occur between bacteria that are members of different genera. Thus, bacteria can pick up new genes from many different sources. Conjugative transposons— integrated elements that excise themselves from the chromosome of the donor bacterium, transfer themselves by conjugation to another bacterium, and integrate into the chromosome of the recipient cell—are the focus of this chapter. Also covered briefly are smaller integrated elements called mobilizable transposons. Mobilizable transposons are triggered to excise by conjugative transposons, which also mobilize them to a recipient cell (Salyers et al., 1999; Salyers et al., 1995b).

How big are the evolutionary "giant steps" represented by acquisition of a conjugative or mobilizable transposon? Conjugative transposons can range in size from 18 kbp to more than 500 kbp. Mobilizable transposons are smaller, usually less than 15 kbp, but, like the conjugative transposons, they can introduce new traits into a bacterial recipient. When one takes into account the fact that there are bacteria with genomes as small as 580 kbp, the potential magnitude of acquisition of a conjugative transposon becomes impressive. Even in the bacteria with genomes as large as 6 to 10 Mbp, these transmissible elements can introduce a significant amount of new genetic information into the organism's genome. Not only are most of these integration events maintained stably by the bacterium that receives them, but

Abigail A. Salyers, Gabrielle Whittle, and Nadja B. Shoemaker, Department of Microbiology, University of Illinois-UC, Urbana, IL 61801.

Microbial Evolution: Gene Establishment, Survival, and Exchange
Edited by Robert V. Miller and Martin J. Day, ©2004 ASM Press, Washington, D.C.

a bacterial strain can acquire multiple conjugative and mobilizable transposons. Thus, transmissible elements may come to constitute a substantial portion of the bacterial genome.

Conjugative transposons and mobilizable transposons were first noticed because some of them carry antibiotic resistance genes, but, like plasmids, they can introduce a much greater variety of new metabolic traits. These traits include the following: virulence genes, genes that allow the bacteria to catabolize new carbon sources such as sucrose or phenolic compounds, genes that allow bacteria to take advantage of new nitrogen sources, DNA restriction-modification systems, and sensitivity to UV. The full extent of the contribution of transmissible integrated elements to bacterial genome evolution is only beginning to be appreciated. The contribution of such elements will be revealed more fully as more genome sequences become available. In addition, conjugative and mobilizable transposons are interesting in their own right. The study of such elements is revealing novel mechanisms of excision, integration, and transfer.

WHAT IS A CONJUGATIVE TRANSPOSON?

Conjugative transposons were first described about 20 years ago. Progress toward understanding their characteristics was slow at first, because there was no way to isolate them physically by a procedure analogous to the plasmid isolation procedure. To make matters worse, these integrated elements were first found in the gram-positive streptococci/enterococci and in the gram-negative *Bacteroides* spp. These bacterial groups did not at that time have the sort of sophisticated genetic systems that would have facilitated the mapping and cloning of these integrated elements. In subsequent years, mostly during the 1990s, large portions of a small group of conjugative and mobilizable transposons have been cloned, and conjugative transposons have been found in genera outside of the streptococcal, enterococcal, and *Bacteroides* species, including *Clostridium, Mesorhizobium, Salmonella,* and *Vibrio* species.

Mobilizable transposons, being more recently discovered, have not yet been shown to be so widely distributed. Despite an increasing number of reports of new conjugative transposons, the number of elements that have been well characterized remains relatively small. This lack of a large database of information on conjugative and mobilizable transposons has led to two problems. One is the lack of a full range of "sequence signatures" that would reveal the presence of a conjugative transposon in a genome sequence. Second, it is now clear that there is a considerable amount of diversity in this group of elements, diversity that has led to some uncertainty as to how to define the category "conjugative transposon."

In the absence of a clear, widely agreed on definition for what is meant by a conjugative transposon, we will start this chapter with a description of the elements that have, for better or worse, come to define the term conjugative transposon. We will then proceed to a description of mobilizable transposons, a category that is, if anything, less clearly delineated. Before embarking on a journey into the teeming swamp of transmissible integrated elements, however, a word about nomenclature is in order. When plasmids and transposons were first described, an orderly nomenclatural system was developed and enforced fairly consistently. Plasmids had names that were usually preceded by a small "p," although the designation that followed was left to the discoverer. Transposons, DNA segments that transfer from one DNA locus to another within the same cell, were designated by the letters "Tn" followed by numbers that were assigned from a central registry. At first, some conjugative transposons were given a "Tn" designation (e.g., Tn916). This proved to be confusing later when it was discovered that conjugative transposons have virtually nothing in common with transposons, except that they integrate into and excise out of DNA. In fact, they have more in common with lysogenic phages, but, so far, no one has had the nerve to suggest that they be given bacteriophage designations. Belatedly, we recommended the use of a

"CTn" designation, which has now been adopted by many, but by no means all, who study conjugative transposons. The subsequent discovery of mobilizable transposons muddied the waters even further. The designation "MTn" has been proposed for mobilizable transposons, but some of the earliest discovered MTns were called NBUs (nonreplicating *Bacteroides* units) or were given a Tn designation. More recently, still other designations such as ICE (for integrative and conjugal elements) have been suggested for integrated transmissible elements, and one group wants to call a type of site-specific integrating element (SXT) a "constin" (Burrus et al., 2002a; Hochhut and Waldor, 1999).

A summary of the features of the integrated elements covered in this chapter is provided in Table 1. These examples have been chosen to illustrate the diversity of transmissible integrated elements.

Section Summary
- CTns carry genes encoding an integrase, excision protein(s), proteins that form the transfer apparatus through which DNA moves from cell to cell, and mobilization proteins that make a single-stranded nick at the *oriT*. Many carry genes that confer traits other than those needed for the CTn to excise and transfer, such as antibiotic resistance genes.
- CTns have been found in many species of bacteria and range in size from 18 to 560 kbp.
- MTns are smaller integrated elements that transfer with the assistance of a CTn.

THE BEST-STUDIED CONJUGATIVE TRANSPOSONS: CTnDOT AND Tn*916*

Steps in Transfer of a CTn
To date, the most extensively studied conjugative transposons are the *Bacteroides* conjugative transposon, CTnDOT, and the gram-positive conjugative transposon, Tn*916*. Tn*916* is the smallest known conjugative transposon, weighing 18 kbp (Scott and Churchward, 1995). CTnDOT is larger (65 kbp). These conjugative transposons differ in more than size. Tn*916* integrates almost randomly in most hosts, whereas CTnDOT integrates site-selectively into at least five to eight sites. Also, as will be seen later, excision and transfer of these conjugative transposons differ both in mechanism and regulation.

What CTnDOT and Tn*916* have in common is the series of steps involved in excision and transfer. These steps are shown in Fig. 1. The first step is excision from the chromosome to form a nonreplicating circle. This circle is nonreplicating in the sense that it does not maintain itself independent of the chromosome after excising and circularizing. That is, it does not become an autonomously replicating plasmid. Yet it does replicate in a sense, because, in the next step, conjugal transfer of a copy of the circular intermediate, a single-stranded copy of the circular form of the conjugative transposon is transferred from donor to recipient via a multiprotein mating bridge. Since the single-stranded copy left behind in the donor and the copy that is transferred to the recipient are next rendered double-stranded, one could argue that this is, in fact, a form of replication, even if it is only a transient one. Finally, the double-stranded circular form integrates into the recipient. Integration and excision are both independent of homologous recombination.

Good experimental evidence exists for all of the steps shown in Fig. 1 except one, the reintegration of the double-stranded circular form into the donor chromosome. A circular suicide plasmid containing the joined ends of the conjugative transposon and the integrase gene can integrate into the chromosome if it is introduced into a bacterial cell, so presumably, the excised circular form should be able to reintegrate into the donor chromosome, but this event has not been shown directly. The problem is that the frequency of excision is low enough (<0.1% in most cases) that it has been difficult to find reintegration events in which the circular form has integrated into a site different than the one it excised from.

TABLE 1 Examples of conjugative and mobilizable transposon families[a]

Element	Size (kb)	First identified in	Marker genes	Integration (*tyr/ser-int*) -excision (*xis*)[b]	Target site(s) (coupling seqs)[c]	Reference(s)
Conjugative transposons (CTns)						
CTnDOT	65.0	*Bacteroides thetaiotaomicron*	*tet*Q, *erm*F	*tyr-int*; *xis* and additional CTn gene products; tetracycline induced excision and transfer	GTAnnTTTGC (5 bp)	Cheng et al., 2001; Whittle et al., 2002
Tn916	18.3	*Enterococcus faecalis*	*tet*M	*tyr-int* family; *xis*; tetracycline induced promoters for transfer genes	AT-rich sites (6 bp)	Clewell et al., 1995
Tn5252	47	*Streptococcus pneumoniae*	*cat*, SOS operon	*tyr-int*; *xis*	72-bp sequence in element end and in primary target site; secondary sites in some hosts missing primary sequence	Alarcon-Chaidez et al., 1997; Kilic et al., 1994
CTn5397	20.658	*Clostridium difficile*	*tet*M	*ser-tndX*, for integration and excision; Tra products ~Tn916.	–GA–target (2 bp –GA–)	Roberts et al., 2001; Wang et al., 2000a
CTn5276	70	*Lactococcus lactis*	*nis*A (nisin), *sac*A (sucrose-6-phosphate hydrolase)	*tyr-int*; *xis*	TTTTTG between ends and in primary target site	Rauch and de Vos, 1994
Site-specific CTns						
CTn-ICE*St1*	34.734	*Streptococcus thermophilus* not shown to transfer yet	Type II restriction system ΦST84 resistance	*tyr-int*; *xis*	27-bp 3′ end of *fda* (fructose-1, 6-diphosphate aldolase)	Burrus et al., 2002b
SXT	99.483	*Vibrio cholerae*	*sul II*, *dhf18*, *strAB*	*tyr-int*; *xis*	17-bp 5′ end of *prfC*	Hochhut et al., 1997
CTnR391 (IncJ)	88.532	*Providencia rettgeri*	*mer* operon (Hg), *aph* (Km), *rumAB* (UV sensitivity)			Beaber et al., 2002

128

Name	Size (kb)	Organism	Cargo genes	Integrase/mechanism	Target site	Reference
CTnscr94	100	*Salmonella enterica* serovar Senftenberg	*scr* (sucrose)	Not identified. Contains regions of identity to 118-kb pathogenicity island in *S. enterica*	3' end Φala-tRNA	Hochhut et al., 1997; Pembroke et al., 2002
Symbiosis islands	502	*Mesorhizobium loti*	Nodulation genes, nitrogen fixation genes	*tyr-int*; *xis*	3' end Φala-tRNA	Sullivan et al., 2002
Pathogenicity islands	10–200	Gram + and Gram + pathogens	Virulence genes	*tyr-int*	3' ends of tRNAs	Hacker and Kaper, 2000
Mobilizable transposons (MTns)						
NBU1	10.276	*Bacteroides uniformis*	None	*tyr-int*; no *xis* homolog	14-bp 3' end of Leu-tRNA	Shoemaker et al., 1996, 2000
NBU2	11.123	*Bacteroides fragilis*	*lin*A homolog	*tyr-int*; no *xis* homolog	14-bp 3' ends of two identical Ser-tRNA genes	Wang et al., 2000b
MTn4555	12.105	*Bacteroides vulgatus*	*cfx*A	*tyr-int*; *xis*	Some strains contain a primary 589-bp site containing 207-bp direct repeats (TnpA dependent); other sites more random (TnpA independent) (6–7 bp)	Tribble et al., 1999a, 1999b
MTn5520	4.692	*Bacteroides*	None	*tyr-int*; *xis*	No known site selectivity; coupling sequences unknown	Vedantam et al., 1999
MTn4451 (~4452- 4453–4454)	6.338	*Clostridium perfringens*	*catP*	*ser-tndX* (~*tndX* on CTn5397); *tnpZ* (~*mobN1* of NBU1-MTn4555 group)	Target site -GA-	Adams et al., 2002; Bannam et al., 1995

[a]Abbreviations: *aph*, kanamycin resistance; *catP*, chloramphenicol resistance; *cfx*A, cefoxitin resistance; *dhf18*, trimethoprim resistance; *lin*A, lincomycin resistance; *mer*, mercury resistance operon; *sulII*, sulfonamide resistance; and *strAB*, streptomycin resistance.

[b]The *tyr-int* is from the tyrosine-integrase of the bacteriophage integrase family, *xis* is small basic protein that facilitates excision as seen for bacteriophage λ, and *ser-int* is the integrase/resolvase of the Tn3 family of integrases that does not require additional gene products for excision.

[c]Coupling sequences are bases found between the ends of the excised circular form of the element that originated from the site it was previously integrated.

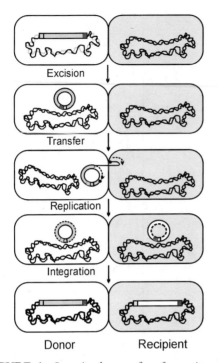

FIGURE 1 Steps in the transfer of a conjugative transposon. The integrated element, represented by the thick bar, excises to form a circular intermediate. The ends are highlighted to show where they are located in the circular form and to emphasize the fact that the transferred element integrates via its ends. A single-stranded copy of the circular form is transferred through a multiprotein mating apparatus that joins the donor and recipient cells. The single-stranded copy becomes double stranded before integrating into the recipient chromosome. (Adapted from Whittle et al., 2002, with permission.)

The Role of Coupling Sequences

The mechanisms of integration and excision of conjugative transposons differ from those proposed for transposons and lysogenic phages, yet there are some intriguing similarities. The proposed model for excision and integration of a conjugative transposon is illustrated in more detail in Fig. 2. During excision, staggered cuts are near the end of the conjugative transposon, at a point that is 5 bp (CTnDOT) or 6 bp (Tn916) from the true end of the element. That is, when a conjugative transposon excises prior to transfer, it brings along small stretches of the donor's chromosomal DNA

that had been adjacent to its ends. These DNA segments are called coupling sequences.

Curiously enough, the coupling sequences from the two ends of the conjugative transposon do not base pair with each other, and a small region of heterology is formed transiently. This region is resolved either by repair enzymes or by replication in favor of one or the other coupling sequences. The result is a mixture of circles, half of which contain one coupling sequence and half of which contain the other one. The best evidence for this unusual excision mechanism comes from experiments in which one-sided PCR was used to amplify one of the strands, followed by two-sided PCR to further amplify the sequence in the coupling sequence region. Also, since integration once more introduces regions of heterology (Fig. 2), the coupling sequences derived from the donor bacterium should end up adjacent to different ends of the integrated conjugative transposon with approximately equal frequencies. This is observed both in Tn916 and CTnDOT.

The steps illustrated in Fig. 1 and 2 have been used as a tentative definition for conjugative transposons. Another defining feature is that the integrases of most of the conjugative transposons, except for the elements in clostridia, have proven to be members of the lambdoid phage integrase family or tyrosine-integrases. The elements from clostridia have a recombinase/resolvase-type integrase (TndX) similar to the Tn3 family of serine-integrases. The CTn tyrosine-integrases do not share much sequence similarity with the phage integrases except at the carboxy-terminal end of the protein. In that region, they have all the conserved residues known to be essential for integration and excision. The excision process requires several accessory proteins encoded by the element for excision. How an enzyme that is more closely related to a phage integrase than to any other known type of integrase can catalyze the reactions shown in Fig. 2 for CTnDOT and Tn916, which differ substantially from the reactions thought to be involved in excision and integration of the lambdoid phages, is an unsolved mystery.

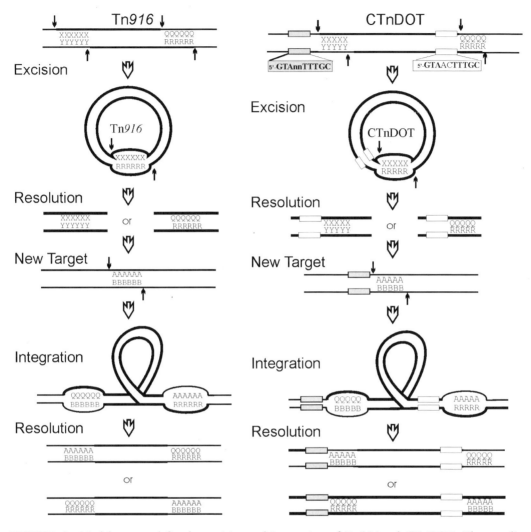

FIGURE 2 Model proposed for the excision and integration of Tn916 and CTnDOT. The coupling sequences, which are chromosomal sequences, are represented by XXX/YYY or QQQ/RRR to indicate that initially they are complementary in sequence. XXX does not base pair with RRR and YYY does not base pair with QQQ. Vertical black arrows indicate the location of the staggered cuts that are produced prior to excision or integration. In the illustration of CTnDOT excision and integration, a thick bar (consensus sequence GTAnnTTTGC) identifies a sequence that is found at one end of CTnDOT immediately adjacent to the coupling sequence and in the site into which CTnDOT integrates. This sequence may account, at least in part, for the fact that CTnDOT integration is less random than integration of Tn916. This model is based on the model described by Scott and Churchward (1995).

The lambdoid phages are limited in their host range not just because the phages need to find the appropriate cell surface receptor but also because of their site-specific mode of integration. The CTns, especially the Tn916 and CTnDOT families of elements, have extensive host ranges. This can be explained in part by the fact that conjugal transfer apparently does not have any receptor requirement, at least not one that has been discernible so far. But perhaps the most significant reason is that integration of these particular conjugative transposons is not site specific, and they can usually find an integration site in almost any host they enter. Several of the CTns described in the table are site specific like the bacteriophages and they do not seem to have the host ranges of the less-site-specific elements. Given this, it is not surprising that Tn916 and related conjugative transposons have been found in many gram-positive bacteria and even in gram-negative bacteria such as the *Neisseria* species. The CTnDOT elements can transfer to all the *Bacteroides* species tested so far and have been shown to be able to transfer to at least some of the other members of the *Bacteroides-Cytophaga-Flavobacterium* phylogenetic group. CTn DOT also transfers itself to *Escherichia coli* under laboratory conditions, although integration into the *E. coli* chromosome is very inefficient. Whether this integration inefficiency is due to poor expression of the CTnDOT integrase gene in *E. coli* hosts or perhaps to a requirement for some *Bacteroides*-specific host factor (or both) remains to be established.

CTnDOT

CTnDOT-type elements were isolated originally from patients with life-threatening *Bacteroides* infections. *Bacteroides* species are gram-negative obligate anaerobes and are prominent members of the microflora of the human colon, where they account for about one fourth of the colonic bacteria. However, if they escape from the colon due to abdominal surgery or other trauma, they can cause life-threatening infections. More recently, members of the CTnDOT family of conjugative transposons have been found to be widespread in nonclinical and clinical isolates. In fact, more than 80% of *Bacteroides* isolates examined in a recent survey (Shoemaker et al., 2001), which included colonic and clinical isolates, carried at least one CTnDOT-type conjugative transposon. The widespread occurrence of CTnDOT is not due to the spread of a clonal population of *Bacteroides* that happens to carry CTnDOT. CTnDOT has been found in virtually all *Bacteroides* species. The extensive transfer of conjugative transposons among

human colonic *Bacteroides* species has revived the old hypothesis that human colonic bacteria might serve as reservoirs of antibiotic resistance genes; that is, bacteria in the human colon acquire, carry, and transmit resistance genes not only among themselves but between themselves and bacteria that pass transiently through the human colon (Whittle et al., 2002).

A genetic map of CTnDOT is shown in Fig. 3A. CTnDOT carries two resistance genes, both of which prevent the action of antibiotics that inhibit ribosome function: the tetracycline resistance gene, *tetQ*, and the eryth-romycin resistance gene, *ermF*. The ErmF protein modifies a ribosomal RNA molecule on the large subunit of the ribosome so that eryth-romycin no longer binds to the ribosome. TetQ is a cytoplasmic protein that somehow protects the ribosome from binding tetracycline. The *tetQ* gene is important not only because it encodes a protein that protects *Bacteroides* ribosomes from tetracycline, but also because of its role in regulation of CTnDOT excision and transfer.

CTnDOT excision and transfer is enhanced by 100- to 1,000-fold by exposure of donor cells to tetracycline. The starting point of the regulatory cascade that is triggered by tetracycline is a three-gene operon in which *tetQ* is the first gene (Fig. 3B). Tetracycline acts through the *tetQ* promoter region, causing increased *tetQ* operon transcription and translation. This operon also contains two genes, *rteA* and *rteB*, that encode proteins with amino acid sequences that suggest they are the sensor and response regulator, respectively, of a two-component regulatory system (Salyers et al., 1995a). The closest database match to RteA is the BvgS of *Bordetella pertussis*, a protein that regulates *Bordetella pertussis* virulence genes. What RteA senses is still unknown. The closest match in the databases to RteB is NtrC, a response regulator that controls the response of *E. coli* to nitrogen starvation. So far, although there is evidence that both RteA and RteB are essential for excision and transfer of CTnDOT, there are no clues as to what signal RteA is sensing. The only thing that is clear at this point is that RteA is not sensing tetracycline. RteB is essential for expression of a third

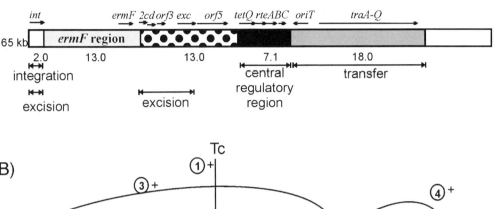

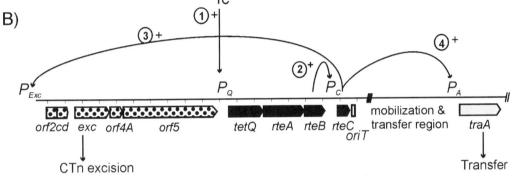

FIGURE 3 (A) Location of genes on CTnDOT that are essential for integration (*int*), excision (*orf2c*, *orf2d*, and *exc*), regulation (*rte*), and transfer (*tra*). The *ermF* region is a 13-kbp composite of transposon and mobilizable transposon genes (Whittle et al., 2001). (B) An expanded view of the excision and regulatory regions of CTnDOT in which the main layers of regulation are illustrated by arrows. Exposure to tetracycline (step 1) leads to increased production of TetQ, RteA, and RteB. RteB, in turn, activates transcription of *rteC* (step 2). RteC activates transcription of genes in the excision region (step 3) and at least some *tra* genes (step 4). Steps 3 and 4 appear to occur simultaneously. + indicates that induction occurred.

regulatory gene, *rteC*, which is encoded by an open reading frame (ORF) located immediately downstream of *rteB*. RteC is essential both for excision and for transfer of the CTnDOT circular intermediate.

A region necessary for excision of CTnDOT is located upstream of the *tetQ-rteA-rteB* operon. In some members of the CTnDOT family, such as CTnERL, this excision region is located immediately downstream of the CTn integrase gene (*intCTnDOT*). In CTnDOT, a 13-kbp insertion, which contains the *ermF* gene, separates the *intCTnDOT* gene from the excision region, so the excision region need not be immediately contiguous to the *intCTn* gene. Excision of phage lambda requires the phage integrase and a small basic protein, called Xis, that helps to drive the integrase

reaction backward in the direction of excision. The excision apparatus of CTnDOT is more complex. Two small *orf*s, which lie immediately downstream of the P$_{exc}$ promoter, *orf2c* and *orf2d,* are both essential for excision. Downstream of these small *orf*s are two larger *orf*s, *orf3* and *exc*. The *orf3* gene can be deleted without affecting excision, so Orf3 is not an essential excision protein. Deletion of *exc*, however, abolishes excision. Exc is a curious protein. It has topoisomerase activity in vitro, but a mutation in the topoisomerase active site that abolishes topoisomerase activity nonetheless is still capable of wild-type levels of excision (Sutanto et al., 2002). From these preliminary studies, it appears that excision of CTnDOT will be very different in mechanism from excision of the lambdoid phages.

Conjugal transfer of the excised circular form of CTnDOT is also regulated by tetracycline and is mediated by a set of transfer genes (*tra* genes) located downstream of *rteC*. The transfer origin (*oriT*) and mobilization genes (*mob* genes) of CTnDOT are also located in this region (Fig. 3). If the *tra* gene region is cloned onto a plasmid, away from other CTn DOT genes, transfer of the plasmid is constitutive. If, however, CTnDOT is provided in *trans*, the plasmid containing the CTnDOT *tra* genes now exhibits tetracycline-regulated, RteC-dependent transfer. The mechanism of regulation is still unclear, but it is obviously complex. Not only is the transcription of at least some of the *tra* genes regulated, probably by RteC, but two small *orfs* located near the end of the *exc* region are also necessary for regulated expression of transfer genes. The complex regulation of transfer gene expression could function to keep transfer genes from being expressed under most conditions. Since many of the transfer proteins are membrane proteins, untimely production of these proteins could be deleterious to the cell that carries CTnDOT. CTnDOT and other conjugative transposons of this family are very stably maintained in cells that carry them. Tight regulation of conjugative transposon activities such as excision and transfer could possibly contribute to the stability of these integrated elements.

The integrase of CTnDOT (*intCTnDOT*) is located upstream of the excision region at one end of CTnDOT (Fig. 3). In contrast to the excision and transfer genes, the *intCTn DOT* gene is expressed constitutively. The integrase and the joined ends of the circular form of CTnDOT are sufficient to mediate integration of the joined ends into a chromosomal site. Integration of CTnDOT is not completely random; there are a limited number of sites per genome. This site-selectivity may be explained by a 10-bp sequence that is located at one end of CTnDOT and matches up with a nearly identical 10-bp sequence adjacent to the integration site of the circular form (Fig. 2 and Table 1).

Recently, an in vitro assay for integration of CTnDOT has been developed. In addition to purified integrase and the DNA substrates (*attDOT*, the joined ends of the circular form of CTnDOT, and *attB*, the chromosomal site into which the circular form integrates), this assay requires purified *E. coli* integration host factor (IHF). The genome sequence of *Bacteroides thetaiotaomicron* 5482, the strain used as a recipient in the in vivo integration assay and the source of *attB* in the in vitro assay, contains several sequences that are distantly related to IHF subunits. So far, none of these IHF homologs seem to have an effect on integration of CTnDOT. Accordingly, although IHF substitutes in the in vitro assay, the real *Bacteroides* host factor may well be quite different at the amino acid sequence level from IHF. The *Bacteroides* host factor that participates in the CTnDOT integration reaction is currently being sought.

Tn*916*

A diagram of the genetic organization of Tn*916* is shown in Fig. 4. Tn*916* carries a tetracycline resistance gene, *tetM*, which, like *tetQ* of CTnDOT, encodes a protein that protects ribosomes from tetracycline binding. It is interesting, although perhaps coincidental, that relatives of Tn*916* that carry resistance genes other than *tetM*, e.g., Tn*1545*, often carry macrolide-lincosamide-streptogramin (MLS)-type *erm* genes. Not all of the conjugative transposons carry antibiotic-resistance genes, but many of the ones that do seem to specialize in the ribosomal protection tetracycline resistance genes like *tetM* and *tetQ* and in MLS-type erythromycin resistance genes. By contrast, genes conferring resistance on β-lactam antibiotics (penicillin, ampicillin) seem to be carried more often on plasmids rather than conjugative transposons. This is by no means an absolute rule, but it is a good bet that a transmissible ribosome protection tetracycline resistance gene or a transmissible MLS *erm* gene in gram-positive bacteria or in *Bacteroides* species is likely to be associated with a conjugative transposon rather than a plasmid.

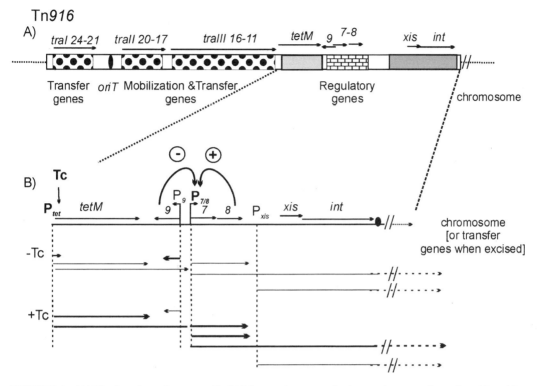

FIGURE 4 (A) The location of genes on Tn*916* that are important for integration, transfer, and excision. Most of these genes are transcribed in the same direction, as indicated by the arrows. (B) A simplified version of how Tn*916* excision and transfer genes are regulated (adapted from Celli and Trieu-Cuot, 1998). Exposure to tetracycline increases the number of transcripts that pass through *tetM, xis,* and *int.* In the circular form of the excised element, these transcripts would cross the joined ends and move through the transfer region (see panel A).

The integrase gene of Tn*916*, *int*, is located at one end of the conjugative transposon oriented toward the end of the integrated element. Immediately upstream of the *int* gene in the same orientation is a gene called *xis* that encodes a protein essential for excision of Tn*916* from the chromosome. As the name suggests, the Tn*916* Xis protein is thought to have a role similar to the role of Xis in phage lambda excision. As previously mentioned, the Int protein of Tn*916* is only distantly related genetically to the λInt but it contains the conserved residues known to be essential for catalyzing the integration/excision reactions.

Upstream of the *int-xis* genes is *tetM* and a cluster of transfer (*tra*) genes. The *oriT* of the circular form of Tn*916* is also in this region. A noteworthy feature of the *tra* gene clusters of Tn*916* and of CTnDOT in Fig. 3 is their small

size. The *tra* regions of most self-transmissible plasmids are at least twice as large. One explanation for the compact size of the conjugative transposon *tra* gene clusters is that the conjugative transposon conjugal transfer systems do not use sex pili. Genes encoding the components of sex pili account for at least half of the genes in the transfer regions of F plasmid, the IncP plasmids, and other well-studied self-transmissible plasmids. No well-characterized conjugative transposon has been shown to have pili genes essential for transfer. This is not very strong evidence for the absence of pili, however, since sex pili are notable for their fragility and can be easily lost from cells being prepared for electron microscopy. Only when the functions of all the conjugative transposon transfer proteins are accounted for will it be clear whether pili or some cell-surface protein complex that serves the same function of binding donor to receptor are involved in the conjugal transfer process.

Although the excision frequency of CTn DOT is enhanced by tetracycline stimulation, there is no apparent effect on the excision frequency of the sequence of the site from which CTnDOT is excising. This is not true of Tn916. Tn916 integrates almost randomly, with a preference for AT-rich sites (Table 1). The frequency of excision of Tn916 from one site, however, can differ as much as 1,000 fold from the frequency of excision from another site. This phenomenon was noted in early articles on the excision and transfer of Tn916, but there is still no explanation for the difference. There is certainly no obvious sequence similarity between high- and low-frequency excision sites.

Integrated elements have an organizational problem. In theory, they could transfer without excising, much the same way as Hfrs transfer. This would result, however, in transfer of a portion of the conjugative transposon that would no longer be able to transfer itself since essential excision and transfer genes had been left behind. Thus, in the interest of preserving the integrity of the conjugative transposon, it would be desirable for transfer to occur only after the element had excised from the chromosome and assumed its circular form. Tn916 has solved this problem in two ways. One way is to decouple the promoter that drives tra gene expression from the tra genes in the integrated form of the conjugative transposon (Fig. 4). Excision of Tn916 is stimulated up to 20-fold by exposure of the donors to tetracyline. As with CTnDOT, this regulation involves the Tn916 tetracycline resistance gene, tetM.

The promoter that will ultimately control expression of the tra genes in the circular form is located in the region upstream of the tetM gene. In the integrated form of the element, transcription from this promoter is directed away from the tra genes into the chromosome. Exposure of donor bacteria to tetracycline causes an upshift in transcription through tetM by a transcriptional attenuation mechanism. Transcription of the region downstream of tetM, which contains xis and int, also increases, allowing excision and circularization to occur. Transcripts originating in the tetM, xis-int region can now continue across the joined ends of the circular form into the transfer genes. Thus, expression of the transfer genes occurs only after excision and circularization of Tn916 to produce the transfer intermediate. A second way Tn916 has of coordinating excision and transfer is that Int protein binds to the transfer origin, a step that may prevent the cleavage at the oriT that initiates transfer of the circular form (Hinerfeld and Churchward, 2001).

Section Summary

- CTns excise from the chromosome to form a covalently closed circle, which is the transfer intermediate. The integration site is regenerated except for a short 4- to 6-bp sequence. Excision and transfer of CTnDOT and Tn916 are regulated by tetracycline, but the mechanisms of regulation are different.

- The excised form of the CTn contains 4- to 6-bp sequences from the chromosome of the donor. These sequences, which are located between the two ends of the circular form of the CTn, are called coupling sequences. Thus, the CTn carries along a small portion of the donor's chromosome when it transfers to a new bacterial host.

- The excision mechanism of Tn916 resembles that of phage lambda and involves a lambda Xis-like protein. Excision of CTnDOT is more complex, involving at least three proteins in addition to the integrase. The CTn-encoded integrase is an essential part of the excision of both Tn916 and CTnDOT.

OTHER SELF-TRANSMISSIBLE INTEGRATED ELEMENTS

Most of the integrated transmissible elements listed in Table 1 have many traits in common with CTnDOT and Tn916 and have thus been called conjugative transposons. Most of them have phage-type or tyrosine-*int* genes, with the exception of the clostridia elements, and many have an *xis* gene that is closely linked to *int*.

Two types of integrated self-transmissible elements that integrate site specifically are CTn*scr94* and two closely related elements called SXT and R391. CTn*scr94* was discovered in a *Salmonella* strain. It carries genes for sucrose fermentation rather than antibiotic resistance genes. CTn*scr94* integrates site specifically in two sites. One site is the *pheV* gene, a gene that encodes a phenylalanine tRNA. Integration in the 3′ end of this gene occurs in such a way that an intact copy of the tRNA gene is preserved. As is evident from the name given to this transmissible element, the discoverers considered it to be a member of the conjugative transposon family. The SXT element is found in many strains of pathogenic *Vibrio cholerae*, including the epidemic strain O139. The SXT element carries multiple antibiotic resistance genes, including genes that confer resistance to sulfonamide, trimethoprim, chloramphenicol, and streptomycin. Like the conjugative transposons, it has a lambdoid-type integrase. Unlike most of the conjugative transposons, except CTn*scr94*, integration of SXT is site specific. The integration site is not a tRNA gene but rather is the *prfC* gene, and the SXT element integrates into the 5′ end, not the 3′ end, of the gene. Although integration occurs independently of homologous recombination, excision may have a partial requirement for homologous recombination to act optimally. R391, which has been found in *Providencia* species and was originally thought to be a plasmid (IncJ), is very closely related to the SXT element except that it carries different resistance genes. Both the SXT element and R391 have transfer genes that are related to the transfer genes of F plasmid. Like CTnDOT, SXT has three regulatory genes. Two of these, *setC* and *setD*, are related to the regulators of flagella assembly, *flhC* and *flhD*. The third gene, *setR*, encodes a protein that is most closely related to a lambdoid repressor, cI repressor.

The discoverers of SXT have argued that the site specificity of the element dictates that it should be put in a separate class from the conjugative transposons. They have given SXT and similar elements the designation "constin." More recently, the discoverers of another site-specific transmissible element, ICE*St1*, which is a 38-kbp element found in *Streptococcus thermophilus*, have made a similar argument for placing this element in a separate class from the conjugative transposons and have given this group of elements the name ICE (for integrative and conjugative elements). Where this battle of the competing nomenclatures will end remains to be seen, but it will probably be decided mainly on the basis of mechanism of excision and integration rather than on site specificity. Whether SXT- and ICE-type elements integrate and excise similarly to CTnDOT and Tn916, by the mechanism illustrated in Fig. 2, remains to be established.

Section Summary
- CTn-like integrated, self-transmissible elements have been found in many different types of bacteria. For example, the SXT element was found in *V. cholerae*. An element called ICE*St1* was found in a *Streptococcus* species.
- Both SXT and ICE*St1* integrate site specifically, unlike CTnDOT and Tn916.

FELLOW TRAVELERS: MOBILIZABLE PLASMIDS AND MOBILIZABLE TRANSPOSONS

Relationships between Conjugative Transposons and the Elements They Mobilize

Many self-transmissible plasmids can mobilize coresident plasmids, called mobilizable plasmids, in *trans*. That is, the conjugative plasmid provides the mating bridge proteins that allow DNA to be transferred from a donor cell to a recipient cell, and the mobilizable plasmid takes advantage of this conduit to a recipient cell by providing the enzyme that makes the single-stranded nick and thereby starts the transfer process that conveys a single-stranded copy of the mobilizable plasmid to the recipient cell. Not surprisingly, associations between mobilizable plasmids and the self-transmissible plasmids that mobilize them are quite specific, since the mobilization proteins of the mobilizable plasmid have to be able to interact with the transfer apparatus provided by the self-transmissible plasmid.

A similar phenomenon is seen in the conjugative transposons. A conjugative transposon can mobilize a coresident plasmid just the same way a self-transmissible plasmid does, by providing the mating bridge through which the plasmid DNA transfers (Fig. 5). At least some conjugative transposons can also do something that conjugative plasmids have so far not been shown to do. They can trigger the excision and circularization of unlinked integrated elements that have been called mobilizable transposons. Most of the mobilizable transposons so far described are not smaller, defective forms

Mobilization of co-resident plasmids *in trans*

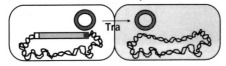

Excision and mobilization of MTns

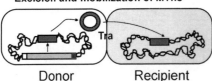

Donor   Recipient

FIGURE 5   Some CTns mobilize elements in *trans*. The mobilizable coresident plasmid provides the protein(s) that nick the plasmid at the transfer origin. The CTn provides the multiprotein mating apparatus through which the single-stranded copy of the plasmid is mobilized (mob) into the recipient cell. Some CTns, such as CTnDOT, also act in *trans* to trigger excision and mobilization of coresident mobilizable transposons (MTns). In this case, the CTn provides factors that stimulate excision and circularization of the MTn and the transfer (tra) functions that allow the single-stranded copy of the MTn to be transferred to the recipient cell. The MTns integrate on their own without help from the CTn.

of the conjugative transposons. They differ in DNA sequence and even in their projected mechanisms of excision and integration.

An example of a mobilizable transposon (MTn) is NBU1, a *Bacteroides* element that is 10.3 kbp. Excision and circularization of NBU1 is triggered in *trans* by the CTnDOT regulatory protein RteB (Fig. 5). The mechanism of this interaction is still not understood. A copy of the circular form of the mobilizable transposon is then transferred in *trans* by CTnDOT similar to a coresident mobilizable plasmid as shown in the top panel. That is, NBU1 provides a protein (MobN1) that nicks at the NBU1 *oriT*, then interacts with the mating apparatus provided by CTnDOT, to transfer a single-stranded copy of NBU1 to the recipient. In the recipient, the recircularized circular form integrates into the donor chromosome. In contrast to CTnDOT, the NBU1 transfer intermediate integrates site specifically in the 3' end of a leucine tRNA gene. NBU1 carries no antibiotic resistance genes, but a related MTn, NBU2, carries at least one.

The smallest known mobilizable transposon is MTn5520, which is 4.7 kbp. It carries genes for excision, mobilization, and integration but no known antibiotic resistance genes. Its site selectivity has not been established. A particularly interesting mobilizable transposon is MTn4555, a 12-kbp MTn that carries a cefoxitin resistance gene. Cefoxitin is an important drug for treating *Bacteroides* infections, so the appearance of this resistance gene on a mobile element was bad news. MTn4555, like most of the conjugative transposons and the other MTns has an integrase and Xis that fall in the lambdoid phage family of integrases and excision proteins, but it has an additional feature. Although it can integrate almost randomly, it carries a gene that encodes a protein, TnpA, which directs integration to a specific site that is found in some hosts. This MTn also has coupling sequences, like the conjugative transposons, a trait that is not shared by most of the other MTns.

Several MTns have recently been found in *Clostridium perfringens* and *Clostridium difficile*.

These MTns carry antibiotic resistance genes and display the degree of diversity we have begun to expect from the mobilizable transposons. Wang et al. (2000a) found that like the CTn5397 isolated from *C. difficile*, the *Clostridium* MTns utilize a large serine integrase of the Tn3 family of transposases called TndX that is sufficient for both integration and excision of the elements. So far, all the information about mobilizable transposons has come from studies of *Bacteroides* and *Clostridium* species, but it seems unlikely that this phenomenon is restricted to a few genera of obligate anaerobes. As was the case with the conjugative transposons, since it is difficult to find transmissible integrated elements except by accident, an awareness of the existence of mobilizable transposons may spark the discovery of similar elements in other types of bacteria as scientists interested in horizontal gene transfer begin to look for transfer elements of this type.

Ecology of Mobilizable Transposons

A few of the MTns discovered to date carry antibiotic resistance genes but most of them seem to encode only those genes necessary for their excision, transfer, and integration. Are the MTns just a curiosity or do they have some practical importance? It is important to note that NBUs, like the CTnDOT-type conjugative transposons have spread widely among different *Bacteroides* species in nature. More than half of the *Bacteroides* isolates screened in a recent survey carried DNA that cross-hybridized with a probe consisting of a highly conserved region of NBU1 that carries the element's mobilization and *oriT*, indicating that all these strains carry an NBU1-type element. In contrast to the CTnDOT elements, which seem to have done most of their transferring among *Bacteroides* species in recent years, carriage of NBU-hybridizing elements seems not to have changed nearly so much from the pre-1970s period to the present. The MTns are starting to pick up antibiotic resistance genes, and we may be seeing the beginning of a trend toward acquisition of resistance genes by MTns, thus increasing the potential for transfer of these resistance genes. In the clostridia, no such extensive surveys have been done, but the increased reporting of MTns in this genus in recent years indicates that they may be equally widespread.

MTns also contribute to the evolution of other elements. This is perhaps most clearly seen from an event or events that took place at some time in the past and created the 13-kbp insertion that is one of the few differences between CTnDOT and members of the family that do not carry *ermF*. The 13-kbp insertion is illustrated in Fig. 3. Clearly, this 13-kbp insert was the result of a series of transposon and MTn insertions to create a composite element that is now found widely in the CTn DOT family. Conjugative transposons also seem to have a proclivity for integrating into other conjugative transposons to form larger hybrid elements, as is seen in the streptococcal conjugative transposon, CTn5253 (CTn5252 plus integrated copy of the Tn916 type element CTn5251) and the *Bacteroides* conjugative transposon, CTn12256 (unnamed CTn with an integrated copy of CTnDOT). Both of these compound elements contain two potentially active conjugative transposons that are now moving as a unit.

Section Summary

- MTns range in size from 4.7 to 12 kbp. NBU1, Tn5520, and Tn4555 are examples of MTns. So far, MTns have been found only in *Bacteroides* and *Clostridium* species.
- MTns, like CTns, have a circular intermediate but rely on a CTn for transfer functions. MTns encode one or more mobilization proteins that make a single-stranded nick at the *oriT* of the circular form and interact with CTn transfer proteins to transfer a single copy of the MTn to the recipient.
- MTns can become part of a CTn or form composite elements and may be making a contribution to the evolution of transmissible elements.

THE END OF THE ROAD

This chapter started with the image of large transmissible elements, such as plasmids and conjugative transposons, as "giant steps" on the evolutionary road. It proceeded with a survey of what is known about how the giant steps caused by conjugative transposons and the slightly smaller steps caused by mobilizable transposons occur. The mechanisms of these excision, transfer, and integration processes, the "how" questions, are clearly interesting ones and are revealing novel processes. Ultimately, however, the most compelling question is the "so what" question: how much of this sort of giant-step change actually occurs in nature and what is its significance for bacterial genome plasticity? Until recently, about the best we could do to answer the "so what" question was to monitor the extent and frequency of gene transfer events by DNA hybridization analysis by using probes that identified known gene transfer agents. Remarkably few studies of this type have been done. In the few studies we have initiated, we have found evidence for very extensive transfer of conjugative transposons and NBUs within the family of *Bacteroides* species. This widespread transfer has had the obvious effect of spreading antibiotic resistance genes, but it is likely that there are other effects stemming from such transfers.

Clearly, it is of practical importance to stem the spread of resistance genes among disease-causing bacteria, but there are some indications that transmissible integrated elements are contributing to other ecological features of bacteria. At least some of the pathogenicity islands that are contributing to the evolution of pathogens may be conjugative transposons. The largest conjugative transposon so far discovered, the 560-kbp conjugative transposon found in *Mesorhizobium loti*, has been called a symbiosis island because it carries nitrogen fixation genes and other genes that contribute to the development of a symbiotic interaction between the bacteria and the plants with which they associate. A conjugative transposon, CTn*5276*, which was found in a *Lactococcus lactis* strain that is used in the cheese industry, carries genes for sucrose fermentation and for an antibacterial protein nisin, which helps to control contamination of dairy products by undesirable microbial species.

Ultimately, a full accounting for the evolutionary and ecological effect of conjugative transposons—how many have entered new bacterial strains and what has been the consequence of their presence—will only be achieved by analysis of genome sequences and by more sophisticated tests of the phenotypic effects mediated by these integrated transmissible elements. Such an analysis will only be possible, however, when sequence motifs that identify possible conjugative transposons and mobilizable transposons, with special emphasis on their ends, have been identified. One sequence signature that seems to identify many, if not all conjugative and mobilizable transposons, is their integrase protein. An integrase gene in the lambda integrase family that is not accompanied by capsid and tail-fiber genes might be attached to a conjugative or mobilizable transposon. The excision genes are proving to be much more variable. Even within the lambdoid phages the amino acid sequence of Xis is too variable to serve as a reliable sequence signature, and this variability is only going to be greater in the conjugative transposons that use this type of excision system. For the clostridia the TndX family of integrases and resolvases may be the target gene since it functions as both integrase and excisase for all the elements described to date.

For elements like CTnDOT and the NBUs, which have different types of excision systems, finding sequence motifs has so far proved to be even more elusive. Yet, some sequence signatures among the transfer genes are beginning to emerge. For example, a CTnDOT transfer protein, TraG, has recognizable homologs among transfer genes of many conjugative transposons and self-transmissible plasmids. This type of gene may not distinguish a conjugative transposon from a conjugative plasmid, but it could signal the presence of a self-transmissible element and provoke further scrutiny of the surrounding DNA sequence.

Up to this point, investigations of conjugative and mobilizable transposons have focused on genetic structures of these elements and mechanisms of excision, transfer, and integration. The question of how these integrated transmissible elements contribute to bacterial evolution and ecology remains to be answered.

So, what are conjugative transposons and mobilizable transposons and where did they come from? Stating that conjugative transposons and mobilizable transposons are integrated elements that are capable of transfer by conjugation begs the question of how these elements arose. Osborn and Boltner (2002) among other scientists who work on these elements have noted that they have phage-like, plasmid-like, and transposon-like properties, although they are clearly not members of any one of these groups. To us, the phage connection is particularly intriguing. The integrase connection is the most obvious one that links phage and conjugative transposons, but are there other possible links? It is interesting to note that the transfer proteins, which connect the cytoplasm of the donor with the cytoplasm of the recipient, have some properties in common with phage capsid proteins; they cover the DNA and help to inject it into the recipient cell. Eukaryotic cells have two types of viruses, viruses that have an extracellular phase and fusogenic viruses that move from cell to cell. Some viruses of eukaryotic cells, such as human immunodeficiency virus, exhibit both modes of spread. The bacterial viruses discovered to date have all been viruses that have an extracellular phase. Is it possible that conjugative and mobilizable transposons are related to the bacterial equivalent of the fusogenic viruses of eukaryotic cells, assuming that such bacterial viruses exist? If so, they have been missed in all investigations of bacteriophages undertaken so far.

Whatever the origin of conjugative and mobilizable transposons, it is clear that they are undergoing a vigorous and continuing evolution that produces elements that are mosaics of DNA segments from phages, plasmids, and transposons. This variability makes it difficult to identify them from sequence data alone, although as mentioned in the previous section, some sequence signatures are beginning to emerge. One lesson from the mosaic nature of many conjugative transposons is that sequence data from portions of an element are not likely to give an adequate picture of the nature and origin of the element unless the sequence data come from the "housekeeping genes" of conjugative transposons: the integrase gene, excision genes, and transfer genes. Even then, given the extensive sequence diversity seen already in this group of elements, analyses of relationships will remain problematic.

Section Summary

- CTns contribute to the evolution of bacteria by conferring new traits such as resistance to an antibiotic, production of an antibiotic or nitrogen fixation, although some CTns seem to be simpler stripped-down elements that contain only the genes necessary for their excision, transfer, and integration.
- Sequences of some CTn genes, especially integrases and transfer genes, may be conserved enough to serve as sequence "signatures" that will make it possible to identify possible CTns in a genome sequence.
- The evolutionary origin of CTns and MTns is still a mystery.

QUESTIONS

1. Why would an integrated element take a small segment of chromosomal DNA with it when it transfers to a new host? Could this property lead to mistakes during excision that alter the length of the CTn?

2. Could CTns be a type of bacteriophage? How could this be proven (or could it), especially if free forms of the "phage" particle are not found?

3. How could analyses of genome sequences be used to ascertain the extent to which CTns or MTns are actually affecting bacterial evolution? Would such analyses require that genome sequences are available for more than one member of a bacterial species?

4. It has been suggested that some of the so-called pathogenicity islands might be transmissible elements such as CTns and MTns. How could this hypothesis be tested?

5. Are virulence genes and antibiotic resistance genes ever transferred together? There are almost no examples of this type of linkage so far. Is this due to a lack of linkage or to the fact that scientists interested in resistance gene transfer do not look for virulence genes and vice versa?

REFERENCES

Adams, V., D. Lyras, K. A. Farrow, and J. I. Rood. 2002. The clostridial mobilisable transposons. *Cell. Mol. Life Sci.* **59:**2033–2043.

Alarcon-Chaidez, F., J. Sampath, P. Srinivas, and M. N. Vijayakumar. 1997. Tn5252: a model for complex streptococcal conjugative transposons. *Adv. Exp. Med. Biol.* **418:**1029–1032.

Bannam, T. L., P. K. Crellin, and J. I. Rood. 1995. Molecular genetics of the chloramphenicol-resistance transposon Tn4451 from *Clostridium perfringens*: the TnpX site-specific recombinase excises a circular transposon molecule. *Mol. Microbiol.* **16:**535–551.

Beaber, J. W., V. Burrus, B. Hochhut, and M. K. Waldor. 2002. Comparison of SXT and R391, two conjugative integrating elements: definition of a genetic backbone for the mobilization of resistance determinants. *Cell. Mol. Life Sci.* **59:** 2065–2070.

Burrus, V., G. Pavlovic, B. Decaris, and G. Guedon. 2002a. Conjugative transposons: the tip of the iceberg. *Mol. Microbiol.* **46:**601–610.

Burrus, V., G. Pavlovic, B. Decaris, and G. Guedon. 2002b. The ICESt1 element of *Streptococcus thermophilus* belongs to a large family of integrative and conjugative elements that exchange modules and change their specificity of integration. *Plasmid* **48:**77–97.

Celli, J., and P. Trieu-Cuot. 1998. Circularization of Tn916 is required for expression of the transposon-encoded transfer functions: characterization of long tetracycline-inducible transcripts reading through the attachment site. *Mol. Microbiol.* **28:** 103–117.

Cheng, Q., Y. Sutanto, N. B. Shoemaker, J. F. Gardner, and A. A. Salyers. 2001. Identification of genes required for excision of CTnDOT, a *Bacteroides* conjugative transposon. *Mol. Microbiol.* **41:**625–632.

Clewell, D. B., D. D. Jaworski, S. E. Flannagan, L. A. Zitzow, and Y. A. Su. 1995. The conjugative transposon Tn916 of *Enterococcus faecalis*: structural analysis and some key factors involved in movement. *Dev. Biol. Stand.* **85:**11–17.

Hacker, J., and J. B. Kaper. 2000. Pathogenicity islands and the evolution of microbes. *Annu. Rev. Microbiol.* **54:**641–679.

Hinerfeld, D., and G. Churchward. 2001. Specific binding of integrase to the origin of transfer (oriT) of the conjugative transposon Tn916. *J. Bacteriol.* **183:**2947–2951.

Hochhut, B., K. Jahreis, J. W. Lengeler, and K. Schmid. 1997. CTnscr94, a conjugative transposon found in enterobacteria. *J. Bacteriol.* **179:** 2097–2102.

Hochhut, B., and M. K. Waldor. 1999. Site-specific integration of the conjugal *Vibrio cholerae* SXT element into prfC. *Mol. Microbiol.* **32:**99–110.

Kilic, A. O., M. N. Vijayakumar, and S. F. al-Khaldi. 1994. Identification and nucleotide sequence analysis of a transfer-related region in the streptococcal conjugative transposon Tn5252. *J. Bacteriol.* **176:**5145–5150.

Osborn, M. A., and D. Boltner. 2002. When phage, plasmids, and transposons collide: genomic islands, and conjugative- and mobilizable-transposons as a mosaic continuum. *Plasmid* **48:**202–212.

Pembroke, J. T., C. MacMahon, and B. McGrath. 2002. The role of conjugative transposons in the Enterobacteriaceae. *Cell. Mol. Life Sci.* **59:**2055–2064.

Rauch, P. J., and W. M. de Vos. 1994. Identification and characterization of genes involved in excision of the *Lactococcus lactis* conjugative transposon Tn5276. *J. Bacteriol.* **176:**2165–2171.

Roberts, A. P., P. A. Johanesen, D. Lyras, P. Mullany, and J. I. Rood. 2001. Comparison of Tn5397 from *Clostridium difficile*, Tn916 from *Enterococcus faecalis* and the CW459tet(M) element from *Clostridium perfringens* shows that they have similar conjugation regions but different insertion and excision modules. *Microbiology* **147:**1243–1251.

Salyers, A. A., N. B. Shoemaker, G. Bonheyo, and J. Frias. 1999. Conjugative transposons: transmissible resistance islands, p. 331–346. *In* J. B. Kaper and J. Hacker (ed.), *Pathogenicity Islands and Other Mobile Virulence Elements*. ASM Press, Washington D.C.

Salyers, A. A., N. B. Shoemaker, and A. M. Stevens. 1995a. Tetracycline regulation of conjugal transfer genes, p. 393–400. *In* J. A. Hoch and T. J. Silhavy (ed.), *Two-Component Signal Transduction*. American Society for Microbiology, Washington, D.C.

Salyers, A. A., N. B. Shoemaker, A. M. Stevens, and L. Y. Li. 1995b. Conjugative transposons: an unusual and diverse set of integrated gene transfer elements. *Microbiol. Rev.* **59:**579–590.

Scott, J. R., and G. G. Churchward. 1995. Conjugative transposition. *Annu. Rev. Microbiol.* **49:**367–397.

Shoemaker, N. B., H. Vlamakis, K. Hayes, and A. A. Salyers. 2001. Evidence for extensive resistance gene transfer among *Bacteroides* spp. and among *Bacteroides* and other genera in the human colon. *Appl. Environ. Microbiol.* **67:**561–568.

Shoemaker, N. B., G. R. Wang, and A. A. Salyers. 1996. The *Bacteroides* mobilizable insertion element, NBU1, integrates into the 3′ end of a Leu-tRNA gene and has an integrase that is a member of the lambda integrase family. *J. Bacteriol.* **178:**3594–3600.

Shoemaker, N. B., G. R. Wang, and A. A. Salyers. 2000. Multiple gene products and sequences required for excision of the mobilizable integrated *Bacteroides* element NBU1. *J. Bacteriol.* **182:**928–936.

Sullivan, J. T., J. R. Trzebiatowski, R. W. Cruickshank, J. Gouzy, S. D. Brown, R. M. Elliot, D. J. Fleetwood, N. G. McCallum, U. Rossbach, G. S. Stuart, J. E. Weaver, R. J. Webby, F. J. De Bruijn, and C. W. Ronson. 2002. Comparative sequence analysis of the symbiosis island of *Mesorhizobium loti* strain R7A. *J. Bacteriol.* **184:**3086–3095.

Sutanto, Y. S., N. B. Shoemaker, J. F. Gardner, and A. A. Salyers. 2002. Characterization of Exc, a protein required for the excision of the *Bacteroides* conjugative transposon, CTnDOT. *Mol. Microbiol.* **46:**1239–1246.

Tribble, G. D., A. C. Parker, and C. J. Smith. 1999a. Genetic structure and transcriptional analysis of a mobilizable, antibiotic resistance transposon from *Bacteroides*. *Plasmid* **42:**1–12.

Tribble, G. D., A. C. Parker, and C. J. Smith. 1999b. Transposition genes of the *Bacteroides* mobilizable transposon Tn*4555*: role of a novel targeting gene. *Mol. Microbiol.* **34:**385–394.

Vedantam, G., T. J. Novicki, and D. W. Hecht. 1999. *Bacteroides fragilis* transfer factor Tn*5520*: the smallest bacterial mobilizable transposon containing single integrase and mobilization genes that function in *Escherichia coli*. *J. Bacteriol.* **181:**2564–2571.

Wang, H., A. P. Roberts, D. Lyras, J. I. Rood, M. Wilks, and P. Mullany. 2000a. Characterization of the ends and target sites of the novel conjugative transposon Tn*5397* from *Clostridium difficile*: excision and circularization is mediated by the large resolvase, TndX. *J. Bacteriol.* **182:**3775–3783.

Wang, J., N. B. Shoemaker, G. R. Wang, and A. A. Salyers. 2000b. Characterization of a *Bacteroides* mobilizable transposon, NBU2, which carries a functional lincomycin resistance gene. *J. Bacteriol.* **182:**3559–3571.

Whittle, G., B. D. Hund, N. B. Shoemaker, and A. A. Salyers. 2001. Characterization of the 13 kb *ermF* region of *Bacteroides* conjugative transposon, CTnDOT. *Appl. Environ. Microbiol.* **67:**3488–3495.

Whittle, G., N. B. Shoemaker, and A. A. Salyers. 2002. The role of Bacteroides conjugative transposons in the dissemination of antibiotic resistance genes. *Cell. Mol. Life Sci.* **59:**2044–2054.

FURTHER READING

The December 2002 issue of the *Journal of Cellular and Molecular Life Sciences* contains a special section on conjugative and mobilizable transposons (Vol. 59, p. 2013–2082). This collection of review articles written by scientists working in the field is a valuable resource for those who would like to delve more deeply into the mysteries of conjugative and mobilizable transposons.

BACTERIOPHAGE-MEDIATED TRANSDUCTION: AN ENGINE FOR CHANGE AND EVOLUTION

Robert V. Miller

9

Transduction is a mechanism for horizontal transfer of genetic material among bacteria that have the common property of being infected by a specific bacterial virus (bacteriophage). The bacteriophage provides the transfer vessel (a capsid) and the mechanism of packaging and delivery of the genetic information. The fate of the transferred genetic information is independent of the process of transduction and depends on mechanisms of recombination and DNA stabilization that you learned about in chapter 2. It is an elegant process tailor-made to stimulate reassortment of genetic information in a "sexless" group of organisms. Transduction has the potential to greatly affect evolution in the bacteria.

To understand transduction, we must first understand the biology of bacterial viruses

Robert V. Miller, Department of Microbiology and Molecular Genetics, Oklahoma State University, 307 Life Sciences East, Stillwater, OK 74078.

called bacteriophages. We will then explore the mechanisms of transduction and discuss some of the evidence that it is more than a simple laboratory phenomenon. Finally, we will discuss the potential of transduction to affect bacterial evolution.

WHAT ARE BACTERIOPHAGES?
Twort and D'Herelle first described bacteriophages independently around 1915. They were conducting experiments in pathogenic microbiology and observed that petri dishes with lawns of bacteria on them showed "holes." When more closely examined these holes or "plaques" were found to contain bacteriophage (bacteria eaters), a term coined by D'Herelle. D'Herelle went on to suggest that these bacteriophages could be used to fight disease by killing the pathogens. Although not frequently attempted in the twentieth century, phage therapy is currently being explored as an alternative to the use of antibiotics for the control of bacterial infections.

Bacteriophage Structure
Bacteriophages, like all viruses, are intracellular parasites. They utilize the metabolic machinery of their hosts for reproduction. All the "life functions" carried out by viruses are carried out while infecting their host cell. However, viruses are transported between hosts by virions. It is these virions that we recognize as viruses and use to classify them.

Virions are made up of a protein shell (the capsid) that encloses and protects the genetic

Microbial Evolution: Gene Establishment, Survival, and Exchange
Edited by Robert V. Miller and Martin J. Day, ©2004 ASM Press, Washington, D.C.

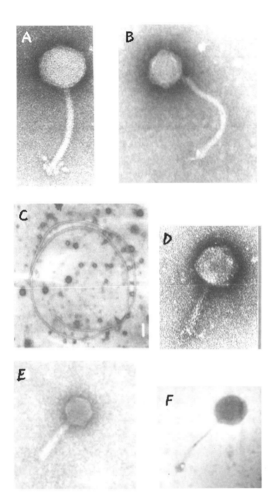

FIGURE 1 Electron micrographs of various bacteriophages. (A, B, D–F) Complex bacteriophages; (C) filamentous bacteriophage. (A, B, D) Bacteriophages isolated from sewage; (C, E, F) those isolated from a freshwater lake.

FIGURE 2 A Buckminster Fuller-like home in Oklahoma. The rendering in panel B emphasizes the geometry of the icosidodecahedron making up the building's roof.

material of the virus (the genome). Depending on the virus, the capsid can take several different geometries (Fig. 1). The first of these is an icosadeltahedral capsid that resembles as multifaceted ball much like a Buckminster Fuller geodesic dome (Fig. 2). The second is a simple helix that leads to a filamentous shape. The third, which is by far the most common shape among bacteriophages that have been identified and studied, is the "complex capsid." This form is made up of a modified icosadeltahedral "head" and a filamentous "tail."

No matter what its shape, each capsid contains a bacteriophage genome made of nucleic acid that contains the genetic information necessary for biosynthesis of phage components and their assembly into new phage particles. While they are often double-stranded DNA, genomes of specific viruses may be single-stranded DNA or even RNA. These unusual types of genomes are found in the icosadeltahedral and filamentous phages. To date, all complex-capsid bacteriophages (Fig. 3) that have been studied contain double-stranded DNA. Because they contain double-stranded DNA, it is the complex viruses that are most likely to be agents of horizontal transfer of bacterial genes. Therefore, this chapter will concentrate on these bacteriophages.

Life Cycles

The life cycles of bacteriophages consist of both biotic and abiotic components. The abiotic portions of the cycles are those that occur while

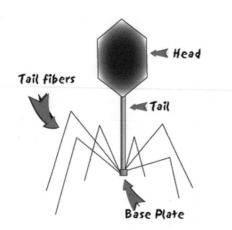

FIGURE 3 A complex bacteriophage. The various parts of the capsid are identified. The nucleic acid (double-stranded DNA in all known cases) is contained in the head. Contact with the host bacterium is initially made with the tail fibers. Irreversible contact is made at the base plate and the DNA is "injected" into the cell through the tail that often contracts. Compare this diagram with the electron micrographs in Fig. 1.

the virus is in the form of a virion and "looking" for a new host to infect. The biotic portions all occur during the infection of a bacterial host cell. These life cycles can be divided into two types. The biotic portion of the first type of life cycle, referred to as "productive," ends with the production of new virus particles. These progeny virions exit the infected cell in one of two ways. The first way leads to the lysis of the host cells with the release of progeny viruses as a burst. In the second way, each virion "buds" though the host cell's membrane as it is matured. Because viruses with complex virions invariably lyse their host with the release of a burst of progeny phage particles, this type of productive cycle, usually called a "lytic" cycle, will be highlighted here.

"Reductive cycles" are the second form of phage life cycles. They do not lead to the production of virus particles. In these cycles, the phage genome enters the host cell but its genetic functions are not expressed. Instead, the phage genome becomes an integral part of the genetic complement of the host. In this form, known as a prophage, the viral DNA is replicated along side other genetic elements in the host and is

segregated along with these host DNA elements into daughter cells at cell division. Because these infections are benign to the host, this type of life cycle is referred to as "temperate." The temperate response can often be maintained for many generations. However, it can be terminated when the bacterial cell is exposed to various environmental stresses including UV irradiation, desiccation, heat, and other conditions that might lead to the death of the host cell. When terminated, the prophage is "activated" and a lytic cycle ensues. Because of this, bacteria that are infected in this manner, are referred to as "lysogens" because they are "lysogenic," that is, they are "capable of being lysed."

LYTIC CYCLE

The lytic cycle of bacteriophages is a well-programmed process (Fig. 4A). It begins with the identification of a host through the interaction between specific phage proteins and "receptors" on the host cell (Fig. 5). It is the uniqueness of these receptors on the bacterial host that determines the host range of the phage. For any given phage the receptor will be one of the specific structures or compounds found on the external surface of the host cell. These include pili, flagella, glycoproteins, and other structures in the cellular envelope. For complex phages, the phage-recognition protein is contained at the base of the tail. For other phages, it is a unique protein usually at one of the vertices of an icosadeltahedral phage or at one of the ends of a helical phage. Bacteria that do not have the phage receptor are said to be "resistant" to infection. The interaction of the host receptor and phage-recognition proteins triggers a process that allows infection of the viral nucleic acid (for all known complex viruses, this is double-stranded DNA) into the cytoplasm of the host bacterium and initiates the timed expression of phage genes that will lead to the production of new phage particles: the phage nucleic acid is replicated; capsid proteins are synthesized and assembled into capsids; and finally, the nucleic acid is inserted into the capsid. As we will see, sometimes the packaging machinery makes a mistake and bacterial DNA is packaged instead

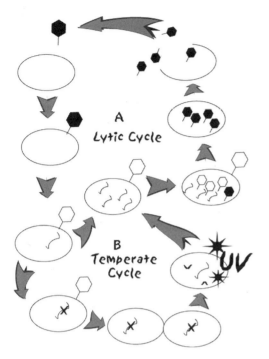

FIGURE 4 The life cycles of a complex bacterio-phage. (A) Lytic life cycle that leads to the production of progeny phages; (B) temperate life cycle in which a prophage is established in the host cell producing a bacterium referred to as a lysogen. The prophage is not transcribed (indicated by the Xs) but replicates with the host cell's genome and is partitioned into its daughter cells. Some prophages are integrated into the host genome (see Fig. 6) while other types are carried as plasmids. Note: The host genome is not illustrated in this figure.

of viral DNA, producing a transducing particle. In any case, the process ends with the lysis of the host and the release of a burst of phage particles.

THE TEMPERATE CYCLE

Many, if not all, lytic bacteriophages have an alternate life cycle that allows them to establish a benevolent interaction with their host. This "temperate" life cycle (Fig. 4B) can produce a long-term interaction lasting many generations of the host bacterium. In this cycle, the bacterio-phage DNA is introduced into the host, but, due to a specific repressor protein, lytic functions are not expressed. During the establishment of the temperate life cycle, called the "establishment of lysogeny," the nucleic acid is frequently inserted into the host cell's chromosome. Over time, the phage genome may evolve into a cryp-tic prophage. These cryptic prophages often provide the cell with new functions that in-crease their virulence or survivability (see chap-ter 17). Some viruses establish prophages that are maintained as plasmids within the host cell.

Lysogenization alters the phenotype of the host in subtle but often significant ways. The most common alteration is the manifestation of immunity to superinfection by the same phage. Because lysogeny is maintained by the produc-tion of a repressor that inhibits expression of the lytic functions of the phage, any infecting genome is immediately repressed. Thus, lyso-

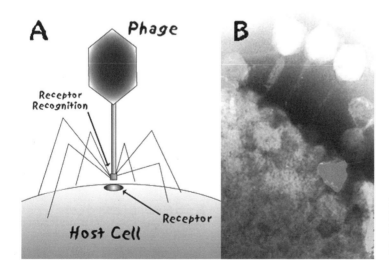

FIGURE 5 Attachment of a bacteriophage particle to the host-cell receptor. (A) Diagram; (B) electron micrograph of phage UT1 infecting *P. aeruginosa*.

gens are not susceptible to the cytopathic effects of infection by phages that respond to the repressor of the resident prophage. This property is called "immunity to superinfection."

Other characteristics that may be induced by the establishment of a prophage include alteration of cellular receptors that change the serotype of the host. These changes in cell envelope components may eliminate or produce receptors for other, unrelated phages, and lysogens will become resistant to some phages that infect nonlysogens of the same species. Alternatively, lysogeny may become sensitive to new phages. Other characteristics may increase the virulence of the host. For instance, the gene that encodes diphtheria toxin is not part of the bacterial chromosome but is a gene expressed for prophages in lysogens. Likewise, cholera toxin is encoded by a prophage that lysogenizes *Vibrio cholerae* (Waldor and Mekalanos, 1996).

Because the concentration of repressor varies from cell to cell, it may fall below the critical level to maintain lysogeny in a small but usually characteristic fraction of cells during each generation of a lysogenized population. In these cells, the prophage is "activated" and a lytic life cycle ensues with the lysis of the cell and a burst of progeny virions. Thus, there is a constant production of virions from such a population.

In addition, various stresses to the host cell can "activate" the population in mass.

In all these cases, the repressor is modified or destroyed, allowing the expression of the lytic functions. The phage genome, if integrated into the host chromosome, is excised and expressed. During the excision process, a portion of the host-cell chromosome adjacent to the integration site may be incorporated into the bacteriophage genome, producing a specialized transducing phage (Fig. 6). After excision, replication of phage DNA and expression of capsid proteins is followed by lysis of the host and the liberation of a burst of virions.

Section Summary

- Bacteriophages are viruses that only infect bacteria.
- The virion is a biologically inert form of the virus used to enable its extracellular survival and to transport the phage genome between hosts.
- Biological activities of the virus all take place while the phage genome is an intracellular parasite of the host bacterium.
- Viral host range is determined by the presence of a "receptor" on the host-cell surface. Cells that do not have this receptor are called "resistant."

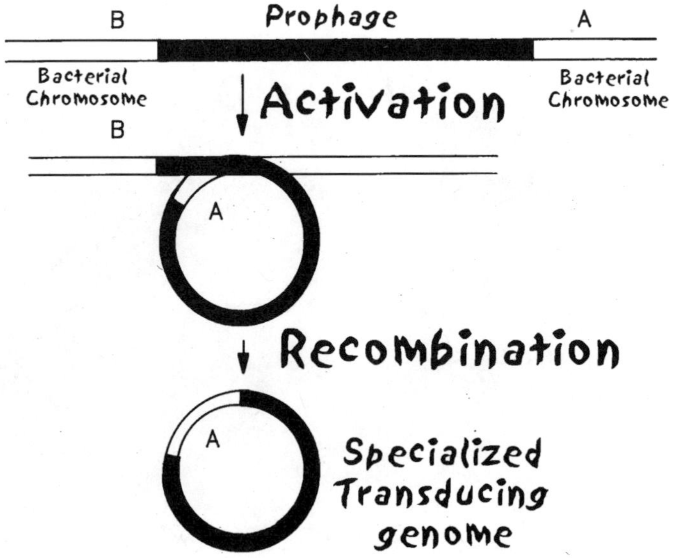

FIGURE 6 Diagram of the "activation" of an integrated prophage into the host genome. When the prophage is "activated" to the lytic cycle by some environmental stressor (perhaps solar UV radiation), it disintegrates as the first step in producing progeny bacteriophages. Sometimes the disintegration is not precise and a "transducing sequence" containing part of the host genome is incorporated. This gene sequence (here "A") is then found in all the progeny specialized transducing particles.

- Virions come in three major types: (i) icosadeltahedral, (ii) helical, and (iii) complex.
- Complex viruses contain double-stranded DNA genomes.
- There are two types of life cycles for complex viruses: (i) the lytic cycle that leads to the lysis of the host cell and production of progeny virions and (ii) the temperate cycle in which a viral genome (prophage) becomes part of the genetic component of the host cell and is replicated and passed on to daughter cells.
- Populations of bacteria containing prophages, known as "lysogens," are "immune to superinfection" by the same phage and release low numbers of virions through "activation" of the prophage in a portion of the lysogenized cells.
- Lysogenized bacteria may exhibit new characteristics directly associated with the presence of the prophage.

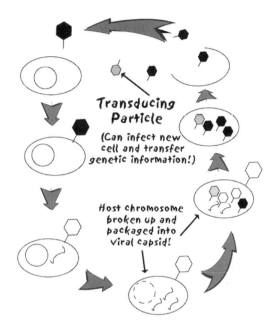

FIGURE 7 Production of generalized transducing particles.

TRANSDUCTION

Transduction is a mechanism for horizontal transfer of genetic material among bacteria that have the common property of being infected by a specific bacterial virus (bacteriophage). Two types of transduction are known and are dependent on the way in which the transducing particles are produced. In each case, bacterial DNA is incorporated into a functional capsid. Since the capsid has the ability to attach to host-cell receptors and inject the nucleic acid it contains into the new host, these "transducing particles" are perfectly capable of transferring DNA from one host cell to another. Clearly, the host range of transduction is the same as the host range of the phage and essentially defines a specific gene pool and breeding population. The only caveat is that there is homology between the transferred DNA and the genome of the recipient to allow integration into the endogenote through recombination.

General Transduction

Generalized transducing particles are made during the lytic life cycle (Fig. 7). They arise when bacterial DNA is packaged into a capsid instead of viral DNA. This process occurs because capsids are filled with DNA by a "head-full" mechanism that, except for a few unique phages, does not depend on a specific DNA sequence. Even in these exceptional phages, only a short recognition sequence is required for a DNA molecule to be packaged. If the packaging machinery attaches to host DNA, either chromosomal or plasmid, portions of the bacterial DNA of approximately the same molecular length as the phage DNA are packaged into the capsid.

Generalized transducing particles do not contain phage DNA but only bacterial. Because packaging is sequential and random, a population of transducing particles will contain all parts of the bacterial chromosome and any plasmids that may have been contained in the host cell. These will be divided into unique and specifically sized portions of molecular size similar to the bacteriophage's genome. Thus, any one transducing particle will only transfer a small portion of the entire bacterial genome. This property has been used by microbial geneticists to identify linkage between bacterial genes. It can also lead to the production of a "mosaic patch" (see chapter 20) in the bacterial DNA sequence.

Because packaging occurs as a "mistake," transducing particles are a very small portion of the total number of phage particles released during a lytic infection. Usually around one in every 10^5 to 10^7 phage virions produced is a transducing particle. However, because they do not contain phage DNA, they do not cause infection or cytopathic damage to the host. Because no phage repressor is made, they do not produce superinfection immunity in the transduced cell. Hence, coinfection of the "transduced cell" by a virion can cause disease and host death. The consequences of transduction are observed most often when either the phage-to-bacterium (PBR) ratio is low or a majority of the potential transductants are lysogens for the transducing phage. Because the infectivity of a transducing particle is contained in the virion, lysogens are equally susceptible to "infection" by a transducing particle, but because they do not allow expression of the viral lytic functions, coinfection does not destroy a potential transductant. Various studies have demonstrated that, in natural environments, lysogens act as effective recipients of transduced DNA. In any case, the transduced DNA must be stabilized by recombination (see chapter 2) into the host genome if chromosomal or established as an extrachromosomal, self-replicating element if a plasmid (see chapter 7) for a productive horizontal gene transfer event to occur.

Because generalized transduction can transfer all the DNA of an infected host (in small pieces), this process is not limited to any specific genes in a bacterial genome. However, because the hosts to which DNA can be transferred are limited to those cells that possess the phage receptor, the population of organisms that contain a specific phage receptor may be considered to form a DNA-exchange group similar to a "species."

Specialized Transduction

Specialized transducing particles are not produced during the initial lytic infection of a host by a bacteriophage. Instead, they come about as the result of the activation from lysogens of an integrated prophage (Fig. 6). Here an illegitimate recombinational event (see chapter 3) occurs during the excision of the prophage from the host genome leading to the formation of a phage "genome" lacking some phage functions but retaining some host functions. Because the host gene sequence is now part of the replicating phage genome, all phage particles produced from this cell will contain the transducing sequence. If the phage genes that are left behind in the host chromosome are not essential to virion production, these transducing virions can be propagated in ensuing cycles of infection. If the genes left behind in the bacterial chromosome are essential, the specialized transducing phage produced will only be capable of replication in the progenitor cell from which it was activated. The particles produced, while infective, will deliver a defective genome to a new host. This genome may establish itself as a cryptic prophage, or it may simply contribute the bacterial genes to the host chromosome through homologous recombination. In these cases, the phage genes are often lost from the daughter cells in subsequent generations.

Because specialized transducing phage are produced by the illegitimate excision (see chapter 3) and packaging, usually by the head-full mechanism, of a phage-host cointegrate, only "specialized" bacterial genes whose loci are adjacent to the prophage integration site can be transduced. In an evolutionary sense, specialized transduction is most likely to contribute to the evolution of unique portions of the bacterial genome. These may be recognized as unique genetic entities such as pathogenicity islands (see chapter 17) and the like.

Section Summary

- Transduction is a bacteriophage-mediated mechanism of horizontal gene transfer among bacteria.
- There are two types of transduction: (i) generalized and (ii) specialized.

- In generalized transduction, (i) transducing particles are made during lytic infection, (ii) they contain only bacterial DNA, (iii) all genes found in the donor cell can be transduced with equal efficiency, (iv) plasmid and chromosomal DNA are transduced with equal efficiency, (v) only about one in 10^5 to 10^7 particles produced in the lytic infection are transducing particles, and (vi) because they contain only bacterial DNA, transducing particles are defective.
- In specialized transduction, (i) transducing particles are made during the activation of an integrated prophage, (ii) they contain a mixture of bacterial and viral DNA, (iii) only genes that are adjacent to the prophage integration site can be transduced, (iv) chromosomal DNA is probably favored because more prophage integration sites are likely to be found in chromosomal than plasmid DNA elements, (v) all progeny particles from a cell producing specialized transducing particles will be transducing particles, and (vi) transducing particles may be effective or defective in producing new transducing particles following infection depending on which viral genes are replaced with host DNA.

TRANSDUCTION IN THE ENVIRONMENT

For many years, microbiologists dismissed transduction as unimportant in natural environments and relegated it as a phenomenon limited to the bacteriology laboratory. This was because they believed that populations of both bacteria and bacterial viruses were too low in the environment to allow transduction to occur. Since the infection process begins with what appears to be a passive interaction between host and predator controlled by Brownian motion alone, scientists reasoned that interactions would be so infrequent as to be unimportant. However, studies during the past 10 years have demonstrated that the numbers of bacteriophages and bacteria in many natural ecosystems are very high. For instance,

bacteriophage numbers often reach concentrations of 100 billion virus particles per milliliter of seawater (Bratbak et al., 1990; Proctor et al., 1993). These observations have caused microbial ecologists to do an about face. Today many microbiologists believe that bacterial viruses are a prime regulator of biomass and carbon and energy flow in natural environments, in particular, in aquatic habitats (Thingstad, 2000). With this realization has come the understanding of the potential of transduction to alter and regulate gene flow and potentially evolution in natural bacterial communities.

In the past decade, transduction has been studied in many natural environments through the use of environmentally incubated microcosms. Studies have included soils and plant surfaces; lakes, oceans, rivers, and sewage-treatment facilities; and internal organs of shellfish and mice. In each case, transduction was detected. Often the frequencies observed were as high and, in some cases, higher than was predicted from laboratory studies.

In other studies, bacteria and bacteriophages have been collected from natural environments (Schicklmaier and Schmieger, 1995) and shown to be, in the first case, excellent recipients of transduced DNA; and, in the second case, excellent vectors of horizontal gene movement among bacterial hosts (Ripp et al., 1994). Other studies have demonstrated that the vast majority, if not all, of bacteria in natural environments are lysogenized by at least one and usually several prophages (Miller et al., 1992). These cells can act as safe havens for viral genomes, allowing them to remain part of the ecosystem for periods greatly exceeding the infective life of a virion. Because of the continuous induction of a fraction of these lysogenic populations, these lysogens act as continuous sources of phage virions and transducing viruses. The remainder of the lysogenized population that is not induced can serve as excellent recipients for transduced genes because they are themselves immune to superinfection. These characteristics ensure that any transduced gene has an excellent opportunity to become part of the gene pool and is available to be acted on by natural selection.

The findings of these environmental studies have led to a model of environmental transduction illustrated in Fig. 8. Simply put, this model states that when a bacterium carrying a new gene enters a habitat, bacteriophages released from established lysogens infect the new cell and create more bacteriophages and some transducing particles. If any of these particles end up containing the new gene, that gene can be passed on to the indigenous bacterial populations.

This model is equally applicable to the transduction of chromosomal (Saye et al., 1990) and plasmid (Saye et al., 1987) DNA. Once established in the indigenous populations, that gene is available as evolutionary fodder. It is clear from the numerous environmental studies on transduction that this form of horizontal gene transfer has a great potential for horizontal transfer of genes in natural populations of bacteria and potentially contributes to their evolution.

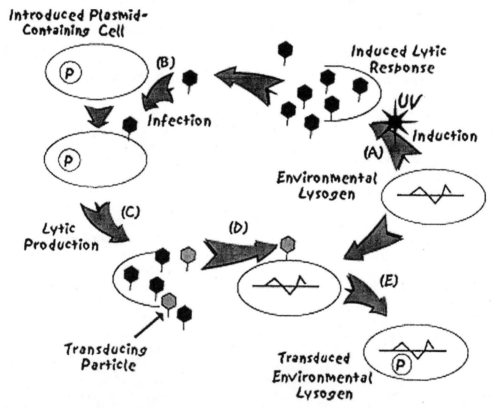

FIGURE 8 A model of the consequences of transduction of genetic markers from a bacterium introduced into an environmental bacterial community. Here the transduction of plasmid DNA is illustrated for simplicity, but chromosomal genes could enter the community's gene pool in a like manner. The following sequence of events is envisioned. (A) Environmental stressors lead to the "activation" or "induction" of a prophage from an environmental lysogen (⌒⌒), resulting in the production and release of bacteriophage virions (♀). (B) One of these virions then infects an introduced bacterium containing a plasmid (P) and propagates lytically in the cell (C) releasing virions and transducing particles (♀). (D) Transducing particles produced in this way are absorbed to members of the resident environmental bacterial community. Following absorption of the plasmid DNA from the transducing particle, the plasmid is established in the environmental lysogen (E) and enters the environmental gene pool that can be transferred by any of a number of horizontal gene-transfer mechanisms (see chapters 7 through 10). By virtue of being a lysogen, the transductant is immune to subsequent lytic infection by virions of the transducing virus. This helps to ensure the survival of the transductant in the environmental community.

Section Summary

- Bacteria and bacteriophages are found in high concentrations in many natural environments.
- Bacteriophages are root regulators of natural ecosystems, affecting both the flow of energy and carbon.
- Transduction has been observed in numerous environments.
- Transduction has the potential to be an important mechanism for horizontal gene transfer in natural environments.

TRANSDUCTION AND EVOLUTION: A MECHANISM FOR ENRICHING GENE POOLS

Transduction is now recognized as a viable horizontal gene-transfer system in many natural environments from soil to rat kidneys. However, it is a reductive process. Although it allows a gene to be transferred to a new cell, the "donor" cell is killed in the process. Because of this, we can ask whether transduction can really be a moving force in the evolution of bacteria. Replicon, Frankfater, and Miller (1995) tackled this question when they developed a model that demonstrated that transduction is indeed a viable force in regulating the diversity of bacterial gene pools even though it is a reductive process.

Replicon and her colleagues used continuous-culture techniques to test their hypothesis that transduction can increase the diversity of bacterial gene pools by allowing genes with negative fitnesses to remain in the gene pool for extended periods. That is, transduction works in opposition to the effects of negative fitness of a gene by increasing the number of individuals in the population that harbor the gene. Thus, the residence time before extension of that gene from the gene pool is increased. In other words, transduction acts to increase genetic diversity and richness of microbial gene pools such that, if conditions change, the number of alternative genotypes available for natural selection is increased and the survivability of the population is increased. Thus, transduction is a driving force of evolution.

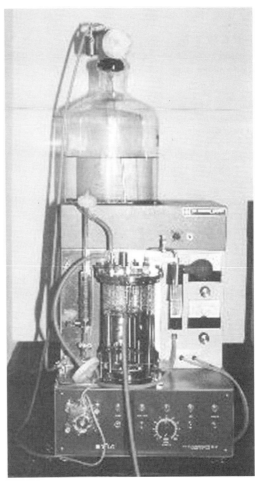

FIGURE 9 A chemostat used for continuous cultivation of bacteria allowing studies of long-term evolution of populations of bacteria. Chemostats are used for many experiments in microbiology and in industry to produce bacteria in quantity.

In their studies, Replicon and coworkers set up chemostats (Fig. 9) that contained different populations of *Pseudomonas aeruginosa* cells (Table 1). Some included a mixed population of bacteria but no bacteriophages. These chemostats served as controls. Two genotypes were present in high and equal numbers. The first genotype, RM2235, contained a Tra⁻, Mob⁻ (cannot be transferred by conjugation) plasmid, Rms149, while the second genotype, RM287, did not. A third genotype was also introduced at a very low level (about 10^{-6} as

TABLE 1 Strains used in continuous–culture experiments[a]

Chemostat	Donors	Recipients	Mock transductants	Transductants
Control (no transduction)	RM2235 (*nalA5, amiE200,* Rms149 [Tra⁻, Mob⁻, Cb^r, Sm^r, Su^r, Gm^r])	RM287 (*rif-901, chl-901*)	RM289 (*rif-901, chl-901,* Rms149)	
Experimental (transduction allowed)	RM2235 (*nalA5, amiE200,* Rms149 [Tra⁻, Mob⁻, Cb^r, Sm^r, Su^r, Gm^r])	RM300 (*rif-901, chl-901,* F116 lysogen)		Produced in the chemostat (*rif-901, chl-901,* F116 lysogen, Rms149)

[a] Genotypes are listed in parentheses (adapted from Replicon et al., 1995).

numerous as cells that had the other two genotypes). This strain, RM289, had the genotype of the second parent but also contained the plasmid Rms149. The authors called this strain a "mock transduction" (MT) because it had the same genotype as the transductants that will be produced in the experimental chemostat populations. While the numerous genotypes shared a common fitness, the MT had a slight (f ≈ −0.02 ± −0.01) negative fitness. The chemostats were run at various turnover times. In each case, the MT was lost from the populations after several generations of growth (Fig. 10).

In the second set of chemostats (experimental), a different set of organisms was introduced (Table 1). Here only the two majority genotypes were introduced. Again, RM2235 containing Rms149 was present to act as a donor of the transduced plasmid DNA. A derivative of RM287 that was lysogenic for the transducing phage F116 (Fig. 11, RM300) was also present. This strain acted as both the source of phage particles and as recipient for transduced DNA (see model in Fig. 8). Again these two strains had equal fitnesses. After these strains were incubated together in the chemostat chamber for several days, the population was found to contain a minority population of RM300 that had been transduced for the plasmid Rms149 as predicted by the model. These transductants (T) were present in approximately the same numbers as the MTs had been at the beginning of the control continuous culture. However,

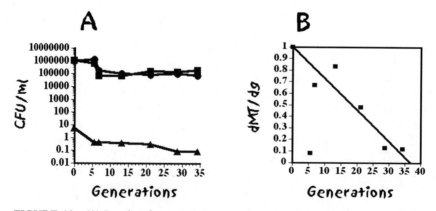

FIGURE 10 (A) Results of a control chemostat experiment in which gene transfer by transduction was not allowed to take place. CFU, colony-forming-units. (B) The mock transductant was lost for the community due to a slight negative fitness coefficient. The change in mock transductants as a function of generation is plotted (dMT/dg). (Adapted from Replicon et al. [1995] with permission.)

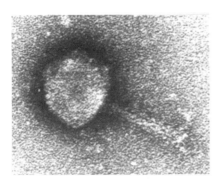

FIGURE 11 Electron micrograph of bacteriophage F116 of *P. aeruginosa*. F116 is a generalized transducing phage with a genome of 85.5 kbp with a G + C content of 61%. This is approximately 2.5% of the *P. aeruginosa* genome and almost exactly the same size as the Tra⁻, Mob⁻ plasmid Rms149. Its head is 65 nm in diameter, and its tail has a 12-nm diameter and is 80 nm long.

when the chemostats were activated the population of Ts was not lost but actually increased in number (Fig. 12). Again, chemostats were run at different turnover times and each time the same results were obtained. Thus, Replicon and her colleagues proved their hypothesis that transduction can act to maintain and even increase the frequency of a gene or set of genes (in this case, the plasmid Rms149) in the gene pool that imposes an otherwise negative fitness.

Section Summary

- The validity of the model of environmental transduction (Fig. 8) can be demonstrated in continuous-culture experiments.
- These experiments show that transduction is a viable source of genetic diversity in a natural bacterial gene pool.
- In bacterial populations, transduction can act to overcome the negative fitness imposed by certain genes that would otherwise lead to extinction of the genotype.

WHERE DO WE GO FROM HERE?

The experiments of Replicon, Frankfater, and Miller clearly demonstrate that transduction not only is an effective way of moving genes horizontally through a bacterial population but also is capable of maintaining a negatively selected genotype in a bacterial gene pool and even increasing its frequency. Bacteriophages are a primary component of all natural ecosystems as are their hosts, bacteria. Sequencing of whole bacterial genomes, as well as environmental studies, have demonstrated that most, if not all bacterial genomes contain prophages as stable components. Bacteriophage transduction and lysogenization have been suggested as frequent precursors to such phenomena as pathogenicity islands (see chapter 17) and

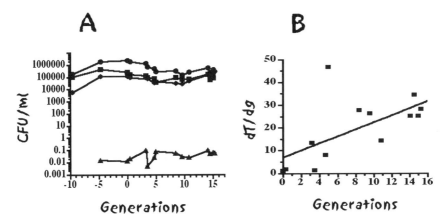

FIGURE 12 (A) Results of a chemostat experiment in which transduction was allowed to occur. CFU, colony-forming units. (B) The transductants increased in frequency in the populations even though they have a slight negative fitness coefficient. The change in transductants as a function of generation is plotted (dT/dg). (Adapted from Replicon et al. [1995] with permission.)

transposons (see chapters 3 and 8). The more we learn about the environmental roles and characteristics of these bacteriophages, the more evident it becomes that they are a potent force for horizontal gene transfer in natural bacterial communities and, hence, are likely to have been a viable mechanism contributing to the evolution of bacterial "species."

All groups of bacteria are known to be infected by some phages. In the correct circumstances, all phages produce generalized transducing particles. All phages are likely to establish lysogeny in their host under certain environmental conditions (Miller and Ripp, 2002; Ripp and Miller, 1997). Thus, the environmental conditions are right for transduction to be a very important mechanism of horizontal gene transfer (Miller, 2001). Transduction has been shown to occur in terrestrial environments (soils and plant surfaces), aquatic environments (lakes, oceans, marine sediments, rivers, epilithon [natural biofilms], and sewage-treatment facilities), and inside eukaryotic organisms (plants, shellfish, and mice). It is now accepted as a true environmental phenomenon and a laboratory one. Unlike other forms of horizontal gene transfer, both chromosomal and plasmid DNA can be transduced with equal efficiencies. Therefore, transduction must be viewed as a potent mechanism for horizontal gene transfer and evolution among bacteria.

QUESTIONS

1. Think about the following statement:

> "Transduction is such an important mechanism of evolution that it is wrong to say that bacteriophage infection is specific for a certain bacterial species. Instead, we should say that bacterial species are specified by which bacteriophages can infect them."

Do you agree or disagree with this statement? Why did you take the stand that you have?

2. You have isolated a new *Vibrio natriegens* phage from seawater. How might you go about showing that it was a generalized transducing phage? Do you think that it would be as easy to show that it was a specialized transducing phage?

Can you think of a way to do that?

3. In an environmental experiment you are tracking chromosomal genes by determining the number of cotransduced alleles. If the phage you are using transduces 2% of a bacterial chromosome whose map is "80-minutes" long, how close (in minutes) should the markers you choose to evaluate be on the chromosomal map?

4. Why do you think that the study of the effects of sunlight on bacteria is important to the study of in situ transduction?

5. Would you expect transduction to be as good a system of gene transfer in the soil as it is in aquatic environments? What factors might affect transduction in the two environments that would either increase or decrease the efficiency when compared with the other?

REFERENCES

Bratbak, G., M. Heldal, S. Norland, and T. F. Thingstad. 1990. Viruses as partners in spring bloom microbial trophodynamics. *Appl. Environ. Microbiol.* **56:**1400–1405.

Miller, R. V. 2001. Environmental bacteriophage-host interactions: factors contribution to natural transduction. *Antonie van Leeuwenhoek* **79:**141–147.

Miller, R. V., and S. A. Ripp. 2002. Pseudo-lysogeny: a bacteriophage strategy for increasing longevity *in situ*, p. 81–94. *In* M. Syvanen and C. Kado (ed.), *Horizontal Gene Transfer*, 2nd ed. Academic Press, San Diego, Calif.

Miller, R. V., S. Ripp, J. Replicon, O. A. Ogunseitan, and T. A. Kokjohn. 1992. Virus-mediated gene transfer in freshwater environments, p. 50–62. *In* M. J. Gauthier (ed.), *Gene Transfers and Environment.* Springer-Verlag, Berlin, Germany.

Proctor, L. M., A. Okubo, and J. A. Fuhrman. 1993. Calibrating estimates of phage-induced mortality in marine bacteria: ultrastructural studies of marine bacteriophage development from one-step growth experiments. *Microb. Ecol.* **25:**161–182.

Replicon, J., A. Frankfater, and R. V. Miller. 1995. A continuous culture model to examine factors that affect transduction among *Pseudomonas aeruginosa* strains in freshwater environments. *Appl. Environ. Microbiol.* **61:**3359–3366.

Ripp, S., and R. V. Miller. 1997. The role of pseudolysogeny in bacteriophage-host interactions in a natural freshwater environment. *Microbiology* **143:**2065–2070.

Ripp, S., O. A. Ogunseitan, and R. V. Miller. 1994. Transduction of a freshwater microbial community by a new *Pseudomonas aeruginosa* generalized transducing phage, UT1. *Mol. Ecol.* **3:**121–126.

Saye, D. J., O. A. Ogunseitan, G. S. Sayler, and R. V. Miller. 1990. Transduction of linked chromosomal genes between *Pseudomonas aeruginosa* during incubation *in situ* in a freshwater habitat. *Appl. Environ. Microbiol.* **56:**140–145.

Saye, D. J., O. Ogunseitan, G. S. Sayler, and R. V. Miller. 1987. Potential for transduction of plasmids in a natural freshwater environment: effect of plasmid donor concentration and a natural microbial community on transduction in *Pseudomonas aeruginosa. Appl. Environ. Microbiol.* **53:**987–995.

Schicklmaier, P., and H. Schmieger. 1995. Frequency of generalized transducing phages in natural isolates of the *Salmonella typhimurium* complex. *Appl. Environ. Microbiol.* **61:**1637–1640.

Thingstad, T. F. 2000. Elements of a theory for the mechanisms controlling abundance, diversity, and biogeochemical role of lytic bacterial viruses in aquatic systems. *Limnol. Oceanogr.* **45:**1320–1328.

Waldor, M. K., and J. J. Mekalanos. 1996. Lysogenic conversion by a filamentous phage encoding cholera toxin. *Science* **272:**1910–1914.

FURTHER READING

Baross, J. A., J. Liston, and R. Y. Morita. 1978. Incidence of *Vibrio parahaemolyticus* bacteriophages and other *Vibrio* bacteriophages in marine samples. *Appl. Environ. Microbiol.* **36:**492–499.

Calender, R. (ed.). 2003. *The Bacteriophages*, 2nd ed. Oxford University Press, Oxford, United Kingdom.

D'Herelle, F. 1949. The bacteriophage. *Sci. News* **14:**a44–a59.

Jiang, S. C., and J. H. Paul. 1998. Gene transfer by transduction in the marine environment. *Appl. Environ. Microbiol.* **64:**2780–2787.

Kokjohn, T. A. 1989. Transduction; mechanism and potential for gene transfer in the environment, p. 73–98. *In* S. B. Levy and R. V. Miller (ed.), *Gene Transfer in the Environment.* McGraw-Hill Publishing Co., New York, N.Y.

Masters, M. 1996. Generalized transduction, p. 2421–2441. *In* F. C. Neidhardt (ed.), *Escherichia coli and Salmonella Cellular and Molecular Biology* American Society for Microbiology, Washington, D.C.

Miller, R. V. 1998. Bacterial gene swapping in nature. *Sci. Am.* **278:**46–51.

Morrison, W. D., R. V. Miller, and G. S. Sayler. 1978. Frequency of F116 mediated transduction of *Pseudomonas aeruginosa* in a freshwater environment. *Appl. Environ. Microbiol.* **36:**724–730.

Ogunseitan, O. A., G. S. Sayler, and R. V. Miller. 1990. Dynamic interactions of *Pseudomonas aeruginosa* and bacteriophages in lakewater. *Microb. Ecol.* **19:**171–185.

Saye, D. J., and R. V. Miller. 1989. The aquatic environment: Consideration of horizontal gene transmission in a diversified habitat, p. 223–259. *In* S. B. Levy and R. V. Miller (ed.), *Gene Transfer in the Environment.* McGraw-Hill, New York, N.Y.

Summers, W. C. 2001. Bacteriophage Therapy. *Annu. Rev. Microbiol.* **55:**437–451.

Suttle, C. A., and A. M. Chan. 1992. Mechanisms and rates of decay of marine viruses in seawater. *Appl. Environ. Microbiol.* **58:**3721–3729.

Syvanen, M., and C. I. Kado (ed.). 2002. *Horizontal Gene Transfer*, 2nd ed., Academic Press, San Diego, Calif.

Twort, F. W. 1915. An investigation on the nature of ultra microscopic viruses. *Lancet* **11:**1241–1243.

Wommack, K. E., and R. R. Colwell. 2000. Virioplankton: viruses in aquatic ecosystems. *Microbiol. Mol. Biol. Rev.* **64:**69–114.

Zinder, N. D., and J. Lederberg. 1952. Genetic exchange in *Salmonella. J. Bacteriol.* **64:**679–699.

TRANSFORMATION

Martin J. Day

10

Change is not made without inconvenience, even from worse to better.

—Richard Hooker

There continues to be a debate on why some organisms have sex, since many do not have a sexual cycle as a key component of their life cycle. It is intriguing to ask why not? One reason might be because there are significant energetic costs to sex. Without any exchange of genes, an asexual organism will reproduce 50% more efficiently than one with an obligate sexual cycle.

But the consequences of lower energy costs are an "apparent" certain extinction as the cell's genome steadily accumulates disadvantageous mutations and eventually acquires a lethal mutation. This effect is termed Muller's ratchet, and it occurs because most mutations are detrimental. Gene transfer is needed to allow these mutated genes to be replaced by "good ones" through recombination. In the right environment combinations of beneficial genes aggregate through processes of transfer, selection, and exchange, which results in the generation of more offspring with successful gene combinations enriched by selection. A dynamic environment requires plasticity in

Martin J. Day, Cardiff School of Biosciences, Cardiff University, Main Building, P.O. Box 915, Cardiff CF10 3TL, Wales, United Kingdom.

Microbial Evolution: Gene Establishment, Survival, and Exchange
Edited by Robert V. Miller and Martin J. Day, ©2004 ASM Press, Washington, D.C.

the phenotypic characteristics of species, and these can be delivered through repetitive cycles of gene exchange to produce diversity in genotypes.

Is there any evidence for gene-exchange mechanisms being selected? It appears that chromosomal gene transfer by phages and plasmids occurs by accident, because the processes do not appear to be governed/regulated in any way by the bacterial chromosome. The selection of chromosomal DNA to transfer by these mechanisms appears to be random. Transformation, however, appears (at least in *Haemophilus* and *Neisseria* species) to be different. Uptake is decided by an uptake signal sequence, and self-DNA is distinguished from DNA from other species. Thus, there is evidence for transformation having evolved as a mechanism of gene exchange.

Transformation in principle is like other transfer processes and these, together with mutation, are processes that result in genetic changes that are unlikely to be beneficial to host cells. However, despite this, there was clearly a reason, even if it no longer exists, for the evolution and establishment of pathways for gene exchange. What drove the evolution of these processes is a matter of conjecture, but the fact that they are so highly evolved and integrated into the life cycles of their host cells reflects their importance to a cell.

AN OVERVIEW OF TRANSFORMATION

What Is It?

Until the 1990s gene exchange between bacteria was considered to be relatively limited and have questionable significance. Apart from a few examples, mainly concerning the spread of antibiotic resistance mediated by plasmids between clinical bacteria, the exchange of DNA between bacteria was generally considered questionable, and its significance was, at best, unproven. This is despite the clarity of the very first genetic transfer experiment done in 1928 by Griffith, a transformation experiment. He injected mice with a mixture of dead pathogenic pneumococci, termed "smooth" because they were encapsulated, with a live avirulent "rough" strain; he found that the mice died soon after through infection with a virulent "smooth" strain. The nonpathogenic strain was transformed into a pathogen by an agent (now recognized as DNA encoding the genes for the capsule carried by the dead strain). The purpose of this chapter is to examine the process of transformation and discuss why this process of gene exchange is of value to those bacteria that participate in it.

In trying to understand how bacteria interact with one another it is important to appreciate that they probably spend their lives in an association with siblings, interacting together with other species and genera in communities. Thus, organisms grow and succeed in associations where they are reliant on one another for mutual metabolic advantage. It is logical for them to also interact genetically. Why? Well one reason could be that certain beneficial phenotypes (e.g., those for antibiotic resistances in a clinical environment) could be spread between members of the community to ensure its continuity. So there is a requirement for genomic modification by cells, but the participating organisms are "doing the changes" blind to the prevailing and shifting environmental conditions. As a result, for many individuals that participate in transformation, their genome changes are more likely to be from better to worse. The fortunate few, which acquire and integrate the right sequences, are either remaining the same or are moving from worse to better. Their situation is, in fact, well described by the term "worse to better" since there is no ideal genotype or phenotype in nature, because the environment changes both predictably and unpredictably, physically and temporally.

The journey taken by a gene during transformation is not an easy one. The sequence can fail at any of the many steps it takes to become established in a host genome. This explains why the frequency of transformation by a gene rarely does better than 10^{-6}, even under laboratory conditions. This number reflects just one successful integration of the gene by recombination in a million opportunities.

Why should a process of exchange, providing an opportunity for the gain of a gene and for the replacement of a deleterious mutation by a beneficial one, be a significant event? In bacterial populations all cells derived from one parent are identical and are thus termed clonal. Because of mutations their genetic diversity is low, but not absent. This is well illustrated in clinical infections, where the infection population is clonal. Regardless of population size, genetic diversity is low. An analysis by Redfield showed that transformation was only really advantageous in small populations because they had low genetic diversity. This advantage would seem to apply directly to all naturally occurring clonal microbial populations.

Simply, transformation is a process by which double-stranded DNA reaches an appropriate site on the cell surface, is taken through to the cytoplasm, aligned to an homologous region in the genome, and integrated via recombina-

tion. This cell is a recombinant and is called a transformant (Fig. 1). Cells capable of this activity are termed competent. The term competence describes a complicated process and the capability is present in various prokaryotes from within eubacteria and archaea. Competence is achieved through the expression of many dedicated proteins, organized into a pathway, to allow the recognition, uptake, and integration of the exogenous (extracellular) DNA sequence into the host genome. It is genetically encoded, highly evolved, and physiologically governed, and, as we are slowly becoming aware, it is far from the rare and trivial process portrayed in many texts. The initial steps in the process, to the point when the DNA enters the cytoplasm, vary across genera and may even vary between some species. In part, this is due to the differences in outer cell surfaces. However, once the DNA is present in the cytoplasm the events that follow and result in integration are probably similar across all bacterial phylogenies. Recombination

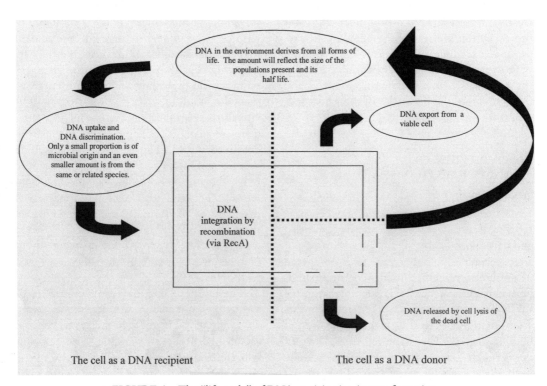

FIGURE 1 The "life cycle" of DNA participating in transformation.

is mediated through sites of homology and allows genes to reassort within one and to be exchanged between two DNA molecules within the cell. Finally, since transformation is a highly evolved process, there has to be or to have been a good selective advantage for its evolution and retention.

Where Does the DNA Come From?

Although DNA is probably spread throughout most environments, its occurrence might well be patchy. The abundance of DNA will reflect the populations and types of organisms currently and previously occupying the niche. DNA is present in the environment due to cell lysis, resulting from predation and cell death, and from spontaneous release of DNA from cells as they grow. These processes of release will probably bias the relative DNA concentrations in an environment and reflect the genomes of successful high-density populations. The more successful gene sequences will have a greater relative concentration in the immediate environment of the more successful populations of cells. Consequently, environmentally successful gene sequences become enriched, in a cloning-like manner, and these consequently become more likely to be transferred between members of the transforming community. If there were repeated selective bottlenecks, then genomes of a given population would tend to be similar for environmentally important selectable loci. An obligatory requirement for these variable loci would be the need to flank them with homologous sequences to allow recombination to continue to drive their integration. Thus, localized uniformity would occur in transformation-dependent populations, since the spread of a few specialized loci/gene sets/alleles would occur and be maintained due to their "selective value."

The spontaneous release/export of DNA by viable cells is a conundrum, as it is an energy-rich molecule. Why "waste" a molecule that has cost so much to construct? Maybe the payback can be explained though an examination of the "selfish DNA" concept (Doolittle and Sapienza, 1980).

Quality of Exogenous DNA for Transformation

In the environment, DNA is subject to a variety of chemical and physical stresses and, therefore, it will have a half-life that reflects these impositions. Thus, its survival or persistence will be determined by a combination of many physical (e.g., humidity and temperature), chemical (e.g., pH and ion types and their concentrations), and biological factors (extracellular nucleases, etc.). In addition, DNA comes in all "shapes and sizes"; it may be in solution, spread over, or discretely bound to surfaces and it will be present in an ecosystem in a nonregular distribution. Providing it remains as a duplex and is not degraded into too short a length, it retains its capability to be taken up by competent bacteria. Degradation in its length will affect its ability to be taken up effectively. Modification of the bases is mutagenic and will reduce its genetic quality, but this will not be "seen" by the competence or recombination processes. This means the genetic value of the sequence will suffer as the sequence ages. The genetic value will be determined only when the sequence is expressed, and a change in phenotype modifies the cell's interaction with its environment.

Uptake

Initially duplex DNA may become passively associated with the cell surface before it is actively bound; single-stranded DNA is not recognized. The steps of uptake are as follows. The DNA duplex loosely associates with specific outer components of the cell termed com (competence). ComA and B are ATP-binding cassettes (known as ABC transporters) and are responsible for transporting a range of molecules into the cell. While bound, the DNA is subject to enzymatic cleavage to reduce its overall length to an appropriate consistent size and then taken in part way through the cell surface layers. At the interface between the cell wall and cell membrane the duplex is reduced to a single strand by a nuclease (known as EndA in *Streptococcus pneumoniae*) and this is transferred the rest of the way into the cell. Its fate is decided here. If it has no homology with any DNA molecule in the cell it will be degraded, but if it has homology, then it will intercalate and hybridize with the homologous genome sequence to form a heteroduplex and be integrated by recombination by RecA, replacing the original sequence.

Recombination, Cell Division, and Expression

Recombination, achieved by the RecA protein, will result in displacement of the parental sequence and insertion of the novel fragment. Consequently, a transformant will always be a recombinant, because the DNA that enters the cell is always linear and needs to be integrated into a replicon to survive. The recombinant will be formed by the intercalation of the transforming single-stranded sequence into the duplex at a region of homology. Recombination will follow, displacing the homologous resident strand with the introduced sequence. The duplex contains, at this point, one "old" strand and one "different in sequence new strand." To allow expression of the information in the new strand the cell genome must undergo replication and segregation into daughter cells. At this point, the daughter cells will be different genetically and able to express the genotypes they have as a phenotype. One daughter will be parental, i.e., not a recombinant. The other daughter will now have a duplex containing the new recombined sequence. Thus, to become phenotypically differentiated, the transformant must grow and divide.

Selection and Transfer Frequency

The transformed cell has a parental sequence on one strand and a complementary recombinant sequence on the other. The cell must divide to produce two daughters, one the recombinant insert and that has a transformant and one that is a chromosomally unaltered daughter cell. Transformants are selected on their phenotype, which determines their rela-

tive growth rate compared with the parental phenotype. However, it is clear, i.e., it is a statistical likelihood, that most of the DNA available for integration will be the same in sequence as the DNA it displaces, so it will not yield a cell with a change in its genome and hence any growth advantage.

Transformation Is a Dependent Chromosomal Gene Expression

Transformation is a complex, well-coordinated, and regulated process occurring after a cell expresses the genes to become competent. This state is defined as the sequential steps of DNA recognition and uptake. In some bacteria competence is an inducible system, dependent on environmental cues to affect the physiology of the host cell and promote expression. In other bacteria, competence is constitutive. In either case, it is a highly evolved process. Transforming DNA is single-stranded DNA transported onto the cytoplasm. Single-stranded DNA is insensitive to restriction and, when it is made into a duplex for integration, it will acquire the modification pattern of the host. So restriction can limit the success of DNA transfer in the other two gene-exchange mechanisms, but does not do so for transformation. It is worth noting that making cells of *Escherichia coli* and other "nontransformable" species competent in the laboratory involves combinations of physical disruption of the membrane/cell wall by enzymatic treatments, divalent ions, heat shock, electroporation, protoplasting, etc. These allow DNA to get into the recipient cell.

Does the cell respond directly and uniquely to the presence of transforming DNA in the cytoplasm so that uptake is for the information held and not for the energy in it as a substrate? So far, it has been shown that RecA is induced by competence in pneumococci. In *Bacillus subtilis* RecA synthesis is associated with competence and DNA damage. In *Haemophilus* spp. and probably also in *Neisseria* spp., competence partially regulates RecA synthesis. However, there are no data yet to confirm this occurs in any other transformation-proficient bacteria.

Section Summary

- Transformation is one of three gene-exchange processes.
- Dead and live cells participate.
- DNA is widespread in the environment.
- Recombination is required to establish a gene in the genome.
- Transformation is a chromosomally encoded process.

WHO DOES IT?

Table 1 provides some idea of the diversity of organisms that can participate in natural transformation. As investigations widen to examine even more diverse bacterial isolates, we find that the number of species that have the capacity is expanding. However, the property is patchy. There is a species in a genus that is readily transformable (e.g., *Pseudomonas stutzeri*) and others that appear not to be (e.g., all the other pseudomonad species). In addition, within a genus recognized as proficient for transformation (e.g., *Acinetobacter* spp.), there are isolates that do so poorly and others that do not at all. Until recently, *E. coli* was only considered to be transformable after quite stringent laboratory manipulation; however, a strain was found to be transformable in a divalent-rich

TABLE 1 Bacteria active in gene transfer by transformation[a]

Species isolated from terrestrial or aquatic habitats	Tf[b]	Phylum based on 16S phylogeny[c]
Protolithotrophic		
Chlorobium limicola	1.0×10^{-5}	Chlorobi
Agmenellum quadruplicatum	4.3×10^{-4}	Cyanobacteria
Anacystis nidulans	8.0×10^{-4}	Cyanobacteria
Nostoc muscorum	1.2×10^{-3}	Cyanobacteria
Synechocystis sp. strain 6803	5.0×10^{-4}	Cyanobacteria
Chemolithotrophic		
Thiobacillus thioparus	$10^{-7} \times 10^{-2}$	Proteobacteria
Heterotrophic		
Mycobacterium smegmatis	$10^{-7} \times 10^{-6}$	Actinobacteria
Achromobacter spp.	$+$[d]	Proteobacteria
Azotobacter vinelandii	9.5×10^{-2}	Proteobacteria
Bacillus subtilis	3.5×10^{-2}	Proteobacteria
Pseudomonas stutzeri	7.0×10^{-5}	Proteobacteria
Deinococcus radiodurans	2.1×10^{-2}	Deinococcus-Thermus
Thermus aquaticus	6.4×10^{-4}	Deinococcus-Thermus
Methylotrophic		
Methylobacterium organophilum	5.3×10^{-3}	Proteobacteria
Archaebacteria		
Methanococcus voltae	8.0×10^{-6}	Euryarchaeota
Clinical isolates of pathogenic species		
Campylobacter jejuni	2.0×10^{-4}	Proteobacteria
Haemophilus influenzae	7.0×10^{-3}	Proteobacteria
Helicobacter pylori	5.0×10^{-4}	Proteobacteria
Neisseria gonorrhoeae	1.0×10^{-4}	Proteobacteria
Neisseria meningitidis	1.1×10^{-2}	Proteobacteria
Staphylococcus aureus	5.5×10^{-6}	Firmicutes
Streptococcus pneumoniae	2.9×10^{-2}	Firmicutes
Streptococcus sanguis	2.0×10^{-2}	Firmicutes
Streptococcus mutans	7.0×10^{-4}	Firmicutes

[a] Modified from Day (2002).
[b] Tf, transformation frequency (chromosomal marker transformants/viable cell).
[c] *Bergey's Manual of Systematic Bacteriology*, 2nd ed.
[d] Transfer recorded but no frequency reported.

aqueous environment (an aquifer). Thus, transformation may be a gene-exchange process in which all bacteria may participate to some degree if the cellular physiology and environmental conditions are appropriate. So it seems the potential of this mechanism for gene exchange is widespread, but that the capability maybe precluded unless special conditions are met.

Section Summary

- Transformation is a widespread but "patchy" gene-exchange process.

DETAILS OF ESSENTIAL STEPS REQUIRED TO PRODUCE A TRANSFORMANT

Serendipity and Recognition

The molecules may drift to the organism, if it is attached in an aquatic environment, or the organism might have motility and locate/interact with a DNA source. DNA must be positioned immediately adjacent to uptake sites on the cell surface. In *B. subtilis* and *S. pneumoniae* there are some 50 sites per cell where the sequence is targeted and transfer across the cell wall membrane is initiated. *Haemophilus influenzae* has just four to eight sites. In addition, the genomes of *H. influenzae* and *Neisseria gonorrhoeae* have uptake signal sequences (USS) that are recognized by these uptake sites. *H. influenzae* has a 29-bp USS consensus (with an 11-bp core, which is more highly conserved; AAGTGCGGTCA) and *N. gonorrhoeae* has a USS consensus GCCGTCGAA. The latter occurs ·frequently as an inverted pair (providing stem-loop configuration potential) downstream of a gene terminus. There is more known about the USS of *H. influenzae* (Smith et al., 1995). This sequence occurs in chromosome, plasmid, and phage genomes. In the chromosome it is relatively randomly located, but for two observations. First, there are 734(+), i.e., orientated in one direction around the genome and 731(−). This organism has a 62% AT genome, so you would expect just eight, not 1,465 sites based on this AT ratio. It is an easy calculation to see

that this accounts for 3.5% of the coding capacity of the genome, so it is not an inconsequential energy load to impose to maintain these sites. Second, 61% of these sequences are found in open reading frames, averaging one every 1,248 bp.

Fragment Size

To become recognized by the receiving cell, DNA must be in a double-stranded form and have a size greater than about 5 kb. The gram-positive bacteria *B. subtilis* and *S. pneumoniae* most efficiently bind duplex DNA of about 10 to 18 kb and transport it within about a minute. Only double stranded DNA is bound and able to move through the outer cell layers. Environments in which DNA becomes dissociated into the single-stranded form (e.g., in alkaline pH), tightly bound (e.g., clay soils), or degraded to shorter lengths due to nuclease activity (e.g., intestines) effectively reduce the likelihood of recognition. Note that RNA is not a substrate for the process, as it is single-stranded.

Pathways of Uptake/Transport

Several genes are required for DNA uptake and transport, and gram-positive and -negative bacteria adopt slightly different pathways, which display different efficiencies. In most bacteria the state of competence is transient, but in *N. gonorrhoeae* and probably *Acinetobacter* spp. it is constitutive, signifying most cells in a population are competent during most of their life cycle. The transient nature of the process indicates that competence is under physiological control. In *Anacystis nidulans* it occurs in the late-exponential phase, and in *B. subtilis* and *P. stutzeri* it occurs at the transition into the stationary phase. Although the process is clearly physiologically governed, the proportion of cells within a population that are competent varies with the species. For example, all *Azotobacter vinelandii* cells exhibit competence, 10 to 25% of *B. subtilis* and *Acinetobacter* sp. cells exhibit competence, and less than 1% of *H. influenzae* and *P. stutzeri* cells exhibit competence.

In gram-positive cells double-stranded DNA binds rapidly without either selection or size discrimination. Thus, DNA from any source, eukaryotic or prokaryotic, jointly competes for the surface uptake sites. Once at the cell surface, there appears to be an ordered process that enzymatically reduces the DNA into fragments of a predetermined optimal size for translocation (averaging 15 to 18 kb). The "newly generated ends" are adjacent to uptake sites/pores on the cell surface; at this point, the sequence begins the first stage of uptake as the newly generated "end" of the sequence becomes bound. In *B. subtilis* and *S. pneumoniae* the DNA becomes inaccessible to DNase after 60 to 90 seconds

The uptake pathways of gram-negative bacteria *H. influenzae* and *N. gonorrhoeae* are similar. Both organisms are selective in the DNA taken up; they identify specific sequences, which coincidently occur far more frequently than would be expected if these occurred randomly. However, *Acinetobacter* sp. does not have a dedicated sequence and so its mechanism is more like the gram-positive. *H. influenzae* has fewer than 10 sites on the cell surface for DNA uptake; these are termed transformasomes. They appear to be specialized DNA-binding regions on the outside of the cell surface.

All uptake appears to be at much the same efficiency of about 6,000 to 12,000 nucleotides (nt) per second.

Transport and Accessory Proteins

The cell wall in most gram-positive bacteria is thick and complex, containing peptidoglycan, teichoic acids, and polyanions. DNA must therefore breach a physical and electrostatic barrier. Analysis shows the mechanisms of DNA transfer across the cell membranes are complex. This is for two reasons. Competence is achieved not by one or two proteins but through the interaction of many. Of these proteins few are uniquely or primarily concerned with DNA transport, and many would be considered as just accessory factors if their absence was not critical to the success of DNA transport.

The PSCT proteins are a group intimately involved in cell surface activities, namely pilus, secretion, competence, and twitching motility. Knockout PSCT genes and transformation activity falls dramatically or is lost completely. Other proteins are those produced from the *com* (for competence) genes in *B. subtilis*, and they have analogues in other species. ComEA is involved in DNA binding and transport across the cell wall, it has an ortholog in *S. pneumoniae*, and similar proteins are found in *N. gonorrhoeae* and *H. influenzae*. ComG is a membrane protein that may increase porosity, provide a channel, or remove a membrane-positive charge so DNA can transverse the membrane. ComC is another DNA-binding protein, with peptidase activity, found in *B. subtilis*. In addition, there are other proteins (accessory factors) that interact with these primary proteins. First, there are enzymes, e.g., in *H. influenzae* the protein Por (periplasmic protein disulfide oxidoreductase), that are essential for transfer. Why are they needed? They seem to ensure the correct tertiary structure of at least one of the competence proteins. Then, there are proteins with structural roles that interact with DNA normally like SSB, which is required for DNA replication. SSB is also essential for transformation because the DNA is single-stranded on entry.

Models for uptake seem to be either selective or nonselective across the Gram reaction divide. How transformation is achieved in the *Archaea* is an open question.

B. subtilis and *S. pneumoniae* form the basis of the models for DNA uptake and integration. Double-stranded DNA binds to a protein anchored in the membrane (ComEA or analogue). A nuclease cleaves the bound DNA and the new terminus is delivered to the membrane transport system (ComEC and FA). It becomes invisible to extracellular nucleases at this point. A nuclease (EndA in Sp) degrades one strand (5′-3′), allowing the 3′ end of a single-stranded molecule to travel into the cytoplasm. Thus, a large group of proteins interact to provide transport for DNA into the cell.

The Cost of Competence

The process has two basic costs. First, there is the energetics of the process and this has not been determined. Transformation requires the input of energy at several steps. Could it be ATP or the proton-motive force that drives it? The energy driving the process may vary between genera if there are major differences in the proteins involved. Second, there is the cost of gene exchange, where no change or an adverse change occurs through integration of the introduced sequence. In the "adverse change" scenario the organism has been penalized both in energy terms and by the fact it is now worse off phenotypically. In the "no change" scenario the organism is unaltered but it has just "wasted" energy.

Section Summary

- Some species select DNA by using uptake signals, others trust to luck.
- Size is important.
- DNA uptake is an ordered process.
- The process has commonality with other uptake systems.
- DNA uptake requires energy.
- Uptake can be selective.

HOW AND WHY DID TRANSFORMATION EVOLVE?

Is It the Same in All Species or May It Have Arisen Independently?

The process of sharing the properties of some of the proteins with others (cell twitching, etc.) suggests that the transfer of DNA across the membrane and cell wall was a problem encountered and solved by other large molecules. The transformation process has just added another subset of DNA-specific proteins to the complex to provide the specificity needed to move DNA into the cell. Aspects of the transformation process in the gram-negative *Acinetobacter* sp. seem to parallel that of the *Bacillus* sp. and not mimic that of the other gram-negative *Haemophilus* sp. and *N. gonorrhoeae*. As a result, are we looking at two independent evolutionary routes or degrees of specialization?

Why Did It Evolve?

Species are reproductively independent populations and have mechanisms for exchanging genes. Why should transformation have arisen as a process for moving genes between such individual cells? If it were for a sex cycle then the process ought, by analogy with higher organisms, to be integrated with the life cycle of, and be regulated by, the organism. By analogy with higher organisms sex allows outcrossing and recombination. This allows for shifts in gene frequencies to occur, through a combination of gene transfer and phenotypic selection imposed by the environment.

Since gene exchange is not an integrated component, perhaps we should look at gene transfer as a means to avoid extinction. All the members of a clone will become extinct except in environments that favor them. So sex seems to reduce species closeness of adaptation to prevent overspecialization and hence avoid the extinction of all clones (Williams, 1975). When applied to bacteria the hypothesis would be that the asexual mode of reproduction allows rapid exploitation of an appropriate environment. Mechanisms that permit the exchange of genes between individuals generate the necessary genetic diversity in these clones, which enables individuals to survive extinction at the next selective challenge.

Role of Competence

Competence, the phase during which the cell takes up exogenous DNA, is a widespread phenomenon among bacterial genera, and many more species remain to be tested. It is clearly a regulatory mechanism to modulate the uptake of DNA. Logically, if it needs to be regulated then there must be a period in the cell-growth cycle when it is not advantageous to have it active relative to one in which it provides an advantage.

There is also a feeling that transformation is probably a more general gene-exchange mechanism than is currently appreciated, since research has shown that more and more isolates possess the capacity. It requires the specific synthesis of at least 12 proteins, which is more than many smaller phages (e.g., MS2) and plasmids (e.g., RSF1010) utilize. So it is similar to other transfer processes in that it is regulated and has evolved to provide a frequency for chromosomal gene exchange that appears similar to the other two mechanisms.

In 1999, Dubnae suggested various reasons for development of transformation, and these are unrelated to its having evolved for DNA exchange. Taking up exogenous DNA avoided the costs of synthesis of precursors in DNA repair, enabling it to be used as either a food source or as a genetic resource. The latter provides a mechanism for generating genetic diversity. Of course it is possible, since these reasons are not mutually exclusive, that each may have contributed to the selective value of the mechanism and thus its widespread adoption. It is also possible that it may have evolved more than once and for different reasons.

The Argument for the Use of Exogenous DNA as a Food Source

Some transformable and many other species excrete nucleases into their immediate environment and possess efficient uptake pathways to scavenge exogenous DNA. This makes the nucleotides produced available for others to use, so selection might have favored the uptake of the larger molecule to be degraded internally. However, in most species one strand is hydrolyzed at the exterior cell surface. This seems to be wasteful, because half the economic value of the process is lost. Since in *H. influenzae* DNA degradation is intracellular, the second strand is not wasted but utilized. But *H. influenzae* does not have an extracellular DNase despite plenty of eukaryote DNA being in the surrounding milieu. In addition, some species are extremely selective in the DNA taken up, reducing the nutritional value of the process, but this ability could have evolved later. Perhaps even more surprising is the release of DNA from cells as they grow.

Thus, this would appear to argue against the idea of transformation evolving to catabolize exogenous DNA as a food source.

The Argument for the Use of Exogenous DNA as a Repair Process

Most DNA in the environment probably comes from the immediate adjacent populations of cells. Since a multitude of enzymes are found in natural environments, the half-life of macromolecules, of which DNA is one, is limited/promoted by many factors through their enzymatic degradation. Consequently, there is believed to be a continual cycling of DNA within the ecosystem. As such, the dominant prokaryotic DNA species prevalent will probably reflect the successful members of the microbial communities present. Thus, cells acquiring disadvantageous mutations have an extracellular resource, DNA from their siblings, to enable mutational repair through uptake and recombination of homologous sequences. Some species become competent as they enter the stationary phase or other stress conditions, and it is in this phase that genomic rearrangements occur due to stress responses. DNA is available to repair damage caused by the induction of these stress responses in the cell. What is the evidence for transformation having a repair role? Little so far, as in *B. subtilis*, DNA repair is induced as part of the competence induction. Competence appears in this organism prior to the stationary phase when its metabolism becomes more aerobic, and then oxidative DNA damage becomes more likely to require repair processes. In *H. influenzae*, competent cells are actually less able to repair UV-induced DNA lesions than are normal cells.

The Argument for the Use of Exogenous DNA as a Fitness Mechanism for Recombination

Although all genetic diversity arises from mutational change, it is recombination that permits the reassortment of genes into novel combinations. For example, in *Helicobacter pylori*, transformation is the only known form of DNA exchange and it is thought that transformation is the mechanism responsible for the panmictic population structure (Suerbaum et al., 1998). It will be interesting to see if *H. pylori* selects by sequence the DNA it takes up. This organism has three distinct gene sets (polymorphisms) at one locus, and these are in equilibrium in the population. Here, transformation serves to reduce overall diversity while allowing recombination to generate local diversity at one site in the genome. So bottlenecks caused by selective sweeps will not reduce the population diversity. Examination of *N. gonorrhoeae*, which has selective uptake of DNA, does show periodic selective sweeps causing founder effects and an advantage of a diversity of genetic backgrounds.

The Argument for the Use of Exogenous DNA as a Molecule for Nucleotide Salvage

Calculations done for *E. coli* indicate that the cost of synthesis of bases for DNA replication consumes about 16% of the cell's total energy budget (Stouthamer, 1979). Consequently, the potential savings of reusing bases from the DNA taken up is significant. In *H. influenzae* the degradation of one strand of the duplex taken up is intracellular, thus providing the cell with a source of nucleotides for replication and repair and saving the cell significant energy. All cells, of course, reutilize the nucleotides from the sequences left after recombination has occurred.

The Argument for Exogenous DNA as a Parasexual Cycle

Transformation enables gene exchange between individual bacteria. The evolution of periods of competence, which coordinately primes the population to be "ready to participate," has selective value. Organisms need to do it together to be effective. Just as with the other gene-exchange processes we see species-specific cues for the enablement of the process. In the laboratory the process provides recombinants at a similar frequency.

Section Summary

- There are several possible routes for the evolution of transformation.

THE EFFICIENCY OF TRANSFORMATION

The Cost of the Process

Competence is regulated by growth conditions, so dissecting the regulatory signals will clarify the control and costs of this metabolic process. The spontaneous release/export of DNA by viable cells is part of an assessment of cost, as it is an energy-rich molecule.

Finally, although transformation is a highly evolved process, it may have evolved under different selective conditions than those that exist now. As a result, we see a process enabling gene exchange using recombination, which may have hijacked an independent and unrelated activity. If this is the case then the cost of the process has redirected the evolution of the activity.

Transfer Frequency or Efficiency

The efficiency of the process can be evaluated from two perspectives, that of the cell and that of the number of DNA molecules required. In laboratory experiments frequency, which is calculated as the number of transformants per recipient cell, can be as high as 10^{-4} per recipient cell. This represents the interaction of one or more DNA molecules with a cell. This is critical when considering the transfer of plasmids by transformation. Transformation will fail unless there are internal sites of homology within one or two copies of the plasmid molecule (linearized at different positions) taken up by the cell, because recombination demands this to reconstitute a replicon. Consequently, plasmid transfer largely depends on the concentration of plasmid DNA in surrounding cells.

Usually in laboratory experiments the numbers of recipient cells are high ($>10^8$) and DNA is homologous and in excess. These conditions are unlikely to occur in many natural environments. Frequencies will also be adversely influenced by the limited periods of competence shown by many transformable bacteria and the finding that in others only a fraction of the population ever becomes competent.

Finally the efficiency of transfer may well be high among those that are competent, even though the transfer frequency may appear to be low when assessed numerically!

Section Summary
- There is a metabolic cost to the organism.
- Transfer is dependent on concentrations of DNA and recipients.

WHY DO IT?

Consequences to Populations: Selective Value

This is a difficult question to address, but it is also one that is relevant to ask about the other mechanisms of transfer, namely conjugation and transduction. It is a mechanism that permits the exchange of genes, and DNA can be provided from living intact cells presumably in any physiological state to cells that are continually or transiently competent. Consequently, each viable cell can be both a donor and a recipient of DNA at the same time. The mechanism provides for both immediacy and long-term exchange too, as DNA in the environment is accessible. If the numbers of transformable bacteria adapted to an environment are amplified through their growth advantage, more homologous DNA will be present also. Consequently, those genes providing phenotypes of growth advantage will be relatively overrepresented in the DNA of an environment. The higher in situ concentrations of these genes would be a leftover from successful organisms, a legacy that can be seen as an accessible historical record of success. It is a legacy in the sense, too, that it will be more likely to be of value to organisms related to the ones who deposited them there.

Differences from Other Transfer Mechanisms

There is an opportunity for both polarity (DNA from dead cells) and reciprocity (DNA release/export) through this transfer mechanism. Transformation is also free from re-striction effects that limit the success of gene integration.

Is this true of transduction? No, in general, transduction requires the death by lysis of the donor host. A few phages are released without the death of their host, but these are not capable of transduction. Phages are affected by restriction.

Is this true of conjugation? Plasmid transfer is affected by restriction, but plasmid gene exchange can be reciprocal. There is no death of the donor. Only this mechanism is capable of transferring high numbers of chromosomal genes.

Section Summary
- We have examined what provides the DNA that is able to participate in transformation.
- We have examined how transformation compares with other transfer processes.

CONTRIBUTION TO TOTAL DNA EXCHANGE

For years people have discussed plasmid transfer and said it is the most important mechanism for gene exchange. Transduction comes next in importance, and transformation is the least significant. Sadly, this is a misunderstanding of the genetic context of the processes. I think it derives, at least in part, from the fact that larger gene packets (more genes, i.e., 10 to 100s) can be moved by plasmids, via conjugation. Transduction and transformation consistently move smaller packets (of the order of 5 to 10 genes). Thus, the process that moves larger packets is viewed as better than one that moves smaller ones. If we first examine this from the viewpoint of frequency, do we know if large packets are transferred more frequently than small packets? No, we do not. What if small packets are exchanged 10-, 100-, or 1000-fold more frequently? It changes the significance of the mechanisms immediately. Second, this analysis assumes that each mechanism is equivalent for all environments. What if each mechanism has a best operational environment? Then this assumption of equivalence falls too. Next, both phage and plasmids can be "moved" by transformation. Their transfer would not be "seen" as a transformation event by experimenters unless they examined the transfer process carefully. It would be assumed that they moved by the expected and relative mechanisms. So one is forced now to question more deeply the relative contributions made by each of the mechanisms. Finally, it is not the transfer event in itself that is important in evolution, but the time and place at which transfer occurs. So, from this perspective, the chances of an evolutionarily significant event are probably higher with a transfer mechanism delivering DNA more often. So mechanisms delivering a quick fix, repairing one or a few genes, can be done by all. The delivery of a pathway or a pathogenicity island, etc., as a large gene cluster really requires a conjugation-related process and will not be achieved by transformation.

This chapter has explained that transformation is a well-evolved and established process for exchanging genes within bacterial populations. Transformation acts in concert with the other processes to enable bacteria to survive, exploit, and evolve to counteract the perturbations of annual cycles. It also allows them to adapt to novel changes in their environments resulting from catastrophes of a local or global scale.

WHERE DO WE GO FROM HERE?

Competence is a widespread phenomenon, and many bacterial species have representatives that exchange genes in this manner. However, competence is not a universal mechanism (or at least we do not think it is) within any species, so one is forced to ask why? Was the ability universal and over evolutionary time has it been lost by some bacteria or has it evolved repeatedly? We need to understand its contribution to the bacteria that have it or have retained it. In achieving this we might then understand why some do not have the ability and what selective pressures were imposed for and against its presence.

There was a range of proposals for the pressures to evolve competence, the ability to take up DNA, and these can be summarized as the following: using it as a food or nucleotide source, for DNA repair, and for generating variability. That we have so many alternatives may indicate that in different genera competence may have evolved for individual reasons, peculiar to a particular group of organisms. Remember, the process makes use of a set of proteins that already have established roles to transport large molecules. Adapting these for DNA for one or another selective reason might provide the same answer since the object is to deliver DNA to the cytoplasm where it can be utilized.

Finally, bacteria have two other transfer mechanisms. Why "invent a third"? If we return to looking at the ecological sites of bacteria and where they live, we might be able to see a reason. Many species of pathogens (*H. influenzae*, *N. gonorrhoeae*, and *H. pylori*) exist in monocultures in host tissue. Consequently, Muller's ratchet will work against them, and they will enter the downward genetic spiral into cell death due to a lethal mutation. However, being able to share genes, and doing so selectively by identifying their DNA specifically, offers a way of avoiding the inevitable and making it not so! Transfer of genes by phage or plasmids probably would not be effective in these niches.

So where do we go from here? We need to look at more isolates from diverse environments and see what happens in nature in a wider way. We need to be able to use genomics to investigate the steps in the process across the Gram reaction divide and beyond into the *Archaea*. We need to get a "feeling" for the process in much the same way as we have for plasmids and conjugation, or phage and transduction. We understand these much better, but there are still lots of questions about their evolution and ecological impact that remain. Answering these for one mechanism will help answer questions about the others.

QUESTIONS

1. In *Haemophilus influenzae*, the genes specifying two resistances (cathomycin/Cm and streptomycin/Sm) are closely linked (Marmur and Lane, 1960). DNA from Cm-r, Sm-s, Cm-s, and Sm-r strains were mixed and heated to 100°C and cooled slowly to allow the DNA to renature. This DNA was used in a transformation experiment with a Sm-s, Cm-s recipient. Some Cm-r, Sm-r transformants were obtained and subcultured normally. Explain why the DNA is cotransferring these linked resistant phenotypes.

2. Explain why
   a. plasmids cannot be transformed unless they are linearized and
   b. the transfer frequency of plasmid borne genes is always lower than that of chromosomal genes

3. A transformation experiment is done with a donor DNA from strain A+B−C+ using the recipient A−B+C−. Selection is made on medium for A+ and the phenotypes of 1,000 transformants are further tested.

Unselected phenotype	Number
B⁻C⁺	48
B⁻C⁻	12
B⁺C⁺	400
B⁺C⁻	540

Then the experiment was repeated selecting for transformants that have the C+ phenotype. The phenotypes of these 1,000 transformant are shown in the table below.

Unselected phenotype	Number
A⁺B⁻	68
A⁺B⁺	340
A⁻B⁻	52
A⁻B⁺	568

What is the order of the three genes A, B, and C?

4. A strain of *Acinetobacter calcoaceticus* was made resistant in the laboratory to four antibiotics (K, kanamycin; S, streptomycin; N, neomycin; and R, rifampicin). These phenotypes are naturally transformable into a strain sensitive to all four antibiotics. The data below show the transfer frequencies per recipient for combinations of the resistance phenotypes.

Calculate which resistances are closely linked and the order of the genes.

Medium + antibiotic(s)	Colony count
+S+N+K+R	0
+S+N+K	0
+S+N+R	411
+S+K+R	0
+N+K+R	21
+N+K	82
+N+R	594
+N+S	425
+S+K	0
+S+R	720
+K+R	20
+R	1,146
+K	1,158
+S	1,162
+N	1,140
	1,000,000,000

5. Is the process of transformation in gram-positive and -negative bacteria fundamentally the same?

REFERENCES

Smith, H. O., J. F. Tomb, B. A. Dougherty, R. D. Fleischmann, and J. C. Venter. 1995. Frequency and distribution of DNA uptake signals sequences in *Haemophilus influenzae* Rd genome. *Science* **269**:538–540.

Stouthamer, A. H. 1979. The search for correlation between theoretical and experimental growth yields. *Int. Rev. Biochem.* **21**:1–47.

Suerbaum, S., J. M. Smith, K. Bapumia, G. Morelli, N. H. Smith, E. Kunstmann, I. Dyrek, and M. Achtman. 1998. Free recombination within *Helicobacter pylori*. *Proc. Natl. Acad. Sci. USA* **95**:12619–12624.

FURTHER READING

Avery, O. T., C. M. Macleod, and M. McCarty. 1944. Studies on the chemical nature of the substance inducing transformation of pneumococcal types. *J. Exp. Med.* **79**:137–158.

Day, M. 2002. Transformation in aquatic environments, p. 63–80. *In* M. Sylvanen and C. I. Kado (ed.), *Horizontal Gene Transfer*. Academic Press, New York, N.Y.

Doolittle, W. F., and C. Sapienza. 1980. Selfish genes, the phenotype paradigm and genome evolution. *Nature* **284**:601–603.

Dubnae, D. 1999. DNA uptake in bacteria. *Annu. Rev. Microbiol.* **53**:217–244.

Garrity, G. M. (ed.-in-chief). 2001. *Bergey's Manual of Systematic Bacteriology*, 2nd ed. Springer, New York, N.Y.

Griffith, M. H. 1928. Significance of pneumococcal types. *J. Hyg. Camb.* **27:**113–159.

Lederberg, E. M., and S. N. Cohen. 1974. Transformation of *Salmonella typhimurium* by plasmid DNA. *J. Bacteriol.* **119:**1072–1074.

Mandel, M., and A. Higa. 1970. Calcium dependent bacteriophage DNA infection. *J. Mol. Biol.* **53:**159–162.

Marmur, J. S., and D. Lane. 1960. *Proc. Natl. Acad. Sci. USA* **46:**453.

Muller, H. J. 1964. The relation of recombination to mutational advance. *Mutat. Res.* **1:**2–9.

Redfield, R. 1988. Evolution of bacterial transformation: is sex with dead cells ever better than no sex at all? *Genetics* **119:**213–221.

Redfield, R. 1993. Genes for breakfast: the-have-your-cake-and-eat-it-too of bacterial transformation. *J. Hered.* **84:**400–404.

Williams, G. C. 1975. Sex and evolution. *In* S. A. Levin and H. S. Horn (ed.), *Monographs in Population Biology*. Princeton University Press, Princeton, N.J.

HORIZONTAL GENE TRANSFER AND THE REAL WORLD

Robert V. Miller and Martin J. Day

||

A few years ago—literally a few years ago—the average microbiologist would have told you that bacteria were clonal and that horizontal gene transfer did not take place in nature. The wisdom of the day told us that genetic exchange was limited to the research laboratory; it was just too uncommon an event to matter in the environment. Of course, conjugation did pass extrachromosomal genes between bacteria, but plasmids were too costly to the cell and they would not maintain them unless dramatic selection criteria were placed on the organism. So few bacteriophages existed in nature (maybe 1 to 10 virions in a milliliter of water) that they could not effect changes in bacterial populations. Besides, the only transducing phages only infected *Escherichia coli* and a few other enteric organisms. The probability of a phage, much less a transducing particle, finding a host in nature was not likely. And transformation, well why even consider trans-

formation? So few bacteria showed competence that it could not even be considered as an evolutionary force in the microbial world. Anyway, free DNA could not exist for more than a few moments in a natural environment!

What has happened between now and then to change this view? There must have been dramatic discoveries, or we would not have included a section on intercellular gene movement in this book.

What happened? Genetic engineering and biotechnology happened! Critics of the use of "genetically engineered microorganisms (GEMs)" in agricultural and environmental settings feared that the genes would "escape to natural organisms." In the United States, these critics demanded that the Environmental Protection Agency and other governmental entities give evidence on the risk of gene exchange among microorganisms. Microbial ecologists and geneticists were forced to look at the natural environment to determine experimentally whether horizontal gene exchange could take place in nature. Supposition was no longer sufficient. More importantly, research funds became available for this type of research for the first time in history. When these scientists looked at nature, they discovered their assumptions were simply wrong!

THE CONDITIONS ARE RIGHT!

Transduction

Many questions about the movement of genetic material among microorganisms originated

Robert V. Miller, Department of Microbiology, Oklahoma State University, Stillwater, OK 74078, and *Martin J. Day*, Cardiff School of Biosciences, Cardiff University, Cardiff CF10 3TL, Wales, United Kingdom.

Microbial Evolution: Gene Establishment, Survival, and Exchange
Edited by Robert V. Miller and Martin J. Day, ©2004 ASM Press, Washington, D.C.

in a concern over the safety of using genetically modified microorganisms for environmental applications. These in turn stimulated a resurgence in environmental research. New and more accurate techniques demonstrated that many old assumptions were wrong. Bergh and colleagues (1989) and Proctor and Fuhrman (1990) were discovering that the number of bacteriophages in a milliliter of seawater was not one or ten but 10^{10} to 10^{14}. Now bacteriophage-host interactions, long thought to have a very low probability of occurring in nature, were seen to be quite common and Proctor and Fuhrman proposed that phage infection was in fact the primary cause of bacterial mortality in marine environments. While the half-life of bacteriophages (Ogunseitan et al., 1990) in fresh and seawater was found to be short (1 to 2 days), Miller and his coinvestigators (1992) and Stotzky's group (Zeph et al., 1988) determined that virions could be stabilized by attachment to suspended particles. Ripp and Miller (1995) also found that this attachment further increased the likelihood that phages and host bacteria would interact. As the realization that the number of bacteriophages in nature was much greater than previously expected was taking hold of the scientific community, Kokjohn and colleagues (1991) demonstrated that, even at low bacteria-to-phage ratios, the likelihood of a phage virion finding and productively infecting a host bacterium in nature was much greater than expected from theoretical biophysical measurements made in the laboratory. They determined that bacteriophage virions could be produced in starved hosts (as they most likely exist in nature), although the latency period was lengthened and the bust size reduced. In addition, it soon became clear that the majority of environmental microorganisms were lysogenic for one or more prophages and that many bacteriophages that produced lytic infections in the well-nourished, laboratory-grown host could establish an unstable but semipermanent relationship with their starved host cells referred to as pseudolysogeny. In this state, the bacterium is neither lysed nor stably lysogenized. Here the genome of the phage seems to be lying inert in the host but is protected from inactivation by environmental conditions that would kill its virion. When the host becomes energized by feeding, the pseudo-prophage can be activated and produce either a lytic infection or establish a true temperate relationship with its host.

Transformation

Lorenz and Wackernagel (1987), Lorenz et al. (1988), and Khanna and Stotzky (1992), working independently, found that DNA could survive for long periods attached to clay and other soil particles. Soon, John Paul and his associates (1987) determined that the half-life of DNA in seawater was at least a day, and the idea that transformation was not possible in the environment because DNA was too unstable was put to rest once and for all. While these investigators were showing that the substrate for transformation did indeed exist in nature, Greg Stuart and Day and Fry, working on opposite sides of the Atlantic, and others, were rapidly discovering that the number of species of bacteria that could acquire a state of competency for the uptake of DNA from the environment was much greater than had been expected previously. It fact it was becoming clear that competency was a widespread phenomenon in many bacteria. Whether a bacterium could become competent for DNA uptake was not the question. The question soon became whether we could discover the correct environmental conditions for the expression of competency. Transformation clearly could take place in many aquatic and terrestrial environments. But did it?

THE PHENOMENA EXIST!

Now that the stage was set, Day and Fry, Lorenz and Wackernagel, John Paul and coworkers, and others set to work to demonstrate that transformation occurred in seawater and fresh waters, as well as in soils (see chapter 10). Miller et al. (1992) and Zeph et al. (1988) soon demonstrated that transduction occurred in water and soil environments (see chapter 9). Conjugation was shown by Barkay and others to be an effective way to rapidly spread genetic elements throughout a bacterial population when the character encoded by this extrachromosomal element afforded a positive selection advantage to the host organism (see chapter 7). They demonstrated that it was not necessary for the entire population to maintain the energy-expensive element, for the population to be protected for that element could rapidly be dispersed in the time of need. These discoveries were soon followed by the studies of Abigail Salyers, June Scott, and others who discovered that transposons were not always confined to intracellular transfer but that some (often carrying drug-resistance markers) had the ability to move among cells much like plasmids (see chapter 8). The consequence of the spread of these antibiotic resistance genes on plasmids, conjugative transposons, and even bacteriophages soon became apparent with the studies of Stuart Levy and others who demonstrated the spread of drug resistance through natural and hospital-associated bacterial populations. Not only could this drug resistance be transferred within a species but it was also able to hop from species to species, genus to genus, and even between highly unrelated bacteria.

WHERE DOES IT NOW APPEAR THAT HORIZONTAL GENE TRANSFER TAKES PLACE?

The simple answer to this question is "everywhere!" We can now confirm horizontal gene transfer in many environments. The data to date suggest that various mechanisms of horizontal gene transfer with various outcomes and consequences can take place in all natural environments, whether aquatic, terrestrial, or biological (Table 1), often in biofilms (Table 2).

Conjugation has traditionally been recognized to occur in the hospital setting. It has now been demonstrated in many environments including freshwater and marine habitats, in terrestrial environments, on the surface of plants and among bacteria associated with insects and mammals, and in sewage-treatment facilities. Its near relative, conjugative transposition, has primarily been demonstrated in intestinal microbes and in various pathogens of mammals including humans.

Transduction has now been observed in many habitats. These include freshwater lakes and streams, sewage-treatment facilities, and estuaries and other ocean habitats. Transduction has been demonstrated in soils, on plants, in the guts of oysters, and in the kidneys and other organs of mice. Insect guts have been shown to act as a hospitable environment for transduction.

Transformation has been shown to occur in freshwater and marine environments and in various soils. The presence of free DNA in many environments suggests that the potential for transformation is widespread. A greater knowledge of the mechanisms and triggers for competence will undoubtedly lead to the identification of transformation as a viable gene-transfer mechanism in even more

TABLE 1 Environments where horizontal gene transfer has been positively observed

Mode of transfer	Environment		
	Terrestrial	Watery	Inside and on other organisms
Conjugation	Soil	Lakes, oceans, marine sediments, river epilithons (biofilms) on river stones, sewage in treatment facilities	Plants, insects, chickens, mice, and humans
Conjugative transposons	*Not yet observed*	*Not yet observed*	Rats and humans
Transduction	Soil	Lakes, oceans, rivers, sewage in treatment facilities	Shellfish, mice, insects, plants
Transformation	Soil	Marine sediments, river epilithon on river stones	Plants, insects, mice

TABLE 2 A summary of the horizontal gene transfer interactions occurring in a biofilm community

Interaction[a]	Density	Location	Half-life
Intracellular[b]			
Transformation	DNA (μg ml^{-1})	Free ↔ attached	Determined by environment
Bacteriophage	0–10^{14} ml^{-1}	Free ↔ attached	Determined by environment
Intercellular-plasmid[c]			
(conjugation)	0–10^9 ml^{-1}	Determined by host	Determined by host

[a] The ability to interact was determined by cell physiology.
[b] Release and uptake, spatially and temporally dissociated events.
[c] Requires contact between donor and recipient.

environmental settings than those in which it has currently been identified.

Besides these mechanisms of gene transfer, others have made suggestions that are less well understood and documented. These include hyphal fusion among the streptomycetes and other filamentous bacteria. There are also some reports of cell fusion in *Escherichia coli* and other enteric organisms. It is likely that we will discover that not only is horizontal gene transfer more widespread than we now can document but that the themes and variations in its mechanisms are much more robust than we can now imagine.

EVIDENCE AND CONSEQUENCES!

In the remainder of this book, you will begin to explore some of the evidence for lateral gene transfer and its consequences for bacterial evolution and diversity. In any case it is now clear that horizontal gene transfer has been and continues to be a strong and important force in the evolution of bacterial genomes. Genomic research (see chapters 17, 20, and 22) has repeatedly demonstrated that bacteria have acquired and maintained prophage as cryptic elements. These have provided bacteria with new and surprising characters from virulence factors to mechanisms of recombination and UV resistance. Milkman and others have clearly demonstrated (see chapter 20) that bacteria, far from being clonal, are a mosaic of genetic sequences clearly showing the history and patterns of intragenic and intergenic recombination.

Our concepts of the importance of transduction and transformation have greatly changed during the past few years. Transduction, once thought to be a phenomenon of the laboratory, has now been recognized as an important environmental phenomenon. In a recent editorial (January 2003) in the *Journal of Bacteriology*, Phil Matsumura (its editor-in-chief), Ry Young, and Lynn Enquist stated:

> The revolutionary findings in genomics have revealed that phages have a more intimate relationship to their bacterial hosts than do eukaryotic viruses. Generalized transduction, probably the main mode of lateral DNA transfer between bacteria, has no real counterpart in eukaryotic biology. Prophages make up a significant fraction of the genomic space of many bacteria, including important pathogens, and often define the pathogenic response, with genes carried on prophages and expressed through phage vegetative growth.

Likewise, we now see that transformation is a significant activity of many bacteria in nature. The discovery that competence is often controlled by quorum-sensing mechanisms has greatly advanced our understanding of the underlying molecular mechanisms of transformation and excited many microbiologists to search for new species with new mechanisms and signals for the initiation of competence and, hence, genetic transformation.

Such insights have excited interest in understanding how bacteria and other microorganisms exchange genetic information in nature. We are only now beginning to scratch the surface of the intricate and devious ways that genetic information has invented to ensure its survival and spread throughout the microbial kingdom. The next decades will be an exciting time of discovery.

REFERENCES

Bergh, Ø., K. Y. Børsheim, G. Bratback, and M. Heldal. 1989. High abundance of viruses found in aquatic environments. *Nature* **340:**467–468.

Khanna, M., and G. Stotzky. 1992. Transformation of *Bacillus subtilis* by DNA bound on montmorillonite and on the transforming ability of bound DNA. *Appl. Environ. Microbiol.* **58:**1930–1939.

Kokjohn, T. A., G. S. Sayler, and R. V. Miller. 1991. Attachment and replication of *Pseudomonas aeruginosa* bacteriophages under conditions simulating aquatic environments. *J. Gen. Microbiol.* **137:** 661–666.

Lorenz, M. G., and W. Wackernagel. 1987. Adsorption of DNA to sand and variable degradation rates of adsorbed DNA. *Appl. Environ. Microbiol.* **53:**2948–2952.

Lorenz, M. G., B. W. Aardema, and W. Wackernagel. 1988. Highly efficient genetic transformation of *Bacillus subtilis* attached to sand grains. *J. Gen. Microbiol.* **134:**107–112.

Miller, R. V., S. Ripp, J. Replicon, O. A. Ogunseitan, and T. A. Kokjohn. 1992. Virus-mediated gene transfer in freshwater environments, p. 51–62. *In* M. J. Gauthier (ed.), *Gene Transfers and Environment.* Springer-Verlag, Berlin, Germany.

Ogunseitan. O. A., G. S. Sayler, and R. V. Miller. 1990. Dynamic interaction of *Pseudomonas aeruginosa* and bacteriophages in lake water. *Microb. Ecol.* **19:**171–175.

Paul, J. H., W. H. Jeffrey, and M. R. DeFlaun. 1987. Dynamics of extracellular DNA in the marine environment. *Appl. Environ. Microbiol.* **53:**170–179.

Proctor, L. M., and J. A. Fuhrman. 1990. Viral mortality of marine bacteria and cyanobacteria. *Nature* **343:**60–62.

Ripp, S., and R. V. Miller. 1995. Effects of suspended particulates on the frequency of transduction among *Pseudomonas aeruginosa* in a freshwater environment. *Appl. Environ. Microbiol.* **61:**1214–1219.

Vettori, C., G. Stotzky, M. Yoder, and E. Gallori. 1999. Interaction between bacteriophage PBS1 and transduction of *Bacillus subtilis* by clay-phage complexes. *Environ. Microbiol.* **1:**347–355.

Zeph, L. R., M. A. Onaga, and G. Stotzky. 1988. Transduction of *Escherichia coli* by bacteriophage P1 in soil. *Appl. Environ. Microbiol.* **54:**1731–1737.

FURTHER READING

Bale, M. J., J. C. Fry, and M. J. Day. 1987. Plasmid transfer between strains of *Pseudomonas aeruginosa* on membrane filters attached to river stones. *J. Gen. Microbiol.* **133:**3099–3107.

Day, M. J. 2002. Transformation in aquatic environments, p. 63–80. *In* M. Syvanen and C. I. Kado (ed.), *Horizontal Gene Transfer*, 2nd ed. Academic Press, San Diego, Calif.

de Lipthay, J. R., T. Barkay, and S. J. Sorensen. 2001. Enhanced degradation of phenoxyacetic acid in soil by horizontal transfer of the *tfdA* gene encoding 2,4-dichlorophenoxyacetic acid dioxygenase. *FEMS Microbiol. Ecol.* **53:**75–84.

Kidambi, S. P., S. Ripp, and R. V. Miller. 1994. Evidence for phage-mediated gene transfer among *Pseudomonas aeruginosa* on the phylloplane. *Appl. Environ. Microbiol.* **60:**496–500.

Levy, S. B. 1998. The challenge of antibiotic resistance. *Sci. Am.* **278**(3)**:**46–53.

Lorenz, M. G., and W. Wackernagel. 1994. Bacterial gene transfer by natural transformation in the environment. *Microbiol. Rev.* **58:**563–602.

Miller, R. V. 1998. Bacterial gene swapping in nature. *Sci. Am.* **278**(1)**:**66–71.

Miller, R. V. 2001. Environmental bacteriophage-host interactions: factors contribution to natural transduction. *Antonie van Leeuwenhoek* **79:**141–147.

Miller, R. V., and S. A. Ripp. 2002. Pseudolysogeny: a bacteriophage strategy for increasing longevity *in situ*, p. 81–91. *In* M. Syvanen and C. I. Kado (ed.), *Horizontal Gene Transfer*, 2nd ed. Academic Press, San Diego, Calif.

Morrison, W. D., R. V. Miller, and G. S. Sayler. 1978. Frequency of F116 mediated transduction of *Pseudomonas aeruginosa* in a freshwater environment. *Appl. Environ. Microbiol.* **36:**724–730.

Novic, R. P., and S. I. Morse. 1967. *In vivo* transmission of drug resistance factors between strains of *Staphylococcus aureus. J. Exp. Med.* **125:**45–59.

Scott, J. R., and G. G. Churchward. 1995. Conjugative transposition. *Annu. Rev. Microbiol.* **49:**367–397.

Shoemaker, N. B., H. Vlamakis, K. Hayes, and A. A. Salyers. 2001. Evidence for extensive resistance gene transfer among *Bacteroides* spp. and among *Bacteroides* and other genera in the human colon. *Appl. Environ. Microbiol.* **67:**561–568.

Stewart, G. J., and C. A. Carlson. 1990. Detection of horizontal gene transfer by natural transformation in native and introduced species of bacteria in marine and synthetic environments. *Appl. Environ. Microbiol.* **56:**1818–1824.

Williams, H. G., M. J. Day, and J. C. Fry. 1992. Natural transduction on agar and in river epilithon, p. 69–76. *In* M.J. Gauthier (ed.), *Gene Transfers and Environment.* Springer-Verlag, Berlin, Germany.

MECHANISMS FOR GENE ESTABLISHMENT AND SURVIVAL

STATIONARY-PHASE-INDUCED MUTAGENESIS: IS DIRECTED MUTAGENESIS ALIVE AND WELL WITHIN NEO-DARWINIAN THEORY?

Ronald E. Yasbin and Mario Pedraza-Reyes

12

ADAPTIVE OR STATIONARY-PHASE-INDUCED MUTATIONS: WHAT'S IT ALL ABOUT?

DIVERSITY AMONG THE SYSTEMS THAT HAVE BEEN UTILIZED TO INVESTIGATE STATIONARY-PHASE-INDUCED MUTAGENESIS

STATIONARY-PHASE CULTURES ARE NOT COMPOSED OF HOMOGENOUS CELLS

IS THERE ANY ROLE FOR DIRECTED MUTAGENESIS IN THE STATIONARY-PHASE MUTATION SYSTEM(S)?

THE ROLE OF MISMATCH REPAIR (MMR) IN "ADAPTIVE" MUTATIONS AND A TRANSIENT HYPERMUTABLE STATE

AN ANCIENT FAMILY OF DNA POLYMERASES (Y SUPERFAMILY) AND POTENTIAL ROLES IN STATIONARY-PHASE-INDUCED MUTAGENESIS

CAN WE SPECULATE ON DIRECTED MUTAGENESIS AND ITS ROLE IN STATIONARY-PHASE-INDUCED MUTAGENESIS?

"WE ARE STANDING ON THE SHOULDERS OF THE GIANTS WHO CAME BEFORE!"

Ronald E. Yasbin, College of Sciences, University of Nevada, Las Vegas, 4505 Maryland Parkway, Box 454001, Las Vegas, NV 89154-4001, and *Mario Pedraza-Reyes*, Institute of Investigation in Experimental Biology, Faculty of Chemistry, University of Guanajuato, Noria Alta s/n, Guanajuato, 36050, Mexico.

The ability of organisms to generate mutations has always been an essential part of the evolutionary process. Ever since biologists (especially evolutionists and geneticists) have been interested in the evolutionary process there have been attempts to delineate the mechanisms responsible for the diversity within a given species. This chapter attempts to raise questions about mutagenesis and how these processes enhance evolution. The focus of the chapter is on the controversial process that has been termed "adaptive" mutagenesis. Although the work described here concentrates on prokaryotic model systems, it is clear (starting with the observations of McClintock on transpositions) that all organisms under stress appear to have genetic mechanisms that permit increase in allelic diversity. We hope that this review adds to the renewed interest in directed mutagenesis, retromutagenesis, transpositional mutagenesis, stationary-phase-induced mutagenesis, and the influences that these processes have had on the evolution of species.

Classically, mutations were supposed to arise during growth, spontaneously, and without any regard to selective pressures. This conclusion not only made sense but it also strongly discounted any Lamarckian concepts that might have lingered. However, from the start of the modern genetics era there have been suggestions and claims that mutations can arise

Microbial Evolution: Gene Establishment, Survival, and Exchange
Edited by Robert V. Miller and Martin J. Day, ©2004 ASM Press, Washington, D.C.

(or at least be influenced) by selection pressures. In 1988, an article by John Cairns and his collaborators forced a rethinking about how spontaneous mutants might arise. This rethinking often involved heated and passionate discussions. Basically, the very tenets of neo-Darwinian theory were being challenged. Most researchers and teachers of genetics began to modify their thinking and lectures to handle the possibility that "adaptive" mutations might occur. This chapter explores some of the possibilities that might exist for departures from the classic mode of thinking associated with the generation of mutations.

ADAPTIVE OR STATIONARY-PHASE-INDUCED MUTATIONS: WHAT'S IT ALL ABOUT?

The very term "adaptive" mutation has been defined somewhat differently by many of the investigators in the field. The definitions range from those mutations that permit organisms to grow and divide in response to natural or artificial selections to those mutational procedures that cause DNA changes after the exposure of the cells to the selection or the selective pressure. In any case, for more than a decade there has been considerable interest in the phenomenon that has been called adaptive or stationary-phase-induced mutagenesis. The result of the mechanism or mechanisms responsible for this phenomenon is the production of mutations that arise in nondividing or stationary-phase bacteria when the cells are subjected to nonlethal selective pressure, such as nutrient-limited environments. While most of the research has involved *Escherichia coli* model systems, similar observations have been made in other prokaryotes as well as in eukaryotic organisms (Foster and Trimarchi, 1994; Kasak et al., 1997).

In the best-studied model system, the F' *lac* frameshift-reversion assay in *E. coli*, stationary-phase mutations that lead to the generation of Lac$^+$ cells can be distinguished from normal growth-dependent spontaneous Lac$^+$ mutations. In this system, bacteria unable to utilize lactose as the sole carbohydrate source are spread on a medium that lacks all other carbo-

hydrate sources. After the second day of incubation, Lac$^+$ colonies begin to appear, and the number of colonies continues to increase daily. Specifically, Lac$^+$ mutations are generated in stationary-phase cells via a molecular mechanism that requires a functional homologous recombination system, F' transfer functions, and a component(s) of the SOS system. Genetic evidence suggests that DNA polymerase III and the SOS-regulated DNA polymerase IV are responsible for the synthesis of errors that lead to these mutations. Furthermore, for the Lac$^+$ mutations, different sequence spectra are generated for the stationary-phase mutations as compared with the types of mutations generated during growth. For instance, a majority of the Lac$^+$ mutations that arise during stationary phase have a -1 deletion at mononucleotide repeats within the target gene. On the other hand, for the spontaneous mutations that arise during growth, various types of mutations occur in seemingly random locations. These characteristics suggested that stationary-phase Lac$^+$ reversions occur via a different molecular mechanism(s) than those reversions, of the same *lac* allele, that are generated during growth. However, there is also evidence that demonstrates that the mutations generated by this *lac* system during stationary phase are the result of amplification of the plasmid borne gene followed by SOS-induced mutagenesis and selection (Harris et al., 1997). These results indicate that, in this system, the mutations are caused by an interaction of recombination functions and the SOS regulon and not a stationary-phase-specific mutagenesis mechanism. Thus, there seems to be a real debate concerning the mechanism(s) responsible for the mutations associated with the Lac system (Galitski and Roth, 1995; Slechta et al., 2002).

Although the very observations of adaptive or stationary-phase mutagenesis could have suggested the existence of a Lamarckian type of genetics, subsequent research has demonstrated that this type of mutagenesis is not necessarily directed only to the selected genes. In fact, some studies have suggested that in a starving or stressed culture a small subpopulation of the cells seems to have an overall increased mutation frequency. Basically, the results mentioned above have been used to generate a proposal that suggests a physiologically stressed bacterial community may differentiate a hypermutable subpopulation and these hypermutable cells generate mutations randomly. If one or more of the mutations help the cell survive longer or grow under the stressful conditions, then the organism will appear to have "adapted" to its environment. However, here too there has been a considerable amount of disagreement as to the existence and importance of a hypermutable subpopulation(s) (Bull et al., 2000; Cairns, 2000; Torkelson et al., 1997).

Section Summary

- Classical theory on how mutations arise is being challenged.
- Stationary-phase-induced or adaptive mutagenesis has been shown to occur in a variety of different organisms spanning the domains of life.
- The actual mechanism responsible for this type of mutagenesis remains to be determined.
- Theory of how mutations arise is under constant revision.

DIVERSITY AMONG THE SYSTEMS THAT HAVE BEEN UTILIZED TO INVESTIGATE STATIONARY-PHASE-INDUCED MUTAGENESIS

The majority of research efforts on stationary-phase mutagenesis have focused on the mechanisms that generate these mutations. The equally exciting elucidation of the mechanism(s) that regulates the activation of this mutation system(s) or even the development of the proposed transitory hypermutable state has not, as of yet, generated the same level of interest. There have been increasing evidence and discussion about the existence of transitory subpopulations in stressed or stationary-phase prokaryotic cultures. Such subpopulations could play an important role in the generation of mutants under stressful conditions like those that exist in stationary-phase cells. While such mutations could be the result of stochastic physiological events, it is intriguing to consider that basic regulatory mechanisms involved in gene expression during stationary phase might control the mutagenic potential of these cells. In fact, using a different model system for the study of stationary-phase mutagenesis, Gómez-Gómez and colleagues (1997) demonstrated that two growth-phase-regulatory proteins (H-NS and σ^S) of E. coli significantly altered the production of adaptive or stationary-phase-induced mutations. Recently, a Bacillus subtilis revertant assay system was utilized to demonstrate that stationary-phase-induced mutagenesis occurs in this gram-positive bacterium (Sung and Yasbin, 2002). In this system stationary-phase mutagenesis was not dependent on a functional RecA protein (i.e., recombination was not required) nor on the presence of the σ^B (a stress-related σ-factor somewhat analogous to σ^S from E. coli) form of the RNA polymerase. However, at least two transcription factors that are involved in the regulation of differentiation and development in this bacterium (ComA and ComK) did influence the eventual production of stationary-phase-induced mutations. Furthermore, the three auxotrophic alleles that were studied did not appear to be reverted during stationary-phase mutagenesis by the same molecular mechanisms. Based on those results, the hypothesis was advanced that during periods of environmental stress, such as found in nutrient-limited stationary-phase cells, subpopulations are differentiated within a culture to generate genetic diversity. Furthermore, it was suggested that within some of these subpopulations mutation frequencies can be increased (hypermutability) by the suppression of DNA-repair systems and/or the activation of mechanisms that would increase the introduction of DNA damage into the genome. However, some or all of the results described could be caused by one or more forms of directed mutagenesis, as described below.

Section Summary

- Adaptive or stationary-phase-induced mutagenesis may be under genetic control and part of eubacterial development.

- The concept of a hypermutable transient subpopulation was raised.
- Evolution of the genetic control of mutagenesis appears to have taken place.

STATIONARY-PHASE CULTURES ARE NOT COMPOSED OF HOMOGENOUS CELLS

There still is considerable interest, with respect to stationary-phase-induced mutagenesis, in whether or not the mutations that were generated were limited to those genes or metabolic processes for which selective pressure was applied. For instance, in the *B. subtilis* system mentioned above (measuring auxotrophic reversion rates of three alleles), does one obtain prototrophic leucine revertants only on media that lacks leucine? In the *E. coli* model, substantial evidence has been provided to indicate that under certain experimental conditions, mutations occur in genes that control metabolic processes unrelated to the initial selective pressure. Although there is still considerable debate, it is argued by some that most, if not all, random adaptive mutations are occurring in a hypermutable subpopulation. In the results obtained with the *Bacillus* auxotrophic reversions described above, 1% of 500 mutant colonies randomly selected and tested contained two different mutations. However, all of the data accumulated, to date, do not prove that hypermutability actually exists in the bacteria that undergo the first mutational event. What is known is that if the cells are undergoing hypermutability, then this increased mutation activity is transitory and not the result of a permanent mutator phenotype. An alternate hypothesis would argue that the "hypermutability" is the result of gene amplification, growth, and the induction of the SOS error-prone repair capacities (lesion by-pass) in those cells that have undergone this process, at least for the plasmid borne F' *lac* frameshift-reversion assay system in *E. coli*. Such an explanation would explain a general hypermutability in the reacting cells but not a situation where a subpopulation is being specifically designated to hypermutate.

The variety of different systems that have been used to examine stationary-phase mutagenesis and the results obtained would tend to argue for the existence of multiple subpopulations rather than for just one hypermutable subpopulation. Specifically, the existence of multiple types of stationary-phase mutagenesis mechanisms is suggested by the collective analysis of the data that have been obtained from studies with *E. coli*, *B. subtilis*, and other model systems. Whether there are multiple subpopulations or only one, a problem remains with how the number of hypermutable cells is limited (remains a subpopulation) and how these cells are prevented from reaching genetic load (the point at which the deleterious mutations are so numerous that the survival of the culture becomes improbable).

Known models for the presence of coexisting subpopulations that communicate with each other, limit population size, and enhance diversity and the ability to survive can be found in the studies of the regulation of sporulation, competence development (the ability to take up and utilize exogenous DNA in a genetic transformation), and the production of secondary metabolites in *B. subtilis*. These studies have revealed the existence of multiple interacting and complex regulatory networks that permit the cells within the culture to sense and adapt to their environments. Specifically, it has been found that with respect to the development of competence (i) only 10 to 20% of the cells in a culture obtain this state, (ii) quorum sensing is an essential component, and (iii) this developmental process does occur in the soil, the natural environment of this bacterium (Dubnau, 1991; Dubnau and Lovett, 2002; Graham and Istock, 1978; Msadek, 1999). Recently, Makinoshima and his colleagues (2002) were able to fractionate subpopulations of an *E. coli* culture similarly to how the competent *B. subtilis* subpopulation can be separated from the noncompetent bacteria of a single culture. These results again suggest the existence of multiple subpopulations within stationary-phase cultures of bacteria. Furthermore, these results would certainly make it seem possible that one or more hypermutable subpopulations could be present in stationary-phase cells. However, although heterogeneity of stationary-phase bacterial cultures has been established, the existence of a hypermutable subpopulation or populations remains a hotly contested topic.

Section Summary

- Genetically "pure" bacterial cultures actually possess a great deal of physiological diversity, especially when the cells are in the stationary phase.
- This concept challenges previously held beliefs that all bacteria, except for the occasional mutant, are identical in physiology because they have the same genetic composition.
- Bacteria can communicate among themselves to achieve diversity within a culture.

IS THERE ANY ROLE FOR DIRECTED MUTAGENESIS IN THE STATIONARY-PHASE MUTATION SYSTEM(S)?

Discussing adaptive mutagenesis and/or directed mutagenesis often invokes images of Lamarckian genetics. However, there is room within neo-Darwinism for both adaptive or stationary-phase-induced mutagenesis and for aspects of directed mutagenesis. As reviewed by Wright (2000), there has been a continuous stream of arguments against every mutation having been completely random during the evolutionary process. Dobzhansky in his seminal publication in 1950 stated that:

> The most serious objection to the modern theory of evolution is that since mutations occur by chance and are undirected, it is difficult to see how mutation and selection can add up to the formation of such beautifully balanced organs as, for example, the human eye.

As pointed out by Wright there are data that strongly support the existence of stress-related "directed" mutagenesis mechanisms. Examples of these stress-related mechanisms for directed forms of mutagenesis depend on the responses of the bacteria to aspects of starvation. Specif-ically, starvation for any number of nutrients leads to the specific transcription of associated regulons. Observations have led to the suggestion that

unprotected single strands, during the transcription process, are more likely to be mutated than are the transcribed strands. Additional evidence has been presented to demonstrate transcription enhancement of mutation levels (Rudner et al., 1999) especially following starvation and the induction of the stringent response. Proposed mechanisms for this transcription-enhanced mutagenesis have ranged from altered susceptibility of bases to damage due to the secondary DNA structures to translesion synthesis by RNA polymerases. With respect to RNA polymerase translesion synthesis, this process has been referred to as "retromutagenesis." The concept is that in nondividing cells (cells subjected to starvation) RNA polymerases could make errors during the transcription process (Rudner et al., 1999; Holmquist, 2002). For instance, in cells starved for histidine, because of a mutation that prevented the production of a functional protein needed for the production of histidine, transcription errors might lead to the translation of mutant proteins that would allow a cell to grow, albeit transiently. These transiently growing cells would then have the opportunity to replicate their DNA and mutate genes in the chromosome. By using the example cited above, a histidine prototroph could be generated by mechanisms associated with the increased mutation frequency occurring in transcribed genes. Bridges and others have used this model to explain "directed" or "cell-selfish" modes of evolution.

Another possible mechanism for a linkage between mutagenesis and transcription would be the transcription-repair-coupling factor (TRCF), also termed the *E. coli* mutation frequency decline gene product (*mfd*). This protein (homologs of which are found in *Eucarya* and in other eubacteria) recruits DNA-repair systems to the sites of damage that are encountered during transcription. This recruitment appears to favor the repair of DNA damage in the transcribed strand (Selby and Sancar, 1993).

In *B. subtilis*, the *mfd* gene product has been shown to be involved in recombination, DNA repair, and carbon catabolite repression mechanisms (Zalieckas et al., 1998). Huang-Mo Sung (personal communication) has discovered that inactivation of the *B. subtilis mfd* gene leads to a dramatic decrease in the number of stationary-phase mutations that are induced. These results suggest that with respect to stationary-phase-induced mutagenesis, the Mfd protein aids the accumulation of mutations in actively transcribed genes. Therefore, at least within the *B. subtilis* model system, the Mfd protein might have an important role in allowing retromutagenesis to occur. Taken collectively, all of the data accumulated to date do provide the potential for some type or types of "directed" mutagenesis to be part of the stationary-phase system(s).

Section Summary
- Spontaneous random mutations could have actually allowed evolution to proceed as rapidly as it has.
- Directed mutagenesis can actually be part of neo-Darwinian evolution.
- Errors in the transcription process can lead to a type of directed mutagenesis that fits well into evolutionary theory.
- The process called retromutagenesis can be invoked to explain a possible form of directed mutagenesis.

THE ROLE OF MISMATCH REPAIR (MMR) IN "ADAPTIVE" MUTATIONS AND A TRANSIENT HYPERMUTABLE STATE

MMR is probably the most important mutation-avoidance system that has been conserved across the domains of life. In addition to mutation avoidance, MMR has been linked to controlling genome rearrangements, gene transfer among species, and the development of cancer (Halas et al., 2002).

MMR was first suggested to be involved in "adaptive" or stationary-phase-induced mutagenesis in the study of the *lac* system in *E. coli*. Here, the types of mutants that gave the *lac*+ revertant genotype were almost all −1 deletions in a region of small mononucleotide repeats. This revertant class was not the predominant type found for the classical spontaneous growth-dependent mutations for this gene. However, this was the category of revertants found in cells lacking a functional MMR system.

Additional suggestive evidence for the role of MMR in the transitory hypermutable state associated with stationary-phase-induced mutations is that the levels of MMR proteins decreased during stationary phase. On the other hand, defects in MMR were shown to increase stationary-phase-induced mutations in the *lac* system, indicating that MMR was functioning in the stressed bacteria. These apparently contradictory results could be explained if one assumes that MMR proteins would be normally reduced or nonfunctional only in the cells that differentiated into the hypermutable state. Rosenberg and her colleagues (1998) tested this hypothesis by overproducing components of the MMR system and determining the level of stationary-phase-induced mutations that arose. The results demonstrated that overproduction of MutL decreased stationary-phase-induced mutations, while the overproduction of MutS (the other major component of the MMR) did not suppress this mutation-generation system. MMR has also been found to play a role in the generation of stationary-phase mutations for the yeast and *B. subtilis* (unpublished observations) systems. Although the mechanism appears complex, the data tend to support the hypothesis that MMR is involved in the stationary-phase-induced mutation process and that a completely functional MMR might be absent or saturated with damages in the subset of cells that differentiate into the hypermutable state. The other possibility is that those cells responsible for producing stationary-phase mutations might have accumulated DNA damages (via several different mechanisms, including the activation of the SOS system) that saturate out the repair capacity of the MMR proteins.

Section Summary
- Altering the proficiency of the mismatch correction mechanism or repair system (as well as other repair systems) can lead to an increased mutation frequency.

- This may be one explanation for a hyper-mutable subpopulation of bacteria.

AN ANCIENT FAMILY OF DNA POLYMERASES (Y SUPERFAMILY) AND POTENTIAL ROLES IN STATIONARY-PHASE-INDUCED MUTAGENESIS

The replicative forms of the DNA polymerases, especially those associated with 5′ to 3′ editing functions, have a high degree of fidelity and, because of this fidelity, often have trouble replicating past lesions or bases that do not provide coding information. Thus, most DNA replication is relatively error-free. However, error-prone forms of DNA replication are known to exist. For instance, in *E. coli*, the *dinB* gene is required for bacteriophage λ untargeted mutagenesis (UTM), an error-prone pathway observed when undamaged λ DNA infects SOS-induced *E. coli* cells. Overexpression of the *dinB* gene confers a mutator phenotype on the cells. However, mutations in the *dinB* gene only caused a modest UV-sensitive phenotype, indicating that this gene product might not play a major role in tolerating DNA lesions introduced by UV irradiation into *E. coli*. In 1999 it was discovered that the purified DinB protein has a template-directed, DNA-dependent DNA polymerase activity and was designated as the fourth DNA polymerase in *E. coli* (DNA Pol IV) (Kim et al., 1997; Friedberg et al., 2002; Wagner et al., 1999).

The DNA damage-inducible UmuD′ and UmuC proteins are required for another type of SOS mutagenesis in *E. coli*. The UmuCD-dependent translesion DNA synthesis (TLS) allows cells to replicate past DNA-damage-induced lesions that would normally block the continuing polymerization by the major replication DNA polymerase (DNA Pol III). In 1999, UmuC or UmuD′$_2$C was discovered to be a template-directed, DNA-dependent DNA polymerase, which was designated as the fifth DNA polymerase in *E. coli* (DNA Pol V) (Reuven et al., 1999).

It has recently become apparent that UmuC is the founding member of a superfamily of novel DNA polymerases that can replicate over lesions or operate on particular classes of imperfect DNA templates. Both DinB and UmuC belong to this superfamily which has been designated as the Y family of DNA polymerases. Genetic evidence suggests that DNA polymerase III and DNA polymerase IV (DinB) are responsible for the synthesis of errors that lead to the stationary-phase-induced mutations in the *E. coli* Lac system that has been described (McKenzie and Rosenberg, 2001). More recently, it has been demonstrated that the product of the *yqjH* gene (a member of the Y superfamily) is involved in at least one aspect of stationary-phase-induced mutagenesis in *B. subtilis*. Thus, these data demonstrate a role for these error-prone DNA polymerases in certain types of stationary-phase mutagenesis in at least two model systems.

Members of this superfamily of polymerases have also been linked to or suggested to be involved in a diverse array of phenomena, including the functioning of the immune system and the generation of neoplastic events (Tang et al., 1999).

Section Summary

- There is an ancient superfamily of DNA polymerases that are error prone (they make mistakes during the replication process).
- These polymerases must play an important role in the survival of species otherwise they would not have been conserved during the evolutionary process.

CAN WE SPECULATE ON DIRECTED MUTAGENESIS AND ITS ROLE IN STATIONARY-PHASE-INDUCED MUTAGENESIS?

This chapter has presented an overiew with respect to the phenomenon that was originally called "adaptive" mutagenesis. While primarily studied in *E. coli*, with use of a plasmid-borne *lacZ* gene, there has certainly been evidence that similar phenomena occur in a variety of different organisms spanning the prokaryotic and eukaryotic domains. While debates continue as to the nature of the mutations and the processes responsible for generating the mutations that arise during the stationary phase, there is very little doubt that mutations do develop in metabolically challenged cells. This result alone forces us to question strict adherence to the classical concepts of how and when mutations are produced. The major and related issues that need to be addressed, at least in the our minds, concern (i) genetic control over stationary-phase-induced mutagenesis, (ii) the possibility of a hypermutable subpopulation(s), and (iii) whether or not this mutagenesis is in any way directed.

As mentioned above, most of the research on "adaptive" or "stationary-phase-induced" mutagenesis has focused on the mechanisms responsible for generating the mutations. It is equally important to understand how this mutagenesis is regulated. Basically, is this phenomenon just a stochastic one or is there actual genetic control? Results from at least two different systems have been presented indicating genetic control of certain aspects of stationary-phase-induced mutagenesis. Clearly, additional investigations are required to determine whether stationary-phase-induced mutagenesis is another component of eubacterial differentiation and development. Such differentiation and developmental processes could involve the generation of distinct subpopulations within a stationary-phase culture, the inactivation or damping down of DNA-repair processes, and/or the enhanced activity of systems designed to allow mutations to accumulate. When one contemplates all the possibilities one very important question is brought to the forefront: how does the culture prevent itself from reaching genetic load? That is, how does a culture that allows, and might even genetically foster, stationary-phase-induced mutagenesis prevent itself from accumulating so many mutations that the viability of the cells is threatened? On the surface, the easy answer would be that the culture differentiates off a subpopulation(s) that will experiment with new mutations (hypermutable). Thus, the majority of the cells are protected from reaching genetic load while a percentage of the population is encouraged to

experiment and possibly find a mutation(s) that would allow better survival under the stressful growth conditions that exist. Although this concept definitely involves a great deal of anthropomorphizing, it is not completely removed from reality. Bacterial models for the generation of subpopulations have already been discussed and it is clear the bacteria within a culture can communicate with each other to achieve diversity and unique physiological characteristics.

Another interesting problem regards how the majority of the cells within a culture would know not to become hypermutable. Our knowledge of quorum sensing would provide models for the activation of genes responsible for hypermutability, competence development, etc. However, mechanisms that limit these phenomena to a subpopulation(s) still have yet to be elucidated. Putting aside questions regarding how subpopulation differentiation is controlled, we can deal with the tremendous survival advantage that the existence of a subpopulation or populations offers a bacterial culture that has been placed under stress. Within a given subpopulation, the mismatch repair system could be turned off or significantly repressed. In such a case, alterations that arise in the DNA would be less likely to be repaired and could result in a hypermutable phenotype for the cells. Similarly, multiple subpopulations could exist with each one being transiently defective in one or more DNA-repair systems. This type of mechanism has been proposed to explain certain aspects of stationary-phase-induced mutagenesis. While this hypothesis may very well be correct, the problem of genetic load still exists. If you have a subpopulation that is randomly producing mutations at a higher-than-normal frequency, what is the chance that a "good" mutation will be achieved before a lethal mutation is generated? This conundrum needs to be addressed if a complete understanding of stationary-phase-induced mutagenesis is to be achieved.

In addition to depressing DNA-repair capacity, subpopulations might differentiate to produce more mutations through the presence of greater quantities of proteins involved in permitting tolerance to DNA errors. An example of this type of mechanism has been discussed and involves enhanced production of members of the Y superfamily of DNA polymerases. On the other hand, one could imagine that the depression of the amount of radical scavengers present in stationary-phase cells also would lead to an increased mutation frequency as has been suggested by Bridges (1995, 1998, 2000). Here too, the "genetic load problem" must be taken into account.

One possible approach to the generation of mutations within a hypermutable subpopulation(s) that would not raise the risk of achieving genetic load could involve an aspect of directed mutagenesis; more specifically, a type of directed mutagenesis that is linked to transcription and translational control mechanisms. This type of hypothesis would involve what has been called retromutations, and the concept advocates that the cells are actually testing (transiently through errors in the mRNA) whether specific mutations impart growth advantage before heritable DNA mutations are generated. Such a scheme would implicate directed mutagenesis (since the genes would have to be actively transcribed) in a process that is still very compatible with neo-Darwinian concepts. Extrapolation of this hypothesis to eukaryotic cells could explain aspects of the mutagenic changes that occur in neoplastic transformations. While a "retromutagenesis" theory appears to fit the criteria necessary to explain stationary-phase-induced mutagenesis, significant research needs to be performed to further establish its existence and importance. However, it must be pointed out that the diversity of results obtained while studying "adaptive" or stationary-phase-induced mutagenesis strongly suggests that a variety of molecular mechanisms are involved. Thus, while a retromutagenesis mechanism might be involved, it probably would not be the only mechanism for stationary-phase-induced mutagenesis. In any case, the existence of stationary-phase-induced mutagenesis represents a starting point for fascinating and important investigations into the mechanisms responsible for the mutations that have most probably driven the evolutionary process.

Section Summary
- There are still many questions that need to be answered concerning mutagenesis and its role in the evolutionary process.
- The phenomenon called adaptive or stationary-phase-induced mutagenesis presents a model for investigating the mechanisms associated with the evolutionary process.

"WE ARE STANDING ON THE SHOULDERS OF THE GIANTS WHO CAME BEFORE!"

The results and hypotheses that are presented in this chapter would never have been contemplated without the brilliant and courageous work of two outstanding scientists, Barbara McClintock and Evelyn Witkin. These two researchers, because of their careful and creative experimentation, were able to challenge the existing concepts on DNA stability and DNA-repair processes, respectively. Their collective works pointed to the enhanced diversity that occurs when organisms were placed under stress. Mention must also be made of Bryn Bridges and his key contributions to the study of the systems that control DNA repair and mutagenesis processes. Without his willingness to challenge accepted beliefs we would be struggling to explain some of the most exciting observations that have occurred in these fields. We strongly recommend that students of all ages review the scientific works and reasoning of these individuals as we attempt to gain a real understanding of how organisms respond to environmental stress without risking the problems associated with reaching genetic load.

QUESTIONS

1. An important question for thought is how have error-prone polymerases been utilized by organisms to provide a selective advantage?

2. What would be the evolutionary benefits of allowing an RNA intermediate in the process of generating mutations? Why would RNA and RNA polymerase be better suited for this type of experimentation than DNA and DNA polymerases?

3. What are the survival benefits of having a heterogeneous population of cells during the stationary phase of growth for bacteria?

4. Which would be more appropriate for generating the mutations necessary for the evolutionary process, mutagenesis during active periods of growth or stationary-phase mutagenesis? Why? How could you test your conclusion?

ACKNOWLEDGMENTS

This work was supported by grant 31767-N from the Consejo Nacional de Ciencia y Tecnología (CONACYT) of México to M.P.-R.; R.E.Y. was supported by grants MCB-9975140 and MCB-0317076 from the National Science Foundation.

REFERENCES

Bridges, B. 2000. DNA polymerases and SOS mutagenesis: can one reconcile the biochemical and genetic data? *BioEssays* **22:**933–937.

Bridges, B. A. 1995. *mutY* 'directs' mutation? *Nature* **375:**741.

Bridges, B. A. 1998. The role of DNA damage in stationary phase ('adaptive') mutation. *Mutat. Res.* **408:**1–9.

Bull, H. J., G. J. McKenzie, P. J. Hastings, and S. M. Rosenberg. 2000. Response to John Cairns: the contribution of transiently hypermutable cells to mutation in stationary phase. *Genetics* **156:**925–926.

Cairns, J. 2000. The contribution of bacterial hypermutators to mutation in stationary phase. *Genetics* **156:**923.

Cairns, J., J. Overbaugh, and S. Miller. 1988. The origin of mutants. *Nature* **335:**142–145.

Dubnau, D. 1991. Genetic competence in *Bacillus subtilis. Microbiol. Rev.* **55:**395–424.

Dubnau, D., and C. M. J. Lovett. 2002. Transformation and recombination, p. 473–482. *In* A. L. Sonenshein, J. A. Hoch, and R. Losick (ed.), *Bacillus subtilis and Its Closest Relative.* American Society for Microbiology, Washington, D.C.

Foster, P. L., and J. M. Trimarchi. 1994. Adaptive reversion of a frameshift mutation in *Escherichia coli* by simple base deletions in homopolymeric runs. *Science* **265:**407–409.

Friedberg, E. C., R. Wagner, and M. Radman. 2002. Specialized DNA Polymerases, cellular survival, and the genesis of mutations. *Science* **296:**1627–1630.

Galitski, T., and J. R. Roth. 1995. Evidence that F plasmid transfer replication underlies apparent adaptive mutation. *Science* **268:**421–423.

Gómez-Gómez, J., J. Blázquez, F. Baquero, and J. Martinez. 1997. H-NS and RpoS regulate emergence of Lac Ara+ mutants of *Escherichia coli* MCS2. *J. Bacteriol.* **179:**4620–4622.

Graham, J. B., and C. A. Istock. 1978. Genetic exchange in *Bacillus subtilis* in soil. *Mol. Gen. Genet.* **166:**287–290.

Halas, A., H. Baranowska, and Z. Policinska. 2002. The influence of the mismatch-repair system on stationary-phase mutagenesis in the yeast *Saccharomyces cerevisiae. Curr. Genet.* **42:**140–146.

Harris, R. S., H. J. Bull, and S. M. Rosenberg. 1997. A direct role for DNA polymerase III in adaptive reversion of a frameshift mutation in *Escherichia coli. Mutat. Res.* **375:**19–24.

Holmquist, G. P. 2002. Cell-selfish modes of evolution and mutation directed after transcriptional bypass. *Mutat. Res.* **510:**141–152.

Kasak, L., R. Horak, and M. Kivisaar. 1997. Promoter-creating mutations in Pseudomonas putida: a model system for the study of mutation in starving bacteria. *Proc. Natl. Acad. Sci. USA* **94:**3134–3139.

Kim, S. R., G. Maenhaut-Michel, M. Yamada, Y. Yamamoto, K. Matsui, T. Sofuni, T. Nohmi, and H. Ohmori. 1997. Multiple pathways for SOS-induced mutagenesis in *Escherichia coli*: an overexpression of *dinB/dinP* results in strongly enhancing mutagenesis in the absence of any exogenous treatment to damage DNA. *Proc. Natl. Acad. Sci. USA* **94:**13792–13797.

Makinoshima, H., A. Nishimura, and A. Ishihama. 2002. Fractionation of *Escherichia coli* cell populations at different stages during growth transition to stationary phase. *Mol. Microbiol.* **43:**269–279.

McKenzie, G. J., and S. M. Rosenberg. 2001. Adaptive mutations, mutator DNA polymerases and genetic change strategies of pathogens. *Curr. Opin. Microbiol.* **4:**586–594.

Msadek, T. 1999. When the going gets tough: survival strategies and environmental signaling networks in *Bacillus subtilis. Trends Microbiol.* **7:**201–207.

Reuven, N. B., G. Arad, A. Maor-Shoshani, and Z. Livneh. 1999. The mutagenesis protein UmuC is a DNA polymerase activated by UmuD', RecA, and SSB and is specialized for translesion replication. *J. Biol. Chem.* **274:**31763–31766.

Rosenberg, S. M., C. Thulin, and R. S. Harris. 1998. Transient and heritable mutators in adaptive evolution in the lab and in nature. *Genetics* **148:** 1559–1566.

Rudner, R., A. Murray, and N. Huda. 1999. Is there a link between mutation rates and the stringent response in *Bacillus subtilis*? *Ann. N. Y. Acad. Sci.* **870:**418–422.

Selby, C. P., and A. Sancar. 1993. Molecular mechanism of transcription-repair coupling. *Science* **260:**53–58.

Slechta, E. S., J. Harold, D. I. Andersson, and J. R. Roth. 2002. The effect of genomic position on reversion of a lac frameshift mutation (*lacIZ33*) during non-lethal selection (adaptive mutation). *Mol. Microbiol.* **44:**1017–1032.

Sung, H.-M., and R. E. Yasbin. 2002. Adaptive, or stationary-phase, mutagenesis, a component of bacterial differentiation in *Bacillus subtilis*. *J. Bacteriol.* **184:**5641–5653.

Tang, M., X. Shen, E. G. Frank, M. O'Donnell, R. Woodgate, and M. F. Goodman. 1999. UmuD'(2)C is an error-prone DNA polymerase, *Escherichia coli* pol V. *Proc. Natl. Acad. Sci. USA* **96:**8919–8924.

Torkelson, J., R. S. Harris, M. J. Lombardo, J. Nagendran, C. Thulin, and S. M. Rosenberg. 1997. Genome-wide hypermutation in a subpopulation of stationary-phase cells underlies recombination-dependent adaptive mutation. *EMBO J.* **16:** 3303–3311.

Wagner, J., P. Gruz, S. R. Kim, M. Yamada, K. Matsui, R. P. Fuchs, and T. Nohmi. 1999. The *dinB* gene encodes a novel *E. coli* DNA polymerase, DNA pol IV, involved in mutagenesis. *Mol. Cell* **4:**281–286.

Wright, B. E. 2000. A biochemical mechanism for nonrandom mutations and evolution. *J. Bacteriol.* **182:**2993–3001.

Zalieckas, J. M., L. V. Wray, Jr., A. E. Ferson, and S. H. Fisher. 1998. Transcription-repair coupling factor is involved in carbon catabolite repression of the *Bacillus subtilis* hut and gnt operons. *Mol. Microbiol.* **27:**1031–1038.

FURTHER READING

Cairns, J., J. Overbaugh, and S. Miller. 1988. The origin of mutants. *Nature* **335:**142–145.

Dobzhansky, T. 1950. The genetic basis of evolution. *Sci. Am.* **182:**32–41.

Grossman, A. D. 1995. Genetic networks controlling the initiation of sporulation and the development of genetic competence in *Bacillus subtilis*. *Annu. Rev. Genet.* **29:**477–508.

Luria, S. E., and M. Delbrück. 1943. Mutations of bacteria from virus sensitive to virus resistance. *Genetics* **28:**491–511.

Modrich, P., and R. Lahue. 1996. Mismatch repair in replication fidelity, genetic recombination and cancer. *Annu. Rev. Biochem.* **65:**101–133.

Newcomb, H. B. 1949. Origin of bacterial variants. *Nature* **164:**150.

Shapiro, J. A. 1998. Thinking about bacterial populations as multicellular organisms. *Annu. Rev. Microbiol.* **52:**81–104.

Sutton, M. D., B. T. Smith, V. G. Godoy, and G. C. Walker. 2000. The SOS response: recent insights into umuDC-dependent mutagenesis and DNA damage tolerance. *Annu. Rev. Genet.* **34:**479–399.

Walker, G. C. 1987. The SOS response of *E. coli*, p. 1346–1357. *In* F. C. Neidhardt (ed.), *Escherichia coli and Salmonella typhimurium*. American Society for Microbiology, Washington, D.C.

Witkin, E. M. 1976. Ultraviolet mutagenesis and inducible DNA repair in *Escherichia coli*. *Bacteriol. Rev.* **40:**869–907.

TEMPORAL SEGREGATION: SUCCESSION IN BIOFILMS

Susse Kirkelund Hansen and Søren Molin

13

It is often claimed that essentially all bacteria in the environment are associated with surfaces, and, consequently, investigations of suspended cells may tell little about bacterial performance in the environment. It is questionable if this statement is correct considering

Susse Kirkelund Hansen and Søren Molin, Center for Biomedical Microbiology, BioCentrum-DTU, Building 301, DTU, 2800 Lyngby, Denmark.

the vast number of bacteria living as planktonic cells in the oceans. However, it is most likely true that the majority of bacterial *activity* is derived from surface-bound communities of bacteria. Therefore, when we assess the diversity of bacteria, the complexity of their gene expression and regulation, and the processes of selection in the evolution of bacteria, we must at some point address these in relation to their sessile mode of life. We must learn how bacteria sense and respond to the complex conditions found in surface-associated communities as they exploit their repertoire of behavioral activities. In particular, the environmental challenges in community life may exceed their individual cellular and physiological abilities. How does their genomic plasticity result in a rapid and effective capability to transcend this challenge? We know a lot about their metabolic diversity and what allows them to switch easily between available nutrients. We also know that intricate and highly effective genetic control systems ensure that this metabolic diversity is optimally exploited by the organisms. Thus, the cells direct their resources into proliferation by down-regulating activities which are not directly necessary under the prevailing conditions. What we do not know is to what extent this metabolic and regulatory flexibility is sufficient to meet the rapidly changing environmental conditions during the development of the surface communities. We also have very limited knowledge about the local and general selective forces operating within biofilm populations, and how these are

Microbial Evolution: Gene Establishment, Survival, and Exchange
Edited by Robert V. Miller and Martin J. Day, ©2004 ASM Press, Washington, D.C.

influencing selection for genetic variations at different levels among the population members. By posing these questions in the context of complex surface-bound bacterial communities, a biofilm, rather than in a monospecies test-tube culture we have a chance to understand how bacterial diversity and adaptability have evolved and keep evolving.

THE BACTERIAL BIOFILM: A PLATFORM TO STUDY ADAPTATION AND EVOLUTION

Biofilm Development

Biofilms are bacterial surface-associated communities attached to solid substrata, growing into a nutrient-containing water phase and embedded in a polymer matrix produced by the bacteria. During the past 10 years or so there has been a rapidly increasing interest in bacterial biofilms. There are several reasons for this interest. First, as emphasized above, active bacterial life outside the laboratory is predominantly associated with biofilms. Second, there are now excellent tools for detailed on-line studies of heterogeneous populations. Finally, the development of multicellular surface populations may share traits with developmental processes common in higher eukaryotic organisms. Recognition of the latter has attracted a large

amount of interest. The challenge is to reveal the molecular mechanism(s) behind the apparently regulated processes which result in the development of structured and multifunctional microbial communities.

Modeling biofilm development has been investigated in a few cases. It is fair to state that, despite the obvious limitations of a few studies, the simplicity of the experimental scenarios and the small number of model organisms the conclusions derived have been very useful and have inspired the field. Pseudomonads and enterobacteria, which are heterotrophic, fast-growing organisms, have been studied in flow-chamber microcosms (Fig. 1). The picture emerging describes a stepwise differentiation process in the development of a biofilm.

1. **Reversible attachment** of planktonic cells to a substratum is based on a combination of random collisions, electrostatic forces, active movement, nutrient attractants bound to the surface, and other factors. This phase may extend from a very short period (seconds) to become a more-or-less permanent state. At any time the attached cells may leave the surface to once again become planktonic or to colonize an alternative location.

2. **Irreversible adherence** of the attached cells is established by cell-surface components,

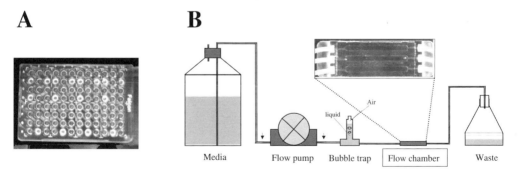

A   **B**

Media   Flow pump   Bubble trap   Flow chamber   Waste

Air
liquid

FIGURE 1   Laboratory biofilm setups. (A) Microtiter plate with biofilms formed in the wells. Visualization of the biofilms is performed after removal of the culture liquid and staining of the attached biomass with crystal violet. (B) The flow-cell biofilm setup as used in the authors' laboratory. The medium containing nutrients for the biofilm is pumped from the reservoir vessel through bubble traps catching air bubbles to the inlet of the flow chamber. The medium runs through the flow-chamber channels (compare blowup), and biofilm is formed on a glass coverslip glued onto the flow chamber. Medium and nonattached cells are finally collected as an effluent in a waste flask. The flow chambers can be placed directly under the microscope for in situ observations.

such as adhesive polymers gluing the cells to the surface. In this phase the cells often lose the ability to move away from the surface, and water movement with or without surfactants or other agents has only marginal effects on loosening the attached cells.

3. Cell proliferation occurs if the water flow contains the necessary nutrients. Initially, growth often occurs as microcolonies as seen on an agar plate. The rate of colony growth is determined by the flow rate and the nutrient concentration in the water phase. In this phase biofilm heterogeneity is developed, as the subsurface will soon be shielded from the stream of nutrients.

4. Biofilm maturation refers to the process that eventually results in the development of a highly structured biofilm in which the biomass is distributed in domains of high cell densities alternating with voids through which water flows. Such biomass domains may extend hundreds of micrometers into the water phase. These are termed variously as "mushrooms," "towers," and "pillars." Internal cavities may be essentially cell-free. During the process of maturation large amounts of extracellular polymeric substances (EPS) may have been excreted from the cells to produce a matrix embedding all or most of the biofilm biomass. The mature biofilm is typically very resistant to most types of antibacterial treatments, including antibiotic addition, and even very harsh treatments aimed at removing the biofilms often fail.

5. Disintegration of the biofilm may happen mechanically by shear forces found in high flow rates or in turbulent conditions. This induces sloughing of large sections of the mature biofilm. Alternatively, sloughing can happen as a result of a conditionally induced change of behavior in one or more subpopulations within the biofilm community. This results in an apparently coordinated removal of large numbers of cells from the biofilm.

Figure 2 visualizes this standardized development scheme.

In both in situ and laboratory experiments biofilm development may follow a route quite different from the one described above. Often the same strain will display different biofilm developmental patterns when the nutritional or physical conditions are varied. Despite the apparent diversity of biofilm development schemes, it is often assumed that the organisms have developed complex regulatory programs to coordinate this process in well-controlled ways. Here we will assume that complexity is the consequence of biofilm heterogeneity, and not the cause. Thus, in this simplified picture we envisage biofilm development to be a self-assembly process, influenced by the external conditions and by a number of genetically determined properties of the biofilm community. As a consequence of this progressive development, the community organisms will become components in an increasingly complex number of heterogeneous niches.

Section Summary

- Biofilms are dominant settings for bacteria in nature, and they may be studied under controlled conditions in the laboratory.
- Biofilm development may be described as a stepwise multicellular differentiation process.
- Structured communities, such as biofilms, create heterogeneous conditions and allow subpopulation development in niches.

ADAPTIVE STRATEGIES IN SURFACE COLONIZATION: GENE REGULATION

The extraordinary capacity of bacteria to coordinate different parts of their biochemical repertoire is evident in biofilm development. The current ideas concerning the individual steps in biofilm development are discussed.

Surface Attachment

The attachment of bacteria to a solid surface has been studied in many ways. Recently, a useful platform for these studies has been a simplified biofilm setup, the microtiter dish array (Fig. 1). Suspended cells are allowed to settle and grow for some time in the wells of the dish, and, after removal of nonattached cells, the attached

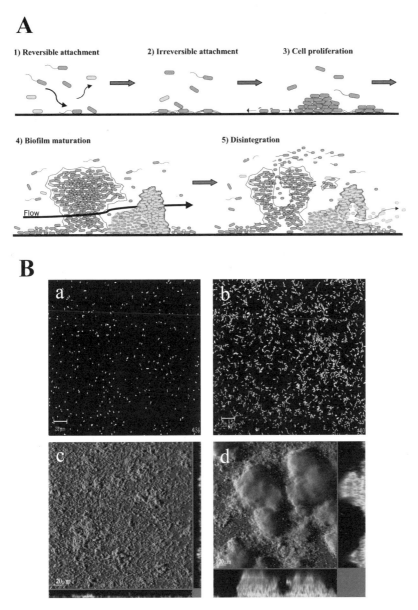

FIGURE 2 Biofilm development. (A) A schematic model of biofilm formation on solid/liquid interfaces. The individual steps in the process are described in the text. The different shapes, greytones and shadings of the microcolonies symbolize the heterogeneity of cell physiology and EPS matrix composition. Planktonic swimming cells are indicated by the presence of polar flagella on the cells, twitching motility is symbolized by surface-attached cells with arrows, and, in all cases, the flow direction is from left to right as indicated in step 4. (B) Biofilm development monitored by scanning confocal laser microscopy (SCLM). Cells of *E. coli* were grown in a flow chamber (see Fig. 1), and the four frames show how the different steps indicated in part A develop over time. In panels a and b, the biofilm is viewed from above, showing the cells attached to the glass surface (a) and the beginning of cell proliferation (b). In panels c and d, the presentations are shadow projections from above showing further cell proliferation (c) and biofilm maturation (d). In the side frames of panels c and d, the biofilms are viewed from the side to indicate the height of the microcolonies. This series of images only represents the first four steps of the model presented in part A. (These images were kindly provided by Janus Haagensen, BioCentrum-DTU.)

biomass adhering to the walls of the wells may be determined by simple crystal violet staining. This format is compatible with analysis of mutant libraries and particularly useful for identification of clones affected in their adhesion to the plastic walls. Factors involved in attachment are surface appendages such as fimbriae, flagella, and curli (Prigent-Combaret et al., 2000). However, none seem to be indispensable, since if a primary attachment factor is missing, alternatives may substitute. If these are not equally effective then mutations may create alternatives. The presence of a noncolonized surface may be considered an available niche for an organism to occupy. Therefore, this niche will be invaded, either by variants in the population or by new organisms. There is evidence that during initial attachment specific genes are induced by the contact through signal transduction, and it is therefore likely that already in this phase the primary colonizing cells express a number of genes which are otherwise silent or poorly expressed in nonattached cells. In a proteomic investigation of biofilm-specific gene induction in *Pseudomonas aeruginosa* several surface-attachment biofilm-specific proteins are synthesized at the time of attachment (Sauer et al., 2002). It is conceivable that the cell-to-cell contact established in microcolonies or during colonization on the substratum may elicit specific gene expression responses, which assist the cells in further development of the biofilm. In a strain of *Pseudomonas fluorescens* there are indications that such cell-to-cell contact induces cellulose polymer production and excretion from the cell poles. This polymer is involved in gluing the cells together, allowing biofilm development.

Biofilm Development

In some ways, the early stages of biofilm growth and development may be compared with the traditional growth curve (Fig. 2). After a lag phase allowing adaptation, the newly attached bacteria begin to grow, and, if the availability of nutrients is not limiting, a phase of exponential growth may be observed. After this initial period of biomass buildup, during which

colonies often appear on the substratum, there is a phase of stationary or slow growth. Due to the difference between a homogeneously dispersed cell culture and a highly structured surface-attached population, the parallel now breaks down. Nevertheless, we may expect that the changes in cellular physiology observed during the normal growth curve, such as shifts in gene expression pattern and in general macromolecular synthesis, i.e., of DNA, RNA, and proteins, will also occur in the course of biofilm development. We would also expect to find cells in different stages of biofilm development, ranging from planktonic, free swimming cells to more sessile cells in late stationary phase. Global gene expression analysis, focusing on the patterns of protein expression (two-dimensional gel electrophoresis) or transcription (DNA microarrays) are being performed on biofilm populations. These investigations aim to understand the developmental process of biofilm formation. So far, no clear consensus picture has emerged, which is most likely due to the heterogeneous nature of global gene expression in biofilms. Additionally, these studies cannot distinguish between simple gene-regulation-induced responses and other more fundamental genetic changes, which in some cases may totally obscure the interpretations made from any such phenotypic analysis. Detailed examinations of the individual cells in such conditions will eventually reveal how the cellular phenotypic properties are distributed in time and space. Then a comparison of such responses within a population will provide information about adaptation through gene expression in response to changing environmental conditions.

Biofilm Dynamics

Many bacterial species are nonmotile and yet form biofilms which resemble those developed by motile bacteria. Motility, where relevant, is nevertheless considered to be an important factor in biofilm development, both in the initial step of attachment (planktonic cells swimming toward the surface) and in subsequent phases of the process. Swimming motility allows the cells to change position in a surface community, and, in fact, significant fractions of surface-associated populations are often found swimming in the liquid surroundings of the matrix-embedded cell clusters of the biofilm. Twitching motility, movements on surfaces mediated by specialized pili, has been implied in formation of microcolonies by providing the cells with a capacity to aggregate. There are, however, also indications that it may provide for colony dissociation and cellular dispersal from colonies (Heydorn et al., 2002). In later stages of biofilm development it is probably a significant feature that the cells are able to leave their location, one of exhausted resources, and swim toward new sites for colonization. Since there is evidence that bacteria may gradually loose their flagella in the course of biofilm development (Garrett et al., 1999), it means that signals are needed to induce their resynthesis before cells may leave the biofilm. Swarming or twitching motility on surfaces is a population expansion mechanism. Cells act in a highly coordinated manner to form a two-dimensional biofilm, allowing them to rapidly occupy available territories before competitors. Finally, the interplay between motility and the chemosensory apparatus results in various types of chemotaxis and is an essential prerequisite for bacteria to position themselves in an environment (Nielsen et al., 2000).

In the course of biofilm development the resident bacteria become surrounded by extracellular polymeric substances (EPS), which help create and stabilize the architecture of the biofilm to resist adverse impacts from water flow. The excreted EPS contains substituted and unsubstituted polysaccharides, proteins, nucleic acids, and other materials, and the matrix may provide protection from harmful chemicals such as toxins and antibiotics and from predation by eukaryotic organisms. In addition, EPS can trap and store nutrients and provides close cell-to-cell interactions at high cell densities. This results in the formation of gradients of nutrients inside and around the microcolonies, resulting in metabolic differentiation (Fig. 3). Therefore, the growth of microcolonies will automatically result in the creation of different

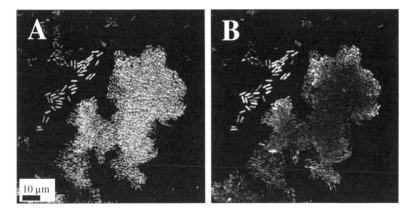

FIGURE 3 Spatial heterogeneity in biofilms. Section through microcolony in flow-chamber biofilm of *P. putida* with an inserted gene fusion between an *rrnB* promoter from *E. coli* and a gene encoding an unstable Gfp fluorescent protein. The two frames show the same microcolony with two different filter sets on the confocal microscope. (A) Staining with a ribosomal probe targeting rRNA in the cells. All cells light up with more-or-less the same intensity, indicating that the cells are distributed throughout the microcolony. (B) Gfp fluorescence expressed in the microcolony. Cells on the surface of the colony express strong green fluorescence, whereas those in the inner parts of the colony are nonfluorescent or show weak fluorescence. The use of an unstable Gfp reporter makes it possible to distinguish between growing and nongrowing cells. (Reprinted with permission from Molin et al., 2000.)

physiological subpopulations according to their position relative to the nutrients in the water phase. EPS may also inhibit cellular motility resulting in a further developed heterogeneity in the population, in which substantial subpopulations most likely suffer from various stress symptoms (nutrient deprivation, pH stress, and osmotic stress). Under such conditions the cellular gene-regulatory repertoire is seriously challenged, and although many bacteria possess a surprising flexibility and adaptability allowing them to prevail under quite extreme conditions, this may not be sufficient to secure their persistence in the long run. To overcome this problem of decreased persistence potential, one strategy could be to disintegrate all or part of the biofilm by inducing a coordinated escape mechanism. The re-formation of flagella, mentioned previously, enables cells to swim away, and the dissolution of the EPS matrix by means of specific extracellular degradation enzymes may be necessary before the cells can be released. This picture of population heterogeneity points toward a phenotypic diversity, even in simple monospecies biofilms. This resembles the phenotypic diversities seen in more complex multispecies populations. Just as we can increase our understanding of the life of complex microbial communities from investigations of controlled monospecies biofilms, we may learn about monospecies biofilms from investigations of multispecies communities. An example of the latter is presented below, in which physiological adaptation to biofilm life of two different species competing for limiting nutrients is described (Fig. 4).

One strain of *Pseudomonas putida* (R1) and one strain of an *Acinetobacter* sp., isolated from an aquifer in Denmark, were investigated in biofilms supplied with benzyl alcohol as the only carbon and energy source. Both organisms were capable of mineralizing this aromatic compound. In *P. putida* the conversion of benzyl alcohol to benzoate was fairly slow, but benzoate was rapidly metabolized to tricarboxylic acid cycle intermediates, whereas in the *Acinetobacter* sp. the conversion of benzyl alcohol to benzoate was very fast, but benzoate

was slowly metabolized, resulting in a continuous release of benzoate to the environment from these cells (Fig. 4A) (Christensen et al., 1998).

When growing together in a biofilm on benzyl alcohol the two organisms interacted in a commensal relationship, and in large parts of the biofilm the *P. putida* cells were found tightly associated with *Acinetobacter* sp. colonies (Fig. 4B). It was further shown that these associated *P. putida* cells were more growth active than nonassociated cells. Thus, based on their metabolic and regulatory properties *P. putida* cells have oriented themselves in the biofilm and optimized their environment to allow their persistence, despite their competitive disadvantage relative to the *Acinetobacter* sp. cells when competing for the primary carbon source. When another strain of *P. putida* (KT2440, derivative of mt-2) lacking the capacity to perform the conversion of benzyl alcohol to benzoate (but which was still an efficient benzoate metabolizer) (Fig. 4C) was substituted for the R1 strain, a very different picture emerged. It was expected that this second strain of *P. putida* would show an even more pronounced association with *Acinetobacter* sp. in the biofilm. However the opposite was observed. The *P. putida* KT2440 cells were nearly always located at some distance from the *Acinetobacter* sp. colonies, even though they depended on the benzoate released from the latter for growth (Fig. 4D). The explanation was that this strain of *P. putida* was adversely affected by a low oxygen concentration caused by the aromatic degradation by oxidative processes in *Acinetobacter*. Thus, for *P. putida* the benefit of released benzoate was counteracted by induced starvation for oxygen. The continued persistence of *P. putida* occurred due to selection of new variants being less affected by the negative oxygen gradient (Fig. 4E). These variants appeared reproducibly and with high frequency and they were stable mutants. In fact, stress-induced generation of genetic variants with subsequent selection of adapted mutants could be an important strategy for bacteria adapting to structured environments such as biofilms (discussed in "Genetic diversification in biofilm populations").

Community development and organization in biofilms is partially determined by the composition of the medium and the respective metabolic properties of the inhabitants. Moreover, the potential to reorganize the population in response to the nutrient gradients arising from growth distribution and the chemotactic repertoire has a significant impact on the eventual distribution of biomass. Severe competitive disadvantages may be overcome through repositioning in the heterogeneous three-dimensional space of a biofilm, and this may explain how multiple species live together despite limited supplies of nutrients. The discovery of population-based gene regulation through cell signaling explains how a population may coordinate phenotypic activities in a synchronous fashion (see chapter 15). The possibility that this mode of cell-to-cell communication for some bacteria can be involved in the process of biofilm development has been suggested. The involvement of population-density-based signaling in the regulation of behavior may also be advantageous in connection with other characteristics of biofilm communities. However, there are reasons to assume that new internal ecological niches develop continuously in complex biofilm communities, which may only be effectively occupied if new variants appear, i.e., selective forces wax and wane over time and constitute changing platforms for the selection of new traits in the population.

Section Summary

- The major cause of biofilm physiological complexity is the development of diverse niches in the community.
- The heterogeneous environment challenges the metabolic and regulatory repertoire of the indigenous populations.
- Growth activity is not the only selective criterion for persistence and competition.
- The development of free niches in biofilms provides new possibilities for incoming and resident organisms.

ADAPTIVE STRATEGIES IN SURFACE COLONIZATION: PHASE VARIATION

When bacteria adhere, grow, and differentiate into mature biofilm communities, they adapt to changes in their environment by modulating gene expression in response to external stimuli and simultaneously making changes to parts of the same environment. Surface-attached bacteria not only have to cope with an array of different biofilm environments, but they often have to respond to sudden shifts in these conditions. For many pathogenic bacteria, a recurring part of their life cycle is that they need to survive in conditions that vary between a low-nutrient external aquatic life and the rich conditions found on infecting a host. These dramatic changes in growth conditions certainly impose severe demands on the gene-regulation capacity of the bacteria. Many microbial pathogens, such as pseudomonads, *Escherichia coli*, and *Vibrio cholerae*, are successful inhabitants of both natural environments and a range of higher organism hosts. They belong to groups of bacteria with large genomes and, consequently, very developed genetic regulatory systems. There are, however, alternative strategies to classical gene regulation that allow for rapid adaptation to changing environments. The ability for these changes is rooted in the genome structure itself in the sense that they involve genomic changes, which may or may not lead to evolutionary developments in the population. One such strategy is phase variation.

Phase Variation: a Genetic Switch

The structural dynamics of bacterial genomes, caused by frequent recombination within and between chromosomes, have made extrapolations about cellular responses to physiological challenges from single cells to populations very difficult. Phase variation is a genetic switch that results in subpopulations with different properties, despite the isogenic nature of the entire population. Phase variation often involves genetic alterations or rearrangements at specific sites in the genome, and it normally occurs

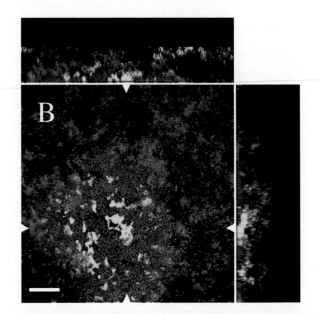

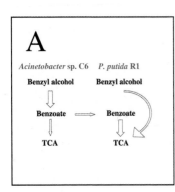

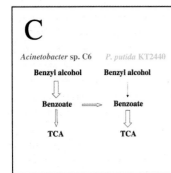

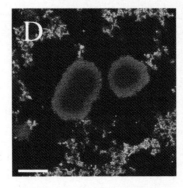

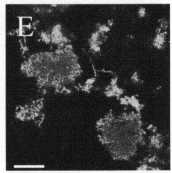

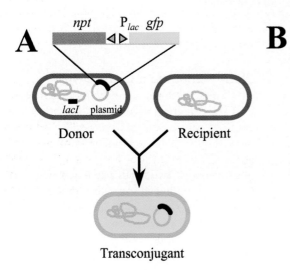

with a fairly high frequency ($>10^{-5}$ per generation) (Henderson et al., 1999). Typical examples of phase variation are site-specific inversion of DNA, leading to the reversible ON $\leftrightarrow$ OFF switching of gene expression, and RecA-dependent homologous recombination, often leading to deletion and ON $\rightarrow$ OFF switching. Gene inactivation may also be caused by reversible insertions of insertion sequences. Transient slipped-strand mispairing during DNA replication involves hot-spot genomic regions with short sequence repeats (SSR), leading to changes in the translational reading frame. High-frequency ON $\leftrightarrow$ OFF switching of phenotype expression creates subpopulations, which may already be expressing the phenotype necessary for survival in a new environment, and such variants will be rapidly selected for if this particular environment is encountered. A population using classical gene-regulatory mechanisms, where one or more genes are either ON (in case of inducing conditions) or OFF (in case of noninducing/ repression conditions), optimizes its genetic expression profile for all cells to match the prevailing conditions. In a stable and relatively homogeneous environment this strategy is appropriate and rational. Phase variation, in contrast, constitutes a strategy where a significant part of the population transiently displays a less optimal gene-expression strategy, in relation to the local conditions, and thus is exposed to the risk of being out-competed. However, if adjacent niches constitute better conditions for this subpopulation, then it is ready and fit to invade and establish itself. Having even a few traits under the control of phase variation ensures a high degree of genetic variation in the population. This can easily be exemplified; six genes under the control of phase variation generate the possibility of $2^6 = 128$ different phenotypes. Thus, phase variation offers colonization diversity and opportunities to some individuals within the population despite the general disadvantage. In microbial biofilms phase variation may be used as a mechanism to allow both for the colonization of surfaces and for the subsequent adaptation to the increasing diversity of niches developing in biofilm communities.

FIGURE 4   Two-species consortia growing as biofilms in flow chambers on a minimal buffered medium with benzyl alcohol as the only carbon and energy source. See text for details. (A) Schematic overview of benzyl alcohol degradation in the biofilm consortium shown in panel B. The arrow size in each step illustrates the degradation rate and the carbon flow between the two consortium members. (B) The two strains are *Acinetobacter* sp. C6 (red cells) and *P. putida* R1 (blue or turquoise cells). Growth-active cells of *P. putida* R1 are turquoise (blue + green) due to the inserted activity reporter cassette containing a ribosomal RNA promoter in front of a gene encoding unstable Gfp protein (compare legend to Fig. 3). The side-frames are *z*-sections of the biofilm in the positions indicated with white arrows. (C) As in panel A, but valid for the consortium shown in panel D. Note that *P. putida* KT2440 scarcely degrades benzyl alcohol and therefore is completely dependent on benzoate excreted from the *Acinetobacter* strain. (D) Section of the biofilm containing the strains of the *Acinetobacter* sp. C6 (red cells) and *P. putida* KT2440 (green cells), a derivative of *P. putida* mt-2. (E) The interactive coupling between *Acinetobacter* sp. C6 and a mutant selected in the panel D consortium. In panel B, *Acinetobacter* sp. C6 and *P. putida* are identified by fluorescence in situ rRNA hybridization using specific probes targeting the organisms. In panels C and D, the biofilm is stained with cyto62 (red), and additionally, *P. putida* was identified by Gfp expression.

FIGURE 8   In situ monitoring of plasmid transfer. (A) Design of a mating-pair system. The donor strain harbors a conjugative plasmid with an inserted fusion between a *lac* promoter and a promoterless *gfp* gene. This strain also has a chromosomal insert of the *lacI* gene expressing a repressor, which prevents expression of Gfp from the plasmid. The recipient strain has no repressor gene. After plasmid transfer the transconjugants will express Gfp (zygotic induction, no repression of the *lac* promoter), thus allowing identification of these as green fluorescent cells. (B) Plasmid transfer in a flow-chamber biofilm. In this demonstration of the mating-pair system described in A, the donor and recipient cells were both *P. putida*. The donor strain harbored the pWWO plasmid (TOL), which is highly efficient for transfer between pseudomonads. Transconjugant cells are green/yellow in the image, whereas no distinction between donors and recipients (all red) can be made in this case. (Reprinted with permission from Molin et al., 2000, and Christensen et al., 1998.)

Phase Variation in Surface Colonization

The development of bacterial biofilms includes several features, which a priori seem like very good targets for phase variation. Cellular adhesion to surfaces, mediated by cell-surface components, constitutes the first step in the process. By keeping surface components or adhesion factors under the control of phase variation, subpopulations are ready to adhere and these will rapidly be selected for. During the growth and differentiation of the surface-associated biomass, it is thought that cell-to-cell adherence may be a key determinant involved in shaping and stabilizing the biofilm structure. It is sufficient that just a part of the population expresses the cell-to-cell adherence factors. Thus, phase variation could easily fulfill this role while the other regulatory systems (i.e., induction and repression) are perhaps more optimal in providing for biomass increase (i.e., the so-called household activities). The next key phase of biofilm maturation is the production of EPS, which eventually provides stabilization and protection from various environmental factors and attacks to the biofilm community. We predict that a distribution of functionality may be beneficial to the population since there is probably no need for all cells in the surface-associated biofilm to be EPS producers. Thus, phase variation control of EPS synthesis could be a rational strategy to produce EPS without letting synthesis be a burden to all individuals in the pop-

ulation. Phase variation in the expression of cell-surface structural components is often easy to detect by screening, since the components frequently give rise to changes in aggregation in liquid culture and colony morphology on plates (Fig. 5). Numerous examples of phase variation involving surface structures in gram-negative bacteria, such as fimbriae, flagella, outer membrane and extracellular proteins, adhesins, lipopolysaccharides (LPS), and extracellular polysaccharides have been described in the literature (Sanchez-Contreras et al., 2002). Some are involved in animal pathogenesis (including biofilm formation in host environments), while others are directly involved in surface colonization in aquatic, marine, and soil environments. In the following examples phase variation is involved in activities relating to complex community lifestyles.

Examples of Phase Variation

E. COLI

One of the best-studied examples of site-specific DNA inversion involves the phase-variable type 1 fimbriae in E. coli, which are important for early steps in biofilm formation in host infection. Phase variation controls expression in several other fimbrial systems found in gram-negative bacteria. Type 1 fimbriae carry adhesion proteins, which enable the bacteria to stick firmly to surfaces comprising mannose residues. However, these fimbriae create problems for such

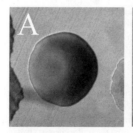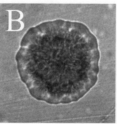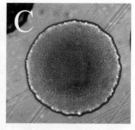

FIGURE 5 Variants with altered cell surface or surface components often give rise to different colony morphologies. Three different colony types of *P. putida* are shown here. (A) The normal smooth wild-type colony growing on a broth agar surface. (B) A variant derived from biofilms; note the wrinkly appearance of the colony. (C) The colony is yet another morphotype, also isolated from a *P. putida* biofilm.

cells penetrating viscous layers, like the mucosa in the large intestine. So in a population of *E. coli* one part is fimbriated and the other is not due to phase variation of fimbrial expression. Thus, there will always be nonfimbriated cells present, which may readily establish in the gut flora mucosa biofilm, and there will be fimbriated cells, which can adhere to epithelial surfaces of the urinary tract. The frequency of switching the fimbriae production "ON" decreases with temperature such that *E. coli* cells in the colder external environment do not express these specific adhesion factors. Hence, cells expressing the host-specific adhesion factors become less frequent outside the 37°C warm environment of the host organism. Phase variation mechanisms are described as essentially random events that generate various combinations of phenotypic variants. The type 1 fimbriae example is therefore interesting, since the directionality of inversion is influenced by temperature and the medium composition. Other examples like this are changing our view of phase variation, since their switch frequencies are conditioned by environmental factors. This modulates the overall adaptability and fitness of a population, and can lead to expression of preferential phenotypes in relation to specific environments. It is therefore very likely that phase variation is an important adaptive strategy in surface colonization, as is illustrated in the following examples.

V. CHOLERAE

V. cholerae O1 El Tor is the causative agent of Asiatic cholera. It is now believed that this bacterium successfully occupies a variety of natural aquatic habitats and resides there between epidemic outbursts. Prolonged incubation of the normal strain in laboratory media such as LB induces diversity in the population, and phenotypic variants with rugose colony morphology appear on plates, contrasting with the smooth colony morphology of the wild type. The rugose phenotype has been shown to be associated with the production of an extracellular polysaccharide (EPS^{ETr}), which significantly enhances biofilm formation (Fig. 6). A chromosomal locus (*vsp*) required for

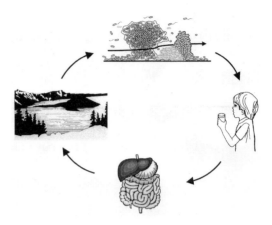

FIGURE 6 The life cycle of *V. cholerae*. This successful pathogen has evolved multiple strategies for adaptation and survival in changing environments. In aquatic environments the EPS producing phase variant (see text) may be a better survivor by forming biofilms on surfaces. When people drink water from sources colonized by *V. cholerae* there is a risk of infections of the gastrointestinal tract, this time selecting pathogenic variants expressing the cholera toxin. Infections with *V. cholerae* cause severe diarrhea, resulting in release of large numbers of bacteria back to the environment.

EPS^{ETr} production and rugose phenotype has been identified together with the response regulator, VspR, belonging to the family of two-component systems (Yildiz et al., 2001). The positive regulator is needed for the expression of EPS^{ETr} and the rugose phenotype, but the exact mechanism of phase variation is virtually unknown. Generating diversity by having a subpopulation ready to colonize surfaces is probably part of a strategy that allows *V. cholerae* O1 to adapt rapidly when challenged with aquatic habitats, a radically different low-nutrient environment compared with the intestine of mammals.

P. AERUGINOSA

Biofilm formation is among the essential infection and survival strategies for the opportunistic pathogen *P. aeruginosa* when it causes persistent infections, and the biofilm developmental cycle (compare Fig. 2) has been modeled extensively from studies of this species. Consider the dramatic shift in lifestyles from adaptation to the nutritional diversity in the natural environment, where competition is fierce and the physicochemical conditions are often extreme, to the challenges in a highly developed host organism. In the latter there may not be many bacterial competitors, but the host defense system will do its utmost to eliminate the invaders. It is not surprising that an arsenal of strategies for effective adaptation has evolved in this species. The examples below show how phase variation may be involved in the regulation of some of the adaptive processes and by globally controlling several related functions.

Development of high-level antibiotic resistance is frequently observed in biofilms. Because of the severity of *P. aeruginosa* infection in humans, in particular in cases of cystic fibrosis (CF), the development of antibiotic resistances that nearly always follow is a constant therapeutic problem. These bacteria proliferate in a biofilm in the lungs of a CF patient, and the chronic state of the infection is determined by this biofilm mode of persistence. For *P. aeruginosa* several hundred-fold increased antibiotic resistance levels have been described, and several physiological explanations have been offered. This could be due to a reduced antibiotic penetration in the biofilm, a change in stationary-phase physiology of biofilm cells, or a shift to a different biofilm mode of activity. An alternative hypothesis is that subpopulations arise in the biofilm with changed physiological properties, including increased antibiotic resistance as a consequence of genetic alterations. In support of this latter hypothesis it has been noted that multiple antibiotic resistant small-colony phenotypic variants of *P. aeruginosa* can be isolated repeatedly from the sputum of CF patients. It seems that antibiotic treatment of *P. aeruginosa* CF infections selects for these resistant small-colony variants (Drenkard and Ausubel, 2002).

One frequent type of variant with a rough small-colony morphology (RSCV) from the clinical isolate *P. aeruginosa* PA14 appears on LB plates supplemented with antibiotics and reverts to the wild-type on culture without antibiotics. In vitro studies of biofilms grown in a continuous flow-chamber system showed that the variant biofilm populations were more resistant to antibiotics than the wild type, and they displayed additional phenotypic changes including enhanced attachment and biofilm formation, pellicle formation in static broth, and cell-surface hydrophobicity. This variation in *P. aeruginosa* PA14 seems to involve a gene participating in the regulation of the phenotypic switch from variant to wild-type configuration. This gene, *pvrR*, shows significant homology to response regulators of two-component regulatory systems. The gene product, PvrR, modulates the phenotypic switch and thereby regulates the biofilm formation. In addition environmental factors interact with components involved in the generation of the variants, thus antibiotics assist in directing the switch in favor of the resistant variants.

In an independent investigation of an environmental strain of *P. aeruginosa* 57PR, Deziel et al. (2001) analyzed similar small-colony variants. When the strain was grown either in continuous-flow biofilm reactors or under static culture conditions, variants were identified. These formed small colonies with rough colony morphology (S variants), whereas normal cells would produce larger flat colonies with an irregular, finely mottled periphery. It turned out that the S variants produced abundant type IV fimbriae, which probably act as strong cell and surface glue enhancing biofilm formation. The S variants are apparently defective in swimming, swarming, and twitching motilities and impaired in chemotaxis. They also seem to increase the production of various secondary metabolites. Although the genetic loci and molecular mechanisms involved are unknown, it is evident that multiple phenotypic traits are involved. It points to the involvement of some kind of regulator gene as the target for phase variation, perhaps as a part of a regulatory cascade. The phenotypic traits of the small-colony variants in *P. aeruginosa* in general are similar but not identical, however. It is therefore likely that the underlying genetic changes are more complex and consist of multiple cases of genetic alterations, which may include elements of phase variation.

P. FLUORESCENS

In the plant rhizosphere some bacteria colonize the surface of roots and simultaneously protect the plant from soil-borne diseases, while other bacteria simply infect the plant host. In both cases this environment selects phase variants, which is not at all surprising since the rhizosphere is a very heterogeneous, frequently changing environment. Phase variation occurs quite often during rhizosphere colonization with the biocontrol strain *P. fluorescens* F113, and the generation of variants has been linked to a gene involved in DNA rearrangements, *sss*, encoding a site-specific recombinase. The frequency of phase variation was shown to decrease significantly in an *sss* mutant colonizing the roots of plants.

Phase variation has been documented in numerous cases of bacterial adaptation to changing environments, and the discussed examples of phase variation also indicate that many of the steps involved in surface community development (biofilms) may frequently be governed by this mechanism. Phase variation has an advantage as a gene-regulation mechanism in some situations and for certain purposes; it ensures that a subpopulation has the gene expression aimed at future available niches, which gives this population a crucial edge when opportunity rises. Subsequently a fine-tuning of global gene expression by sensing the surrounding conditions is necessary to persist in a new niche. Hence, phase variation should be viewed as an important adaptive mechanism that complements the sophisticated gene-regulatory schemes bacteria may employ to occupy new niches. The finding that phase variation may control the activities of regulatory genes resulting in phenotypic switch events involving many genes simultaneously is in accord with the idea that this mechanism represents an essential aspect of adaptive physiological control exerted at the population level. The fact that environmental factors may influence the direction of the phase switch by adding even higher levels of control makes this adaptive mechanism very powerful.

Section Summary

- Phase variation creates physiologically different subpopulations in clonal populations.
- Important biofilm characteristics are controlled by phase variation.
- Phase variation may interfere with global regulator activity and, in this way, control the activity of an array of physiological traits in the population.
- The direction of phase variation switches may be influenced by the external environment.
- Phase variation may result in fast adaptation to new niches.

GENETIC DIVERSIFICATION IN BIOFILM POPULATIONS

In a series of observations made by Jim Shapiro in the mid-1980s and later by John Cairns in 1988, it appeared that bacteria might preferentially mutate under starvation conditions to variants that survive and even proliferate under the adverse conditions. This created turbulence among geneticists; was Lamarckism returning through the back door? DNA-repair enzymes were early suspects as key factors in the mechanism where random mutations in the replication fork of residual chromosomal replication runouts could be fixed by growth whenever a beneficial change occurred. This explained why nonbeneficial mutations did not show up in the growing population with the same frequencies as beneficial ones. Another hypothesis suggested the presence of high-frequency mutating subpopulations and specific induction of mutator phenotypes. The phenomenon seemed to be driven by various mutational mechanisms and is referred to as "stationary-phase mutation" or "adaptive mutation," and has been seen when bacteria are placed under various kinds of stress conditions. The types of mutations reported include point mutations, recombination between similar DNA sequences, horizontal gene exchange and transposon-mediated insertions and deletions. The relationship between these mutational events is complex and remains confusing, and a number of different mechanisms may exist. The following discussion is not a survey of these mechanisms, but is our biased choice of examples relevant to biofilm community development.

E. coli lac⁻ Reversion Studies

Brisson conducted very comprehensive studies of induced mutations in starving and stressed bacteria in *E. coli*, and one is the reversion of a *lac*+1 frameshift allele. The selection/screening system was based on the assumption that mutants already present in the growing population (before transfer to the starvation condition) will grow up under the selective conditions after 1 to 2 days of incubation. However, Lac$^+$ revertants appearing on the selective solid lactose minimal medium after more than 2 days of incubation on the plates occur due to exposure to the selective conditions. The mechanism, denoted "adaptive mutation," is different from those mechanisms producing spontaneous mutations in growing cells. It now appears that these Lac$^+$ revertants arise from a small subpopulation that has undergone a transient genome-wide hypermutation. The state is assumed transient since it was not a heritable phenotype. Heritable hypermutation phenotypes (or mutator bacteria) are usually associated with cells deficient in the mismatch repair system, which normally contributes to genetic stability (see below). A reason for the hypermutation state could be a transient limitation of the mismatch repair system (MMR). By generating a transient hypermutation state, the cells benefit from a fast adaptation to a challenging environment without paying the long-term cost of continuously creating deleterious mutations, which occur more frequently than the adaptive ones.

The SOS signal transduction pathway is responsible for the induction and control of the adaptive mutations in the Lac system. The SOS response was first characterized as a repair system allowing cells to deal with DNA damage (e.g., after UV radiation), and a central aspect of this pathway is the derepression of more than 20 genes under the control of the LexA repressor. In studies of the role of SOS induction of adaptive mutations in the Lac

system, the involvement of DNA recombination and repair proteins and of a special error-prone DNA polymerase IV, belonging to the DinB/UmuC superfamily, has been demonstrated (McKenzie and Rosenberg, 2001). Since homologues of the SOS system and the other factors have been found in most prokaryotes, it is possible that adaptive mutations may be widespread in the microbial world. The role of starvation or stress-induced mutation mechanisms in cells adapting to life on surfaces still needs to be investigated. The present knowledge about these systems and their induction in nongrowing or slow-growing cells strongly suggests that genetic variation could be generated in biofilm subpopulations as a result of induced mutation mechanisms. The dense microcolonies found in biofilms are niches that develop severe gradients and nutrient limitation, which could lead to SOS induction. This relationship remains to be experimentally confirmed.

Induced Mutations in Structured Environments

Bacterial agar surface colonies of *E. coli* have been used as models for growth in structured environments, and they do resemble dense biofilm microcolonies, since they eventually give rise to nutrient gradients and to the accumulation of metabolic intermediates and signaling molecules. The mechanism generating mutations in aging colonies has been described by Taddei and coworkers (1997) and denoted ROSE (Resting Organisms in a Structured Environment). The ROSE mechanism is also SOS induced and similar to the mechanism of the Lac reversion system mentioned above. SOS induction in aging colonies depends on the synthesis of cAMP, a metabolite accumulating in starving bacteria which controls several operons subjected to catabolite repression. In a study the frequency of Rif^R mutations was found to increase linearly with time during colony development, whereas the same bacterial strain in aging liquid cultures (an unstructured environment) did not. Thus, the inducing factors of the ROSE mechanism are linked to structured environments and are not just starvation inducible. Therefore, gradients (nutrients, oxygen, cAMP, or other metabolites) and osmotic and/or water stress could play roles in a mutation-generating mechanism(s). It is also intriguing that the inducer of the SOS system, cAMP, is over-produced in resting cells and excreted to the environment. So in biofilms this nucleotide may act like an intercellular mutation-induction signal, coordinating a transient high-rate mutability in a part of the population.

Several different *Pseudomonas* species show genetic diversity if placed in structured environments under nonfavorable conditions, and eventually specialized and more adapted mutants appear (Rainey and Travisano, 1998). In a few studies the type and location of the specific mutations have been identified, but so far the underlying mutation mechanisms have not been determined. This is also the case with the mucoid conversion of *P. aeruginosa* as outlined below.

Due to an inheritable genetic disorder CF patients are prone to infections in the respiratory tract. The opportunistic pathogen, *P. aeruginosa*, readily infects their lungs, and the conditions in the lungs induce this organism to undergo a genetic change from an acute, controllable state to a persistent, untreatable state involving biofilm formation of a mucoid variant phenotype. The mucoid state is caused by overproduction of alginate, a capsular polysaccharide that protects the bacterial cells from oxidative stress, immune system attacks, and antibiotic therapeutic treatments. The mucoid conversion is often associated with mutations inactivating the *mucA* gene, which encodes an anti-σ-factor counteracting the AlgT σ-factor. The latter is required for expression of the alginate biosynthetic operon. The loss of MucA protein results in the derepression of AlgT, which leads to high-level expression of the alginate biosynthetic genes and alginate production. Mucoid variants isolated from the lungs of CF patients very often have specific mutations in the *mucA* gene, either a deletion of a guanosine residue in a string of five Gs, or

insertion of a copy of 8 bp. There therefore seem to be hot spots for mutations in the two respective positions, which result in enhanced probabilities of mutational events. In both cases, the mutations cause premature translational terminations due to out-of-frame switches. In vitro investigations of *P. aeruginosa* flow-chamber biofilms have indicated that the oxidative stress caused by the host immune response could be one of the factors that generates and/or selects the mucoid mutants. It also appears that the mutations in *mucA* induced by oxygen stress develop with even higher frequencies in biofilm-associated cells than in suspended cells. See the analogy with the ROSE mechanism.

Mutator Bacteria: Do They Play an Important Role in Biofilm Adaptive Developments?

It is often assumed that the genomic mutation rate of organisms has evolved to be as low as possible because new mutations are often lethal and bacteria harbor numerous genetic systems that act to prevent the appearance of mutations. This view is challenged by the finding of naturally occurring bacteria with greatly increased mutation rates, the so-called mutator bacterial strains. Mutator strains have inactivated genes, the products of which are involved in eliminating mutations (*mut*-genes), and the most frequently occurring are those deficient in the mismatch repair system (*mutS, L, U, H*). The elevated mutation rate is not an advantage per se, but the mutators are selected for because of the adaptive mutations they generate, resulting in "hitch-hiking along the evolutionary highway." Since it is still expected that mutants with increased mutation rates are problematic in the long run, one interesting question is under what conditions these mutants flourish, and if and when they disappear.

P. aeruginosa causing chronic lung infections in CF patients has been a model system to document the significance of mutator strains in a population. One study found 11 of 30 CF patients colonized with a mutator strain that persisted for years (Fig. 7). As mentioned previously, *P. aeruginosa* develops a biofilm in the

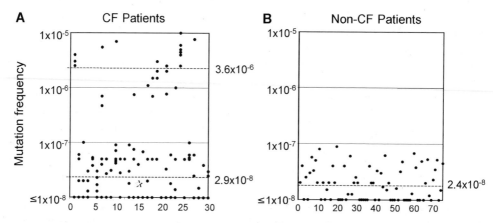

FIGURE 7 The occurrence of mutator strains in populations of *P. aeruginosa*. Clones of *P. aeruginosa* were collected from two groups of patients: (A) cystic fibrosis patients with chronic infections of *P. aeruginosa* and (B) patients not having cystic fibrosis with *P. aeruginosa* lung or blood infections. For each of the isolated clones the frequency of mutation to rifampicin resistance was determined by plating. It is clear that in the CF patients a significant subpopulation of bacteria are mutating with very high frequencies, whereas in the normal patients the determined mutation frequencies are in a range that is normally observed in growing bacteria (standard PAO1 strain included in frame A and indicated by an asterisk). The dashed lines represent the mean of mutator and nonmutator groups. (Reprinted with permission from Oliver et al., 2000.)

lungs of CF patients, and isolates from chronic infections display a wide spectrum of colony variants, including mucoid and small-colony variants, which suggests that diversification through mutations could be an important factor contributing to mutant strain fitness in the lungs of CF patients. The progressive deterioration of lungs in CF patients reflects a highly structured environment during chronic infection. The host immune attack and aggressive antibiotic treatment contribute further to a continuous alteration of the lung and the continuous presence of selection pressures fixes the mutator population. Not all bacteria are mutators due to the cost of carrying reduced repair capacity, which eventually leads to the accumulation of numerous mutations in the genome, which may be neutral or adaptive in one environment, but most likely deleterious in a different environment. Competition experiments in complex environments (the mouse gut) have demonstrated the short-term advantage of the mutator phenotype, in contrast to a long-term disadvantage of high mutation rates once adaptation has occurred. Still, natural mutator strains are

not at all rare, often accounting for a few percent of tested isolates, and this is consistent with evolutionary theory and computer simulations, which predict the evolution of high mutators under certain conditions.

A related question concerns the frequency and role of mutators in environmental biofilms, where conditional changes also occur quite frequently, although often less drastically. We have addressed this question by examining whether a *mutS* mutator variant would be more fit in continuous-flow biofilm development than in the otherwise isogenic wild-type strain of *P. putida* mt-2. The mutator strain developed a biofilm faster, it was more fit in competition assays, and it was more resistant to different kinds of imposed stresses including antibiotic treatment. Thus, a short-term advantage of a high mutation rate is a faster adaptation potential. What remains to be seen is the long-term effects of being a mutator in this system, and if mutator variants would eventually appear spontaneously and be fixed in the biofilm.

To what extent surface populations rely on diversity generators in adaptive processes is an interesting question for the future. The frequencies with which variants appear in structured communities suggest that the conditions favoring induced mutations are probably common, and the mutation frequencies most likely exceed what is normally found. In this way, genomic plasticity seems to add to the adaptability displayed by surface-associated bacterial populations, and the probable outcome may not only be improved growth characteristics in certain environments, but also genetic changes allowing subpopulations to enter free ecological niches present in structured communities. The possible selective benefit of mutator subpopulations is more difficult to assess, and, in any event, it is difficult to see how such variants could be beneficial over longer periods. It may be speculated that severe challenges to the population, such as is made by antibiotic addition, can only be met by the rapid development of resistant cells; if not, the population will become extinct. Under conditions where the population is continuously confronted with new and harsh challenges, such as in the course of an infection in a multicellular organism, there may be a more-or-less constant selection for increased numbers of mutators to ensure the necessary adaptive changes within the population. The increase in mutation frequency obtained in mutator strains could be the last defense of the population, when all other adaptive strategies, gene activity regulation, phase variation of gene expression, and induced mutations, fail.

Section Summary
- The continuous change of the environmental conditions within many surface-associated communities creates selective forces for novel phenotypes.
- Adaptive mutations induced by the environment seem to be particularly suited to the biofilm environment during its development and persistence.
- The occurrence of stress conditions in structured communities stimulates mutant induction and adaptive phenotype selection.
- Mutator strains are a transient source of rapid mutant production.

- In communities with constantly changing adverse conditions, mutators may be found as more permanent community members.

HORIZONTAL GENE TRANSFER
The asexual reproduction mode of bacteria appears to prevent genome diversification by recombination between different members of the same species. However, half a century ago it was discovered that bacteria do exchange genetic material. The significance of horizontal gene transfer is rapidly becoming clearer as we examine genomic sequences of bacteria. There are three major routes of gene transfer in the bacterial world:

- Plasmid mobilization/conjugation, requiring cell-cell contact
- Phage transduction, most often only specific phage-host organism combinations
- DNA transformation, requiring cellular competence and some degree of DNA homology.

Plasmid Transfer
Conjugation of plasmids from donor cells to plasmid-free recipients requires cell-to-cell contact, which may be mediated by specific conjugation pili or other surface-bound appendages (gram-negative bacteria; chapter 20) or by pheromone-induced cell aggregation (gram-positive bacteria; chapter 7). In all cases efficient plasmid transfer depends on high cell densities, such as we find in surface-associated biofilms. We may therefore a priori consider biofilms to be hot spots for plasmid transfer, since it is in these natural environments that bacteria are metabolically active. A positive correlation between the biofilm mode of growth and high rates of conjugation has been documented in recent years by direct in situ visualizations based on zygotic induction of Gfp expression. Thus, it is clear that plasmid transfer in biofilms is high. However, due to the heterogeneous distribution of cellular metabolic activity and the reduced mixing of cells in these structured communities, plasmid

transfer will only involve part of the recipient population (Fig. 8).

Conjugative plasmids may reach sizes approaching those of small bacterial chromosomes, and, thus, this type of horizontal transfer has great potential for producing direct phenotypic impacts on recipients. In addition within a population individual plasmids may be present in high/low numbers in subpopulations. So together the chromosome and plasmids constitute a genomic information set corresponding to several diverse bacterial genomes. Since no homologous recombination is necessary for establishing and maintaining plasmids in the genome of an organism, the evolutionary potential of the conjugation process is very high. In rapidly changing environments the genetic information within the different plasmids may have ecological significance at different times, imposing selection on particular subpopulations. Plasmids are flexible elements contributing to the phenotypic repertoire of a population. It is important to recognize too that there is continuous trafficking of genetic material between chromosomes and extrachromosomal elements. Transposons, various types of functional "islands" of gene sequences, etc., all contribute. This underlines the two driving forces in bacterial development and evolution: the cellular and the population selection platforms.

Phage Transduction

Bacterial viruses, phages, are obligate parasites on bacteria, which either infect, reproduce, kill, and lyse their hosts (virulent phages), or insert themselves as prophages in the host genome (temperate phages), until they at some later time induce a lytic life cycle. Phages generally have narrow host cell specificity, and require a specific receptor on the cell surface for infection to take place (see chapter 9). Aberrant excision may lead to insertion of a small host chromosomal segment in the phage genome or, in some cases, a headful of chromosomal DNA instead of phage DNA. This phage may later infect a

new cell and thereby introduce the bacterial genomic DNA for insertion in the new host. Many, if not most, bacteria harbor inserted prophages, of which some may be inducible. Although the relationship between the environmental conditions and prophage induction is not well understood, it is induced when the bacterial host cell is stressed. In biofilms some cells will always exist under physiologically stressful conditions, and free phage particles are seen. All members of the cell line harboring a specific prophage may release phages, but due to the immunity encoded by the prophage, infection requires the presence of nonlysogenic cell lines carrying the right phage receptor. Thus, phages may contribute to the genetic development of biofilm communities by transduction and may also play another important role in community development through their predator activity. Potentially phages can eliminate entire subpopulations of a species in a mixed-species biofilm community and thus open up novel niches.

DNA Transformation

The direct uptake by bacteria of DNA from the environment has for a long time been known to occur with high efficiency in certain species exhibiting specialized competence for this process. In these cases the process of transformation is a well-regulated and precisely defined activity and constitutes an important part of the natural life cycle of several organisms (see chapter 10). There are examples of induction of competence and subsequent transformation of cells by DNA in the environment. There is also lot of free DNA "out there" in nature. So biofilms are potential hot spots for transformation. The EPS matrix of biofilms comprises significant amounts of both proteins and nucleic acids in addition to polysaccharides, and large amounts of DNA are released in biofilms of P. aeruginosa, which influences the structural development of this particular community. It is possible that the biofilm-associated DNA is simply released from the lysis of dead cells in the community. It is also possible that death and cell lysis is an

---

FIGURE 8 *See page 200.*

important element in the biofilm developmental cycle (apoptosis). Thus, DNA may be released from membrane-surrounded vesicles or as a consequence of phage-induced lysis of bacteria. So when there are domains or conditions in biofilms when bacteria become induced to competence, the presence of free DNA could contribute to acquisition of new genetic information.

Horizontal gene transfer has contributed much to bacterial diversity, and the occasional promiscuity across widely separated phylotypes has challenged the species concept in the bacterial world. The relationship between these processes and the distribution of ecological niches in structured communities is a target for future investigations, as is the relationship between the development of biofilm communities and the stimulation of conjugation, induction of lysogenic phages, and the development of competence. The potential interactions between horizontal DNA-transfer processes and induction of SOS-induced mutations constitute a hot spot of interest in the evolution of structured-surface communities.

Section Summary

- All types of gene transfer occur in biofilms.
- Horizontal gene transfer may be one key determinant in the generation of substantial genetic alterations.
- Plasmids may provide their host cells with flexible metabolic potential.

WHERE DO WE GO FROM HERE?

Bacterial evolution has been a focus point in microbiology and evolutionary biology since the first days of bacterial genetics. The simplicity of bacteria, their fast growth (at least as far as the more common model organisms are concerned), and the ability to perform controlled and reproducible growth experiments have stimulated microbiologists to track genetic changes and interpret the occurrence of these in relation to the chosen selective forces. The chemostat offers opportunities to monitor over extended periods (hundreds of genera-

tions) the rise and disappearance of subpopulations with new or improved capacities to exploit the growth-limiting nutrients in the reactor. Much has been learnt about competition between individual cells and how the environment selects for improved growth properties.

In the classical chemostat there is generally one ecological niche present, that defined by the rate-limiting nutrient. After a period of growth in the chemostat a specific irritating problem always shows up: biofilm development on the walls of the reactor. A substantial amount of scientific creativity has over the years been invested in preventing and removing this heterogeneity from disturbing the controlled environment in the reactor. In retrospect, these biofilms should have been examined with enthusiasm, because their ready and predictable occurrence clearly indicates there is an alternative growth strategy for bacteria. The reactor wall constitutes a free ecological niche in addition to the one defined by the nutrient in the bulk water, and very soon this free niche will consist of a large number of subniches, in which subselective forces operate. So, eventually, the population in the water phase may represent a numerically minor part of the "zoological garden" thriving in the surface-attached communities. The first conclusion is therefore that the bacterial biofilm is a fundamental platform for adaptation and evolution, and the performance and evolutionary strategies of bacteria must be assessed in relation to their life there.

These chemostat investigations on bacterial evolution show that the organisms located in the wall growth would have little chance competing with the planktonic chemostat population under conditions where no biofilm formation were possible. The microbial diversity, both phenotypically and genetically, in the biofilm population is far greater than that in the planktonic phase due to the auto-induced complexity of the ecological niches present in the surface community. The second conclusion, therefore, is that, in biofilms, the inhabitants exploit their metabolic and regulatory repertoire, thus optimizing their chances of occupying niches as they arise. The continuous creation of alternative subpopulations by phase variation provides individuals able to invade a new niche.

The balanced exponential growth in the chemostat ensures that the metabolic and regulatory activities in all cells are directed optimally to occupy the aqueous niche. This balance is achieved by the quality, quantity, and flow rate of the limiting nutrient(s). Any change in phenotype which results in improvement in nutrient uptake and usage is achieved by mutation. This is, of course, why the chemostat has been so popular and effective in bacterial genetics and evolution studies. In the biofilm, conditions are far from stable, and exponential balanced growth rarely, if ever, occurs. However, the community development will inevitably create opportunities and threats to which the population cannot respond, and local or even total elimination may only be avoided if new genetic variants appear on the scene. The third conclusion thus is that multiple ecological niches develop in structured communities and severe challenges to the population call for repertoire transcending variations to secure survival and persistence. Increased mutation rates under these adverse conditions seem to provide the necessary dynamics.

The presence of extrachromosomal elements in bacteria is considered a burden to the cell. Replication, transcription, and translation of plasmid-encoded genes represent an energy cost. The loss of this genetic information is predictable unless the environment selects for its presence. In structured communities the potential for gene transfer is much higher due to the population densities. However, since growth activity and rates are not the only criteria for selection in a biofilm, the spread of information and the recombinational possibilities offered may in fact overrule the energy burden. The final conclusion to be drawn therefore is that evolution of fundamentally new capacities requires large genetic changes, and the potential for transfer of complex genomic information is high in biofilm communities. It is therefore possible that gene transfer enables cells to have complex life cycles and to survive rapid shifts between extreme environmental conditions.

The investigations of the adaptive and evolutionary potentials of bacteria in the context of complex communities in biofilms and other structured environments have only just started. Exciting work remains to be done, and with the tools and methods that are now available due to the rapid technological development of the past decade, there is really no excuse to avoid asking these difficult questions. Bioinformatics almost daily brings new genome sequences and improved tools for their interpretation. Data from elegant experiments done in molecular biology, bioimaging, and physiological studies of complex ecosystems provide the potential for understanding the driving evolutionary forces in our planet's biosphere.

QUESTIONS

1. How may bacteria know that they are in a biofilm (in contrast to being planktonic)?

2. Which types of genes may be expected to be "biofilm-specific"?

3. Which benefits may bacteria obtain from life in biofilms?

4. Under which conditions could gene regulation by phase variation be advantageous compared with signal transduction-mediated regulation?

5. How will you explain increased antibiotic resistance in biofilms?

REFERENCES

Christensen, B. B., C. Sternberg, J. B. Andersen, L. Eberl, S. Moller, M. Givskov, and S. Molin. 1998. Establishment of new genetic traits in a microbial biofilm community. *Appl. Environ. Microbiol.* **64:**2247–2255.

Christensen, B. B., J. A. Haagensen, A. Heydorn, and S. Molin. 2002. Metabolic commensalism and competition in a two-species microbial consortium. *Appl. Environ. Microbiol.* **68:**2495–2502.

Deziel, E., Y. Comeau, and R. Villemur. 2001. Initiation of biofilm formation by *Pseudomonas aeruginosa* 57RP correlates with emergence of hyperpiliated and highly adherent phenotypic variants deficient in swimming, swarming, and twitching motilities. *J. Bacteriol.* **183:**1195–1204.

Drenkard, E., and F. M. Ausubel. 2002. Pseudomonas biofilm formation and antibiotic resistance are linked to phenotypic variation. *Nature* **416:**740–743.

Garrett, E. S., D. Perlegas, and D. J. Wozniak. 1999. Negative control of flagellum synthesis in Pseudomonas aeruginosa is modulated by the alternative sigma factor AlgT (AlgU). *J. Bacteriol.* **181:** 7401–7404.

Giraud, A., M. Radman, I. Matic, and F. Taddei. 2001. The rise and fall of mutator bacteria. *Curr. Opin. Microbiol.* **4:**582–585.

Heydorn, A., B. Ersboll, J. Kato, M. Hentzer, M. R. Parsek, T. Tolker-Nielsen, M. Givskov, and S. Molin. 2002. Statistical analysis of *Pseudomonas aeruginosa* biofilm development: impact of mutations in genes involved in twitching motility, cell-to-cell signaling, and stationary-phase sigma factor expression. *Appl. Environ. Microbiol.* **68:**2008–2017.

McKenzie, G. J., and S. M. Rosenberg. 2001. Adaptive mutations, mutator DNA polymerases and genetic change strategies of pathogens. *Curr. Opin. Microbiol.* **4:**586–594.

Nielsen, A. T., T. Tolker-Nielsen, K. B. Barken, and S. Molin. 2000. Role of commensal relationships on the spatial structure of a surface-attached microbial consortium. *Environ. Microbiol.* **2:**59–68.

Oliver, A., R. Canton, P. Campo, F. Baquero, and J. Blazquez. 2000. High frequency of hypermutable *Pseudomonas aeruginosa* in cystic fibrosis lung infection. *Science* **288:**1251–1254.

Prigent-Combaret, C., G. Prensier, T. T. Le Thi, O. Vidal, P. Lejeune, and C. Dorel. 2000. Developmental pathway for biofilm formation in curli-producing *Escherichia coli* strains: role of flagella, curli and colanic acid. *Environ. Microbiol.* **2:**450–464.

Rainey, P. B., and M. Travisano. 1998. Adaptive radiation in a heterogeneous environment. *Nature* **394:**69–72.

Sanchez-Contreras, M., M. Martin, M. Villacieros, F. O'Gara, I. Bonilla, and R. Rivilla. 2002. Phenotypic selection and phase variation occur during alfalfa root colonization by *Pseudomonas fluorescens* F113. *J. Bacteriol.* **184:**1587–1596.

Sauer, K., A. K. Camper, G. D. Ehrlich, J. W. Costerton, and D. G. Davies. 2002. *Pseudomonas aeruginosa* displays multiple phenotypes during development as a biofilm. *J. Bacteriol.* **184:** 1140–1154.

Taddei, F., J. A. Halliday, I. Matic, and M. Radman. 1997. Genetic analysis of mutagenesis in aging Escherichia coli colonies. *Mol. Gen. Genet.* **256:**277–281.

Whitchurch, C. B., T. Tolker-Nielsen, P. C. Ragas, and J. S. Mattick. 2002. Extracellular DNA required for bacterial biofilm formation. *Science* **295:**1487.

Yildiz, F. H., N. A. Dolganov, and G. K. Schoolnik. 2001. VpsR, a member of the response regulators of the two-component regulatory systems, is required for expression of vps biosynthesis genes and EPS(ETr)-associated phenotypes in *Vibrio cholerae* O1 El Tor. *J. Bacteriol.* **183:**1716–1726.

FURTHER READING

Brisson, D. 2003. The directed mutation controversy in an evolutionary context. *Crit. Rev. Microbiol.* **29:**25–35.

Cairns, J., J. Overbaugh, and S. Miller. 1988. The origin of mutants. *Nature* **335:**142–145.

Christensen, B. B., C. Sternberg, J. B. Andersen, R. J. Palmer, Jr., A. T. Nielsen, M. Givskov, and S. Molin. 1999. Molecular tools for study of biofilm physiology. *Methods Enzymol.* **310:**20–42.

Costerton, J. W., P. S. Stewart, and E. P. Greenberg. 1999. Bacterial biofilms: a common cause of persistent infections. *Science* **284:**1318–1322.

Henderson, I. R., P. Owen, and J. P. Nataro. 1999. Molecular switches: the ON and OFF of bacterial phase variation. *Mol. Microbiol.* **33:**919–932.

Høiby, N., H. K. Johansen, C. Moser, Z. Song, O. Ciofu, and A. Kharazmi. 2001. Pseudomonas aeruginosa and the in vitro and in vivo biofilm mode of growth. *Microbes Infect.* **3:**23–35.

Molin, S., A. T. Nielsen, B. B. Christensen, J. B. Andersen, T. R. Licht, T. Tolker-Nielsen, C. Sternberg, M. C. Hansen, C. Ramos, and M. Givskov. 2000. Molecular ecology of biofilms, p. 89–120. *In* J. D. Bryers (ed.), *Biofilms II: Process Analysis and Applications.* Wiley-Liss, Inc., New York, N.Y.

O'Toole, G., H. B. Kaplan, and R. Kolter. 2000. Biofilm formation as microbial development. *Annu. Rev. Microbiol.* **54:**49–79.

Radman, M., I. Matic, and F. Taddei. 1999. Evolution of evolvability. *Ann. N. Y. Acad. Sci.* **870:**146–155.

Shapiro, J. A. 1984. Observations on the formation of clones containing *araB-lacZ* fusions. *Mol. Gen. Genet.* **194:**79–90.

Wingender, J., T. R. Neu, and H.-C. Flemming (ed.). 1999. *Microbial Extracellular Polymeric Substances.* Springer-Verlag, Berlin, Germany.

SPATIAL SEGREGATION:
THE DEEP SUBSURFACE STORY

David L. Balkwill

14

David L. Balkwill, Department of Biomedical Sciences,
Florida State University College of Medicine, Tallahassee, FL
32306.

Subsurface environments can be defined as any terrestrial environments that are situated beneath the topsoil zone of the Earth's crust. It is now well established that microorganisms are widely distributed in these environments, even at considerable depths. At present, very

Microbial Evolution: Gene Establishment, Survival, and Exchange
Edited by Robert V. Miller and Martin J. Day, ©2004 ASM Press, Washington, D.C.

little definitive information exists on the evolution of microorganisms in the terrestrial subsurface. Yet, the information we do have indicates that there may be interesting aspects of microbial evolution that are peculiar to the subsurface, primarily because of the ways in which the process is likely to be affected by the physical and chemical nature of the environment. For example, the microbial communities in some types of subsurface environments appear to be spatially segregated from those in surface environments or even those located nearby in the subsurface, implying that these segregated communities have been evolving independently for potentially very long periods of time. Other aspects of the terrestrial subsurface, such as very low nutrient levels that result in very slow growth or dormancy of subsurface microbes, might also influence the evolution and genetic structure of microbial communities in deep environments. This chapter focuses on the physical and chemical characteristics of the terrestrial subsurface that are most likely to influence microbial evolution, with emphasis on the spatial segregation issue, and suggests possible directions for future research.

SUBSURFACE MICROBIOLOGY

Emergence of Subsurface Microbiology in the Early 1980s

As noted above, subsurface environments are defined as any terrestrial environments that are situated beneath the topsoil zone (or A horizon) of the Earth's crust. They include saturated environments, such as aquifers and the groundwater that flows through them. They also include unsaturated environments, such as the vadose zone that is situated between the water table and the overlying topsoil zone in many areas.

Until recently, it was widely believed that subsurface environments were almost devoid of microbial life. Studies carried out in the early 1900s indicated that the numbers of microbes detectable with methods available at that time decreased rapidly with increasing depth beneath the topsoil zone. As a result, microbiologists largely ignored the terrestrial subsurface for many years. Reports describing the presence of microbes at considerable depths appeared sporadically in the literature over the next several decades, but these generally were ignored or dismissed as invalid because of concerns that the samples may have been contaminated with organisms from the surface. In the early 1980s, however, a group of scientists funded by the U.S. Environmental Protection Agency showed conclusively that substantial numbers of culturable microorganisms (primarily bacteria) were present in several shallow aquifers. Other groups working in several countries then reported similar findings, and this triggered a new interest in subsurface microbiology that has expanded steadily since that time.

Shift in Focus from Shallow to Deeper Environments

Most studies in the early 1980s focused on relatively shallow subsurface environments that were situated less than 30 meters below land surface. Shallow aquifers were of much interest because the groundwater in them serves as the primary source of fresh water for drinking and irrigation in many parts of the world. Subsurface microorganisms, through their metabolic activities, might affect the chemical quality of this important water source. After it was established that microbes were widespread in shallow subsurface systems, the focus shifted to deeper environments.

Widespread Occurrence of Microbes in Terrestrial Subsurface

Research sponsored by the U.S. Department of Energy during the late 1980s discovered abundant and diverse microbial communities in Atlantic Coastal Plain sediments of the southeastern United States to depths of 500 meters below land surface. A wide variety of different types of subsurface environments have been examined since then, ranging in depth to 3.4 km below land surface, and microorganisms have been detected in almost all of these environments. At present, we do not know the maximum depth at which viable microbes occur within the Earth. High temperature becomes the limiting factor at some point. The currently accepted maximum temperature for microbial life (approximately 115°C) theoretically would be reached at 6 to 7 km below land surface, depending on the local geology, so we might expect to find some microorganisms at least to that depth range. Research carried out to date has shown that microorganisms are very widespread in the terrestrial subsurface. They have been found in saturated and unsaturated environments, and in many types of unconsolidated sediments (sands, gravels, clays, etc.) and rocks (shales, volcanic tuffs, sandstones, etc.).

Physiological Diversity of Subsurface Microorganisms

Microbial numbers in the subsurface often are high enough to influence the chemical nature of the environment, if they are metabolically active. Moreover, subsurface microbial communities are often diverse and include a wide variety of physiologically distinct forms (various types of heterotrophs and chemolithotrophs, aerobes and anaerobes, etc.). There is a potential for many types of microbially mediated chemical transformations to take place in most subsurface environments, depending on which members of the microbial community are active at any given time. As a result, there has been much interest in how subsurface microbes might affect (or, through human manipulation, might be made to affect) the fate of both organic and inorganic contaminants in groundwater and other subsurface environments. In recent years, interest has also increased in how microorganisms affect the geochemistry of the subsurface through oxidation and reduction of metals, formation and weathering of minerals, etc.

Section Summary

- Subsurface environments are any terrestrial environments situated beneath the topsoil zone of the Earth's crust.
- The field of subsurface microbiology emerged in the early 1980s and has grown steadily since then as increasingly deeper environments have been studied.

- Microorganisms are present in many different types of subsurface environments.
- Microbes can occur at considerable depths (up to several kilometers below land surface).
- Subsurface microbes can influence the chemistry of their environments.
- Subsurface microbial populations are physiologically diverse.

THE SUBSURFACE ENVIRONMENT

Wide Variations in Chemical and Physical Environmental Characteristics

Terrestrial subsurface environments vary widely in physical and chemical characteristics. Moisture levels range from fully saturated to exceedingly dry. There is very little transport of water or nutrients through some types of very dry unsaturated systems, so the microorganisms in these environments must be able to survive prolonged periods of starvation and/or desiccation.

Porosity and Permeability: Movement of Water, Nutrients, and Cells

The flux (and, thereby, the concentrations) of nutrients and the movement of microorganisms themselves through the subsurface are also influenced by the porosity and permeability of the environment. In this regard, unconsolidated subsurface sediments range from coarse sands or gravels through which groundwater, nutrients, and microbial cells may be transported relatively quickly if the system is saturated, to extremely fine, dense clays through which there is little or no movement of anything. Rock environments also vary in the extent and rate at which water, nutrients, and microbial cells can move through them, although they are typically less permeable than most unconsolidated sediments.

Implications of Very Low Nutrient Concentrations

In general, concentrations of nutrients, especially organic nutrients, are much lower in the subsurface than in many surface environments, even in saturated systems. As a result, microbes in the subsurface must be adapted for slow growth and survival under low-nutrient conditions. Relatively deep subsurface environments are microaerophilic or fully anoxic. Microbes living in these environments must be able to utilize fermentative processes or other forms of metabolism that do not depend on oxygen (e.g., the use of alternate electron acceptors such as nitrate or Fe^{3+} for respiratory metabolism). Temperature also becomes a significant factor in deeper subsurface environments, limiting the range of microbial species that can survive and function there. (The deepest environments examined in detail to date range from 55 to 65°C, placing them at the low end of the thermophilic range for bacteria.) High hydrostatic pressure could also become a factor at great depths in the terrestrial subsurface, but there is little or no information on this in the current literature.

Harsh Nature of Subsurface as a Habitat for Microbes

In general, subsurface environments can be thought of as "harsh" habitats for microorganisms in comparison with many surface environments. Those studied to date might not be considered "extreme" in terms of their physical characteristics, but they probably qualify as "extreme" examples of low-nutrient environments.

Vertical and Spatial Heterogeneity in Physical and Chemical Characteristics

It is important to realize that there often is a great deal of vertical and horizontal spatial heterogeneity in the physical and chemical characteristics of a subsurface environment. This heterogeneity occurs on many scales and sometimes is reflected by a similar level of microbial heterogeneity. For example, within a single geological formation that appears fairly uniform to the eye, there may be variations in the chemical environment that result in variations in the composition of the microbial community over distances ranging from micrometers to meters. On a larger scale, physical and chemical heterogeneity sometimes results in the juxtaposition of environments that differ greatly in properties that affect microbial survival and growth. For example, pockets of very dense, impermeable materials such as clay lenses may be situated within layers of comparatively porous or transmissive materials, such as sandy aquifer sediments, or vice versa. Physical and chemical heterogeneity sometimes is responsible for the compartmentalization of environments into physically isolated units within the subsurface. A single geological stratum and the microorganisms living in it might be physically isolated from the overlying and/or underlying strata (and their resident microbial communities) by layers of relatively dense, impermeable material.

Section Summary

- Subsurface environments vary widely in their physical and chemical characteristics.
- The rate at which water, nutrients, and microbes can move through subsurface environments is highly variable because the physical characteristics of these environments (e.g., porosity) are highly variable.
- Most subsurface microbes are adapted for slow growth because the concentrations of nutrients are very low.
- Subsurface microbes may have to survive periods of starvation or desiccation.
- There is much spatial heterogeneity in physical and chemical characteristics within the subsurface.
- Some subsurface microbes may be physically isolated from other microorganisms because they are surrounded by layers of dense, impermeable material.

ORIGINS OF SUBSURFACE MICROORGANISMS AND THE SPATIAL SEGREGATION ISSUE

The concept of physical isolation or compartmentalization of environments raises some interesting questions regarding the origins of microorganisms in the subsurface. Why are they now so widespread throughout the subsurface? How did they get there? How long have they been there?

Long-Term Survival In Situ versus Recent Transport from the Surface

It is possible that microorganisms in the subsurface are the survivors or progeny of the organisms that were present in sediments or other materials when they were initially buried thousands to hundreds of millions of years ago. If this is the case, the microbes encountered now probably have been resident in the subsurface for very long periods. Of course, it is also possible that microorganisms are transported into the subsurface from surface environments by the downward movement of meteoric water (water originating from the surface, e.g., rainwater) or other means, over comparatively short periods. In that case, subsurface microorganisms would be unlikely to differ greatly from those at the surface.

Importance of Physical Environmental Characteristics and Geological History

The extent to which either of the above-mentioned possibilities regarding the origins of subsurface microbes may be true depends mostly on the physical characteristics and geological history of a particular subsurface environment. For example, an aquifer that is recharged from the surface and that is composed of porous sand might receive a continual input of surface microorganisms in the recharge water, even at considerable depths. In contrast, transport of surface microbes into some types of subsurface geological formations probably is greatly limited or even entirely prevented by impermeable overlying layers of clay or hard rock. The situation might be less clear in many subsurface environments, however. For example, when microorganisms are found in rocks that are of volcanic origin, they must have been transported into those rocks after they cooled down to temperatures that could support life, but when did transport first

occur and is it still occurring? If it is not still occurring, when did it stop? Unless the geological history of the formation in question is unambiguous (see the contrasting examples discussed in the next sections), there is no easy way to answer such questions.

Spatial Segregation from the Surface and within the Subsurface: Implications for Subsurface Microbial Evolution

The origin question also raises some interesting issues regarding the evolutionary history and genetic composition of microbial communities in the subsurface. With an environment that has been physically separated from the surface for a very long time, for example, one would expect that the microbes have evolved along a line distinct from that seen for similar (or closely related) organisms on the surface. The subsurface microorganisms in this case have been spatially segregated from the environmental conditions and other selective factors that influence evolution on the surface. They also have been segregated from the pool of genes in the surface microorganisms that might otherwise be exchanged through lateral gene transfer or other means. Spatial segregation could even be an important issue within some kinds of geological formations, especially those with very low porosity, in which the movement of groundwater, chemicals, and microbial cells is greatly restricted. In environments like these, microbial communities that are separated only by relatively small distances might be evolving along distinct lines because there is little or no communication between them. Spatial segregation, then, could be one of the most important factors that affect subsurface microbial evolution and the genetic composition of subsurface microbial communities.

Section Summary
- Microbes may have resided in certain subsurface environments for long periods.
- The physical traits of the subsurface might determine how long the microbes have resided therein.

- Spatial segregation most likely affects evolution and genetic composition of subsurface microbial communities.

ATLANTIC COASTAL PLAIN AQUIFER SEDIMENTS: AN AMBIGUOUS SITUATION

Physical Nature of Coastal Plain Sediments at the Savannah River Site

Much of the early microbiological research on relatively deep subsurface environments was carried out in Atlantic Coastal Plain sediments at or near the U.S. Department of Energy's (DOE) Savannah River Site (SRS) in Aiken, SC. The subsurface at this site consists of several layers of saturated sandy sediments that function as regional aquifers (Fig. 1). This system of aquifers extends down to the native bedrock, 250 to 600 meters below land surface, depending on the location within the site. Depth increases with increasing distance from the recharge zones (see Fig. 1). The distinct aquifer formations under the SRS are separated by layers of relatively dense clay (aquatards) that restrict but do not fully prevent the vertical movement of groundwater from one formation to another. As a result, most groundwater movement through the system is lateral, with the water originating at the recharge zones, where the formations intersect with the surface (several kilometers away from the areas where the subsurface samples examined in the DOE-funded research were obtained).

Age of Sediments versus Age of Groundwater: Segregation versus Transport

The sediments in the deepest aquifer formation at the SRS, the Middendorf, are thought to be 66 to 100 million years old. They would have been marine sediments at the time they were buried and almost certainly would have contained bacteria and, perhaps, other microorganisms. It is possible, then, that the Middendorf formation contains microorganisms that have been segregated from surface environments for a very long time. However, the age of the groundwater in this system was estimated to be less than 10,000 years, so it is also possible that microbes have been transported

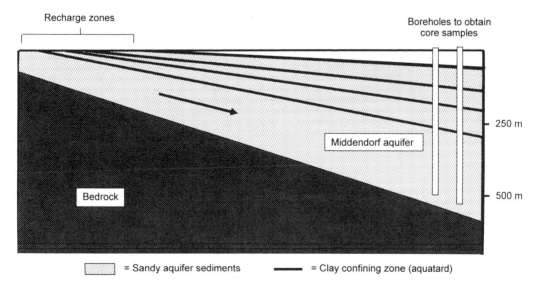

Recharge zones

Boreholes to obtain core samples

Middendorf aquifer

Bedrock

250 m

500 m

▨ = Sandy aquifer sediments   ▬▬ = Clay confining zone (aquatard)

FIGURE 1 Simplified representation of Atlantic Coastal Plain sediments at the U.S. Department of Energy's Savannah River Site in Aiken, SC. Regional aquifers composed mostly of sands are separated by relatively low-permeability aquatards composed mostly of clay. Arrow indicates principal direction of groundwater flow through this system. Depths below land surface (at right) are approximate.

into the Middendorf from the surface in relatively recent times. On the other hand, surface microorganisms might be filtered out or they may die off before they can reach the depths at which the Middendorf aquifer was sampled. The Atlantic Coastal Plain sediments at the SRS are a good example of a subsurface environment whose geological history and hydrological characteristics make it difficult to say whether the microbes therein are likely to have been spatially segregated from surface environments and surface microorganisms for long periods.

Lack of Evidence for Spatial Segregation from 16S rRNA–Based Phylogeny

Bacterial strains were isolated from the various aquifers at the SRS and from different depths within some aquifers as part of the DOE research effort at this site. These isolates were characterized by phylogenetic analysis of their 16S rRNA gene sequences to obtain information on their evolutionary relatedness to each other and to previously described bacteria (Balkwill et al., 1997a). Many of the isolates were placed within existing genera, including several that are commonly found in soils or other surface environments, but most of them appeared to be novel species of those genera. These findings could be taken as evidence that the subsurface microbial communities at the SRS have evolved somewhat independently of surface communities, but it is rather weak evidence at best. In all likelihood, one would find new species of culturable bacteria in almost any environment (e.g., a typical topsoil) that was examined as thoroughly as were the sediment samples from the SRS.

Differences were noted in the genus- and species-level compositions of the microbial communities in different aquifers at the SRS, so the clay confining zones (aquatards) mentioned earlier may afford some degree of spatial segregation between adjacent aquifer layers. However, the differences between the microbial communities could also be the result of different nutrients, nutrient concentrations, groundwater chemistry, or other factors. More-

over, some genera, such as *Arthrobacter* and *Acinetobacter*, were isolated from several of the aquifers at the SRS and thus appeared to be present throughout much of the subsurface vertical profile at this site. When phylogenetic trees were generated from the 16S rRNA gene sequences of the *Acinetobacter* isolates, the isolates from different aquifers did not form distinct clusters, as might be expected if they have been evolving independently as a result of spatial segregation. Similar results were obtained for the *Arthrobacter* isolates, so, in general, phylogenetic analyses of 16S rRNA gene sequences provided no convincing evidence that subsurface microbial communities at the SRS have been segregated from those at the surface or those in adjacent formations for an extended period.

Evidence for Spatial Segregation in Distinct Evolutionary History of Aromatic Degraders

One group of bacterial isolates from the deepest aquifer at the SRS might provide a slightly different perspective on the question of spatial segregation in the subsurface at this site. Several isolates from a depth of approximately 500 meters in the Middendorf formation were found to degrade a wide range of aromatic and polyaromatic compounds. Analysis of 16S rRNA gene sequences, biochemical properties, metabolic traits, and other characteristics have shown that these isolates represent three new species within the genus *Sphingomonas* (Balkwill et al., 1997b). (*Sphingomonas*, a genus that is characterized by the presence of sphingolipids that are not present in other bacteria, was recently reclassified into four closely related genera of bacteria that contain these lipids.) Several *Sphingomonas* strains with degradative abilities very similar to those of the subsurface strains have been isolated from surface soils in recent years. However, the subsurface isolates are phylogenetically distinct from the topsoil strains, being most closely related (according to their 16S rRNA-based phylogeny) to species that cannot degrade aromatic compounds. In addition, the degradative genes in the subsurface strains are located on a large plasmid, whereas those in most topsoil-derived aromatic-degrading sphingomonads are located in the chromosome (Kim et al., 1996). The degradative plasmid of one of the subsurface strains (*Sphingomonas aromaticivorans*, now *Novosphingobium aromaticivorans*, strain F199) has been fully sequenced in the DOE Microbial Genome Program (Romine et al., 1999). Analysis of the sequence data implies that the arrangement of degradative genes in the subsurface strain is quite distinct from that seen in topsoil isolates. The subsurface aromatic degraders, then, seem to have a distinct evolutionary history, which might reflect an extended period of isolation at depth.

A possible explanation for the seemingly contradictory or inconsistent findings regarding spatial segregation in the aquifer sediments at the SRS is that these sediments actually contain a mixture of organisms. Perhaps some of the extant bacteria have been segregated from surface conditions for an extended time (possibly millions of years), whereas others entered the sediments more recently (within the past 10,000 years) via transport from the surface. Additional studies, perhaps focusing on the phylogeny of genes other than 16S rRNA, would be required to resolve this complex situation, a potentially challenging opportunity for future research in this area.

Section Summary

- It is difficult to know how long microbes have resided in some types of subsurface environments because relatively old sediments may be permeated with much newer groundwater.
- It is difficult to know whether microbes have been physically segregated from surface microbes in some types of subsurface environments because the physical characteristics of these environments do not exclude recent transport of microbes from the surface.
- Phylogenetic analyses of microbes from the subsurface may give ambiguous results regarding possible spatial segregation from the surface.

RINGOLD FORMATION SEDIMENTS AT THE HANFORD SITE: A CASE FOR SPATIAL SEGREGATION WITHIN THE SUBSURFACE

Physical Nature of Ringold Formation Sediments at the Hanford Site

Between 1990 and 1996, the DOE research program on subsurface microbiology carried out several studies designed to understand how microbial numbers, diversity, physiological traits, and community structure correlate with physical and chemical conditions in the subsurface. One such study also provided information on how spatial segregation may have affected evolutionary change and shaped the genetic structure of subsurface communities, thereby creating patterns of genetic variability. This study focused on microbial communities living in a series of saturated, unconsolidated sediments situated 173 to 223 meters below land surface at DOE's Hanford Site in south central Washington State. These sediments are located in the Ringold formation, which overlies the Columbia River Basalt Group at the Hanford Site. A stratigraphic column showing the layers of different types of sediment materials that were studied and the areas from which core samples were obtained is shown in Fig. 2. The top of the interval that was sampled consists of dense, fine-grained lacustrine (i.e., lake) sediments (from 173 to 185 meters depth), under which there is a 5-cm-thick layer of volcanic tuff (the tephra layer). Two complete fluvial sequences are situated between the tephra layer and the underlying basalts, each starting with gravel at the base and grading up to sand and, finally, a paleosol (buried soil). The paleosol layers are situated at 185 to 193 meters (the upper paleosol) and at 211 to 225 meters (lower paleosol), with layers of fluvial sands and then coarse-grained fluvial gravel underlying each of them. However, the lowermost fluvial sand and gravel layers were not examined.

16S rRNA and *recA* Sequence Phylogenies

Core samples were obtained from approximately 40 depths within the aforementioned

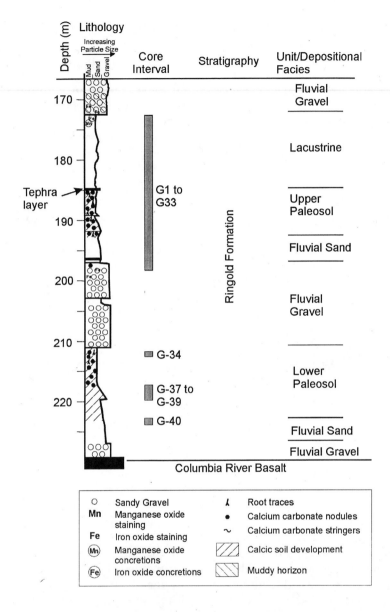

FIGURE 2 Lithographic and stratigraphic column of the Ringold Formation sediments examined at the Hanford Site. The areas from which core samples were taken are indicated, along with their respective sediment sample identities. Reprinted from *Applied and Environmental Microbiology* (van Waasbergen et al., 2000) with permission of the publisher.

subsurface sediment interval at the Hanford Site (see Fig. 2) and aerobic, chemoheterotrophic bacteria were isolated from each of these samples. Preliminary characterization of the isolates by phylogenetic analysis of partial 16S rRNA gene sequences indicated that many of them are likely to be *Arthrobacter* species. The *Arthrobacter* isolates were singled out for further study because they were detected at more depths than was any other species, they were often the most abundant form present, and they have been found in several other subsurface environments, including the SRS Atlantic Coastal Plain aquifers described earlier. *Arthrobacter* species are also quite common in soils and other surface environments. Nearly complete nucleotide base sequences for the 16S rRNA and *recA* genes were then determined for each of the *Arthrobacter* isolates and analyzed to study the evolutionary relatedness and genetic structure of the *Arthrobacter* communities in distinct strata (van Waasbergen et al., 2000). (For the purposes of this research, distinct strata were defined as sedimentary layers characterized by their own unifying lithologic composition.) Phylogenetic trees for the 39 isolates and 17 related type strains were produced by analyzing the 16S rRNA and *recA* gene sequences with maximum parsimony, distance matrix, and other standard methods.

The utility of 16S rRNA gene sequences as a phylogenetic tool is well established. RecA is a relatively conserved bacterial protein involved in DNA repair and recombination processes. RecA gene sequences were studied in addition to 16S rRNA gene sequences so that the resulting phylogenies could be compared and used to confirm one another, as has been the case in several other studies. In addition, *recA* is a protein-coding gene and is thus likely to be more variable at the nucleotide level than the 16S rRNA gene because the codon degeneracy in the genetic code allows for more fluctuation within protein-encoding genes. As a result, it was expected that the *recA* phylogenies might provide higher resolution of the more closely related isolates than would 16S rRNA-based phylogenies.

Spatial Segregation: Development of Genetically Distinct Populations in Different Strata

A phylogenetic tree generated by analysis of 16S rRNA gene sequences is illustrated in Fig. 3. The subsurface *Arthrobacter* strains cluster into several monophyletic groups that most likely correspond to distinct species or slightly

higher-order groups, based on their positions in the tree relative to the type strains that were included in the analysis for comparison. For most of the sediment strata described above, the majority of the isolates from a given stratum grouped in a single monophyletic cluster. This was true for the uppermost two meters of the lacustrine layer (referred to as the "upper lacustrine" in Fig. 3), the rest of the lacustrine layer, the upper and lower paleosol layers, and the fluvial sands. The monophyletic clusters show that the population of *Arthrobacter*-like bacteria in each of these stratigraphic units is generally similar (i.e., most or all members very closely related) throughout the unit itself but distinct from the populations in adjacent units. The lacustrine, fluvial sand, and paleosol units all have an important physical characteristic in common. They all have very low porosity and hydraulic conductivity, and this inhibits the movement of groundwater, nutrients, and bacterial cells through these layers. For all intents and purposes, then, the lacustrine, upper paleosol, and fluvial sands together form a low-permeability hydraulic barrier between two highly permeable gravel layers: the fluvial gravel overlying the sampled interval of sediments and the fluvial gravel unit between 197 and 210 meters depth. The latter fluvial gravel unit was the only comparatively porous, transmissive sediment examined in this study. In contrast to those from the other strata, the *Arthrobacter*-like isolates from this fluvial gravel layer were distributed among several lineages in the phylogenetic tree, rather than forming a monophyletic cluster (see Fig. 3).

The results of the *recA* sequence analysis were largely in agreement with those of the 16S rRNA gene sequence analysis (tree not shown). The most interesting difference was that isolates from the upper lacustrine and lower paleosol layers clustered together in the 16S rRNA tree (see Fig. 3) but separated into two distinct monophyletic groups in the *recA* tree. As was expected, then, the *recA* sequence analysis could better resolve the relationships among this set of isolates. The *recA* data also provide additional evidence that distinct pop-

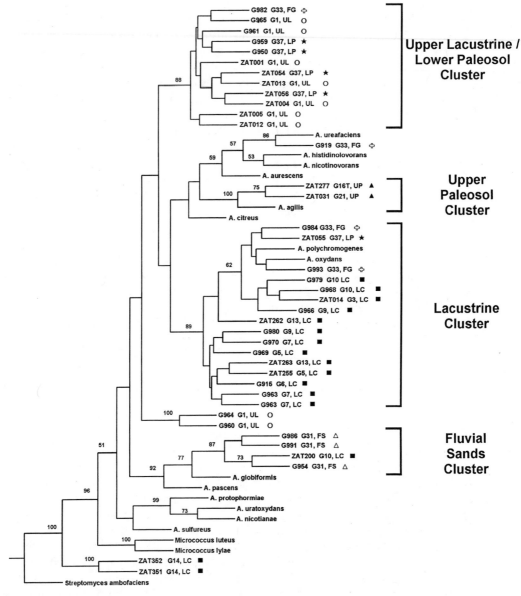

FIGURE 3 Phylogenetic tree based on parsimony analysis of 16S rRNA gene sequences for deep-subsurface and type strains of *Arthrobacter* species. For the deep-subsurface strains, the sediment sample identifier and stratum from which they originate are indicated. FG and ✛, fluvial gravel; FS and △, fluvial sand; LC and ■, lacustrine; LP and ★, lower paleosol; UL and ○, upper lacustrine; UP and ▲, upper paleosol. The tree is rooted using *S. ambofaciens* as an outgroup. Numbers at the nodes indicate the percentages of occurrence in 100 bootstrapped trees; only values that are 50% or greater are shown. Reprinted from *Applied and Environmental Microbiology* (van Waasbergen et al., 2000) with permission of the publisher.

ulations of *Arthrobacter* strains were isolated from different strata.

The results of the Hanford Site sediment study can be interpreted as good evidence that long-term spatial segregation has influenced the evolution and genetic structure of the microbial communities in the subsurface at this site. Permeability and hydraulic conductivity throughout much of the vertical interval of sediments examined are low enough to prevent the movement of groundwater, nutrients, and microorganisms between adjacent strata. This apparently has led to the development of genetically distinct populations of *Arthrobacter* strains in most strata, as evidenced by the clustering patterns in phylogenetic trees.

The *Arthrobacter* strains found in each isolated stratum were very closely related; in fact, virtually the same genotype was found throughout a given stratum. In the lacustrine sediment, for example, genetically very closely related strains were isolated from samples spanning a nine-meter-thick interval. The very low permeability of the isolated strata is just as likely to prevent migration of water, nutrients, and microbes *within* a particular formation as between the adjacent formations. Therefore, the presence of the same genotype throughout an entire stratum probably is best explained by the long-term persistence of a single genotype that was present throughout the formation when it was more porous (i.e., during burial or shortly thereafter). Environmental conditions that became uniform throughout the stratum after burial and subsequent compaction of the sediments might then select for the persistent genotype over time, allowing it to become the numerically predominant member of the population that is seen today. The findings for the lacustrine sediments in this study might seem to argue against this theory in that the uppermost portion of the lacustrine layer (upper lacustrine in Fig. 3) and the rest of the lacustrine yielded distinct populations of *Arthrobacter* strains. However, physical and chemical analyses indicated that the upper lacustrine layer is oxidized, whereas the rest of the formation is not. In addition, the concentration of total organic carbon (TOC) throughout most of the lacustrine sediment is 25 to 40 times higher than in the upper lacustrine. Apparently, these differing conditions selected for distinct dominant genotypes of *Arthrobacter* from among those that were present throughout the lacustrine sediments before they were buried and compacted.

Probable Effects of Selection after Burial and Compaction of Sediments

It seems likely that most of the stratigraphic units examined at the Hanford Site became isolated shortly after they were buried and that the *Arthrobacter* populations detected today have resided in those formations since their deposition six to eight million years ago. At the present time, however, there is no way to determine exactly how long these populations have been the dominant genotypes in their respective strata.

In contrast to the impermeable strata, there would be much less spatial segregation of microbial genotypes in the fluvial gravel layer because of the greater flow of groundwater and nutrients. Under these circumstances, there could be continual fluctuations in environmental conditions over time. Microbial species may also migrate into and out of the formation over time. Either or both of these factors could have prevented the development of a monophyletic dominant population in the fluvial gravel formation.

Unanswered Questions

Thus far, the study of the Hanford Site sediments is the only one to look in detail at how the physical characteristics of the subsurface environment may affect the genetic composition and evolution of the microbial communities therein. It leaves many unanswered questions that constitute interesting opportunities for further investigation. For example, would a similar relationship between permeability of the sediments and the resident microbial populations hold true for other genera (besides *Arthrobacter*) at this site? Do such relationships hold true in other types of environments at other sites? How have these relationships been affected in subsurface environments that have been altered or disrupted by humans (e.g., contaminated environments)? These are just some of the questions that could be addressed in future studies.

Section Summary

- Phylogenetic analysis of 16S rRNA gene sequences may provide evidence for long-term spatial segregation of microbes in some types of subsurface environments.
- Phylogenetic analysis of *recA* gene sequences (instead of 16S rDNA sequences) can provide increased sensitivity for detection of long-term spatial segregation.
- Physical characteristics like permeability may affect spatial segregation of microbes in the subsurface by determining whether microbes, water, and nutrients can move from one location to another.
- Phylogenetic analyses of gene sequences indicate that very low permeability probably resulted in long-term spatial segregation, so that microbial communities in different subsurface sediments at the Hanford Site evolved independently.

ULTRA-DEEP ENVIRONMENTS: AN UNUSUAL OPPORTUNITY?

Rationale for Studying Ultra-Deep Environments

Depending on the local geology and hydrology, the likelihood that subsurface microbial communities have been spatially segregated from surface microorganisms and their gene pools might be expected to increase with increasing depth. One might also expect the time over which the subsurface communities have been segregated to increase with depth. To study the effects of very long periods of isolation on the genetic structure and evolution of subsurface microorganisms, then, it might make sense to examine some of the deepest environments from which samples can be obtained. The drawback of this approach is that opportunities to acquire samples from great depths are rare and typically involve considerable expense, especially when they can only be obtained by drilling down from the surface. Still, there are some preliminary indications that it would be worthwhile to take advantage of any opportunities that do become available.

In 1992, researchers affiliated with the DOE subsurface microbiology program obtained samples of siltstone and soapstone from a depth of approximately 2.7 km below land surface at a drill site in Virginia. The rock samples were located within a geological formation known as the Taylorsville Triassic Basin. Geologic evidence suggests that microorganisms inhabited the basin between 200 and 140 million years ago, when the greatest penetration of meteoric water into the basin is thought to have occurred. It is unlikely that microorganisms have been introduced into the basin since that time because the groundwater flow has always been preferentially channeled through the overlying permeable sediments in the area. Moreover, the microbes would have to be transported through approximately 2.5 km of sedimentary rocks with very low porosity and permeability to reach the basin. In all likelihood, then, any microorganisms detected in the Taylorsville Triassic Basin now have survived there for a very long time.

Bacillus infernus: Evidence of Long-Term Spatial Segregation from the Surface?

Only a few bacterial strains were cultivated from the basin samples, nearly all of which were found to be novel species of bacteria. The most interesting of these was a new species of *Bacillus*, which was named *Bacillus infernus*. *B. infernus* was found to be a strict anaerobe (the only strict anaerobe in the genus when this novel species was described) that can grow on formate or lactate with Fe(III), MnO_2, trimethylamine oxide, or nitrate as an electron acceptor (Boone et al., 1995). It also grows fermentatively on glucose, is halotolerant (growth up to 6 M Na^+), and is thermophilic (optimum growth at 61.4°C). It is tempting to postulate that the physiological traits of *B. infernus* that are unusual (or even unique) for a species of *Bacillus* are the result of long-term segregation from other *Bacillus* species and from the environmental conditions that have influenced the evolution of this genus in surface environments.

Deep Microbial Communities That Are Independent of Input from the Surface

Another potentially interesting target for research on the effects of long-term segregation of microbes from the surface might be subsurface microbial communities that appear to be fully independent of any input of energy or nutrients from the surface (even over extended periods). For example, Karsten Pedersen and his coworkers at the University of Göteborg in Sweden have proposed that some microbial communities in deep granitic groundwaters within the Fennoscandian shield are dependent on hydrogen-oxidizing bacteria for primary production rather than on solar energy. Stable-isotope studies indicate that the groundwater in this environment is highly unlikely to be of meteoric origin. The same is true for some very deep formation waters (3.2 to 3.6 km below land surface) that can be accessed in South African gold mines. Analyses of directly extracted nucleic acid from one of these water sources have detected the presence of bacteria with very-deep-branching lineages (based on 16S rRNA-based phylogeny) that extend nearly back to the origin of the bacterial domain. However, none of the unusual organisms have been cultured or further characterized yet. Moreover, additional research is needed to ascertain with any certainty just how long any of these surface-independent microbial communities have been segregated from surface input.

Section Summary

- There is evidence to indicate that microbes in ultra-deep environments may have been physically segregated from surface microbes for very long times.
- Microbes and microbial communities in

ultra-deep environments may have evolved unique combinations of traits in response to long-term spatial segregation.
- If so, ultra-deep environments might provide especially good opportunities to study evolution of subsurface microbes.

OTHER FACTORS THAT MIGHT INFLUENCE MICROBIAL EVOLUTION IN THE SUBSURFACE

Lateral Gene Transfer: Unlikely to Occur in Low-Permeability Environments

Lateral gene transfer most likely has a very significant effect on the evolution and genetic structure of microbial communities in some environments (see chapters 7 through 11). However, it is not yet known whether this phenomenon takes place in terrestrial subsurface environments. Lateral gene transfer would appear to be very unlikely in dense, impermeable materials like the lacustrine sediments beneath the Hanford Site (see above). Microbial cells that are not in direct contact with other cells would not have any way to come into contact with other cells because they are unable to move through the sediment. As a result, conjugation would be infrequent at best. Transformation and transduction seem equally unlikely because neither groundwater nor chemical substances are moving through the sediment to any appreciable extent either.

Lateral Gene Transfer: Does It Occur in High-Permeability Environments?

Lateral gene transfer might be more feasible in relatively porous, transmissive sediments such as the sandy aquifers at the SRS or the fluvial gravel layer at the Hanford Site, presumably because water, chemical substances, and possibly microbial cells can move through the system. However, even in these relatively porous environments, lateral gene transfer might be limited or prevented entirely by factors other than porosity and permeability. For example, cell densities might be too low to permit any type of gene transfer in some kinds of subsurface environments. Also, it is generally believed that most microbes in the subsurface are attached to solid surfaces. This appears to be the case because cell counts for groundwater typically are one to three orders of magnitude lower than cell counts for the associated sediments or rocks. Does attachment of cells to surfaces interfere with or promote mechanisms of gene transfer and, if so, to what extent? Given the complex chemistry of the mineral surfaces in some types of subsurface environments (e.g., clays), what happens to plasmids or other pieces of free DNA that are released from cells? Can they travel any distance through the environment, or are they rapidly adsorbed onto mineral surface and/or denatured? Are transducing phages more likely to move through the subsurface without being damaged than free DNA? At present, there are no definitive answers to these questions.

Possible Evidence from Antibiotic Resistance Traits of Subsurface Microorganisms

When isolates from the Ringold Formation sediments at the Hanford Site (see above) were screened for antibiotic resistance, 58% of them were found to be resistant to at least one of eight antibiotics tested (Table 1). The percentage of strains resistant to any single antibiotic ranged from 0% for kanamycin and rifampicin to 34% for nalidixic acid (Table 2). Moreover, 14% of the isolates displayed multiple resistance (defined arbitrarily here as resistance to three or more of the eight drugs tested). The presence of multiple resistance suggests that lateral gene transfer may have occurred because this mechanism is known to produce multiple resistance among bacteria in clinical settings, primarily

through transfer of plasmids. As noted earlier, however, the very low permeability of most of the Hanford Site sediments makes it unlikely that gene transfer has occurred since the sediments were buried and compacted. The resistance characteristics seen in the isolates now are therefore more likely to have been present in the bacterial populations when the sediments were originally deposited. In contrast to those from the Hanford Site sediments, a larger percentage of bacterial isolates from the deepest comparatively porous, sandy aquifer at the SRS (Middendorf aquifer, see above) displayed multiple resistance (36% versus 14%; Table 1). The percentage of isolates that were resistant to each of the eight drugs examined was higher as well (Table 2). Could the higher levels of resistance and multiple resistance in the more porous sediments at the SRS be taken as evidence that lateral gene exchange occurs more readily in this system? Alternatively, could they indicate that the bacteria in the Middendorf aquifer are less segregated from those at the surface than are the bacteria in the Hanford sediments? These are interesting questions.

Very Slow Growth and Dormancy: Implications for Evolution of Subsurface Microbes

The extremely low nutrient levels that prevail in most subsurface environments might be yet another factor that strongly influences the evolution of subsurface microbes. Microbes in the subsurface must be adapted for long-term survival under very low-nutrient conditions that most likely include extended periods of starvation. The mechanisms that subsurface microorganisms use to cope with very low nutrient

TABLE 1 Percentages of subsurface bacterial isolates from different sample sites possessing resistance to antibiotics

Sample site	No. of isolates	% of strains resistant to the following no. of antibiotics:								
		0	1	2	3	4	5	6	7	8
Savannah River Site (Atlantic Coastal Plain sediments)	50	0	24	40	20	6	6	0	4	0
Hanford Site (Ringold Formation sediments)	38	42	26	18	8	3	3	0	0	0

TABLE 2 Percentages of subsurface bacterial isolates from different sites possessing resistance to specific antibiotics

Sample site	No. of isolates	% of strains resistant to the following antibiotics[a]:							
		Amp	Ery	Gen	Nal	Str	Kan	Rif	Tet
Savannah River Site (Atlantic Coastal Plain sediments)	50	36	10	16	56	22	2	18	76
Hanford Site (Ringold Formation sediments)	38	21	8	11	34	3	0	0	24

[a]Amp, ampicillin (50 μg/ml); Ery, erythromycin (30 μg/ml); Gen, gentamicin (30 μg/ml); Nal, nalidixic acid (50 μg/ml); Str, streptomycin (50 μg/ml); Kan, kanamycin (50 μg/ml); Rif, rifampicin (50 μg/ml); Tet, tetracycline (12.5 μg/ml).

levels are not yet fully understood. However, it is generally recognized that their metabolic and growth rates must be exceedingly slow. The microbes in a typical subsurface environment may persist for very long periods solely on endogenous sources of energy, in which case DNA replication and subsequent cell division probably would be very rare events. It is also possible that certain types of subsurface microorganisms persist for long periods in a dormant or nearly dormant state, in which they have little or no detectable metabolic activity. How might dormancy or very slow growth rates affect microbial evolution in the subsurface? Does microbial evolution occur more slowly here than in most surface habitats? Does evolution stop for all intents and purposes in dormant or very inactive subsurface microbial communities? These questions remain unanswered, but they merit further investigation because slow growth and/or dormancy could affect subsurface microbial evolution as strongly as spatial segregation apparently does.

Section Summary
- The effects of very slow growth or dormancy on subsurface microbial evolution are not yet known.
- Lateral gene transfer could affect the rate and direction of subsurface microbial evolution if it can occur under subsurface environmental conditions.
- Analysis of antibiotic resistance traits of subsurface microbes might provide preliminary evidence for gene transfer in some kinds of subsurface environments.

- Questions about the importance of lateral gene transfer, slow growth, and other factors in the evolution of subsurface microbes represent good opportunities for future research.

SUMMARY OF THE EFFECTS OF SPATIAL SEGREGATION ON MICROBIAL LIFE
Very little is known in detail about the evolution of microbes in terrestrial subsurface environments, but there are good indications that this is a potentially quite interesting topic that presents many possible directions for future research. The physical and chemical nature of the environment most likely has a pronounced effect on the evolution and genetic composition of subsurface microbial communities. Sometimes the physical nature of the environment leads to spatial segregation, such that subsurface microbes may be physically isolated from those at the surface (and their gene pool) for very long periods. In these situations, subsurface and surface communities are likely to evolve along very different tracks because of their isolation and the different environmental conditions to which they are exposed over time. Spatial segregation appears to be a factor within the subsurface as well, when the microbial communities in adjacent strata become trapped within their respective strata for extended periods. At least one study has shown that phylogenetically distinct communities can develop in closely juxtaposed strata when low porosity prevents the movement of water, nutrients, or microbial cells between these strata. Still, there are many questions about the

evolution of subsurface microorganisms that remain to be addressed. To what extent is spatial segregation a factor in environments that have not yet been studied in any detail, especially environments with very different physical and chemical characteristics? Is spatial segregation the most important factor in subsurface microbial evolution? What about the fact that subsurface microorganisms grow very slowly and might not reproduce for long periods? Most likely, there is no one factor that is most important in all types of subsurface environments. The issue of lateral gene transfer also needs to be resolved. Does this phenomenon even occur in the terrestrial subsurface? If so, under what conditions does it occur and how important is it relative to other factors such as spatial segregation? These are just some of the questions that could be addressed in future research on the evolution and genetic structure of subsurface microbial communities.

QUESTIONS

1. What evidence is there that microorganisms in some types of subsurface environments have been spatially separated from each other and/or from surface microbes for long periods?

2. What evidence is there that long-term spatial separation may have influenced the evolution and genetic composition of subsurface microbial communities?

3. What are some of the most difficult problems associated with studying the evolution of subsurface microbial communities?

4. What are the key unresolved questions regarding the evolution of microorganisms in terrestrial subsurface environments?

5. What types of samples from the subsurface might provide the best opportunities to obtain more definitive information on the evolution of subsurface microorganisms?

REFERENCES

Balkwill, D. L., G. R. Drake, R. H. Reeves, J. K. Fredrickson, D. C. White, D. B. Ringelberg, D. P. Chandler, M. F. Romine, D. W. Kennedy, and C. M. Spadoni. 1997a. Taxonomic study of aromatic-degrading bacteria from deep-terrestrial-subsurface sediments and description of *Sphingomonas aromaticivorans* sp. nov., *Sphingomonas subterranea* sp. nov., and *Sphingomonas stygia* sp. nov. *Int. J. Syst. Bacteriol.* **47:**191–201.

Balkwill, D. L., R. H. Reeves, G. R. Drake, J. Y. Reeves, F. H. Crocker, M. B. King, and D. R. Boone. 1997b. Phylogenetic characterization of bacteria in the Subsurface Microbial Culture Collection. *FEMS Microbiol. Rev.* **20:**201–216.

Boone, D. R., Y. Liu, Z. Zhao, D. L. Balkwill, G. R. Drake, T. O. Stevens, and H. C. Aldrich. 1995. *Bacillus infernus* sp. nov., an Fe(III)- and Mn(IV)-reducing anaerobe from the deep terrestrial subsurface. *Int. J. Syst. Bacteriol.* **45:**441–448.

Kim, E., P. J. Aversano, M. F. Romine, R. P. Schneider, and G. L. Zylstra. 1996. Homology between genes for aromatic hydrocarbon degradation in surface and deep-subsurface *Sphingomonas* strains. *Appl. Environ. Microbiol.* **62:**1467–1470.

Romine, M. F., L. C. Stillwell, K.-K. Wong, S. J. Thurston, E. C. Sisk, C. Sensen, T. Gaasterland, J. K. Fredrickson, and J. D. Saffer. 1999. Complete sequence of a 184-kilobase catabolic plasmid from *Sphingomonas aromaticivorans* F199. *J. Bacteriol.* **181:**1585–1602.

van Waasbergen, L. G., D. L. Balkwill, F. H. Crocker, B. N. Bjonstad, and R. V. Miller. 2000. Genetic diversity among *Arthrobacter* species collected across a heterogeneous series of terrestrial deep-subsurface sediments as determined on the basis of 16S rRNA and *recA* gene sequences. *Appl. Environ. Microbiol.* **66:**3454–3463.

FURTHER READING

Amy, P. S., and D. L. Haldeman (ed.). 1997. *The Microbiology of the Terrestrial Deep Subsurface.* CRC Lewis Publishers, New York, N.Y.

Bachofen, R. (ed.). 1997. Proceedings of the 1996 Internal Symposium on Subsurface Microbiology (ISSM-96). *FEMS Microbiol. Rev.* **20:**179–638. (Special issue.)

Fredrickson, J. K., and M. Fletcher (ed.). 2001. *Subsurface Microbiology and Biogeochemistry.* John Wiley & Sons, New York, N.Y.

Ghiorse, W. C. (ed.). 1989. Special issue on deep subsurface microbiology. *Geomicrobiol. J.* **7:**1–130.

Ghiorse, W. C., and J. T. Wilson. 1988. Microbial ecology of the terrestrial subsurface. *Adv. Appl. Microbiol.* **33:**107–172.

Pedersen, K. 2002. Igneous rock aquifers microbial communities, p. 1661–1673. *In* G. Bitton (ed.), *Encyclopedia of Environmental Microbiology*, vol. 3. John Wiley & Sons, New York, N.Y.

Wilson, J. T., J. F. McNabb, D. L. Balkwill, and W. C. Ghiorse. 1983. Enumeration and characterization of bacteria indigenous to a shallow water-table aquifer. *Ground Water* **21:**134–142.

ARE YOU OUT THERE? INTERCELLULAR SIGNALING IN THE MICROBIAL WORLD

Mike Manefield, Sarah L. Turner, Andrew K. Lilley, and Mark J. Bailey

15

Until recently bacterial intercellular signaling was considered to be an obscure phenomenon occurring in a limited number of organisms to mediate multicellular behaviors. It is now apparent that intercellular signaling is not uncommon among the bacteria and regulates phenotypes of medical, agricultural, industrial, and domestic significance. Despite an explosion in attention to the phenomenon, the evolution of intercellular signaling mechanisms (ISMs) remains poorly understood, as does its role in mediating ecological interactions with both prokaryotes and eukaryotes. For the purpose of this chapter the wealth of information on this topic has been distilled to emphasize recurring themes in bacterial signaling rather than the distinguishing features of individual systems.

AN INTRODUCTION TO INTERCELLULAR SIGNALING

The molecular machinery underlying intercellular signaling differs depending on the signaling organism. There are, however, three components common to all intercellular signaling systems. Clearly, the signaling molecule is the primary component. Although no standard definition exists, a bacterial intercellular signaling molecule can be defined as any molecule produced by a bacterium that exits the cell and elicits a specific and beneficial transcriptional response in receptive bacterial cells. A number of bacterial metabolites fit this description (Fig. 1). These include N-acyl-L-homoserine lactones (AHL) first discovered in *Vibrio fischeri* and produced by various gram-negative bacteria, furanosyl borate diester (AI-2) first discov-

Mike Manefield, Sarah L. Turner, Andrew K. Lilley, and Mark J. Bailey, CEH Oxford, Mansfield Road, Oxford OX1 3SR, United Kingdom.

Microbial Evolution: Gene Establishment, Survival, and Exchange
Edited by Robert V. Miller and Martin J. Day, ©2004 ASM Press, Washington, D.C.

ered in *Vibrio harveyi* and produced by a diverse range of bacteria, signaling peptides produced by many gram-positive bacteria, butyrolactones produced by *Streptomyces* species and 3-hydroxy-palmitic acid methyl ester (PAME) produced by *Ralstonia solanacearum*.

The two remaining components common to all bacterial intercellular signaling systems are the genes encoding signal generation and those encoding the signal response. Signal generation genes may encode for an enzyme or enzymes responsible for signal synthesis or translate directly into peptide-based signals. The genetic material required for signal generation may also encode proteins for signal modification and/or export. The genetic material required for a response to a signal may encode a single signal-responsive transcriptional regulator protein or a network of signal transduction proteins. In some cases signal response machinery also includes proteins for transportation of the signal into receiving cells.

The range of bacterial phenotypes regulated by intercellular signaling is fascinating in its diversity. Within the sizeable and motley collection of proposed signaling-regulated phenotypes, a number of common themes exists. The production of extracellular degradative enzymes, thought to allow access to nutrients contained in host tissues, is one of the most common intercellular signaling-regulated phenotypes. For example, various *Staphylococcus* species regulate expression of genes encoding enzymes with protease and lipase activity via a peptide-based signaling system. Similarly, *Aeromonas*, *Pseudomonas*, *Erwinia*, and *Serratia* species produce extracellular degradative enzymes in response to AHL accumulation. Additionally, extracellular enzyme production is regulated by PAME in *R. solanacearum*. The unrelated evolution of these three distinct signaling mechanisms, each regulating the production of invasive extracellular enzymes in different organisms, is suggestive of a strong selection for the regulation of an ecological function (resource acquisition from host tissues) by intercellular signals.

Another phenotype commonly regulated by intercellular signals is the production of anti-

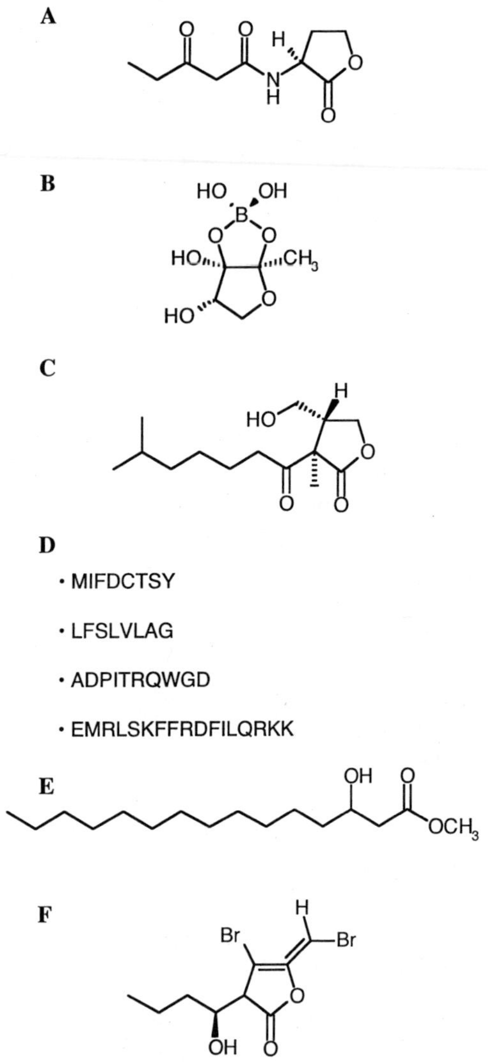

FIGURE 1 Structures of well-characterized signaling molecules. (A) *N*-3-oxohexonoyl-L-homoserine lactone (OHHL), which regulates expression of bioluminescence in *V. fischeri*. (B) Furanosyl borate diester (AI-2), which regulates expression of bioluminescence in *V. harveyi*. (C) A γ-butyrolactone (A-factor), which regulates antibiotic production in *Streptomyces griseus*. (D) Amino acid sequences of peptide signals. In the order listed they are AIF, which regulates antibiotic production in *Staphylococcus aureus*; cAD1, which regulates conjugation in *Enterococcus faecalis*; ComX, which regulates competence in *Bacillus subtilis*; and CSF, which regulates competence in *Streptococcus pneumoniae*. (E) PAME, which regulates virulence factor production in *R. solanacearum*. (F) A halogenated furanone produced by the red marine macroalga *D. pulchra* that acts as an AHL antagonist.

bacterial compounds, thought to be a means of excluding competing microbes from a niche. For example, synthesis of the antibiotics phenazine and carbapenem is AHL dependent in certain *Pseudomonas* and *Erwinia* species, respectively. Streptomycin and related antibiotics are produced in response to butyrolactones in *Streptomyces* species. Lantibiotic production is regulated by intercellular signaling peptides in *Lactococcus* and *Bacillus* species (Kleerebezem et al., 1997). Again, these three separately evolved systems have married a common ecological function (resource defense) with regulation by ISMs.

A third common theme among phenomena regulated by intercellular signaling is that of gene transfer. For example, conjugation of the tumor-inducing Ti plasmid from *Agrobacterium tumefaciens* to plasmid-free recipients is AHL regulated. Analogously, the conjugation of plasmids encoding antibiotic resistance is regulated by signaling peptides in certain species of *Enterococcus*. Finally, *Streptococcus* and *Bacillus* species induce competence for genetic transformation in themselves through genesis and response to peptide-based signals. This constitutes a third example whereby analogous regulatory mechanisms have proved independently to be the most successful means of controlling analogous bacterial behaviors. One must conclude that probing the environment with intercellular signals before committing resources to particular activities offers substantial fitness benefits.

Of all the ISMs currently known the AHL-dependent systems are the most thoroughly characterized (Fig. 2). AHL-mediated gene expression has been reported in both terrestrial and marine environments and is most often involved in bacterial interactions with higher organisms (plants and animals) often of a symbiotic or pathogenic nature. Further, AHLs regulate as diverse a range of phenotypes as that found in the entire gambit of intercellular signaling-regulated phenotypes. For these reasons AHL-mediated gene expression can be considered representative of intercellular signaling systems and will thus be used here as a model to consider the means by which ISMs became established and survive and

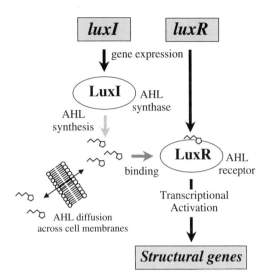

FIGURE 2 *N*-Acyl-L-homoserine lactone (AHL)-mediated gene expression in *V. fischeri*. The *luxI* gene encodes the AHL synthase protein LuxI. AHLs diffuse across cellular membranes. Above a certain concentration AHLs bind to the LuxR signal receptor protein (encoded by the *luxR* gene) which activates transcription of target (structural) genes, thereby expressing the signaling regulated phenotype. In the case of the marine symbiont *V. fischeri* the specific AHL is *N*-3-oxohexonoyl-L-homoserine lactone and the structural genes are those encoding luminescence (*luxCDABE*).

to appraise the impact of signaling capabilities on genome evolution.

Section Summary

- Intercellular signaling mechanisms (ISMs) are composed of a signal synthase, an intercellular signal and a signal responder.
- These fundamental components of ISMs have evolved several times independently and have converged functionally to regulate exoenzyme production, antibiotic production, and gene transfer.

THE ADVANTAGES OF INTERCELLULAR SIGNALING

Quorum Sensing

What advantages do ISMs offer bacteria? More specifically, with respect to our model system, what is the advantage of AHL-mediated gene expression? The common answer to this question is that the AHL-based ISM, and any other

ISM for that matter, confers the ability to sense and respond to cell density. The cell density sensing or "quorum sensing" hypothesis proposes that the concentration of an intercellular signal, produced by a growing population and accumulating in the surrounding extracellular environment, is proportional to the cell density of that population (Fig. 3A and B). Thus, the ability to respond to a threshold concentration of AHL equates to the ability to change behavior at a specific cell density.

This hypothesis is based on laboratory experiments using pure cultures in liquid media. In such artificial systems, where environmental parameters remain relatively constant, there is little doubt that AHL concentration is dependent on and thus reflective of cell density. Therefore, in the laboratory, phenotypes responsive to a specific concentration of signal are consistently expressed at a specific cell den-

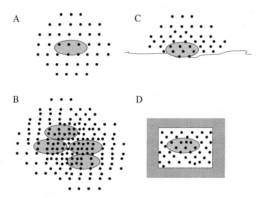

FIGURE 3 Theoretical, diagrammatic representation of the accumulation of intercellular signals in different environments. The ratio of signal molecules (black dots) to cells (grey ovals) is constant in each frame. Target gene expression occurs when signal concentrations in each cell reach a certain level (for example, eight black dots per cell). (A) A single cell producing a diffusible metabolite with no external limitation to diffusion (a planktonic marine cell for example). (B) A population of cells producing a diffusible metabolite with no external limitation to diffusion (a floc for example). (C) A single cell producing a diffusible metabolite with diffusion limited by the surface to which it is attached (on a marine surface for example). (D) A single cell producing a diffusible metabolite with diffusion limited by the space in which the cell is confined (a light organ or xylem for example).

sity. The quorum-sensing hypothesis, however, has never been tested in the environment.

Why has the quorum-sensing hypothesis never been demonstrated in the environment? First, the concept of cell density in a laboratory culture (number of cells per volume) is distinct from that in the more environmentally relevant and difficult-to-measure context of a biofilm or colony (number of cells per surface area). Second, direct methods to measure AHL concentration (not based on bioassays) have been slow to develop even for laboratory-based experiments, let alone those conducted in the environment. Measuring cell density and AHL concentration in the environment represents a major technical challenge to the field. We are left to consider hypothetically whether AHL concentration will reflect cell density in the environment.

Changes in environmental parameters that affect the synthesis, diffusion, or stability of AHLs will disrupt the relationship between AHL concentration and cell density. How responsive are these three factors to environmental change? First, while it is generally believed that AHL synthesis is constitutive, it seems unlikely that expression of AHL synthase genes or the activity of the synthase enzyme remains constant under all environmental conditions. It is clear, for example, that AHL synthesis is dependent on the concentration of plant metabolites in *A. tumefaciens* and on AHL concentration itself in *V. fischeri*. Second, the rate at which AHLs diffuse out of the cytoplasm of producing cells and away from cells in a producing population has important consequences for the behavior of quorum-sensing models (Nilsson et al., 2001). Movement of AHLs out of a cell is likely to depend on the state of the cytoplasmic and/or outer membranes, which adapt responsively to environmental conditions. The movement of AHLs away from the producing population of cells will also depend on the environmental context. For example, the presence of a physical barrier to AHL diffusion results in induction of the regulated phenotype at lower densities. This concept is illustrated in Fig. 3C and D. Finally, although there is little information on the persistence of AHLs in the environment, they are known to degrade both chemically, at high pH, for example, and biologically, through the action of lactonases (Dong et al., 2001).

Let us briefly consider why bacteria might express the phenotypes mentioned above (extracellular enzymes, antibacterial compounds, and gene transfer mechanisms) at specific cell densities. A popular explanation with respect to invasive phenotypes is based on the concept of a minimum infective dose. A commonly used analogy to describe this adjunct to the quorum-sensing hypothesis is that of an army amassing in stealth before launching an attack. For example, the plant pathogen *Erwinia carotovora* partially regulates the expression of plant-cell-wall-degrading enzymes in response to a certain concentration of AHL. Here, according to the quorum-sensing hypothesis, the AHL-dependent regulatory system ensures the invasive phenotype is only expressed at a specific cell density. The rationale is that the plant immune system, which responds to tissue damage, is not capable of resolving *E. carotovora* infections at (or above) this cell density. Rephrased, this adjunct hypothesis states that an *E. carotovora* strain expressing exoenzymes independently of cell density cannot mount a successful infection. This hypothesis was tested indirectly when AHL⁻ mutants of *E. carotovora* were inoculated on to the leaves of a plant genetically modified to produce AHL (Fray et al., 1999). Thus, *E. carotovora* was expressing its invasive phenotype at low cell densities. The finding that an infection was successfully mounted throws into some doubt this stealth-invasion model.

Why produce antibacterial compounds at high cell densities? In this context the adjunct hypothesis is that low-density populations are not capable of generating an effective dose of antibacterial, while high-density populations are. Antibacterial production at low cell densities therefore taxes available resources without the associated benefit of resource defense. In considering this adjunct hypothesis one must weigh the cost of the ISM against the cost of antibiotic production. Expressed another way, ISM regulated antibiotic production will only evolve according to this concept if it is more taxing to produce an antibiotic than it is to produce and respond to a signal.

Why initiate gene transfer mechanisms in response to cell density? As above, the adjunct hypothesis is simple. Conjugation is likely to be more successful in a high-density population because the number of donors and potential recipients will be higher, thus enhancing the frequency of transfer. To further illustrate the point with the non-AHL ISM-regulated example of transformation, a high-density population will generate more naked DNA than a low-density population, which in turn will have more chance of being taken up in a high-density population because there are more potential recipients.

A little-discussed alternative to these three adjunct hypotheses is that the accumulation of an intercellular signal staves off the expression of auxiliary phenotypes while essential functions amounting to replication are established. From this angle the quorum-sensing mechanism can be seen as a means of allocating resources to activities at the heart of survival. The focus then shifts from expressing signaling-regulated phenotypes at the appropriate time to not expressing them at inappropriate times. Based on the likely vulnerability of the relationship between intercellular signal accumulation and cell density to shifting environmental parameters it seems more likely that it is a coarse mechanism evolved to miss a target rather than to hit one. Figure 4 illustrates this subtle but important difference between ISMs activating gene expression when a certain density (a quorum) of cells is achieved and ISMs stalling gene expression at low cell densities as a resource management strategy.

Conclave Sensing

The influence of environmental parameters on the accumulation of an intercellular signal and therefore on the expression of signaling-dependent genes might suggest that the signaling mechanism acts to probe the environment.

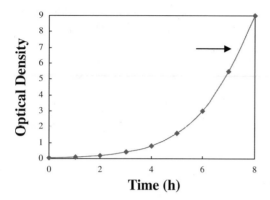

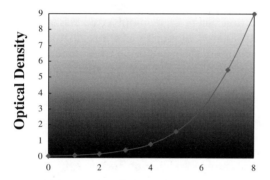

FIGURE 4 The importance of cell density related to a bacterial growth curve (diamonds). (A) The traditional view of cell density sensing, known as quorum sensing, suggests that ISMs have evolved to induce gene expression at a specific cell density, e.g., at an optical density of 7 (arrow). (B) An alternative view of cell density sensing suggests that ISMs have evolved to prevent gene expression at low cell densities when resources are best utilized for primary functions, such as cell division, rather than secondary functions, such as antibiotic or exoenzyme production. The intensity of the shading represents evolution's insatiable demand for replication, which is imagined to abate as cell numbers increase. The intensity of shading, therefore, also inversely represents the availability of resources for the expression of auxiliary phenotypes. This model, in which ISMs are not required to identify specific cell densities, is consistent with the disruptive influence of environmental parameters on the fidelity of the cell-density-sensing mechanism.

If the signal does not accumulate then the conditions are inappropriate for expression of the signaling-regulated phenotype. Because the environmental conditions must be appropriate for quorum sensing to function, the ISM must first and foremost be an environmental

sensing device. Of the environmental parameters signals could be used to measure, space seems the most obvious. As mentioned above, the presence of physical barriers to AHL diffusion results in the accumulation of signal disproportionately to cell density. Thus, the signaling mechanism may have evolved as a means of detecting spatial confinement (compartment or conclave sensing), functionally analogous to the use of sonar by bats to establish spatial relationships. The confined environments in which ISMs are found lend credence to the proposal. Examples include *V. fischeri* in the light organs of marine fish and squid, *Pseudomonas aeruginosa* in the human lung (Davies et al., 1998), and *R. solanacearum* in the plant xylem.

Orchestration

A final concept related to the advantage of intercellular signaling is that of coordination or orchestration. It is difficult to imagine that a diffusible intercellular signal, whatever its function, will not stimulate the population to conduct a united transcriptional response. Orchestration, therefore, is an inevitable consequence of ISMs. Orchestrated responses are the common element of all bacterial multicellular behaviors, the evolutionary advantages of which are discussed in detail elsewhere. An example not conventionally viewed from this angle is that of AHL-regulated expression of the *car* operon (*carABCDEFGH*) which encodes the production of and resistance to the antibiotic carbapenem in *E. carotovora*. The need to produce ecologically functional concentrations of the antibiotic was discussed above in the context of quorum sensing. This focused on antibiotic production. The evolution of an ISM regulating the *car* operon in the context of orchestration requires the focus to shift to the antibiotic resistance genes. Is it possible that the orchestrated expression of carbapenem resistance throughout the population is more important than the cell-density-dependent production of carbapenem? Death is certainly a stronger selection pressure than resource loss.

Advocates of the quorum-sensing hypothesis suggest that ISMs link gene expression in individual cells with the cell density of the population of which they are a part. In this hypothesis the ability of intercellular signals to sense environmental parameters and to orchestrate gene expression is thought to be secondary to the ability to sense cell density. It is clear, however, that the sensing of environmental parameters and the orchestration of gene expression are inevitable consequences of the ISM and that cell density sensing is the only fallible component of the system. That is to say that you cannot have quorum sensing without environmental sensing and orchestration, but you can have environmental sensing and orchestration without quorum sensing. In conclusion, the ISM confers on bacteria the ability to orchestrate the expression of particular genes under appropriate environmental conditions. The quorum-sensing component is likely to be secondary, serving as a mechanism of temporal resource allocation by staving off the expression of auxiliary phenotypes at low cell densities.

Section Summary
- ISMs orchestrate gene expression allowing bacterial cells to cooperate.
- ISMs probe the environment allowing bacteria to sense spatial parameters.
- ISMs manage resource allocation by preventing the expression of auxiliary phenotypes at low cell density.

THE EVOLUTIONARY HISTORY OF INTERCELLULAR SIGNALING

How Are Distinct ISMs Distributed among the Bacteria?

The fact that distinct groups of bacteria have converged on functionally analogous mechanisms (ISMs) to coordinate particular behaviors indicates the strong selective advantage that such mechanisms must confer in the environment. Figure 5 illustrates the distribution of the better characterized ISMs throughout the bacteria. It reveals that peptide- and butyrolactone-based ISMs have only been identified in the phylum Firmicutes and that

known examples of AHL-based ISMs are restricted to the α, β, and γ classes of the phylum Proteobacteria. From this it appears that ISMs are not widely distributed among bacteria, but we must exercise caution with this conclusion in the name of undersampling. This caveat is important and will be emphasized throughout the following discussion relating ISMs to phylogeny.

Because we are using the AHL ISM as a model system it is worth further scrutinizing its distribution among the classes, orders, families, and genera of the Proteobacteria. Table 1 lists the proteobacterial orders according to class and denotes the number of families and genera in each order harboring known AHL-producing species. This survey reveals that only 4% of all proteobacterial genera have been found to contain AHL-producing species. Again, this suggests that the phenomenon is less common than many investigators believe, but again, a comprehensive, systematic survey of Proteobacteria for AHL production has never been undertaken.

What Can the Phylogeny of AHL Receptor Proteins Tell Us about the Evolution of ISMs?

As a result of much attention being paid to the AHL ISM, many sequences of its regulatory proteins are available. These sequence data can be interrogated by phylogenetic analysis, thus revealing the evolutionary relationships between homologues. Such an approach allows us to consider whether the distribution of the components of the AHL ISM is a result of recent horizontal transfers or evolution through lineages derived from a last common ancestor (LCA). In short, it provides an evolutionary context in which to consider the selection pressures responsible for generating the AHL ISM.

As illustrated in Fig. 2, the most basic AHL ISM has two genetic components, a gene encoding an AHL synthase protein (LuxI homologue) and a gene encoding an AHL receptor protein (LuxR homologue). More often than not *luxI* and *luxR* homologues appear adjacent to each other, although their

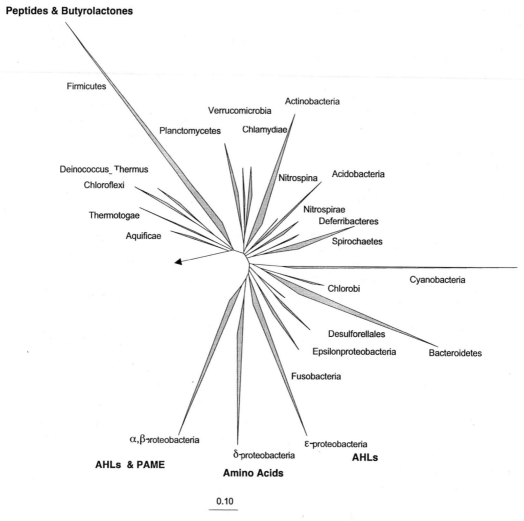

Peptides & Butyrolactones

Firmicutes

Actinobacteria

Verrucomicrobia

Planctomycetes   Chlamydiae

Deinococcus_ Thermus

Chloroflexi

Nitrospina   Acidobacteria

Thermotogae

Nitrospirae
Deferribacteres

Aquificae

Spirochaetes

Cyanobacteria

Chlorobi

Desulforellales

Epsilonproteobacteria   Bacteroidetes

Fusobacteria

α,β-proteobacteria

ε-proteobacteria

δ-proteobacteria   **AHLs**

AHLs & PAME

Amino Acids

0.10

FIGURE 5 Distribution of distinct intercellular signaling mechanisms throughout the bacteria. AHLs have only been found in the α, β, and γ classes of the phylum Proteobacteria. Peptide- and butyrolactone-based signaling systems have only been found in the phylum Firmicutes. PAME has only been found in the genus *Ralstonia*.

orientation is varied. These regulatory components are found encoded on plasmids and chromosomes. Previous phylogenetic analyses have shown that in the many cases where the two genes are paired in this way their evolution has followed an identical path. Because of this, phylograms of LuxI proteins mimic those of LuxR proteins. Therefore, an analysis of the LuxR protein alone provides all the evolutionary information we require.

Relatively recently, a thorough analysis of available LuxI and LuxR sequences was con-ducted. It was concluded that the LuxI and LuxR homologues have been associated with their host lineages for a long time and diverged in parallel with other host genes. They also conclude that discrepancies in branching order have derived from horizontal gene exchanges. Because new AHL ISMs are still being discovered with some regularity, and because new genome sequences are becoming available with increasing regularity, it is worthwhile updating this phylogenetic analysis and challenging the conclusions drawn previously.

TABLE 1 Phylogenetic association of proteobacterial AHL systems

Class	Order (*n*)	Families with AHL system/total families	Genera with AHL system/total genera
Alphaproteobacteria	*Rhodospirillales*	0/2	0/24
	Rickettsiales	0/3	0/17
	Rhodobacterales	1/1	2/23
	Sphingomonadales	0/1	0/9
	Caulobacterales	0/1	0/4
	Rhizobiales	3/10	5/56
	All orders (6)	4/18 (22%)	7/133 (5%)
Betaproteobacteria	*Burkholderiales*	2/5	2/33
	Hydrogenophilales	0/1	0/2
	Methylophilales	0/1	0/3
	Neisseriales	1/1	1/14
	Nitrosomonadales	1/3	1/4
	Rhodocyclales	0/1	0/8
	All orders (6)	4/12 (33%)	4/64 (6%)
Gammaproteobacteria	*Chromatiales*	0/2	0/30
	Acidithiobacillales	0/2	0/2
	Xanthomonadales	0/1	0/9
	Cardiobacteriales	0/1	0/3
	Thiotrichales	0/3	0/15
	Legionellales	0/2	0/3
	Methylococcales	0/1	0/6
	Oceanospirillales	0/2	0/12
	Pseudomonadales	1/2	1/28
	Alteromonadales	0/1	0/12
	Vibrionales	1/1	2/6
	Aeromonadales	0/2	1/7
	Enterobacteriales	1/1	6/41
	Pasteurellales	0/1	0/6
	All orders (14)	4/22 (18%)	10/180 (6%)
Deltaproteobacteria	All orders (7)	0/18	0/55
Epsilonproteobacteria	All orders (1)	0/2	0/6
Proteobacteria	All orders (34)	12/72 (17%)	19/446 (4%)

Acquiring the Data Set of LuxR Homologues

Because of their long-term associations with their hosts LuxR homologues have diverged considerably. The extent of sequence divergence is so great that parts of the sequence alignments are unreliable because of the high level of amino acid substitutions. Unfortunately, removal of these regions from the phylogenetic analysis reduces the number of residues considered, and hence the phylogenetic signal, but it can increase the robustness of the data due to the removal of noise. An additional problem encountered when collecting sequences for inclusion in the protein alignments was estimating

what were likely to be true functional LuxR homologues in the databases. Thus, while we are confident that broad conclusions can be drawn from this analysis, they are somewhat speculative and will hopefully be augmented in the future by additional sequence data.

Database searches for LuxR homologues using the PHI BLAST algorithm yield an abundance of sequence homologues due to the overlap between AHL-receptor proteins and those belonging to the family of response regulators in two component phospho-relay cascades. The LuxR homologues used to construct the phylogenetic tree shown in Fig. 6 were collected using only the N-terminal domain of the LuxR sequence in the search. Truncation and point mutation analyses have indicated that this domain is responsible for AHL binding. The ignored C-terminal domain contains the helix–turn–helix motif common to many DNA-binding regulatory proteins, including response regulators.

In contrast to LuxR database searches, searches for LuxI homologues generate a relatively clean data set, including all described AHL synthases and only a limited number of additional sequence homologues. Despite limiting the LuxR search to the AHL-binding domain we acquire many more putative AHL-binding proteins than we do AHL synthase proteins. This finding is surprising given that the described LuxI and LuxR homologues are generally paired and may reflect the fact that AHL ISMs are almost always discovered by screens for AHL production rather than AHL response. Mining genome sequence data sidesteps this limitation. Although many of the LuxR

sequence homologues may not, in reality, be AHL-binding proteins, there is precedent for the maintenance of functional LuxR homologues in the absence of AHL synthases. For example, *Escherichia coli* does not produce AHLs but encodes the functional AHL-dependent regulatory protein SdiA. The evolution of organisms that can respond to AHLs without incurring the cost of producing them is discussed further below.

Following the LuxR Evolutionary Trail

A phylogenetic tree (Fig. 6) describing the divergence of LuxR homologues was generated from full-length sequences (excluding gaps) using ClustalX. Figure 6 is simplified to emphasize the clustering patterns and distribution of LuxR homologues. In agreement with previous findings, the phylogram indicates that the LuxR homologues have become highly diverged, separated by long branch lengths, indicative of a

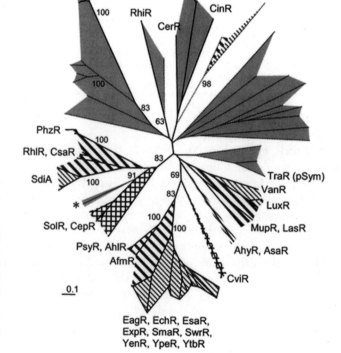

FIGURE 6 Neighbor-joining phylogeny of LuxR homologues with sequence clusters shaded according to taxonomic grouping: alphaproteobacteria, grey; betaproteobacteria, square cross-hatching; and gammaproteobacteria, light and heavy (*Pseudomonas* spp.) diagonal lines. Known functional AHL receptors are labeled and indicated by solid lines. Sequences that have only been identified by sequence homology are indicated by dotted lines. Bootstrap support (>60%) is indicated. Asterisk denotes *B. melitensis* sequence.

long existence or a recent selection for rapid divergence. That the branching order broadly parallels that of housekeeping genes suggests the former is more likely.

The clustering patterns indicate that the LCA of the gammaproteobacteria had at least two distinct LuxR homologues. This conclusion is derived from the observation that there are at least two distinct (separated by high bootstrap support, >60%) clusters of gammaproteobacterial sequences (Fig. 6, one containing SdiA and the other containing ExpR) and that the relative branching order in each cluster parallels the branching order of chromosomal housekeeping genes such as 16S rDNA. The cluster containing LuxR also shows the same branching order and might suggest that the LCA contained a third system, but this cluster is not well supported by bootstrap values (so no clear conclusions can be drawn). Figure 6 also hints that these distinct ancestral LuxR homologues may have been present before the beta- and gammaproteobacteria diverged. This suggestion is based on betaproteobacterial sequences lying basal to the γ-sequences of the SdiA and ExpR clusters; however, this is also somewhat speculative because there is only one known betaproteobacterial sequence (CviR) at the base of the ExpR cluster. Again, we predict that this reflects the paucity of betaproteobacteria that have been screened for AHL ISMs.

Earlier investigators were unequivocal about the position of the CviR sequence, because statistical tests confirmed that it could not be grouped with the other betaproteobacterial cluster (containing SolR) without a significant increase in tree length. This result is consistent with CviR having been acquired by *Chromobacterium violaceum* by horizontal gene transfer (HGT) from a gammaproteobacterium, but it is also consistent with the independent, parallel divergence of paralogues (duplicated genes) within the same cell as we propose has occurred in the gammaproteobacteria. More widely accepted tests for evidence of HGT involve anomalous genetic signatures such as differences in GC content, dinucleotide frequency, and codon usage between the host genes and the horizontally acquired genes (Koonin et al., 2001). However, these tests only detect relatively recent HGTs. The branch lengths connecting the CviR sequence to the ExpR gammaproteobacterial cluster are very long, suggesting that if this grouping is the result of HGT it must have occurred a long time ago and that genetic signatures are likely to have been completely erased by amelioration (acquisition of the new host's DNA signatures). Additionally, these tests require substantial sequence data from both the putative donor and recipient in the HGT event, which is not available for *C. violaceum*.

It has also been suggested that HGT was necessary to explain the anomalous positions of two *Pseudomonas syringae* homologues (PsyR and AhlR) at the base of the ExpR cluster and the separation of the SdiA sequences away from the other gammaproteobacterial sequences. It was proposed that the *P. syringae* sequences had been acquired from a *Serratia-*, *Erwinia-*, or *Yersinia*-like organism and that *E. coli* and *Salmonella enterica* serotype Typhimur-ium had acquired their LuxR homologues from a pseudomonad. These are effectively reciprocal horizontal exchanges. Gene transfer, acquisition, and fixation is a complex and unlikely process. It seems more parsimonious to assume that the last common ancestor of the gammaproteobacteria had two AHL quorum-sensing systems and that different lineages have lost or retained different copies through evolutionary time.

Thus far we have neglected appraisal of LuxR homologues belonging to the alphaproteobacteria (represented by the *Rhizobiaceae* and *Rhodobacter sphaeroides*). This, in part, stems from the fact that there appear to be many LuxR homologues in the four published rhizobial genomes, in addition to those identified in molecular studies. This is clearly seen in Fig. 6 where alphaproteobacteria dominate the phylogram; however, the majority of homologues in the alphaproteobacteria sectors have not been shown to bind AHLs or elicit regulatory changes in response to AHLs.

The better-studied systems in this group of bacteria are the TraR-dependent systems associated with the tumor-inducing plasmids of agrobacteria, RhiR of the rhizobial plasmid pRL1JI, and the chromosomal CinR- and RaiR-dependent systems found in both *Rhizobium leguminosarum* and *Rhizobium etli*. One problem with trying to identify LuxR/I pairs in rhizobia is that many of the known pairs of homologues associated with the plasmids of these bacteria are not adjacent unlike in the majority of systems encoded on chromosomes.

Of the unpaired rhizobial LuxR homologues in Fig. 6, the most striking, insofar as it does not group with any other rhizobial sequences, is the *Brucella melitensis* (BMEI1758) sequence that nestles among the SdiA and CepR clusters, indicated by an asterisk (Fig. 6). Both of these clusters contain described AHL intercellular signaling systems, and it seems likely that this LuxR homologue can bind an AHL and possibly act as a gene regulator. This sequence might represent an ancient HGT, since no sequences that cluster with it were identified among the other rhizobial genomes. However, the role of this putative gene/protein needs to be established experimentally before conclusions can be drawn. Another feature of the alphaproteobacterial LuxR homologues is that sequences within the different clusters appear to be more diverged than the beta- or gammaproteobacterial clusters, even though they are all from taxonomically closely related genera. This is especially noticeable for the TraR and RaiR homologues in Figure 6, both of which are components of functional quorum-sensing systems.

The presence of unpaired LuxR homologues in genome sequences is not restricted to the rhizobia. Both *P. aeruginosa* and *Vibrio cholerae* have putative LuxR homologues, in addition to those of the described systems, in their annotated genomes. Additionally, unpaired LuxR homologues have been identified in *Erwinia* species (CarR) and *S. marcescens* (CarR). These observations support the hypothesis of multiple LuxI homologue losses among different bacterial lineages and, further, support the idea that multiple AHL quorum-sensing systems were present in the LCA of these bacteria.

Section Summary

- Distinct ISMs belong to distinct taxa reflecting their independent evolution.
- The phylogeny of AHL ISMs suggests that the system was present at least in duplicate in the LCA of the beta- and gammaproteobacteria.
- The fact that there are beta- and gammaproteobacteria without the AHL ISM suggests that evolution discards the ISM on occasion.
- There are more AHL receptors than there are AHL synthases, suggesting that signaling cheats are tolerated by evolution. This may be a path to losing the ISM.

INTERCELLULAR SIGNALING IN BACTERIAL POPULATION ECOLOGY

Intercellular Signaling and Multicellularity

Microbiologists distinguish between two useful viewpoints that cohabit within their discipline. The first emphasizes the importance of studying microbes in pure cultures because only in this way can properties such as virulence be unambiguously attributed to them. This viewpoint is born out of the work of Robert Koch during the late 1800s and has been buoyed by the remarkable success of medical microbiology. The second viewpoint acknowledges the importance of coordinated bacterial interactions among themselves and with their environment. This viewpoint is born out of the work of Sergei Winogradsky and underpins contemporary microbial ecology.

The relatively recent discovery of bacterial ISMs has explained how bacteria can attain sophisticated levels of interaction and coordination. The coordination of bacterial behaviors by ISMs generates what some consider to be prokaryotic multicellularity. In this, the evolution of ISMs can be seen as part of a powerful trend to the higher orders of life through increasing tissue differentiation. Note that we are interested here in bacterial behaviors that are performed collectively by virtue of their coordination by an ISM. This is distinct from bacterial cells exhibiting uncoordinated yet collective behaviors, such as the movement of a population up a nutrient gradient.

Investigations of coordinated bacterial behavior mediated by ISMs have led to new perspectives on the population ecology of bacteria. The following discussion considers the implications of ISM-regulated coordinated behaviors on three levels. First, we consider in depth how intercellular signaling within species can extend the range of phenotypes and niches available to bacteria at the cost of adaptive flexibility. Second, we appraise the evidence for intercellular signaling between species in mixed communities and consider its potential impact on microbial community dynamics. Third, we consider how bacterial ISMs may be exploited by higher organisms in interdomain interactions.

Intraspecies Intercellular Signaling

As described above, a wide variety of behaviors such as resource acquisition, resource defense, and gene transfer may be coordinated through intercellular signaling. Behaviors requiring coordination by ISMs are beyond those practical for single cells. They can result in different degrees of intercellular association ranging from orchestrated gene expression among otherwise unaffiliated sister cells to the construction of physical structures requiring morphological differentiation and commitment to physical affiliations between sister cells. As hinted at above the evolution of ISMs may have been an essential precursor for the relatively recent evolution of organisms considered by convention to be truly multicellular.

Streptomyces soil bacteria, for example, form highly associated populations. They forage in soil, growing as branching filaments that generate dense mycelial masses. These mycelial masses differentiate with cells committed to various behaviors including production of aerial filaments, sporulation, self-lysis (providing nutrients to sporulation), and production of antibiotics. *Streptomyces* species therefore have a biphasic lifestyle, one phase of which is strikingly multicellular (mycelial mass), while the other is distinctly unicellular (spores). Not sur-

prisingly, coordination of the highly associated multicellular phase is achieved via ISMs.

ISMs controlling differentiation consign constituent mycelial cells to different tasks and fates. An interesting product of this behavior is that competition between mutant genotypes in the resulting clonal populations will be between mycelial masses rather than within them. If mycelial masses were formed by aggregation of *Streptomyces*, then we might expect variants to arise that competed to produce spores at the expense of other genotypes.

Myxococcus xanthus is another bacterium that has life-cycle stages that are markedly multicellular. These are soil bacteria that move by gliding and search for bacterial prey in swarms from which digestive enzymes are collectively released and within which the takeup of nutrients is shared. In conditions of declining nutrients and sufficient cell numbers, *M. xanthus* cells aggregate to form fruiting bodies within which some cells sporulate. Again, this highly coordinated behavior is orchestrated by ISMs. The collective or social nature of *M. xanthus* behavior is underlined by the sacrifice of some cells in lysis and the participation of only a subset of the cells in sporulation, a theme we will return to shortly.

The partially multicellular lifestyles of *Streptomyces* and *Myxococcus* species constitute extreme examples of ISM coordinated behavior in which cells are highly committed to the architecture of the population. While this ISM-controlled commitment to the multicellular phase of the lifestyle taps into the advantages of coordinated cooperative behavior, it sacrifices the flexibility of the unicellular lifestyle.

Is multicellularity the only game in town? The discovery of many distinct ISMs encourages the view that coordinated behavior is integral to the bacterial way of life rather than an option in adaptation (Davey and O'Toole, 2000). Our contention is that organisms encoding ISM-regulated coordinated behaviors are unlikely to be fitter than uncoordinated organisms in all conditions encountered. That is, collective behaviors, just like other phenotypes, are likely to evolve in response to specific conditions and will therefore not be selected for universally. From this, one might expect that a signaling organism will lose its coordination under conditions where the benefits of the coordinated phenotype are not realized.

An excellent test and illustration of this comes from studies into the loss of collective behaviors by *M. xanthus* (Velicer et al., 1998). *M. xanthus* cultures were grown for 1,000 generations in shake flasks, thereby disrupting their social behaviors and selecting for faster-growing strains. In these cultures new variant dominant lineages evolved with substantial loss of collective behaviors. These included 7 of 12 lineages that lost social motility while retaining the capacity for motility as individuals, 9 of 12 lineages that lost fruiting body formation (after 72 days starvation), and 7 of 12 lineages that lost sporulation ability. Where these traits were not lost they were reduced or impaired. Although it was not established whether any of these mutations were in components of the ISMs involved, the findings still serve to illustrate the influence of the environment on selection for coordinated or uncoordinated behaviors.

The changes observed in *M. xanthus* behavior could have been simply stochastic mutations with neutral drift in the absence of selection for collective behavior. However, it may be expected that expression of coordinated behaviors imposes costs that in asocial conditions would result in counterselection of ISMs and their regulated phenotypes. The observed repeated, independent, and rapid loss of social traits in *M. xanthus* indeed suggested that there is selection against their maintenance under these conditions. The costs of coordination include the resources involved in the ISM and in expression of the social phenotype itself. Further, as more traits become regulated by ISMs, opportunity costs such as the reduced flexibility associated with specialization may be incurred.

Are some of the players cheaters? Many behaviors regulated by ISMs generate benefits that are shared by groups of bacteria, such as the release of nutrients and antibiotic antago-

nism to competitors. Collective expression of traits by groups of cells regulated by intercellular signaling is vulnerable to cheaters, which may benefit while avoiding some of the costs. When, for example, a group of bacteria use ISMs to regulate the release of hydrolytic enzymes, the benefits (nutrients) will be available to all members of the group, whether or not they contributed to the release of enzyme. Such cheater strains may be expected to occur as occasional mutants and to increase in frequency within groups at the expense of non-cheaters. The increasing frequency of cheater genotypes in several cycles of colonization by new groups can reduce the overall group efficiency and potentially displace the genotype for collective behavior. This leads us to the questions why do selfish genotypes not predominate and how can collective, especially apparently altruistic, behaviors persist?

A possible solution comes from kin selection theory, which has been particularly successful in explaining this type of behavior in higher organisms (Griffin and West, 2002). This approach proposes that some sacrifices made by individuals can improve their "indirect fitness." Indirect fitness is acquired when the costs of social activity to an individual are more than compensated for by the benefits to others of the same genotype. In other words, these coordinated behaviors can persist when the benefits of their sociality are disproportionately distributed to their social genotype as opposed to the cheater types. Higher organisms can achieve this by social types recognizing one another and cooperating to the exclusion of asocial types. There is no evidence, however, that bacteria are able to recognize asocial (nonsignaling) types, forcing us to look for other mechanisms to protect signal-producing cells from nonproducing cheats.

AHL–mediated gene expression has been modeled in the context of group selection for altruistic traits. The model is based on the familiar scenario of groups of bacteria using an ISM to regulate the release of extracellular enzymes, which in turn generates a shared nutrient resource. In the model, cells from four genotypes competed in successive rounds of patch colonization. The competing genotypes were:

1. cells with the full AHL ISM, both producing AHL and responding to it with release of enzyme

2. cells with no production of AHL and no response to it

3. cells that do not produce AHL but do release enzyme in response to it, and

4. cells that produce AHL but do not release enzyme in response to it.

A key parameter investigated for its effect on maintenance of the AHL ISM was the mean number of cells founding a group or colony (m). When m was low the equilibrium frequency of the complete AHL ISM (the first genotype listed above) was high. As the value of m rose, the equilibrium frequency fell sharply. Therefore, the numbers of cells founding microcolonies is expected to have a strong influence on the prevalence of signaling types. More generally viscous distribution of populations (low dispersal) can create the conditions in which cells are predominantly interacting with their own genotype. It was also found that a wide range of conditions (parameter space) gave rise to stable mixes of social and asocial genotypes. These results suggest that we should expect in nature to find social (signaling) types coexisting with asocial cheating types. The presence of functional LuxR homologues in non–AHL-producing species (mentioned above) supports this as does the finding that the number of LuxR sequence homologues listed in current databases is in excess of the number of LuxI homologues. Brookfield's model (see "Further reading," below) predicts that the fourth genotype listed above will persist at the lowest frequencies. In accordance with this prediction, there are no known examples of AHL-producing organisms that do not encode an AHL receptor protein.

These issues have also been addressed in an experimental study in which the loss of collective behaviors by *M. xanthus* strains was correlated with cheater behaviors. Mutant strains that were defective for fruiting body formation were taken from the *M. xanthus* lineages evolved in the asocial conditions of liquid cultures. Several of these were competed against the progenitor wild type and produced spores in excess of their frequency in the population. Some of these cheater strains were shown also to reduce the overall efficiency of the group as their frequency rose. In further studies single mutations were shown to be sufficient to establish cheater strains of *M. xanthus*. These results indicate that cheating or asocial behavior may be common among *M. xanthus* in nature.

Interspecies Intercellular Signaling

We have seen now that intercellular signaling between conspecific cells allows bacteria to speciate further with the evolution of new behaviors that exploit the benefits of multicellularity. Now we shall briefly consider whether different bacterial species can communicate via ISMs. More than 30 different bacterial species are known to use AHLs as intercellular signaling molecules. There is little doubt that different AHL-producing species cohabit in the environment. There is also little doubt that AHLs produced by one bacterial species can induce expression of AHL-regulated genes in other bacterial species. For these reasons crosstalk between distinct bacterial species in the environment seems plausible, at least when considered from this basic molecular biology perspective. The best evidence for interspecies signaling in the environment comes from the demonstration that naturally coexisting bacterial species on wheat or tomato roots induce AHL-mediated gene expression in nonisogenic species. It has been proposed that interspecies signaling directly affects the outcome of competitive interactions in the rhizosphere, although this remains to be demonstrated.

Furanosyl borate diester or AI-2 is a second signaling molecule produced by a range of different bacterial species. It has been demonstrated that AI-2 producers can induce luminescence in *V. harveyi*. Based on this and its widespread distribution, it has been proposed that AI-2 is a universal bacterial dialect that provides bacteria with the ability to monitor niche occupancy in a communal bacterial census. This proposal is dependent on each bacterial species, some of which will not be AI-2 producers, generating the same amount of AI-2. If different species generate different amounts of AI-2 then the signal will induce gene expression at different cell densities depending on the identity of the producer, which the signal does not reveal, and can therefore not function as a universal quorum-sensing device. An alternative argument for signaling cross talk is that AI-2 could act as a trigger to induce behaviors that are complementary between two different bacterial species, both benefiting from activities that are worthless when not in combination. This concept, however, is also flawed because the signal will also induce the complementary phenotypes when the two are apart. Further, we believe that interspecies communication via universal intercellular signals is too vulnerable to subversion to mediate any long-term ecological interactions. As with the use of universal signals in the ecology of higher organisms such as insects, the evolution of broad bacterial dialects must be considered unlikely. In fact, interspecies intercellular signaling is more likely to act as an evolutionary mechanism selecting against the proliferation of intraspecies ISMs.

Interdomain Intercellular Signaling

Given that interactions with higher organisms are potent drivers for bacterial evolution (see chapter 19) and that ISMs often mediate interactions between bacteria and higher organisms, it is worth considering the impact of intercellular signaling on bacterial evolution via such interactions. To this end we will consider the fascinating possibility that a marine alga has evolved a means of interfering with AHL-mediated gene expression.

Delisea pulchra is a benthic macroalga commonly found in subtidal marine environments along the southeastern coast of Australia. Although the surfaces of marine algae are vulnerable to bacterial colonization and disease, *D. pulchra* displays a remarkable ability to defend itself from such threats. The defense has been attributed to the production of metabolites known as halogenated furanones, which exhibit striking structural similarities with AHLs functioning in an ISM in bacteria (Fig. 1).

Since this observation was made halogenated furanones have been shown to inhibit the expression of many AHL-regulated phenotypes (Givskov et al., 1996). Halogenated furanones are thought to inhibit AHL-mediated gene expression by binding to the AHL receptor protein and triggering its proteolytic degradation. The evolution of AHL antagonists by *D. pulchra* is thought to be an evolutionary response to the pressure of bacterial colonization and phytopathogenicity. Thus, it constitutes an example of bacterial ISMs directing eukaryotic evolution. Further, the environmental presence of halogenated furanones, which have other biological effects such as oxidative stress, is, in turn, a selection pressure likely to yield evolutionary responses from other cohabiting bacteria. While no evidence for this response has been sought, it seems somewhat inevitable, and thus constitutes a mechanism by which bacterial intercellular signals can influence bacterial evolution indirectly via interactions with higher organisms.

Section Summary

- Bacteria can achieve new behaviors and therefore access new niches through ISM-mediated coordination within species. This can be seen as multicellular behavior.
- Interspecies signaling is more vulnerable to subversion than intraspecies signaling and is therefore unlikely to mediate stable cooperative interactions between species.
- Interspecies signaling can mediate antagonistic interactions, which will select against proliferation of ISMs.

SUMMARY OF SIGNALING IN THE MICROBIAL WORLD

In this chapter we have introduced bacterial ISMs and viewed them from three different perspectives. Initially, we used the outlook of molecular microbial ecology to consider how ISMs physically relate bacteria to their environment and to each other. Next, we considered ISMs from a phylogenetic angle to gather insight into how they have evolved. Finally, we considered ISMs from the standpoint of population ecology to understand what ecological processes they mediate and what forces shape their evolution.

We believe that intercellular signaling, as a field of study, is still in its infancy. As a result we must view this discussion with an open mind and be ready to adapt our conclusions as new information becomes available. This integrated vision of ISMs will be nourished by new experimental data from molecular microbiologists, bioinformaticists, and population ecologists. For example, the ability to measure signal concentration and cell density in the environment will provide further insight into what environmental factors lead to accumulation of the signal. Additional sequence data and a comprehensive screen of bacterial taxa for signals are required to fortify or amend our phylogenetic conclusions. Finally, additional long-term experiments in the vein of Velicer et al. (2000) will be crucial to understanding how ISMs proliferate or decline in bacterial populations and communities. Regardless of perspective we must not underestimate the ability of bacteria to control their behavior, their environment and maybe even their own evolution through intercellular signaling.

QUESTIONS

1. In what environmental contexts is it an advantage for bacterial cells to respond to the density of their own populations?

2. How can the various hypotheses describing the advantages of intercellular signaling be tested?

3. What are the two possible hypotheses explaining the incongruous positions of certain LuxR homologues in the LuxR phylogenetic tree?

4. If intercellular signaling mechanisms are vulnerable to subversion by cheats then how do they persist?

5. How might bacterial intercellular signal-ing affect the evolution of genomes in higher organisms?

REFERENCES

Davey, M. E., and G. A. O'Toole. 2000. Microbial biofilms: from ecology to molecular genetics. *Microbiol. Mol. Biol. Rev.* **64:**847–867.

Davies, D. G., M. R. Parsek, J. P. Pearson, B. H. Iglewski, J. W. Costerton, and E. P. Greenberg. 1998. The involvement of cell-to-cell signals in the development of a bacterial biofilm. *Science* **280:**295–298.

Dong, Y. H., L. H. Wang, J. L. Xu, H. B. Zhang, X. F. Zhang, and L. H. Zhang. 2001. Quenching quorum-sensing-dependent bacterial infection by an *N*-acyl homoserine lactonase. *Nature* **411:**813–817.

Fray, R. G., J. P. Throup, M. Daykin, A. Wallace, P. Williams, G. S. Stewart, and D. Grierson. 1999. Plants genetically modified to produce *N*-acylhomoserine lactones communicate with bacteria. *Nat. Biotechnol.* **17:**1017–1020.

Givskov, M., R. de Nys, M. Manefield, L. Gram, R. Maximilien, L. Eberl, S. Molin, P. D. Steinberg, and S. Kjelleberg. 1996. Eukaryotic interference with homoserine lactone mediated prokaryotic signalling. *J. Bacteriol.* **178:**6618–6622.

Griffin, A. S., and S. A. West. 2002. Kin selection: fact and fiction. *Trends Ecol. Evol.* **17:**15–21.

Kleerebezem, M., L. E. N. Quadri, O. P. Kuipers, and W. D. de Vos. 1997. Quorum sensing by peptide pheromones and two-component signal-transduction systems in Gram-positive bacteria. *Mol. Microbiol.* **24:**895–904.

Koonin, E. G., K. S. Makarova, and L. Aravind. 2001. Horizontal gene transfer in prokaryotes: quantification and classification. *Annu. Rev. Microbiol.* **55:**709–742.

Nilsson, P., A. Olofsson, M. Fagerlind, T. Fagerstrom, S. Rice, S. Kjelleberg, and P. Steinberg. 2001. Kinetics of the AHL regulatory system in a model biofilm system: how many bacteria constitute a "Quorum"? *J. Mol. Bol.* **309:**631–640.

Velicer, G. J., L. Kroos, and R. E. Lenski. 2000. Developmental cheating in the social bacterium *Myxococcus xanthus*. *Nature* **404:**598–601.

FURTHER READING

Brookfield, J. F. Y. 1998. Quorum sensing and group selection. *Evolution (Lawrence, Kans.)* **52:**1263–1269.

Dunny, G. M., and S. C. Winans (ed.). 1999. *Cell-Cell Signaling in Bacteria.* ASM Press, Washington, D.C.

Gray, K. M., and J. R. Garey. 2001. The evolution of bacterial LuxI and LuxR quorum sensing regulators. *Microbiology* **147:**2379–2387.

Nealson, K. H., T. Platt, and J. W. Hastings. 1970. Cellular control of the synthesis and activity of the bacterial luminescent system. *J. Bacteriol.* **104:**313–322.

Shapiro, J. A., and M. Dworkin (ed.). 1997. *Bacteria as Multicellular Organisms.* Oxford University Press, New York, N.Y.

Whitehead, N. A., A. M. L. Barnard, H. Slater, N. J. L. Simpson, and G. P. C. Salmond. 2001. Quorum-sensing in Gram-negative bacteria. *FEMS Microbiol. Rev.* **25:**365–404.

Williams, P., N. J. Bainton, S. Swift, S. R. Chhabra, M. K. Winson, G. S. Stewart, G. P. Salmond, and B. W. Bycroft. 1992. Small molecule-mediated density-dependent control of gene expression in prokaryotes: bioluminescence and the biosynthesis of carbapenem antibiotics. *FEMS Microbiol. Lett.* **79:**161–167.

GENE ASSOCIATIONS IN BACTERIAL PATHOGENESIS: PATHOGENICITY ISLANDS AND GENOMIC DELETIONS

Eshwar Mahenthiralingam

16

BACTERIAL PATHOGENS
Nonpathogens
Opportunistic Pathogens
Primary Pathogens
Virulence

ACQUISITION OF BACTERIAL VIRULENCE GENES BY HORIZONTAL TRANSFER: PAIs
PAIs Encode Virulence or Pathogenicity-Enhancing Genes
PAIs Consist of DNA Regions That Differ in G+C Content and Codon Usage from the Core Genome
PAIs Are Often Large Genomic Regions
tRNA Genes and Small Directly Repeated (DR) Sequences Are Frequently at the Boundaries of PAIs
PAIs Often Carry Genes Associated with Mobile Genetic Elements
PAIs Are Often Unstable Genomic DNA Regions

TYPE III SECRETION SYSTEMS: AN EXAMPLE OF VIRULENCE GENE CLUSTERS CARRIED ON PAIs

HORIZONTAL GENE TRANSFER AND THE EVOLUTION OF VIRULENCE IN PATHOGENIC *E. COLI* STRAINS: A GENOMIC PERSPECTIVE

LOSS OF DNA AS A MECHANISM OF PATHOGEN EVOLUTION
Black Holes in *Shigella* Species and Enteroinvasive *E. coli*
The Leprosy Bacillus: Genome Decay, Pseudogenes, and Homologies to Eukaryotic Genes

GENOMIC PROCESSES INVOLVED IN BACTERIAL PATHOGEN EVOLUTION

SUMMARY

Eshwar Mahenthiralingam, Cardiff School of Biosciences, Main Building, Museum Avenue, PO Box 915, Cardiff University, Cardiff CF10 3TL, Wales, United Kingdom.

Many bacteria live in what we consider the natural environment. This includes terrestrial and marine habitats and niches, such as animals and plants that live throughout both environments. Organisms evolve a genetic content of genes suited to survival, growth, and replication in their chosen environment. Many microorganisms are not restricted to a single growth environment and are capable of survival in many different niches. In contrast, certain microorganisms, such as intracellular symbionts or obligate intracellular pathogens, have chosen to live within a single, very restricted environment, inside the cells of a host organism. In this chapter we will examine how bacteria have altered their genome content to better survive and replicate within a disease environment. We will examine what kinds of genes they have acquired or lost and some of the major genetic mechanisms behind the evolution of their ability to cause disease.

To incorporate and express virulence genes and adapt their overall genetic content to disease states, bacterial pathogens have evolved by using mechanisms of mutation and gene transfer. The acquisition, recombination, duplication, inactivation, and deletion of genes are common events occurring throughout evolution. In this chapter, we will specifically examine two of the latter evolutionary pathways and their relationship to the development of bacterial pathogenesis: (i) gene acquisition and (ii) gene deletion. Both processes have been responsible for generating sequence diversity and driving evolution of bacterial infectious diseases.

Microbial Evolution: Gene Establishment, Survival, and Exchange
Edited by Robert V. Miller and Martin J. Day, ©2004 ASM Press, Washington, D.C.

BACTERIAL PATHOGENS

The overwhelming desire of humans to combat infectious disease and discover new methods to prevent infection has driven a considerable proportion of research on prokaryotes. Many routine techniques of bacterial culture and identification were pioneered during the study of bacterial infectious diseases. Bacterial pathogens were also among the first organisms to be subjected to complete genome sequence determination with the gram-negative *Haemophilus influenzae* leading the way in 1995. Currently, of the 146 complete microbial genome sequences available at the National Center for Biotechnology Information databases (http://www.ncbi.nlm.nih.gov/PMGifs/Genomes/), more than half are bacteria species that are pathogenic to plants, animals, and humans. This focus of research and application of genomic

methods to agents of infectious disease has led to considerable increases in our knowledge of bacterial genetics.

What constitutes a bacterial pathogen? In simplest terms, pathogens possess the ability to cause damage in a host organism (Fig. 1). This definition and others described below were drawn from a review of host-pathogen interactions by Casedevall and Pirofski (2000). Key definitions and concepts of bacterial pathogenesis are illustrated in Fig. 1. The definition of a pathogen as a microbe capable of causing damage to a host leads to the term pathogenicity, the ability to cause damage and infection. As you can imagine, pathogenicity is a qualitative trait consisting of a spectrum of different disease states. Bacteria may cause infections that lead to no damage to the host (asymptomatic). Or infections may lead to morbidity and ulti-

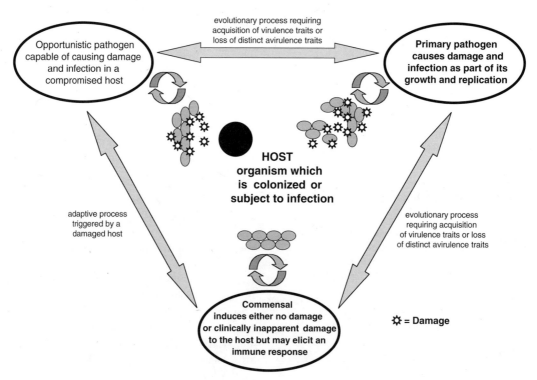

FIGURE 1 Concepts of pathogen–host interactions in bacterial infectious diseases. Definition of commensal bacteria, opportunistic pathogens, primary pathogens, and host organism are illustrated in relation to disease and pathogenicity. The relationship between each class of bacteria and the colonized host organism is indicated by the arrows. Definitions are drawn from Casadevall and Pirofski (2000).

mately death of a host in virulent diseases. As a result of this continuum of pathogenic states, bacterial pathogens have been divided into three classes: (i) nonpathogens, (ii) opportunistic pathogens, and (iii) primary pathogens.

Nonpathogens

Nonpathogenic bacterial species are not capable of causing damage to a host organisms. Nonpathogens may reside freely in natural environments such as those defined above. Other nonpathogenic bacterial species live on or within a given host organism and are known as commensals. Examples of harmless commensal bacterial species that choose to live on or within a human host are: (i) gram-negative nonpathogenic *Escherichia coli* strains that live within our intestines or (ii) gram-positive *Staphylococcus aureus* species that live on our skin. However, commensalism itself may be considered a state of infection that results in no damage to the host (Fig. 1), and considering commensalism in this way leads logically to the next category of infectious organism we will consider.

Opportunistic Pathogens

Opportunistic pathogens possess the ability to cause host damage and subsequently disease; however, this is not the natural form of growth and replication. Opportunistic pathogens only cause damage when the host organism is compromised or injured in some way. Because of their close association with host organisms many commensal bacterial species are capable of opportunistic infection and may progress from a state of harmless colonization of host tissues to virulent disease (Fig. 1). For example, *S. aureus* may cause infections if the host skin becomes damaged or bruised. *Pseudomonas aeruginosa* is another good example of an opportunistic pathogen. It may reside freely within the natural environment, and it is often carried by humans as a commensal organism. However, *P. aeruginosa* is a major cause of opportunistic infections, frequently causing wound and burn infections in compromised hospitalised patients. *P. aeruginosa* is also the major cause of infection and death in patients

with cystic fibrosis, whose lungs fail to adequately clear this microorganism. Another good example of an opportunistic bacterial pathogen, but one that is not carried as a commensal and normally lives in the natural environment, is the gram-negative *Burkholderia cepacia*. Like *P. aeruginosa*, *B. cepacia* has the ability to cause devastating respiratory infections in patients with cystic fibrosis and invasive systemic infections in other compromised individuals. So opportunistic pathogenicity represents an adaptation from a state of commensalism or growth in the natural environment (Fig. 1), but what about bacterial species that depend on causing damage to a host organism as part of their normal life cycle?

Primary Pathogens

Finally, primary pathogens (also known as frank pathogens) are those that cause disease as part of their natural survival and life cycle (Fig. 1). Certain pathogens, such as *Salmonella* and *Vibrio* species, may be capable of surviving freely within the natural environment or be carried without signs of infection by certain hosts. However, when given the opportunity, they will cause virulent epidemic infections, resulting in disease outbreaks, rapid multiplication, and transmission to further uninfected hosts. These pathogenic species are also known as facultative pathogens. Other primary pathogens such as *Mycobacterium tuberculosis* only reside within an infected host and are often described as obligate pathogens. Tuberculosis infection may be asymptomatic, where the organism persists for many years without showing signs of infection. However, more often than not, tuberculosis infection is virulent with a gradual progression in disease, where the host becomes damaged and the viable numbers of the pathogen increase.

Virulence

The degree to which a pathogen causes damage to the host during disease is described as its virulence. Virulence is considered to be a quantitative trait and may be measured using assays examining cellular invasion, cell lysis, growth within an animal model, or the dose required to produce death in an animal model. Infection often requires multiple steps, such as adhesion to and colonization of host tissues, evasion of host defenses, invasion of host cells, replication, and finally escape and transmission to a new host. Each stage in the disease process may require expression of specific virulence factors. Hence, virulence is nearly always multifactorial with factors such as toxins, proteases, adhesins, and invasins all contributing to pathogenesis at different stages of infection. Bacterial pathogens may infect many different host organisms, but for the purposes of this review we will only examine examples of infection in humans. Since many areas of the body colonized and subsequently infected by pathogens are the same for different infections, several common mechanisms of pathogenesis have evolved. Many of the genes involved in these mechanisms of pathogenesis are conserved and have spread among different species by horizontal gene transfer.

Section Summary

- Microbial pathogens all possess the ability to cause infectious disease (pathogenesis) but may exhibit different degrees of virulence.
- There are several states of bacterial infection, including commensalism, opportunistic pathogenesis, and primary pathogenesis.
- Bacterial infections may use one or several host organisms as a growth environment.

ACQUISITION OF BACTERIAL VIRULENCE GENES BY HORIZONTAL TRANSFER: PAIs

Many pathogenic traits are encoded on mobile genetic elements. Broad host-range plasmids, such as the R (resistance) plasmids that encode antimicrobial resistance genes, were one of the first groups of mobile elements found to mediate horizontal gene transfer among bacterial pathogens. There are also examples of bacteriophages carrying pathogenicity-related genes. For example, the diphtheria toxin of *Corynebacterium diphtheriae* is encoded on the genome of a temperate phage.

However, in the early 1980s it was discovered that many bacterial virulence determinants were encoded on the bacterial chromosome as clusters of genes that demonstrated features associated with their acquisition by horizontal gene transfer. The term "Pathogenicity Island" was coined in the 1990s by Jorg Hacker and his colleagues, who were examining a large cluster of unstable regions encoded on the chromosome of *E. coli* strain 536, which caused urinary tract infections. Excision of these regions occurred from tRNA-specific loci and loss of the DNA was associated with reduction in hemolytic activity, serum resistance, adherence, and virulence in mouse models of infection. Because of their functional role during infection, these clusters of genes were described as pathogenicity islands (PAIs). PAIs were subsequently found to occur in many members of the *Enterobacteriaceae* that are pathogenic to humans. The distribution of similar virulence gene clusters now appears to be very broad, with PAIs described in several gram-negative genera and gram-positive bacteria. It has been found that PAIs may not be exclusively encoded on the bacterial chromosome, but PAIs also form components of large virulence plasmids or bacteriophages. Pathogenicity islands are perfect examples of how gene acquisition by horizontal transfer can generate sequence diversity in bacterial pathogens and enable them to evolve new virulence traits.

The term pathogenicity island is used widely in the literature to describe regions of bacterial DNA which encode factors important for infection and disease. However, several specific criteria may be considered to define a region of DNA as a PAI; these are illustrated graphically in Fig. 2. The concepts and defining features of PAIs are described below. PAIs:

- Encode virulence- or pathogenicity-enhancing genes, a universal trait of PAIs
- Consist of DNA regions that differ in G+C content and codon usage from the core genome
- Occupy relatively large genomic regions
- Associate with transfer RNA genes and

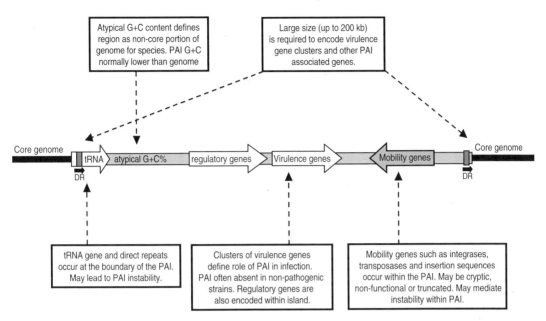

Atypical G+C content defines region as non-core portion of genome for species. PAI G+C normally lower than genome

Large size (up to 200 kb) is required to encode virulence gene clusters and other PAI associated genes.

Core genome   tRNA ⟩ atypical G+C% ⟩ regulatory genes ⟩ Virulence genes ⟩ ◁ Mobility genes   Core genome

DR   DR

tRNA gene and direct repeats occur at the boundary of the PAI. May lead to PAI instability.

Clusters of virulence genes define role of PAI in infection. PAI often absent in non-pathogenic strains. Regulatory genes are also encoded within island.

Mobility genes such as integrases, transposases and insertion sequences occur within the PAI. May be cryptic, non-functional or truncated. May mediate instability within PAI.

FIGURE 2 The concepts that define a bacterial pathogenicity island. A model pathogenicity island is illustrated using the criteria described in the text. The PAI (large grey box) is embedded within an area of core genome (black line). Genes and features characteristic of PAIs are indicated by the arrows, and a brief explanation of each defining feature is provided in the text boxes.

are flanked by small directly repeated (DR) sequences

· Encode genes/factors associated with mobile genetic elements

· May be unstable DNA regions

PAIs Encode Virulence or Pathogenicity-Enhancing Genes

In simple terms, PAIs can be considered to occur in disease-causing strains and to be absent in avirulent strains of the same or closely related bacterial species. PAIs possess multiple genes encoding virulence factors, and Kaper and Hacker (1999) state that this term should not be applied to a region encoding a single virulence gene. However, because bacterial virulence is normally multifactorial, the simple presence or absence of a PAI may not always be appropriate. For example, for a species to achieve its most virulent phenotype, a given strain may require the presence of several pathogenicity islands, with each PAI contributing an enhancement of virulence phenotype. Acquisition of a novel PAI by a

bacterial strain that is already pathogenic may confer an additional virulence trait, such as invasion and the ability to survive and multiply within host cells, thus enhancing the pathogenic ability of the organism. Hence, the most appropriate criteria to consider is that PAIs encode clusters of virulence genes.

In the classic examples of PAIs natural deletion or site-directed mutagenesis of PAI-encoded genes leads to strains that are less pathogenic. For example, deletion of the *E. coli* 536 pathogenicity islands I and II leads to loss of hemolytic activity, serum resistance, adherence, and virulence in mouse models of infection as described above. Other examples of specific virulence traits that are encoded on PAIs include pili and other adherence factors, bacterial toxins, iron-uptake systems, and type III or type IV secretion systems (see "Type IV secretion"). Many PAIs that fit the classical criteria listed above have been found in members of the *Enterobacteriaceae*, and selected examples associated with virulent strains within their respective species are shown in Table 1.

TABLE 1 Examples of classical pathogenicity islands and their features

Organism	Name	Size (kb)[a]	G+C content (PAI/genome)	Mobility genes present	Boundary sequence	Associated tRNA gene	Stability	Virulence functions
Uropathogenic *E. coli*	PAI I	70	41/51	Cryptic integrase	DR	Selenocysteine	Unstable	Hemolysin secretion
	PAI II	190	41/51	Cryptic integrase	DR	Leucine	Unstable	Hemolysin and P-fimbriae
Enteropathogenic *E. coli*	LEE	35	39/51	None	None	Selenocysteine	Stable	Type III secretion
Shigella flexneri	SHI-1	51	Mosaic/51	IS elements	IS	None	Unstable	Enterotoxin, proteases
	ipa-mxi-spa	37	35/48[b]	IS elements, plasmid host	IS	None	Stable	Type III secretion and invasion
Yersinia species	Yop virulon	47	44/45[c]	IS, plasmid host	None	None	Stable	Type III secretion

[a] All PAI described carry more than 20 genes.
[b] G+C content of the large virulence plasmid pWR501, which carries the *ipa-mxi-spa* pathogenicity island; plasmid G+C content is the same as *Shigella* chromosome G+C content.
[c] G+C content of the large virulence plasmid pCD1; overall plasmid pCD1 is lower in G+C content than the *Yersinia* genome, which is 48% G+C.

The PAIs of uropathogenic (UPEC) *E. coli* strain 536 were the first to be studied in detail, leading to the coining of term pathogenicity island (Table 1). PAI I of UPEC encodes a hemolysin virulence determinant that is responsible for lysis of host cells during infection. PAI II encodes both a hemolysin and P-fimbriae genes, which encode adherence factors enabling UPEC to colonize and infect the urinary tract. Both pathogenicity islands are large (70 and 190 kb for PAI I and II, respectively) and each is flanked by two short, directly repeated DNA sequences. In PAI I, 16-bp direct repeats occur within the tRNA gene encoding selenocysteine, which is known to act as the site of integration for the *E. coli* bacteriophage ΦR73. PAI II is associated with an 18-bp direct-repeat sequence that forms part of the tRNA gene for leucine; this region of the *E. coli* genome can also act as a site of integration for a bacteriophage, phage P4. The presence of repeated DNA around these regions of the chromosome enable recombination to occur, resulting in loss of the PAIs at a frequency of about 1 in 50,000 clones. One copy of each direct repeat is retained after deletion. As stated above, UPEC *E. coli* strains that have lost PAI I or II are less virulent, defining the role of the DNA region in disease.

Enteropathogenic *E. coli* strains were also found to possess a large cluster of virulence genes involved in the invasion of epithelial cells of the small intestine, causing infant diarrhea. This region was designated the locus of enterocyte effacement (LEE) by James Kaper and his group (Hacker and Kaper, 2000). LEE was large (35 kb) and carried a type III secretion system, which enabled invasion and destruction of enterocytes.

Several PAIs have been characterized in members of the genus *Salmonella*. *Salmonella* species cause diarrheal diseases in humans and animals. *Salmonella* pathogenicity island I (SPI-1; Table 2) encodes a type III secretion system, which enables invasion and survival of *Salmonella* within host cells during the early stages of infection. Recent examination of the complete genome sequence of *Salmonella enterica* serovar Typhi demonstrated the presence of five known pathogenicity islands, SPI-1 through SPI-5. In addition, at least five further regions with characteristic features of pathogenicity islands were found and designated as SPI-6 through SPI-10 (Parkhill et al., 2001; Fig. 3).

TABLE 2 Defining features of *Salmonella enterica* serovar Typhi PAIs

PAI[a]	G+C content relative to genome	Associated flanking sequences	Mobility genes	Regulatory genes present
SPI-1	Low	None	None	*araC* family, transcriptional regulators
SPI-2	Low/mosaic	tRNA Val	None	*luxR*, *merR*
SPI-3	Low	tRNA SelC	IS, transposase	Transcriptional regulators
SPI-4	Low	None	None	None
SPI-5	Low	tRNA Ser	IS, transposase	None
SPI-6	Mosaic	tRNA Asp	Transposase and integrase core	*lysR* family
SPI-7	Low/mosiac	tRNA Phe	Bacteriophage, integrase, IS, transposases	None
SPI-8	Low	tRNA Phe	Integrase pseudogene	*luxR*
SPI-9	High	10Sa RNA[b]	Adjacent to lysogenic phage	None
SPI-10	Low/mosaic	tRNA Leu	Phage integrase transposases	*araC* family

[a]Virulence functions of each PAI are given in Fig. 3.
[b]10Sa is a region encoding a small-subunit regulatory RNA.

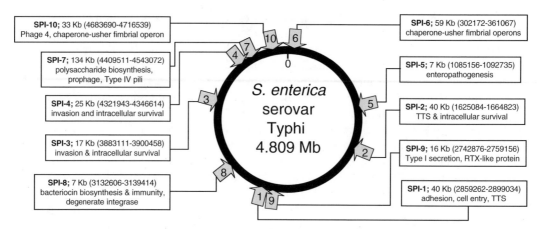

FIGURE 3 Distribution of PAIs in *Salmonella enterica* serovar Typhi. The large black circle represents the circular chromosome of serovar Typhi. The relative position of each PAI to base zero is indicated by the arrows and linked to a text box providing the name and pertinent features of each PAI. The data presented were adapted from Parkhill et al. (2001) or retrieved from the genome sequence file, GenBank accession no. AL513382.

All these regions encode genes that have been implicated in virulence and pathogenesis of *Salmonella* species and illustrate the impact of genome sequencing on the discovery of virulence-related genes and PAIs (Table 2). The serovar Typhi PAIs also demonstrate another feature of pathogenicity islands and their encoded gene clusters: the presence of regulatory genes within the island (Table 2). Transcriptional regulators and two-component response regulators are commonly encoded on PAIs to sense host factors and regulate expression of virulence genes within the island. The presence of multiple PAIs within a single pathogenic species such as serovar Typhi also clearly illustrates the concept that bacterial virulence and pathogenicity are multifactorial, with many sets of virulence genes playing a role in the evolution of a complete disease phenotype (Fig. 3 and Table 2).

The final examples of pathogenicity islands described here are drawn from another human enteric pathogen, *Shigella flexneri*. *Shigella* species are a major health problem causing more than 100 million cases of dysentery each year. Disease is characterized by invasion of the bacterium into epithelial cells of the colon, where they replicate, causing cell death and inflammation that leads to the bloody diarrhea

associated with disease. Like many species of pathogens, *Shigella* species not only possess PAIs encoded on their chromosomes, but also harbor a large virulence plasmid that encodes numerous virulence genes and an additional pathogenicity island. *Shigella* pathogenicity island I (SHI-1) was discovered by Rajakumar and colleagues (1997) and at the time was designated the *she* PAI. This region of the *Shigella* chromosome was known to encode several major virulence factors including the *Shigella* enterotoxin, ShET1, and ShMu protein, a putative hemagglutinin and mucinase. Two genes, *set1A* and *set1B*, encoded subunits that formed the complete ShET1 enterotoxin. Each of these toxin subunit genes is encoded within a third oppositely orientated open reading frame, *she*, which encodes the ShMu virulence factor. This overlapping arrangement of genes and the presence of toxin genes led Rajakumar and colleagues to investigate whether the region formed part of a larger PAI. The approach they used was to mark the *she* gene region by insertion of a tetracycline resistance cassette and then select for tetracycline-sensitive revertants by plating on fusaric acid–containing agar (a counter selection for tetracycline resistance). Derivatives surviving this selection demonstrated the loss of about

51 kb of DNA which spanned the *she* region. Insertion-sequence-like sequences were found adjacent to the deleted region. When the authors probed other enterotoxin-producing strains of *Shigella* they found that most of the genes from the deletable region were present, suggesting that the encoded genes formed part of a large pathogenicity island (Table 1).

As explained above, *Shigella* species also carry a large virulence plasmid, which in *S. flexneri* strain 5a is called pWR501 and is 210 kb. This plasmid encodes multiple virulence factors, which include an additional enterotoxin, ShET2; a type I secretion system, VirA-VirG region; and the *ipa-mxi-spa* locus, a region of DNA that has all the classical features of a pathogenicity island (Table 1). This pathogenicity island encodes a type III secretion system, which, on contact with host cells, secretes bacterial effector molecules directly into them. These effector molecules stimulate cytoskeletal rearrangements that lead to internalization of *Shigella* and allow the organism to replicate intracellularly.

PAIs Consist of DNA Regions That Differ in G+C Content and Codon Usage from the Core Genome

All bacteria have a mean DNA base composition that is characteristic of their species and is expressed in terms of the mean genomic G+C content. The use of this genetic feature as a specific taxonomic criterion is described in chapter 22. Examination of the G+C content of a region of DNA is also often used as a means to determine whether it has been acquired via horizontal gene transfer and therefore atypical for the host organism. Pathogenicity islands have a G+C content that is generally distinct from the chromosomal average for the host species. For the majority of PAIs identified to date, this value is lower than the genomic G+C content. The archetypal PAIs of uropathogenic *E. coli* possess a G+C content 10% below that of the mean genome G+C content of 51% (Table 1). Examination of complete genome sequence data also vividly demonstrates how G+C content fluctuates through-

out bacterial genomes and is significantly reduced in regions encoding PAIs. Plots of G+C content for *Salmonella* pathogenicity islands 2 and 9 of serovar Typhi are shown in Fig. 4. The mean G+C content of the serovar Typhi genome is 52%, whereas the content of the SPI-2 and SPI-9 pathogenicity islands varies from 40% to 57%. Most PAIs have a G+C content that is lower than the core regions of the bacterial genome. The lower G+C content of SPI-2 is clearly visible across the genomic region encoding this PAI (Fig. 4A). The reduced G+C content of PAIs is indicative of the different codon usage found within the PAI-encoded genes. PAI-gene codon usage will be more characteristic of the ancestral species from which the PAI was acquired during evolution of a pathogen.

However, for many PAIs not all genes within their boundaries demonstrate a G+C content lower than the genome average. For example, within the 3′ region of serovar Typhi pathogenicity island SPI-2 are a cluster of genes that encode a tetrathionate sensory and utilization system that all have a G+C content that is either around or higher than the *Salmonella* chromosomal average (see Fig. 4A). SPI-6 and SPI-7 also show a G+C content that varies greatly across the PAI, including both high and low G+C regions (Table 2). This kind of fluctuation within a PAI suggests that additional genes from *Salmonella* or other species may have been incorporated during PAI evolution, ultimately producing a mosaic of genes (Fig. 4). A gene mosaic is also seen in the *Shigella* SHI-1 PAI, which encodes enterotoxins and proteases (Table 1).

Shigella species encode PAIs on both the chromosome and a large virulence plasmid. What is the G+C content of PAI regions carried on plasmids in relation to the overall plasmid G+C content? For the *Shigella ipa-mxi-spa* PAI region carried on plasmid pWR501, G+C content is 35% compared with the average plasmid G+C content of 48% (which is close to the average genome G+C content, 50%, of the *Shigella* chromosome; Table 1). So the distinct G+C content of PAIs holds true

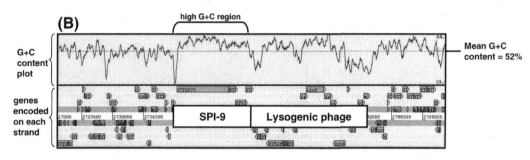

FIGURE 4 G+C content of serovar Typhi PAIs 2 and 9 (SPI-2 and SPI-9). Annotated genome sequence data (GenBank accession no AL513382) was viewed using the computer program Artemis (Rutherford et al., 2000). Each plot shows the encoded genes and G+C content of the genomic region (indicated on the right of the figure). The low G+C content (typical for most PAI) of SPI-2 is shown in A. In addition, SPI-2 demonstrates a mosaic structure with G+C content at the 3′ end of the PAI, carrying tetrathionate genes, similar to the genome average. The atypical high G+C content of SPI-9 is shown in B; the lysogenic phage inserted adjacent to SPI-9 is also shown.

for these virulence gene clusters even when they are encoded on mobile genetic elements.

Are PAIs always lower in G+C content with respect to the chromosomal mean? A limited number of genomic islands implicated in pathogenesis possess a G+C content higher than the genome average. For example, serovar Typhi SPI-9 demonstrates a G+C content of 57% compared with the average of 52% seen across the genome (Fig. 4B and Table 2). In light of these limited exceptions to the low G+C content of PAIs, the most appropriate concept to remember in defining PAIs is that they frequently show a G+C content that is atypical for the host bacterial species.

PAIs Are Often Large Genomic Regions

In 2000, Hacker and Kaper reviewed the pathogenicity islands from many species of bacteria.

Thirty-seven of known size were examined, the largest was 190 kb, and the smallest was 4 kb; and the average size was 43 kb. This size suggests that PAIs on average carry clusters of between 30 and 40 genes. Examination of the serovar Typhi genome sequence indicates an average size of 38 kb for the 10 PAIs identified. In total, these PAIs account for at least 8% of coding capacity of the genome of serovar Typhi, demonstrating what a significant impact horizontal gene transfer and PAI acquisition may have on genome size.

tRNA Genes and Small Directly Repeated (DR) Sequences Are Frequently at the Boundaries of PAIs

The presence of tRNA genes at the boundaries of genomic regions encoding virulence genes was one of the first hallmark features identified in the characterization of the PAIs in uropathogenic *E. coli*. Hacker and Kaper have subsequently noted that more than 75% of the PAIs identified are also associated with tRNA genes. This statistic is also borne out by examination of the genome of serovar Typhi where eight of the ten PAIs are associated with tRNA genes or other subunit RNA genes (Table 2). Bacterial transfer RNA genes are often a target site for the integration of foreign DNA, such as bacteriophages or episomal plasmids (see chapters 7, 8, and 9). Extrachromosomal genetic elements often encode partial or complete tRNA genes and direct repeats are present in the 3′ region of tRNA loci and other small regulatory RNAs. Therefore, homologous recombination between these repeats encoded on mobile elements and the corresponding chromosomal tRNA genes may be a mechanism by which PAI integration is achieved. Normally, after recombination, the tRNA loci will be left intact, a feature of many PAIs (Fig. 2 and Table 1). The size of DRs associated with the boundaries of PAIs varies from as little as 9 bp to 135 bp. A minority of PAIs, such as the *Salmonella* SPI-1, lack the presence of any associated tRNA or DR sequence (Table 1).

Single tRNA loci may also act as target sites for different types of PAI. For example, *selC*, encoding the selenocysteine-specific tRNA is associated with three different *E. coli* PAIs (including PAI-1 and LEE; Table 1). Serovar Typhi PAIs SPI-7 and SPI-8 are both associated with phenylalanine-specific tRNA loci (Table 2). PAIs are also frequently unstable (see below) and those associated with tRNA loci have also been shown to be capable of loss by deletion. This may lead to truncation and mutation of the associated tRNA. Hence, only multicopy tRNA genes such as phenylalanine-specific ones, those which have redundancy due to codon wobble, such as leucine-specific ones, or those encoding nonessential tRNAs, such as selenocysteine-specific tRNAs, generally form the target sites for unstable PAIs (Table 1 and 2). Finally, although insertion sequences and transposons are commonly found within PAIs (Fig. 2), they do not normally flank them in the same way as tRNA or DR loci.

PAIs Often Carry Genes Associated with Mobile Genetic Elements

The atypical G+C content of PAIs suggests that they have been acquired as a result of horizontal gene transfer; hence, it is not surprising that many PAIs also carry genes that are associated with mobile genetic elements such as insertion sequences, transposons, phages, and plasmids. The PAIs in uropathogenic *E. coli* strain 536 both carry cryptic integrases (Table 1). The *Shigella* SHI-1 chromosomal island is flanked by IS elements, while the type III secretion system *ipa-mxi-spa* PAI is associated with a large virulence plasmid and also flanked by multiple ISs (Table 1). Many of the mobility genes present within PAIs are often cryptic and not capable of expression, suggesting that their stabilization within pathogenic bacterial strains has been achieved by mutation. Examination of the serovar Typhi PAIs demonstrates that only SPI-1, SPI-2, SPI-4, and SPI-9 do not carry genes associated with mobile genetic elements, while the remaining six PAIs all possess sequences related to IS, transposases, or phages (Table 2). Even in SPI-9, although it does not possess mobility genes within its boundaries, it is located immediately adjacent to a prophage in the serovar Typhi genome (Fig. 4).

PAIs Are Often Unstable Genomic DNA Regions

Instability was noted very early in the characterization of uropathogenic *E. coli* PAI I and PAI II. These were the prototypes for the definition of PAIs. These PAIs deleted at a frequency of 1 in 10,000 clones, leading to the formation of strains attenuated in specific aspects of virulence. Deletion of the UPEC PAIs is linked to recombination between to the two flanking DR sequences. Hence, recombination can act as mechanism for both PAI insertion and deletion. In uropathogenic *E. coli* strain 536, deletion of PAIs occurs by a *recA*-independent pathway, suggesting that a novel or PAI-specific recombinase facilitates the recombination (Hacker et al., 1997). Some PAIs appear to have become stabilized by the host species and do not demonstrate instability, but they possess many of the other classical features associated with these regions. The LEE pathogenicity island of enteropathogenic *E. coli* and *Salmonella* pathogenicity island SPI-2 are stable in individual pathogenic strains. The identification of several PAIs occurred as a result of virulence gene or phenotype instability and the absence of PAI DNA in non-pathogenic strains. Therefore, instability and uneven distribution of virulence-determinant DNA between strains of a single species or closely related species can be a useful experimental aid in the discovery of PAIs.

Section Summary

- PAIs always carry clusters of virulence genes.
- PAIs may possess additional features such as (i) an atypical G+C content in comparison with the host genome, (ii) a large size, carrying multiple structural and regulatory genes, (iii) the possession of flanking sequences such as tRNA genes or DR sequences, and (iv) being associated with classical mobile elements such as insertion sequences, bacteriophages, and plasmids.
- The unusual structure and content of virulence genes within a PAI suggest they have been recently acquired in most bacterial species as a result of horizontal gene transfer.

TYPE III SECRETIONS SYSTEMS: AN EXAMPLE OF VIRULENCE GENE CLUSTERS CARRIED ON PAIs

We will now examine in more detail a specific cluster of virulence genes that are frequently carried on PAIs, the type III secretion systems.

Approximately 20% of all the proteins expressed by bacteria are destined for a location associated with or outside of the cytoplasmic membrane. Hence, secretion of proteins is a vital physiological process for both gram-negative and gram-positive bacteria. The presence of both an inner and outer membrane in gram-negative bacteria adds an extra level of transport complexity for proteins that are des-

tined for export outside of the cell. Consequently, it appears that gram-negative bacteria have evolved at least four pathways of protein secretion, which are outlined below:

1. Type I secretion. This pathway is possessed specifically by a number of gram-negative bacteria and is responsible for protein transport across both inner and outer membranes. The process occurs in a single step, does not rely on the secreted proteins possessing a signal sequence, and requires ATP as an energy source. Examples of proteins secreted by type I systems are the *E. coli* hemolysin and *Bordetella pertussis* adenylate cyclase toxin.

2. Type II secretion. The general secretory pathway. This is the basic pathway that delivers proteins across the prokaryotic cytoplasmic membrane. The general secretory pathway is sufficient for protein export in gram-positive bacteria, but only results in secretion into the periplasmic space in gram-negative bacteria. The pathway recognizes a conserved secretory signal sequence at the amino terminus of proteins destined for export. The signal sequence is cleaved during transport, and secretion requires the hydrolysis of ATP to provide energy.

3. Type III secretion. This pathway is similar to type I secretion in that it is specific to gram-negative bacteria and exports proteins across both inner and outer membranes in a single step. Type III secretion systems are more complex than type I systems, however, requiring many more components to build the complex secretion system. In addition, type III systems couple secretion with pathogenesis and are frequently carried on pathogenicity islands. The biology and evolutionary significance of type III secretion systems will be expanded below.

4. Type IV secretion. The prototypic type IV transporter, the VirB system, was described in *Agrobacterium tumefaciens*. This system specifically exports the tumor-inducing (Ti) DNA across the bacterial membrane and into plant cells, where it integrates within the plant genome. The VirB system can also direct conjugal transfer of certain plasmids between *A. tumefaciens* and other bacteria. Other type IV secretion systems that possess genetic homology to the VirB system have been found in pathogens such as *B. pertussis*, but transport toxins and not DNA. These data and the close homology between type IV components and conjugal transfer genes suggest that type IV secretion systems may have evolved from a conjugal transfer system.

The type III secretion system in the *Yersinia* Yop virulon is well documented (Cornelis and van Gijsegem, 2000), and so I will examine this further. It was discovered during the study of the pathogens, *Yersinia pestis*, *Y. pseudotuberculosis*, and *Y. enterocolitica*. *Y. pestis* is the causative agent of plague, a virulent systemic human disease acquired after individuals being bitten by infected fleas. *Y. pseudotuberculosis* and *Y. enterocolitica* infections cause diarrheal diseases and are acquired after ingestion of contaminated food. Guy Cornelis and colleagues (1998) investigated a group of secreted proteins, Yops, which were released into *Yersinia* culture supernatants when the bacteria were grown without calcium ions (inducing what was known as the low-calcium response or *lcr* genes). The Yop secretion pathway was encoded on a large virulence plasmid and was studied using a combination of genetics and microbiology, uncovering a highly complex pathway encoded by a cluster of genes. The type III secretion system is responsible for Yop formation and translocation of toxic effector molecules into host cells during infection and is conserved in all three *Yersinia* pathogen species. *Y. enterocolitica* and *Y. pestis* were also subsequently found to possess chromosomally encoded type III secretion pathways.

Homologous secretion systems were subsequently discovered in many other gram-negative pathogens that caused disease in both plants and animals. A common feature of all these systems is that protein secretion occurs from the bacteria directly into the host cell. Regulation of secretion is often highly complex with secretion only triggered by specific signals occurring during infection, such as contact between bacteria and host cells. Highly conserved proteins form the secretion apparatus and translocate specific effector molecules into host cells. These effector proteins may then act in a number of different ways. The structure and genetic organization of the *Yersinia* Yop system is discussed below to illustrate commonalities among gram-negative type III secretion systems.

The proposed structure of the *Yersinia* Yop secretion system together with clustered organization of component type III secretion genes is illustrated in Fig. 5. The proteins involved in forming the type III Yop secretion channel were termed Ysc for Yop secretion. Several proteins, YscD, -R, -S, -T, -U, and -V appear to form a secretion channel within the inner membrane (Fig. 5). YscN is homologous to ATPase/translocases and is thought to provide energy for the secretion pump. The inner membrane complex of type III secretion proteins is also highly homologous to those that form the basal body of the flagellum in motile gram-negative bacteria, with both gene sequence and order being highly conserved (Fig. 5) in both loci. The outer membrane channel is formed by YscC, which belongs to the secretin family of transporters, which are involved in a variety of secretion systems. YscC forms a ring-shaped pore structure in the outer membrane and works with several other proteins to facilitate Yop secretion (Fig. 5). Secretion via this membrane-spanning complex is regulated by many different proteins, including YopN, LcrG, and TyeA, which may form a "stop-valve" for the type III secretion system.

On contact with eukaryotic cells, specifically the macrophage in the case of *Yersinia* pathogenesis, the type III secretion apparatus is opened and energized to facilitate secretion of the Yop effector molecules into the host cell. YopB and YopD form a pore within the host-cell membrane after export from the bacterium and then translocate the effector Yop molecules directly into the macrophage cytoplasm (Fig. 5). Within the bacterial cytoplasm

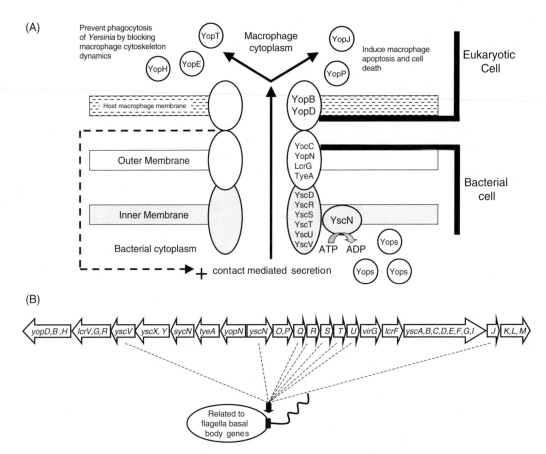

FIGURE 5 Gene order and organization of the type III secretion system of *Y. pestis* encoded on plasmid pCD1. (A) A model of the type III secretion system is shown spanning the inner and outer bacterial membranes and inserting into the cytoplasmic membrane of a host cell (the macrophage). Component proteins for each membrane domain are indicated and correspond to the gene names shown in B. (B) The genes encoding the major components of the type III secretion are shown as arrows indicating their direction of transcription. Shaded arrows indicate individual genes drawn as a single cluster for the purposes of the figure. Genes that are homologous to those which encode the gram-negative flagellum basal body are shown.

many of the immature Yop effector molecules are bound by specific chaperones that are released during secretion, enabling the Yops to carry out their functions without damaging the bacterium. The presence of such secretion-specific chaperones is another characteristic feature shared by other type III secretion systems. Once within the macrophage, the effector molecules YopT, -E, and -H appear to interfere with phagocytosis of *Yersinia*. They block various aspects of cytoskeleton dynamics and actin filament formation, preventing uptake and killing of the pathogen by the macro-

phage. YopJ and YopP effectors go one step further and induce apoptosis and death in the macrophage. By specifically targeting the host cells that protect against bacterial infection in this way, *Yersinia* species are able to undermine the host defenses, multiply, and cause systemic disease.

The arrangement of the secretion-translocation component genes and proteins is conserved in most type III secretion systems and present as compact clusters of genes (Fig. 5). The secreted effector molecules and regulatory components are often diverse and dependent on the pathogen. The *Yersinia* Ysc system is closely related to the *P. aeruginosa* Psc system in terms of gene organization and overall sequence conservation, despite the evolutionary distance between these two bacteria. However, the effector molecules of the *P. aeruginosa* type III secretion system are a group of exotoxins and exoenzymes, of which only ExoS protein shows some homology with YopE, the *Yersinia* protein that blocks phagocytosis by host macrophages. Other examples of homology include the type III secretion system encoded on serovar Typhi SPI-1, which is closely related to the plasmid-borne *ipa-mxi-spa* system of *Shigella* species. The *E. coli* LEE type III secretion system is also very similar to that encoded by serovar Typhi SPI-2.

The conservation of gene content and gene order of type III secretion systems suggests that they derived from a common ancestor and transferred among species by horizontal gene transfer. The propensity of type III secretion systems to be encoded on either plasmids or as part of pathogenicity islands also points to their evolutionary spread as part of bacterial mobile elements. The components of the inner-membrane secretion system may have evolved from the gene clusters that encode the flagellum basal body since they are highly related in sequence and frequently conserved in organization (Fig. 5). However, each type III secretion system has been adapted during the evolution of pathogens to secrete a variety of different effector molecules directly into cells of the host, causing damage and facilitating infection.

Section Summary

- Gram-negative bacteria possess at least four different pathways for protein and macromolecule secretion across their inner and outer membranes.
- Type III secretion systems are frequently encoded on PAIs.
- Type III secretion systems play a major role in bacterial pathogenesis.
- Type III secretion systems secrete toxic effector proteins from a pathogen directly into host cells.

- The genes and proteins of type III secretion systems are conserved across many different bacterial species.

HORIZONTAL GENE TRANSFER AND THE EVOLUTION OF VIRULENCE IN PATHOGENIC *E. COLI* STRAINS: A GENOMIC PERSPECTIVE

In this section we will examine the entire genome of a bacterium, *E. coli*, to illustrate how horizontal gene transfer (see chapters 7 through 11) has contributed to the evolution of pathogenesis in pathogenic strains.

Much of our current knowledge of bacterial genetics derives from the intensive study of the nonpathogenic *E. coli* laboratory strain K-12. There are also several pathogenic lineages within this species, such as uropathogenic *E. coli*, discussed above in relation to the discovery of PAIs, and enterohemorrhagic *E. coli* (EHEC), which cause a dysentery-like disease. EHEC secrete toxins that are homologous to the *Shigella* enterotoxins, damage the intestinal lining by the production of a variety of other virulence factors, but rarely invade host epithelial cells. Recent comparison of the K-12 genome sequence and that of the enterohemorrhagic *E. coli* O157:H7 revealed considerable new information on the evolution of this species. Phylogenetic examination of each strain suggests that they derived from a common ancestor about 4.5 million years ago. Genome sequence comparison demonstrated that nearly all the subsequent evolution of each strain was driven by horizontal gene transfer, and, in strain O157:H7, many of these events were associated with the acquisition of virulence genes. Several of these horizontally acquired virulence mechanisms are illustrated in Fig. 6.

The size difference of each *E. coli* genome immediately indicates that gene acquisition has been most active in evolution of virulence in strain O157:H7. Blattner and colleagues (1997) found that the enterohemorrhagic *E. coli* genome is 5.5 Mb, nearly a megabase large than the K-12 genome (4.6 Mb). However, each strain shares a common backbone of about 4.1 Mb of DNA carrying homologous genes. Colinearity of genes is preserved for nearly all

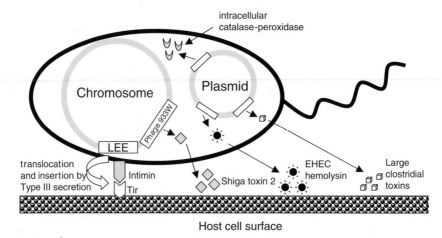

FIGURE 6 Horizontally acquired virulence factors that contribute to the pathogenesis of enterohemorrhagic *E. coli* O157:H7. Virulence genes and their location are shown by the open blocks on either the chromosome or large virulence plasmid of strain O157:H7. The pathogenicity island, LEE, and toxin-producing phage are encoded on the chromosome. The large virulence plasmid encodes multiple virulence factors including a hemolysin and clostridial-like toxins. All the latter horizontally acquired virulence genes encode traits that result in the damage of the host gut epithelium.

this backbone DNA, with only a 422-kb inversion being present near the replication terminus of each genome. Most of the differences between the two *E. coli* strains are associated with the presence or absence of clusters of genes. Clusters specific to the pathogenic strain O157:H7 were designated O islands, while gene clusters specific to *E. coli* K-12 were designated K islands. O islands formed a total of 1.34 Mb of DNA, with each segment of DNA comprising clusters of genes of up to 88 kb. In contrast, the amount of K–island DNA specific to strain K-12 was much smaller, at only 0.53 Mb. Perna and colleagues (2001) used the term "islands" for these clusters of strain-specific DNA as an extension of the term pathogenicity island; not all of the O islands or K islands carried virulence-associated genes, and hence their description as genomic islands was more appropriate. Several genomic islands encoded new metabolic capabilities demonstrating that other phenotypic traits of strains might evolve through horizontal transfer events that mirror those associated with pathogenicity islands.

As you would expect, many of the large O islands carried genes implicated in virulence of EHEC O157:H7 and several of these possess features associated with pathogenicity islands. For example, the PAI locus of enterocyte effacement (LEE; Table 1) and Shiga-toxin-producing lysogenic phage 933W had been previously associated with virulence in animal studies (see Fig. 6). Other O islands demonstrate considerable homology with known PAIs, such as O island 122, which encodes a type III secretion system closely related to the *Shigella ipa-mxi-spa* host invasion locus. Four other large O–islands encoded virulence factors, phage integrases and were adjacent to tRNAs, such as serine-, phenylalanine-, and selenocysteine-specific transfer RNAs, indicating that they may be putative pathogenicity islands. Other smaller O islands carried other virulence-associated genes, such as fimbrial biosynthesis operons, iron uptake systems, and adhesins. Many O islands also carried other genes not linked to virulence, such as carbohydrate-transport systems, antibiotic efflux, and aromatic compound degradation. Most of these O islands demonstrate atypical G+C content in keeping with their recent acquisition by horizontal gene transfer.

E. coli O157:H7 also carries a large, 92-kb, virulence plasmid, pO157, which has also been completely sequenced (Burland et al., 1998) and shows that several virulence gene clusters are encoded on this F-like plasmid. During the evolution of strain O157:H7, plasmid pO157 was probably acquired via conjugation. However, in its current state, pO157 is known to be nonconjugative. DNA-sequence analysis of the plasmid demonstrated that large portions of the putative transfer region had been replaced by insertion sequences leading to the lack of conjugal phenotype. Acquisition of this plasmid during the evolution of strain O157:H7 provided the organism with several new virulence factors (Fig. 6). The plasmid also encodes a hemolysin (specific to EHEC) that is part of an operon responsible for its secretion and insertion into mammalian cells during infection. A novel catalase-peroxidase is present on the plasmid, conferring additional protection to host defenses during infection. Other plasmid-encoded virulence factors include various proteases related to known virulence genes in other pathogenic *E. coli* strains. Finally, sequence analysis of pO157 demonstrated the presence of a novel toxin-encoding gene. *E. coli* O157:H7 infection produces lesions of intestinal damage that are similar to those observed during *Clostridium difficile* infections. The novel plasmid-encoded EHEC toxin was very similar to the large clostridial toxins, which mediate that kind of cellular destruction (Fig. 6).

Overall, comparative analysis of *E. coli* K-12 and O-157:H7 genomes demonstrated that most of the key differences between the strains were attributable to horizontal gene transfer events, in particular, the acquisition of new virulence traits in O157:H7. Many of the acquired virulence genes are shared by other pathogenic *E. coli* lineages, such as the enteropathogenic *E. coli*, and lead to the question of whether these pathogenic *E. coli* lineages derived from a common pathogenic clone? Recent population studies suggest that the pathogenic *E. coli* lineages appear to have been evolving independently with little exchange of genes with one another (Ochman and Jones, 2000). Acquisition of virulence appears to have occurred relatively recently with independent pathogenic *E. coli* lineages picking up the same virulence factors in parallel. Overall, as a model of horizontal gene transfer and the acquisition of virulence, pathogenic *E. coli* strains are excellent examples.

Section Summary

- Whole-genome comparison is a rapid and detailed means of examining the evolution of a species.
- The evolution of pathogenic *E. coli* strains demonstrates a bias toward acquisition of genes by horizontal gene transfer.
- In addition to finding PAIs within the genome of strain O157:H7, other non-virulence-associated gene clusters were found and designated as genomic islands.
- Genomic islands demonstrate features such as atypical G+C content and presence of mobility genes that are indicative of their acquisition by horizontal gene transfer.

LOSS OF DNA AS A MECHANISM OF PATHOGEN EVOLUTION

The preceding section describes many examples to show that the horizontal transfer of genes leads to the evolution of pathogenic bacterial strains and species. However, evidence of a complementary pathway of gene deletion has also emerged. Certain genes may be either (i) detrimental to the pathogenic lifestyle of a bacterium or (ii) may not be required and hence redundant in the disease environment in which the pathogen lives. Over time these genes may be lost or become inactive as pathogens adapt to their environment. We will consider two examples of gene loss in bacterial evolution that illustrate the following concepts.

- Pathoadaptive changes may result in the deletion of DNA.
- Obligate intracellular pathogens are subject to reductive evolution.
- Loss of DNA forms a parallel evolutionary pathway to gene acquisition.

Black Holes in *Shigella* Species and Enteroinvasive *E. coli*

The family *Enterobacteriaceae* consists of several very closely related species such as *Escherichia*, *Shigella*, and *Salmonella*, and during the past 50 years, clinical microbiologists have developed various biochemical and phenotypic tests to distinguish them. If modern taxonomic criteria used to define bacterial species (such as the level of whole-genome DNA-DNA hybridization being greater than 70%; see chapter 21) were applied to this group of enteric bacteria, many of the species would be difficult to distinguish. The four species within the genus *Shigella*, *S. flexneri*, *S. sonnei*, *S. boydii*, and *S. dysenteriae*, possess greater that 90% homology by DNA-DNA reassociation with *E. coli* K-12. Therefore, *E. coli* and *Shigella* could be considered a single species. In addition, genetic mapping and sequence analysis of chromosomes of *E. coli* and *Shigella* demonstrate that they possess a high degree of colinearity, with the organization and arrangement of the majority of genes being identical. The high degree of sequence homology also enables the recombination of genes transferred between these two species by conjugation, transduction to occur at high frequency.

However, the pathogenicity and epidemiology of *E. coli* and *Shigella* species is radically different. *Shigella* species are major pathogens that cause virulent dysentery in humans, while, with the exception of a few pathogenic strains, *E. coli* is generally considered to be a harmless human commensal. Enteroinvasive *E. coli* (EIEC) is a pathogenic clone that can be considered a genetic hybrid between *E. coli* and *Shigella*. EIEC carry virulence plasmids that encode toxins related to those encoded on the virulence plasmid of *Shigella*. As a result of toxin expression and the presence of other virulence factors, EIEC also cause an invasive enteric disease that is indistinguishable from *Shigella* dysentery. The commonality between these two pathogens was further examined by Anthony Maurelli and colleagues in 1998. They noted that *Shigella* and EIEC also share several biochemical features and, in particular, the absence of lysine decar-

boxylase activity. In contrast, more than 90% of *E. coli* strains are lysine decarboxylase positive. This observation led Maurelli et al. (1998) to postulate that the absence of lysine decarboxylase activity may be important to the virulent lifestyle of EIEC and *Shigella* species. They uncovered a large genomic deletion encompassing the lysine decarboxylase gene locus that was shared by both *Shigella* and EIEC and enhanced their pathogenicity. Their work clearly demonstrates a role for gene loss in the evolution of bacterial virulence.

Initially, Maurelli et al. attempted to introduce the *E. coli* gene encoding lysine decarboxylase, *cadA*, into *S. flexneri*, and found that this was not possible by transduction. Complementation of the *cadA* genotype by this means required homologous recombination between common regions of DNA, which they assumed would be present flanking the absent *cadA* gene. Failure to transduce the *E. coli cadA* gene suggested that *S. flexneri* possessed a large genomic deletion in this region of its genome. Introduction of the *E. coli cadA* gene was finally achieved by transformation of a cloned copy carried on a plasmid vector. The resulting transformant was designated *S. flexneri* BS529 (pCAD+) and was able to express lysine decarboxylase activity. Tests examining its invasive properties demonstrated that possession and expression of the *E. coli cadA* gene did not alter this aspect of *Shigella* virulence, with the transformant being highly invasive.

Because *Shigella* species also produce several enterotoxins, which play a major role in their invasive and inflammatory phenotype, Maurelli et al. next investigated toxin production and activity. *S. flexneri* produces at least two kinds of enterotoxin. One type, *Shigella* enterotoxin 1, ShET1, is chromosomally encoded on a PAI that is present in all strains of *S. flexneri* (Table 1). *Shigella* enterotoxin 2, ShET2, is encoded on a large virulence plasmid that is found in all strains of *Shigella* species and also EIEC. Toxin activity was assayed in a rabbit model of infection, where fluid accumulation was measured within the ileal intestinal loop. The presence of a functional *cadA* gene in

S. flexneri BS529(pCAD+) significantly reduced toxin activity. The toxicity of both the chromosomally encoded ShET1 and plasmid-encoded ShET2 was reduced. In addition, toxin activity in wild-type *S. flexneri* culture supernatant could be inhibited by the addition of the culture supernatant from the *cadA*⁺ transformant. These data indicated that the toxin inhibition mediated by introduction of the *cadA* lysine decarboxylase gene was acting directly on the toxin protein rather than at the level of toxin gene expression.

Maurelli and colleagues next determined the nature of the inhibitory factor. They postulated that it was either the lysine decarboxylase protein encoded by *cadA*, or a product of the catabolic reaction it catalyzes. Because lysine decarboxylase is a cytoplasmic protein not usually found in culture supernatants, they investigated cadaverine, a product of lysine decarboxylation, which is known to be secreted from bacteria with lysine decarboxylase activity. Cadaverine was present in the culture supernatant of the *S. flexneri* *cadA*⁺ transformant, but absent in *Shigella* species and EIEC culture supernatants. The levels of cadaverine secreted by the *S. flexneri* transformant were equivalent to that found in nonpathogenic *cadA*⁺ *E. coli* strains. The addition of purified cadaverine to entertoxin-containing culture supernatants also inhibited fluid accumulation in the intestinal models. Pretreatment of these tissues with cadaverine prior to the addition of the enterotoxin-containing supernatants demonstrated that cadeverine appeared to function by protecting the tissues rather than by directly acting on the enterotoxin molecules. Overall, Maurelli et al. had defined a mechanism whereby virulent *Shigella* sp. and EIEC strains had enhanced the activity of their enterotoxins by deleting *cadA* and not producing cadaverine, which they found was a potent inhibitor of toxin activity.

Finally, the exact size of the genomic deletion around *cadA* was determined by a combination of mapping using pulsed-field gel electrophoresis and Southern hybridizations with gene probes known to be encoded around *cadA* in *E. coli* K-12. It was found that up to 90 kb of DNA was deleted dependent on the *Shigella* sp. or EIEC strains examined. The deleted region was not exactly the same and the end points of the deletion varied from strain to strain. Also, one *E. coli* K-12 gene, *proP*, encoding a proline transporter, was found to be present in the middle of the deleted region for all *Shigella* spp. and EIEC strains investigated, suggesting that this gene was retained because it was essential or beneficial to these pathogens. The *cadA* region was not found to encode any phage-like attachment sites that may have played a role in its deletion during a phage excision event.

It had been previously noted that nonpathogenic *E. coli* strains also contain a cryptic prophage DLP12, which inserts at a tRNA gene *argU*, carried just at the boundary of the *cadA* deletion in EIEC. This 21-kb prophage is also known to be absent in *Shigella* spp. Its absence also constitutes a mechanism by which the pathogen has enhanced its virulence potential by gene loss. The prophage encodes a protease activity that degrades the *Shigella* outer membrane protein, VirG, a virulence factor encoded on the large plasmid that is crucial for cellular invasion and spread of the pathogen. The *cadA* deletion was unique in being much larger than the prophage-mediated gene deletion and in not being associated with a mobile genetic element. The comparison of *E. coli* K-12 and *E. coli* O157:H7 genome sequences performed by Perna et al. also demonstrated that the nonpathogenic K-12 strain possessed 528 genes that were absent in the enterohemorrhagic pathogen. Many of these absent genes were associated with large genomic K islands that were inserted at the same site on the co-linear chromosome as pathogenic O islands in strain O157:H7. These data indicate that both gene acquisition by horizontal transfer and gene deletion act in parallel during the evolution of pathogenesis in virulent strains. The absence of lysine decarboxylase activity in *Shigella* spp. and EIEC is an excellent example of the enhancement of virulence by gene loss.

The Leprosy Bacillus: Genome Decay, Pseudogenes, and Homologies to Eukaryotic Genes

Mycobacterium leprae is the causative agent of leprosy, an age-old human disease. Leprosy has been very difficult to study because *M. leprae* can only replicate within cells of a human host or a surrogate host such as the nine-banded armadillo. Its growth rate is also tortuously slow, taking up to a year to reach levels where sufficient bacterial growth is available for study. As a pathogenic bacterium, *M. leprae* can serve as a useful model in understanding the evolution of bacteria toward an obligate intracellular lifestyle. Many of the enteric pathogens described throughout this chapter possess stages of infection where they invade host cells, multiply, cause cellular destruction, and transmit to new cells or a new host. However, *M. leprae* is an obligate intracellular pathogen and apart from transmission, which is believed to occur via a respiratory route, it spends its entire life growing within host cells.

M. leprae is a *Mycobacterium*, a genera that consists of phenotypically diverse species, including pathogens such as *Mycobacterium tuberculosis* (the causative agent of tuberculosis) and free-living environmental species such as *Mycobacterium fortuitum*, which rarely cause infections. Genome-sequence analysis has been applied to both *M. leprae* and *M. tuberculosis*. Comparison of each genome revealed several interesting features of *M. leprae* that reflect its adaptation to a parasitic intracellular lifestyle. In contrast to *M. leprae*, *M. tuberculosis* can be grown in laboratory culture on defined minimal media, but both pathogens grow intracellularly during infection.

The genome of *M. tuberculosis* is 4.4 Mb and consists of approximately 4,000 genes with a G+C content of 65.6%; this G+C content and genome size is typical of most of the other species of mycobacteria (Cole et al., 1998). In contrast, the *M. leprae* genome was considerably reduced in size at 3.2 Mb and possessed a lower G+C content of 57.8%. In addition, only half of the *M. leprae* genome was found to contain protein-coding sequences. More than

a quarter of the *M. leprae* genes identified were homologous to other known or functional genes in *M. tuberculosis*, but many of the *M. leprae* homologues contained mutations that made them inactive and led to their designation as pseudogenes. Of the potentially active genes found in *M. leprae*, only 1,400 were common to both *M. leprae* and *M. tuberculosis*. Although significant gene loss had occurred in *M. leprae*, gene inactivation to form pseudogenes also appeared to be major mechanism of reductive evolution to which it has been subjected as an obligate intracellular pathogen.

What types of genes were lost during the evolution of *M. leprae*? Despite being identified over a century ago, scientists have failed to grow *M. leprae* in laboratory culture outside of an animal host. Determination of the genome sequence of *M. leprae* demonstrated the absence of many metabolic genes, providing clues as to why it could not be cultured in vitro. *M. leprae* has only two lipase genes in comparison with the 22 possessed by *M. tuberculosis*. Because the degradation of host-derived membrane lipids is an important source of energy for intracellular mycobacterial pathogens, the limited pathways for scavenging lipids in *M. leprae* may considerably restrain its growth potential. Also, simple carbon sources such as acetate and galactose cannot be utilized because of the absence of vital metabolic genes in *M. leprae*. Other carbon- and nitrogen-metabolizing pathways are also restricted because of an absence of key genes and regulators. Overall, both aerobic and anaerobic metabolism was highly restricted in *M. leprae* by gene loss and inactivation. In contrast, many essential anabolic pathways are present in *M. leprae*. It possesses nearly all the genes and pathways necessary for the biosynthesis of amino acids, nucleotides, vitamins, and cofactors, suggesting that adaptation to states of auxotrophy were not prominent in its evolution to the host environment.

A feature of the *M. leprae* genome, which was also unusual for a bacterium, was the high proportion of apparently noncoding DNA, which accounted for 23.5% of its genome. For nearly all the bacterial species that have been sequenced to date, gene density is greater than 90% of the genome (leading to the basic rule of thumb that 1.1 kb of bacterial genomic DNA is equivalent to one gene). The presence of significantly greater noncoding DNA in *M. leprae* is a trait more in common with eukaryotic genome organization. In eukaryotes, noncoding DNA may serve regulatory functions; it is not known whether the noncoding DNA in *M. leprae* serves a similar role. Alternatively, the extensive presence of noncoding DNA may have arisen from drastic mutation of *M. leprae* genes resulting in them being unrecognizable. Another link to eukaryotic organisms was found when Cole and colleagues (2001) examined the tRNA synthetase genes of *M. leprae*. The prolyl-tRNA synthetase, encoded by the gene *proS*, was more closely related to eukaryotes than other mycobacteria or prokaryotes. The *proS* gene of the intracellular pathogen, *Borrelia burgdorferi*, was also found to be more closely related to eukaryotic organisms than to bacteria. In addition to the lack of mycobacterial homology, the position of the *proS* locus within the genome was different than the arrangement of genes in the same region of the *M. tuberculosis* genome. These data suggest that both *M. leprae* and *B. burgdorferi* may have acquired the *proS* gene as a result of a horizontal gene transfer event that took place between each bacterium and their eukaryotic hosts.

How has the *M. leprae* genome evolved to produce so many pseudogenes, deletions, and rearrangements in comparison with *M. tuberculosis*? Although we have discussed horizontal gene transfer extensively as a means of genome evolution in this chapter, changes in gene sequence due to base mutation during DNA replication forms the baseline mechanism of natural evolution in all organisms (see chapter 1). Two features of *M. leprae* may have accelerated its natural rate of mutation leading to the high proportion of pseudogenes within its genome. Its intracellular lifestyle within macrophages leads to considerable exposure to oxidative radicals that damage DNA. Second, vital proof-reading activity is absent from its DNA polymerase III, which may contribute further to the incorporation of mutations during DNA replication.

Rearrangement and deletion of DNA appears to form the other major pathway of genome evolution in *M. leprae*. Many of the boundaries of regions that have been rearranged or deleted are associated with tRNA genes and repetitive DNA sequences. Mycobacteria are known to possess a high frequency of illegitimate recombination and, in *M. leprae*, this has produced a mosaic structure of genes around the genome produced by multiple recombination events during its evolution. Finally, little genomic change in *M. leprae* appears to have been driven by the acquisition of foreign DNA. *M. tuberculosis* contains two prophage elements and multiple insertion sequences, whereas *M. leprae* contains only three genes with some homology to bacteriophage DNA and 26 transposase-like gene fragments. Horizontal transfer of prokaryotic genes would be expected to be limited in an organism with an obligate intracellular lifestyle. It is also interesting that the one predicted transfer event, acquisition of *proS*, appears to have derived from the transfer of a host tRNA synthetase. Hence, most of the evolution observed in *M. leprae* has been reductive, driven by its continuing adaptation to a highly conserved intracellular environment.

Section Summary

- The highly conserved intracellular niche of *M. leprae* has had a profound influence on the genomic evolution of this pathogen.
- Considerable reductive evolution occurs in intracellular bacterial pathogens.
- Gene loss is mediated by both deletion and gene inactivation, which results in pseudogenes.
- Gene acquisition by horizontal transfer may be limited by the physical limitations of an intracellular lifestyle, but gene acquisition from the host may occur in some instances.

GENOMIC PROCESSES INVOLVED IN BACTERIAL PATHOGEN EVOLUTION

I have described both additive and reductive mechanisms of evolution in bacterial pathogens. These mechanisms act to create genome sequence diversity that contributes to the ability of a pathogen to survive, grow, and replicate during infection. Mira and colleagues (2001) have elegantly reviewed the processes that contribute to the evolution of bacterial genomes, and their findings are used to summarize and bring perspective to the gene associations in bacterial pathogenesis.

Bacterial genomes do not vary greatly in size, especially when they are compared with eukaryotic species. Most bacterial genomes are between 1 and 10 Mb, while eukaryotes may carry between 5 and 3,000 Mb of DNA. In addition to their tenfold limitation in genome size, bacteria also demonstrate very high coding densities with, on average, greater than 90% of the genome occupied by genes. Hence, large bacterial genomes possess proportionally larger numbers of genes. In contrast, eukaryotes may differ 100-fold in genome size but carry the same number of genes. The processes that contribute to this efficiency in the maintenance of genome size and content have been widely debated. Mira et al. examined the genome sequences of more than 40 prokaryotic species to shed light on the mechanisms of bacterial genome evolution. Their findings in the context of the bacterial pathogenesis discussed here are illustrated in Fig. 7. By comparing the number of pseudogenes with their functional counterparts in a variety of different bacterial genomes, Mira et al. (2001) predicted the number of deletions and insertions per pseudogene. In every bacterial genome compared, deletions occurred on average twice as frequently as insertions; hence, they proposed that the major force driving the evolution of bacterial genomes is gene loss and deletion (Fig. 7).

Gene loss may occur by at least two mechanisms. One or several genes may be lost during large deletion events facilitated by recombination (Fig. 7). The loss of the lysine decarboxylase gene and several associated gene loci in *S. flexneri* illustrates how the deletion of a large region of DNA enhanced the pathogenicity of this species. A second mechanism of gene loss is gene inactivation and the formation of pseudogenes (Fig. 7). Examination of *M. leprae* demonstrated that pseudogene formation (and deletion) was a prominent genomic event in the adaptation of the leprosy bacillus to an obligate intracellular lifestyle. However, in the context of other intracellular pathogens and symbionts, such as *Mycoplasma* or *Buchnera* species, *M. leprae* is unusual in that it possesses a very large number of pseudogenes and a high proportion of noncoding DNA. Mira and colleagues suggest that this may be because *M. leprae* is at an early stage of reductive evolution from its tuberculosis-like ancestor. Although it is an ancient human disease, leprosy is very new in terms of the 4-billion-year evolutionary period of bacteria, a fact that lends weight to this hypothesis. What drives this bias toward reductive evolution? Among intracellular pathogens such as *M. leprae*, there appears to be decreased selection to maintain gene function. The host intracellular environment is generally less dynamic and can provide many essential factors required for growth. Hence, many genes encoding phenotypes for these characteristics in such bacteria are redundant. Over time, these redundant genes acquire mutations becoming inactive pseudogenes, which eventually may be completely deleted (Fig. 7).

The argument above for a deletional bias in the evolution of pathogenic bacterial genomes also makes sense if one considers all the examples of gene acquisition by horizontal transfer that are illustrated here in regard to pathogenesis and throughout this book in association with other bacterial traits. Horizontal gene transfer is a continual process occurring in nearly all, if not all bacterial species, with only *M. leprae* and other intracellular bacteria showing limited gene acquisition via this pathway. However, bacterial genome size is strictly maintained for most species, suggesting that a parallel process of gene loss must balance the additive evolutionary process of gene acquisition. New virulence genes may be assimilated by bacterial pathogens

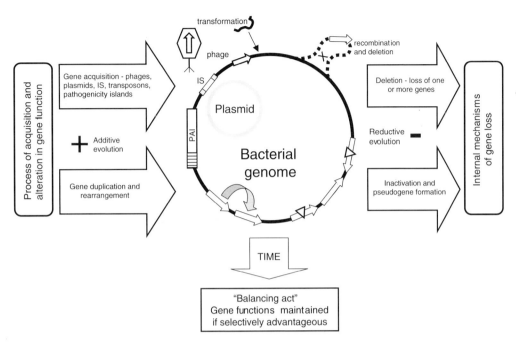

FIGURE 7 The major mechanisms involved in the evolution of bacterial pathogen genomes. A bacterial genome is indicated by the black circle. Processes of additive evolution are indicated on the left, while the complementary mechanisms of reductive evolution are indicated on the right. Genes, gene clusters, and mobile genetic elements associated with evolution pathogenesis are shown as arrows or labeled features.

if they contribute to the process of disease. Phages, plasmids, insertion sequences, transposons, and pathogenicity islands may all contribute to this process of additive evolution, but it must be balanced by a simultaneous process of gene loss to maintain the compact genomes possessed by bacteria (Fig. 7).

Pathogenicity islands and their associated genes, such as type III secretion systems, demonstrate how many bacterial species have acquired and evolved virulence functions using the same basic mechanism of horizontal gene transfer followed by mutation and adaptation to the lifestyle of the host bacterium. The insertion and deletion of pathogenicity islands as complete blocks of genes suggest that they may act as discrete molecular units within the genome of bacterial pathogens. Hacker and Kaper noted that the bacterial species in which pathogenicity islands are prevalent are also those which participate extensively gene transfer via plasmids, phages, and conjugative transposons. In contrast, they noted that PAIs were rare in species that were naturally competent, being able to take up DNA by natural transformation (see chapter 10). Naturally competent bacterial species demonstrate many instances of sequence variation which have resulted from the recombination of small DNA fragments within genes. This provides another mechanism by which pathogens may evolve virulence functions with a good example being the evolution of penicillin-binding proteins in *Streptococcus pneumoniae* (see Chapter 10).

Comparison of pathogenic and nonpathogenic *E. coli* strain genomes clearly demonstrates a mosaic-like genome structure of genomic islands occurring within a common species-specific backbone of DNA. This kind of organization (extensively observed with regard to pathogenicity islands) has now also been noted for clusters of genes encoding resistance, degradative, metabolic, and a whole variety of other functions. Perna and colleagues described such gene clusters as genomic islands. Many PAIs and genomic islands possess features that indicate their acquisition by horizontal gene transfer, such as atypical G+C content, flanking boundary sequences like tRNAs, and direct repeats or the presence of mobility genes. These findings have led to the emergence of the genetic concept that core DNA encodes basic housekeeping functions and traits specific to a bacterial species, while accessory DNA, such as PAIs and genomic islands, may encode more diverse functions that have been acquired via horizontal gene transfer. Ultimately, it is the environment that maintains the genome size and content of a given bacterial species (Fig. 7). Pathogens require virulence genes to cause cellular damage to provide resources for growth and to exploit new niches within a host organism. This results in disease of the host. Many gene associations have been identified in relation to bacterial pathogenesis and each provides a mechanism by which sequence diversity may be generated.

Section Summary

- Prokaryotic genomes are compact and show a limited size range when compared with eukaryotic organisms. The factors maintaining the size and coding efficiency of bacterial genomes were identified.
- Gene inactivation and deletion appear to be the most active pathway in the evolution of bacterial genome size.
- Gene acquisition among bacterial pathogens enabled the rapid evolution of pathogenic clones.
- Pathogenicity islands contributed significantly to the evolution of bacterial diseases.
- Genomic islands have added to the evolution of many other nonpathogenic bacterial functions.
- Identification of pathogenicity islands and genomic islands in bacteria led to the concepts of core DNA and accessory DNA.

- Environmental selection for gene function is ultimately responsible for the maintenance of genome size and content.

SUMMARY

Bacterial pathogens have evolved to cause disease in various host organisms. They may live exclusively on or within these hosts, or be capable of growth and replication in several environments, including the natural environment and states of disease. Pathogens possess multiple genes that encode virulence factors used to damage or infect the host. The evolution of bacterial virulence depends on a number of mechanisms, including gene acquisition by horizontal gene transfer and gene loss by deletion or inactivation. Pathogenicity islands are good examples of gene acquisition, carrying clusters of genes that enhance virulence. They possess atypical features that indicate their recent evolutionary acquisition by most pathogens. Many pathogenicity islands encode type III secretion systems that secrete pathogenic proteins from the bacterium directly into host cells, mediating invasion and tissue destruction. PAIs were originally discovered by researchers investigating specific bacterial virulence mechanisms; however, the comparison of whole genomes of pathogens with nonpathogenic strains has revealed the presence of many different kinds of genomic islands encoding clusters of genes. Not all of these regions encode virulence traits, leading to the emerging concept of discrete genomic islands. These genomic islands encode accessory functions required by bacteria for unusual interactions and possess many features of pathogenicity islands, such as atypical G+C content and the presence of mobility-associated genes. Finally, in addition to gene acquisition, a complementary pathway of gene loss exists in the evolution of bacterial pathogenesis. This may involve the deletion of genes that interfere with specific pathogen functions required for disease. Genes that are redundant in conserved habitats such as the intracellular environment may also be lost by deletion or gene inactivation. In summary, both additive and reductive evolutionary pathways have contributed to the evolution of bacterial pathogens.

QUESTIONS

1. Give an example of bacterial pathogen evolution by the acquisition of virulence genes.

2. Give an example of a bacterial pathogen evolution by loss of genes.

3. What are the features of bacterial pathogenicity islands?

4. Discuss with examples the concept of bacterial genomic islands.

5. Horizontal gene transfer is a well-described mechanism of bacterial pathogen evolution, yet most bacteria maintain a strict genome size. Describe the evolutionary processes involved in maintaining this balancing act.

REFERENCES

Blattner, F. R., G. Plunkett III, C. A. Bloch, N. T. Perna, V. Burland, M. Riley, J. Collado-Vides, J. D. Glasner, C. K. Rode, G. F. Mayhew, J. Gregor, N. W. Davis, H. A. Kirkpatrick, M. A. Goeden, D. J. Rose, B. Mau, and Y. Shao. 1997. The complete genome sequence of *Escherichia coli* K-12. *Science* **277**:1453–1474.

Burland, V., Y. Shao, N. T. Perna, G. Plunkett, H. J. Sofia, and F. R. Blattner. 1998. The complete DNA sequence and analysis of the large virulence plasmid of *Escherichia coli* O157:H7. *Nucleic Acids Res.* **26**:4196–4204.

Casadevall, A., and L. A. Pirofski. 2000. Host-pathogen interactions: basic concepts of microbial commensalism, colonization, infection, and disease. *Infect. Immun.* **68**:6511–6518.

Cole, S. T., R. Brosch, J. Parkhill, T. Garnier, C. Churcher, D. Harris, S. V. Gordon, K. Eiglmeier, S. Gas, C. E. Barry III, F. Tekaia, K. Badcock, D. Basham, D. Brown, T. Chillingworth, R. Connor, R. Davies, K. Devlin, T. Feltwell, S. Gentles, N. Hamlin, S. Holroyd, T. Hornsby, K. Jagels, B. G. Barrell, et al. 1998. Deciphering the biology of *Mycobacterium tuberculosis* from the complete genome sequence. *Nature* **393**:537–544.

Cole, S. T., K. Eiglmeier, J. Parkhill, K. D. James, N. R. Thomson, P. R. Wheeler, N. Honore, T. Garnier, C. Churcher, D. Harris, K. Mungall, D. Basham, D. Brown, T. Chillingworth, R. Connor, R. M. Davies, K. Devlin, S. Duthoy, T. Feltwell, A. Fraser, N. Hamlin, S. Holroyd, T. Hornsby, K. Jagels, C. Lacroix, J. Maclean, S. Moule, L. Murphy, K. Oliver, M. A. Quail, M. A. Rajandream, K. M. Rutherford, S. Rutter, K. Seeger, S. Simon, M. Simmonds, J. Skelton, R. Squares, S. Squares, K. Stevens, K. Taylor, S. Whitehead, J. R. Woodward, and B. G. Barrell. 2001. Massive gene decay in the leprosy bacillus *Nature*. **409**:1007–1011.

Cornelis, G. R., A. Boland, A. P. Boyd, C. Geuijen, M. Iriarte, C. Neyt, M. P. Sory, and I. Stainier. 1998. The virulence plasmid of *Yersinia*, an antihost genome. *Microbiol. Mol. Biol. Rev.* **62**:1315–1352.

Cornelis, G. R., and F. Van Gijsegem. 2000. Assembly and function of type III secretory systems. *Annu. Rev. Microbiol.* **54**:735–774.

Hacker, J., G. Blum-Oehler, I. Muhldorfer, and H. Tschape. 1997. Pathogenicity islands of virulent bacteria: structure, function and impact on microbial evolution. *Mol. Microbiol.* **23**:1089–1097.

Maurelli, A. T., R. E. Fernandez, C. A. Bloch, C. K. Rode, and A. Fasano. 1998. "Black holes" and bacterial pathogenicity: A large genomic deletion that enhances the virulence of *Shigella* spp. and enteroinvasive *Escherichia coli*. *Proc. Natl. Acad. Sci. USA* **95**:3943–3948.

Mira, A., H. Ochman, and N. A. Moran. 2001. Deletional bias and the evolution of bacterial genomes. *Trends Genet.* **17**:589–596.

Ochman, H., and I. B. Jones. 2000. Evolutionary dynamics of full genome content in *Escherichia coli*. *EMBO J.* **19**:6637–6643.

Parkhill, J., G. Dougan, K. D. James, N. R. Thomson, D. Pickard, J. Wain, C. Churcher, K. L. Mungall, S. D. Bentley, M. T. Holden, M. Sebaihia, S. Baker, D. Basham, K. Brooks, T. Chillingworth, P. Connerton, A. Cronin, P. Davis, R. M. Davies, L. Dowd, N. White, J. Farrar, T. Feltwell, N. Hamlin, A. Haque, T. T. Hien, S. Holroyd, K. Jagels, A. Krogh, T. S. Larsen, S. Leather, S. Moule, P. O'Gaora, C. Parry, M. Quail, K. Rutherford, M. Simmonds, J. Skelton, K. Stevens, S. Whitehead, and B. G. Barrell. 2001. Complete genome sequence of a multiple drug resistant *Salmonella enterica* serovar Typhi CT18. *Nature*. **413**:848–852.

Perna, N. T., G. Plunkett III, V. Burland, B. Mau, J. D. Glasner, D. J. Rose, G. F. Mayhew, P. S. Evans, J. Gregor, H. A. Kirkpatrick, G. Posfai, J. Hackett, S. Klink, A. Boutin, Y. Shao, L. Miller, E. J. Grotbeck, N. W. Davis, A. Lim, E. T. Dimalanta, K. D. Potamousis, J. Apodaca,

T. S. Anantharaman, J. Lin, G. Yen, D. C. Schwartz, R. A. Welch, and F. R. Blattner. 2001. Genome sequence of enterohaemorrhagic *Escherichia coli* O157:H7. *Nature.* **409:**529–533.

Rajakumar, K., C. Sasakawa, and B. Adler. 1997. Use of a novel approach, termed island probing, identifies the *Shigella flexneri she* pathogenicity island which encodes a homolog of the immunoglobulin A protease-like family of proteins. *Infect. Immun.* **65:**4606–4614.

Rutherford, K., J. Parkhill, J. Crook, T. Horsnell, P. Rice, M. A. Rajandream, and B. Barrell. 2000. Artemis: sequence visualization and annotation. *Bioinformatics* **16:**944–945.

FURTHER READING

Brogden, K., J. Roth, T. Stanton, C. Bolin, F. C. Minion, and M. J. Wannemuehler (ed.). 2000. *Virulence Mechanisms of Bacterial Pathogens,* 3rd ed. ASM Press, Washington, D.C.

Kaper, J. B., and J. Hacker (ed.). 1999. *Pathogenicity Islands and Other Mobile Virulence Elements.* ASM Press, Washington, D.C.

Salyers, A. A., and D. Dixie (ed). 2001. *Bacterial Pathogenesis: A Molecular Approach,* 2nd ed. ASM Press, Washington, D.C.

Winstanley, C., and C. A. Hart. 2001. Type III secretion systems and pathogenicity islands. *J. Med. Microbiol.* **50:**116–126.

STRATEGIES IN ANTAGONISTIC AND COOPERATIVE INTERACTIONS

Angela E. Douglas

17

Most microbiological research is conducted on microorganisms in axenic culture. Indeed, axenic cultivation is traditionally perceived as a prerequisite for a microbiological study. Although it is a very successful approach that has yielded many fundamental insights, it is a distortion of microbiological reality. Outside of the laboratory, microorganisms do not live in isolation but in multispecies assemblages involving many other microorganisms and often plants or animals. This can be illustrated

by the microbiology of forest soil. In an important experiment, Torsvik and colleagues (1990) showed that each gram of bulk soil that they examined contained approximately 10^9 metabolically active bacterial cells, comprising representatives of at least 13,000 taxa. If the bacteria were uniformly distributed in the soil, each bacterial cell would be no more than 12 μm from its neighbors, many of which are different taxa. In fact, most members of the soil microbiota are in much closer proximity than this simple calculation would suggest, because many microorganisms live in aggregations, often of multiple species. The conclusion, that microorganisms live in a complex web of biotic interactions, is not peculiar to forest soils; it applies to all habitats. The purpose of this chapter is to address the nature of the interactions.

Biotic interactions are classified as either antagonistic, with a negative impact on at least one of the participants, or cooperative (or mutualistic), with a positive impact on all the participants. Negative and positive impacts, equivalent to "harm" and "benefit," respectively, can be identified by comparing the number or proliferation rate (or some other index of fitness) of microorganisms in the presence and absence of the interaction. Most interactions can be identified unambiguously as either antagonistic or cooperative, but the outcome of certain interactions can be ecologically variable (e.g., dependent on environmental conditions) and evolutionarily labile (i.e., responsive to selection).

A. E. Douglas, Department of Biology (Area 2), University of York, P.O. Box 373, York, YO10 5YW, United Kingdom.

Microbial Evolution: Gene Establishment, Survival, and Exchange
Edited by Robert V. Miller and Martin J. Day, ©2004 ASM Press, Washington, D.C.

As an approach to explore the significance of biotic interactions on the biology and evolution of microorganisms, this chapter considers three topics in detail: the production of compounds with bacteriocidal properties; interactions between bacteria and bacteriophage; and metabolic cooperation, both among microorganisms and with "higher organisms" (i.e., animals and plants).

CHEMICAL WARFARE: CHEMICALS IN ANTAGONISTIC INTERACTIONS

Numerous structurally diverse compounds with biological activity are synthesized by microorganisms, and many of these compounds appear to function as mediators of antagonistic interactions. Antibiotics and bacteriocins are among the best characterized of these compounds at the chemical and genetic levels, and they are considered here.

Antibiotics

Antibiotics are compounds that, at low concentrations, inhibit the growth of different taxa of microorganisms. Antibiotics are chemically very diverse but, in terms of their mode of action, they can be classified into four broad groups: those that interfere with (i) bacterial cell wall synthesis (e.g., the penicillins and other β-lactams), (ii) membrane integrity (e.g., polymyxin), (iii) nucleic acid metabolism (e.g., rifampicin), and (iv) protein synthesis (e.g., the tetracyclines). Despite the immense clinical importance of antibiotics during the past 50 years, the significance of these compounds to the organisms that naturally produce them is poorly understood. It remains as a reasonable working hypothesis that antibiotic production mediates competition between the producers and other bacteria in natural environments and, linked to this, that it is particularly advantageous to slow-growing microorganisms. Antibiotic production may play a vital role in shaping certain microbial communities.

Most antibiotic producing microorganisms have been isolated from soil, raising the possibility that the soil habitat has favored evolution of this trait. One possible reason is that soils are highly structured environments; the pores between soil particles approximate to multiple separate (or poorly connected) microhabitats with different abiotic characteristics and arrays of resources. In such environments, an antibiotic-producing bacterium may, by chance, be at a locally high density relative to nonproducers and, consequently, outcompete the nonproducers through antibiotic production, even though the latter may have higher growth rates. The notion that antibiotic production is promoted in structured environments is supported by mathematical models that generate persistence of antibiotic producers in structured environments even when antibiotic production is costly, but not in homogenous (well-mixed) environments.

Antibiotic-producing bacteria are also exploited by higher organisms as a means of chemical defense. Two striking examples are considered here, in fungus-growing ants and in parasitic nematodes.

Ants of the tribe Attini cultivate a single species of fungus in their nests as a source of food. The ant-fungus association is potentially susceptible to a virulent fungal pathogen, an ascomycete of the genus Escovopsis. Invasion by Escovopsis sp. is prevented, at least in part, by an actinomycete Streptomyces sp., which grows profusely on specific regions of the ant body (Fig. 1) and produces diffusible compound(s) that suppress spore germination and hyphal growth of Escovopsis sp. The chemical identity of the antibiotic compound(s) produced by Streptomyces sp. remains to be established.

The microbial production of antibiotics is crucial to the life cycle of nematodes that parasitize insects. These nematodes, which are generally known as entomopathogenic nematodes, bear specific bacteria in the related genera of Xenorhabdus and Photorhabdus. When the nematode gains entry to an insect host, it releases the bacteria, which proliferate rapidly, causing septicemia and killing the insect. The insect carcass supports the growth and reproduction of the invading nematode, whose offspring acquire an inoculum of the bacteria, so completing the life cycle (Fig. 2). The usual

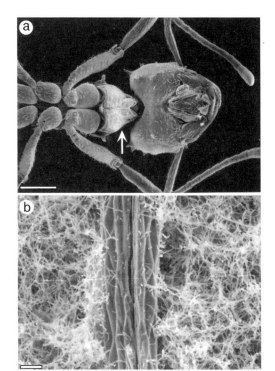

FIGURE 1 The association between the actinomycete *Streptomyces* sp. and fungus-growing ants. (a) Scanning electron micrograph of ventral view of the ant *Acromyrmex octospinosus*, showing the *Streptomyces* sp. infection on the insect exoskeleton (arrow) on a body segment posterior to the head (scale bar = 500 μm). (b) Scanning electron micrograph of *Streptomyces* sp., showing characteristic growth pattern on the insect cuticle (scale bar = 10 μm). Reproduced from Currie et al. (1999).

hosts of these nematodes are soil-inhabiting insects, and the insect carcass is, in principle, an excellent source of nutrients for microorganisms in the soil. In practice, however, the nematode-infested carcass does not putrefy and this is because the bacteria produce potent antibiotics that protect the carcass against generalized microbial attack. The bacteria are also luminescent and produce a red pigment (see Fig. 2).

Bacteriocins

Bacteriocins are proteinaceous toxins synthesized by bacteria (*Eubacteria* and *Archaea*). They kill only closely related bacteria, and the very narrow taxonomic range of bacteriocin activity indicates that these compounds are important as mediators of intraspecific interactions.

The best-studied bacteriocins are the colicins produced by *Escherichia coli*. At least 25 different colicins have been described. Colicins are produced when intraspecific competition for resources is high (e.g., low-nutrient conditions). They are released into the environment, generally by lysis of the releasing cell, mediated by a lysis protein, and transported into susceptible cells, where they degrade the DNA, inhibit protein synthesis, or destroy cell membrane integrity (varying with colicin type), and kill the cell. Colicin-producing cells are not killed by their own colicin because they constitutively

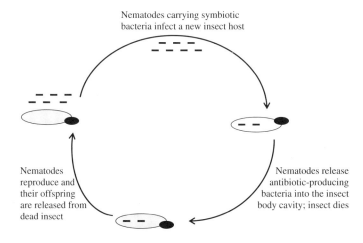

Nematodes carrying symbiotic bacteria infect a new insect host

Nematodes reproduce and their offspring are released from dead insect

Nematodes release antibiotic-producing bacteria into the insect body cavity; insect dies

FIGURE 2 The relationship between antibiotic-producing bacteria of the genera *Xenorhabdus* and entomopathogenic nematodes. (a) The life cycle of the nematode. (b) Hydroxy-stilbene antibiotics produced by *Xenorhabdus luminescens*. Figure reproduced from Nealson et al. (1988).

produce a protein ("immunity protein") that binds to, and inactivates, the colicin. The three proteins essential for colicin function (i.e. the colicin, the lysis protein, and the immunity protein) are coded by three tightly linked genes borne on a plasmid. Resistance to the colicin can also evolve by changes to the cell wall or cell membrane that prevent the uptake of the colicin.

Colicin production in natural *E. coli* populations has been studied extensively. The identity of colicins varies widely between populations and can also change over time in one population, as a result of invasion by cells with a different colicin profile. For example, in the study of Gordon et al. (1998) on *E. coli* in mice, a total of eight colicin types were identified and about half of the bacterial population could produce at least one colicin (Fig. 3a). The incidence of *E. coli* strains resistant to the colicins varied from about 25% to 95%, varying among the colicin types (Fig. 3b).

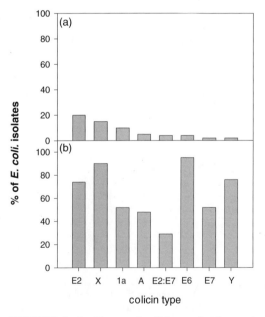

FIGURE 3 Incidence of colicin production and resistance in *E. coli* isolated from feral mice. The percentage of *E. coli* strains tested that produced (a) and were resistant (b) to each of eight colicins is shown. Redrawn from Riley and Gordon (1999).

The population dynamics of colicin-producing, colicin-susceptible, and colicin-resistant cells have been modeled, and coexistence of the different classes of bacteria has been found to be promoted in structured environments (e.g., Pagie and Hogeweg, 1999). These results offer remarkable parallels with the conclusions obtained for coexistence of antibiotic-producing and antibiotic-susceptible bacteria (see "Antagonistic interactions"), and illustrate that the spatial structure of an environment can have a major impact on the outcome of interactions between organisms.

There is considerable variation among the many colicins in the amount of colicin produced per cell and the incidence of cell lysis, but both experimental studies and modeling reveal that this variation has little impact on the patterns of colicin production in natural populations. Further studies suggest that the relative abundance of various colicin types may be influenced by factors unrelated to intraspecific competition. For example, colicin Ib has been reported to protect *E. coli* cells from bacteriophage and to mediate competition (although the mechanism remains uncertain); and this colicin may be favored in environments with high phage prevalence.

One final issue arises. As considered above, bacteriocins play a role in protecting resources from closely related competitors, and, for *E. coli* and many other bacteria, the resource is a living host. In principle, the host could benefit from the bacteriocin production by its resident bacteria where alternative bacterial genotypes are more virulent than the resident strain(s). Furthermore, such an interaction could evolve into bacterial-mediated competition between hosts with different bacterial complements. This is displayed by some protists, although bacteriocins are apparently not implicated in the interactions. The best-studied system is the association between ciliates of the *Paramecium aurelia* complex and various bacteria, including members of the genus *Caedibacter* (also known as kappa, lambda, etc. in the traditional literature), which mediate killing of susceptible paramecia lacking the bacteria. Most research has been done on *Caedibacter taenospiralis* (kappa) in *Paramecium tetraurelia*. *C. taenospiralis* displays two alternative forms: "bright," which are refractile by phase-contrast microscopy, and "nonbright." The refractile structure in brights appears as a coiled ribbon, known as the R body, but its structural organization is not fully established. Killing occurs when bright *C. taenospiralis* cells are released from their host and phagocytosed by susceptible paramecia, followed by the uncoiling of the R body. Paramecia bearing nonbrights have no killing activity but are resistant to the killer activity. In this system, the killer activity has been attributed to virus particles associated with the R body, but the detail of the virus-bacterium-paramecium interaction remains to be established. It has been shown experimentally that bacteria with killer activity enable paramecia to out-compete sensitive paramecia.

Numerous protists bear intracellular bacteria, but the interactions in most of these associations have not been studied. Further research on these systems may reveal systems of killer activity, analogous to the *Caedibacter/Paramecium* relationship, but based on bacteriocins.

Section Summary

- In general, antibiotics mediate interspecific competition and bacteriocins mediate intraspecific competition among microorganisms.
- Coexistence of antibiotic/bacteriocin producers and nonproducers is promoted in structured environments.
- Antibiotic production by microorganisms is exploited by higher organisms as a defense against microbial attack, and bacteria have also been implicated in intraspecific competition among protists.

GENOMIC SUBVERSION BY BACTERIOPHAGE

Bacteriophage are viruses of bacteria. They are not living organisms because they are absolutely dependent on the enzymes and subcellular structures of their hosts for propagation. Bacteriophage have been described as bacterial

plasmids that have acquired the capacity to survive (but not reproduce) outside their hosts (chapter 9).

Antagonistic Interactions

Most interactions between phage and their hosts are unambiguously antagonistic. One group of phage, called virulent phage, kill their bacterial host. Phage bind to the bacterial surface, inject their genetic material into the bacterial cell, and redirect the cellular machinery to the production of phage progeny, which are liberated, usually by lysis and death of the host cell. The number of phage particles produced by an infected bacterial cell varies widely with species and conditions; for example, several hundred infective phage and a comparable amount of incomplete (and noninfective) phage are produced within 25 minutes of infection of *E. coli* by phage T4.

The interaction between phage and bacterial hosts is predicted to result in antagonistic coevolution, i.e., the selection of resistant bacterial cells, which, in turn, results in selection for increased infectivity of phage, then increased bacterial resistance and so on. Antagonistic coevolution between bacteria and phage can be investigated readily because the studies can be conducted with bacteria and phage that are genetically uniform at the start of the experiment and maintained under uniform conditions. Any change in resistance or infectivity during the experiment can therefore be attributed to mutation and selection. In addition, interactions between the phage and bacteria from different time points in the experiment can be examined. For example, the evolutionary increase in bacterial resistance throughout each generation can be quantified by testing resistance to the ancestral phage (the genotype at the start of the experiment), and the evolutionary increase in phage infectivity can be determined by testing the infection of the ancestral bacterial genotype.

Using this approach, Buckling and Rainey (2001) investigated the antagonistic coevolution of the bacterium *Pseudomonas fluorescens* and the phage SBW25Φ2 over 50 serial transfers (roughly equivalent to 400 bacterial gener-

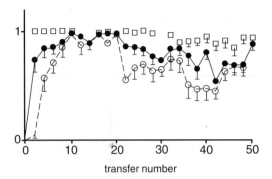

FIGURE 4 Antagonistic coevolution between the bacterium *P. fluorescens* and a bacteriophage. The proportion of resistant bacteria is shown on the *y* axis. The bacteria resistant to the ancestral phage (open squares), contemporary phage (closed circles), and phage from two transfers in the future (open circles) in 12 replicate experiments (mean ± standard error) are displayed. Figure reproduced from Buckling and Rainey (2001).

ations). As illustrated in Fig. 4, the proportion of bacteria resistant to the ancestral phage stock rose rapidly to unity by the first time of assay, indicative of very effective selection against susceptible cells, and the resistance to the ancestral stock remained high throughout the experiment. Resistance to "contemporary" phage (i.e., phage at the same generation in the experiment as the bacteria) also rose initially, but never exceeded resistance to the ancestral phage, and resistance to phage collected from the experiment two transfers in the future was always lower than resistance to the contemporary phage. Furthermore (not shown in the figure), the resistance of bacteria to phage at two transfers in the past was significantly higher than resistance to the contemporary phage. These results indicate that the infectivity of the phage was coevolving with bacterial resistance.

Such coevolutionary changes are unlikely to proceed indefinitely, at least under natural conditions, because the ever-increasing resistance/infectivity is anticipated to have negative pleiotropic effects, resulting in reduced bacterial growth rates or depressed progeny number per infecting phage, especially under nutrient-limited conditions. In addition, the coevolving bacteria and phage would be susceptible to invasion by strains with different resistance or infectivity characteristics.

"Less Antagonistic" Interactions

Many phage, known as temperate phage, have two alternative pathways: the lytic pathway, in which they produce multiple progeny and usually lyse the host cell (the pathway displayed by virulent phage, described above), and the lysogenic pathway, in which viral DNA is integrated into the bacterial chromosome or transformed to a low-copy-number plasmid and then copied passively with each round of bacterial chromosomal replication and cell division. The phage may switch from the lysogenic condition to the lytic pathway at any time. This twofold strategy enables a single infective phage to exploit many more hosts (i.e., the infected host and all its progeny) than a virulent phage with the obligate lytic pathway, albeit over a longer timescale. Bacterial cells bearing a lysogenic phage are protected from superinfection by related phage. For example, when the phage lambda is integrated into the bacterial chromosome (in this condition, it is called a prophage), just one phage gene *d* is expressed, coding for a protein that represses transcription from all genes in the lytic pathway, and this protein also represses incoming lambda phage. The prophage is expected to be under selection pressure to minimize the costs of its maintenance to the host, as this would increase the proliferation rate of the bacterial host and phage. Under these conditions, the selective interests of the phage and its bacterial host overlap, even though a lineage of bacterial cells bearing a lysogenic phage is living in the shadow of the "sword of Damocles" because the lytic pathway could be induced at any time.

Selection pressure to ameliorate (i.e., reduce) the costs imposed by phage on its host has been demonstrated for the f1 filamentous phage, whose interactions with host cells are similar, but not identical, to temperate phage. Bacteria infected with f1 are (like the prophage of lambda) resistant to superinfection and transmit the resident viral DNA through the generations but, unlike temperate phage, the bacterial cells are not killed when phage progeny are produced and released. f1-infected *E. coli* cells grow more slowly than uninfected cells. To investigate the selection pressures on the phage, Bull and colleagues (1991) used two strains of f1 that differed in their cost to the host, i.e. the extent to which they depressed host growth rates. When these strains were grown together in competition, under conditions that prevented the production of phage progeny but permitted the transmission of phage DNA to daughter cells, the strain that was less costly to the host cells increased in number relative to the more damaging strain. This advantage of the less costly strain was lost under culture conditions that permitted the production and release of phage progeny.

These experiments illustrate that it is not always in the interests of a phage to be "nasty" to its host. Specifically, when the interaction between the offspring of one participant and the offspring of its partner is assured, the selective interests of the partners tend to converge, and antagonistic interactions are selected against. We will return to this issue in "Genomic consequences of cooperation for microorganisms."

Section Summary
- Bacteriophage (viruses of bacteria) are small genomes that subvert the cellular machinery of their bacterial hosts to produce phage progeny.
- Phage and their bacterial hosts display antagonistic coevolution involving escalating bacterial resistance and phage infectivity.
- When phage are transmitted vertically exclusively, strains with small negative effects on the growth rates of their bacterial hosts are at a selective advantage, suggesting that antagonistic interactions may be reduced by partner fidelity.

METABOLIC COOPERATION

Cooperation among Microorganisms
Metabolic cooperation between microorganisms is known as syntrophy, which can be defined as two or more microorganisms contributing different elements of a common metabolic pathway, resulting in the net synthesis or degradation of specific compounds, to the advantage of all participants. An example of syntrophy is the degradation of the persistent herbicide atrazine by the syntrophic soil bacteria *Clavibacter* spp. and *Pseudomonas* spp. via products of overlapping sets of genes (Fig. 5).

Syntrophy is promoted by close proximity between the interacting microorganisms, which minimizes the distance over which compounds diffuse between cells (the time required for diffusion increases with the square of distance). Proximity is obtained by two alternative routes: (i) microbial consortia, in which the interacting cells are in direct cell-to-cell contact, often held together by extracellular material, such as slime or mucus; and (ii) intracellular symbiosis,

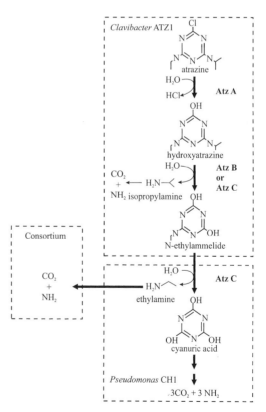

FIGURE 5 Catabolism of the *s*-triazine herbicide atrazine by a consortium of *Clavibacter* spp. and *Pseudomonas* spp. Reproduced from de Souza et al. (1998).

in which one microorganism is located within the cell of the other microorganism. Consortia usually involve either multiple bacterial taxa, or protists and bacteria, and the usual partners of intracellular symbioses are bacteria within protists. Intracellular symbioses between bacteria are exceptionally rare, but have been described in the cells of certain insects, where multiple cells of a gammaproteobacterium are borne within the single cell of a betaproteobacterium.

The promotion of metabolic cooperation by close proximity is illustrated by the study of Christensen and colleagues (2002) on the bacteria *Pseudomonas putida* and *Acinetobacter* spp. Both bacteria can utilize benzoyl alcohol as sole carbon and energy source, but *Acinetobacter* spp. can utilize this substrate more efficiently than *P. putida*, and *P. putida* can metabolize benzoate, an intermediate in the degradation of benzoyl alcohol, more efficiently than *Acinetobacter* spp. When the bacteria were cocultured on a surface, the two bacteria formed stable mixed assemblages, with *P. putida* cells accumulating on the surface of microcolonies of *Acinetobacter* spp. to give a mixed population in the ratio of 1:5 cells of *P. putida* to *Acinetobacter* spp. However, when the bacteria were cocultured in suspension in a chemostat (where cell-to-cell contact is largely precluded), the population of *Acinetobacter* spp. increased much more rapidly than *P. putida*, generating a ratio of 1:500 cells of *P. putida* to *Acinetobacter* spp. and a lower total bacterial biomass and consumption of benzoyl alcohol than in the surface cultures. The interaction between the bacteria on the surface-attached bacteria was cooperative: benzoyl alcohol was consumed predominantly by *Acinetobacter* spp. and the intermediate degradation product, benzoate, diffused rapidly from *Acinetobacter* spp. to the nearby *P. putida* cells, which completed the degradation. In suspension, however, the interaction was primarily competitive (i.e., antagonistic) because the large distance between the cells of the two bacteria prevented the efficient cross-feeding of benzoate.

The selective advantage of cooperation through syntrophy is particularly great in hypoxic and anoxic environments (i.e., environments with low or no free oxygen) because anaerobic metabolism yields little energy and requires high metabolic efficiency. Syntrophies involving the interspecies transfer of hydrogen and sulfur are ecologically important, and are now considered in turn.

For hydrogen, the metabolic "problem" is that the fermentation of short-chain fatty acids, primary alcohols, and certain aromatic hydrocarbons is endergonic (i.e., energy-consuming reaction) under standard partial pressure of hydrogen (pH_2) but exergonic at very low pH_2 (Fig. 6a). The "solution" is syntrophy with hydrogen-scavenging bacteria, such as homoacetogens, methanogens, and sulfate-reducing bacteria, which mediate the net consumption of hydrogen (Fig. 6b) by exergonic reactions. The metabolic advantage of interspecies hydrogen transfer is illustrated by studies on the syntrophic conversion of ethanol to methane by the fermenter *Desulfovibrio* spp. and methanogen *Methanobacterium* sp. In mixed cultures in a chemostat, cells allowed to form consortia produced methane at higher rates and hydrogen at lower and more stable rates than cells maintained separately in suspension, and attained higher total biomass. In these consortia, the syntrophic partners are predicted to grow at equal rates and possibly to share energy of the overall bioenergetic metabolism equally. The implication is that the pH_2 equilibrates at the level that yields equal amounts of energy to each of the partners.

For sulfur, obligately anaerobic sulfate-reducing bacteria face the metabolic "problem" of access to sulfate, especially in freshwater environments where the concentration of inorganic sulfur compounds, especially sulfate, is low. The "solution" is syntrophy with sulfide-oxidizing bacteria, either bacteria performing anoxygenic photosynthesis (e.g., green sulfur bacteria or purple sulfur bacteria) or aerobic colorless sulfur bacteria. Sulfate-reducers commonly cooccur with phototrophic sulfur bacteria in the chemocline of lakes, and with colorless sulfur bacteria in microbial mats, which typically show steep gradients of sulfide and oxygen. The conditions permitting the

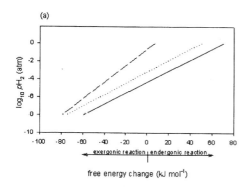

(a)

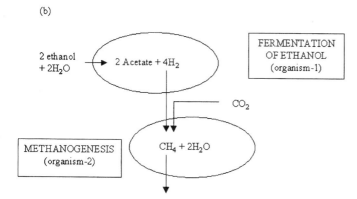

(b)

FIGURE 6 The metabolic basis of interspecies hydrogen transfer. (a) Free energy change associated with the oxidation of the primary alcohols ethanol (unbroken line), butyrate (dotted line), and propionate (dashed line) as a function of the partial pressure of hydrogen. (b) The dissimilation of ethanol by a consortium of a fermenter and methanogen. Redrawn from Fenchel and Finlay (1995) with permission from Oxford University Press.

coexistence of anaerobic sulfate-reducers and aerobic sulfur bacteria have been explored using chemostat cultures. For example, cocultures of *Desulfovibrio* spp. and *Thiobacillus* spp. proliferate stably under oxygen tensions sufficient to complete *Thiobacillus*-mediated oxidation of sulfide to sulfate, but not at higher oxygen levels, which are deleterious to *Desulfovibrio* spp.

Morphologically distinctive consortia and intracellular symbioses are observed routinely in certain habitats, including anoxic freshwaters, animal guts, sewage sludge, and even deep-sea environments. The participating organisms of some consortia are unknown, and many are known by binary names in quotation marks (e.g., "*Chlorochromatium aggregatum*") because associations have no standing in the formal Linnaean nomenclature for single organisms. Application of molecular techniques has, however, revealed unique forms in some consortia. For example, the tiny bacterium (0.4-μm diameter) associated with an *Ignicoccus* sp. iso-

lated from a submarine hot vent is so divergent from all other known bacteria that it has been assigned to a novel phylum, the Nano-archaeota. The nature of the metabolic interaction (if any) in most of these consortia is obscure, and plausible, but erroneous, metabolic scenarios have been published. For example, it has been suggested that the phototrophic consortium "*Chlorochromatium aggregatum*" between a single "inner" motile, colorless bacterium and a layer of "outer" rod-shaped photosynthetic green sulfur bacteria cycled sulfur, i.e., the inner bacterium is a sulfate reducer, whose product, sulfide, is utilized as reductant by the outer phototrophs. Recent metabolic and phylogenetic data indicate that the inner bacterium is not a sulfate-reducer.

Cooperation between Microorganisms and "Higher" Organisms

The various associations considered in "Cooperation among microorganisms" illustrate that many instances of metabolic cooperation are underlain by differences in the metabolic repertoire among the participating organisms. This inequality is particularly marked for associations involving eukaryotes because, as a group, the eukaryotes are metabolically impoverished relative to the bacteria. The lineage that gave rise to the eukaryotes lacked the capacity to respire aerobically, photosynthesize and fix nitrogen, and, in addition, the animals cannot synthesize 9 of the 20 amino acids which make up protein (the "essential" amino acids) and have specific vitamin requirements. On multiple evolutionary occasions, eukaryotes have acquired these metabolic capabilities by entering into intimate associations with microorganisms. As examples, photosynthetic algae have been acquired by both fungi (the lichens) and various animals (e.g., corals); many leguminous plants have acquired nitrogen fixation by forming root associations with nitrogen-fixing rhizobia; and specific bacteria provide certain insects with essential amino acids (e.g., the bacteria *Buchnera* in aphids) and vitamins (e.g., the bacteria *Wigglesworthia* in tsetse flies). Two associations early in the evolutionary history of eukaryotes had a truly revolutionary impact on the eukaryotes; the first involved bacteria with aerobic respiration, which evolved into mitochondria, and the second with cyanobacteria, which evolved into plastids. All the various microorganisms and the derived organelles cooperate, in that they share the products of their metabolic capabilities with the eukaryotic partner.

The metabolic basis of cooperation has been studied in rhizobia associated with legumes. Rhizobia are members of the alphaproteobacteria and comprise multiple genera, including *Bradyrhizobium* and *Rhizobium*. In the symbiosis, they are usually located in plant cells, separated from the plant cytoplasm by a membrane of plant origin. In this condition, they fix nitrogen at high rates and release much of the nitrogen to the plant cell in the form of ammonia. This is mediated by two special features of rhizobia:

• Regulation of the activity of ammonia-assimilating enzymes. The principal ammonia-assimilating enzyme in both the rhizobia and plant cells is glutamine synthetase, which incorporates ammonia into glutamate, to form glutamine (Fig. 7). The activity of the enzyme in the plant cell cytoplasm surrounding the rhizobial cells is very high, thereby maintaining a very low ammonia concentration, but the activity of the enzyme in the rhizobial cells is undetectable, such that the rhizobia have essentially no capacity to assimilate the ammonia that they fix. The consequence is a very steep concentration gradient of ammonia from the rhizobial cell contents to the plant cytoplasm (e.g., from 12 mM to <10 μM, respectively, in soybean nodules) and this promotes the passive diffusion of ammonia from rhizobia to the plant cytoplasm.

• Regulation of expression of nitrogen-fixing (*nif*) genes. In nonsymbiotic nitrogen-fixing bacteria (e.g., *Klebsiella*), the expression of the nitrogen-fixing (*nif*) genes is repressed by fixed nitrogen, including ammonia. If the *nif* genes of rhizobia were regulated in this way, the depressed glutamine synthetase activity and high rhizobial ammonia content (see

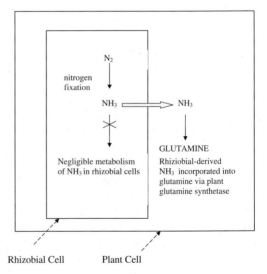

FIGURE 7 The metabolic basis for the transport of ammonia (grey block arrow) from symbiotic rhizobia to their plant host.

above) would lower nitrogen-fixation rates. This does not occur because of a fundamental difference in the pattern of *nif* regulation between rhizobia and non-symbiotic bacteria. In rhizobia, the expression of *nif* genes is promoted by low oxygen tensions, conditions that occur in a root nodule, and it is completely independent of the nitrogen supply.

These data suggest that differences in the regulation of expression of key genes between rhizobia and nonsymbiotic nitrogen-fixing bacteria can account for the release of ammonia from rhizobia to the plant. At present, there is insufficient evidence to assess whether the release of nutrients such as essential amino acids and vitamins from other symbiotic microorganisms has a comparable basis.

A question that arises directly from the special metabolic traits of cooperative microorganisms in symbioses is their evolutionary origin. Have they evolved from pathogens by progressive amelioration of virulent traits? A link between some rhizobia and pathogens is suggested by the close phylogenetic relationship between certain rhizobial genera (e.g., *Rhizobium* and *Sinorhizobium*) and overt pathogens, such as members of the genus *Agrobacterium*, which cause plant tumors. However, *Bradyrhizobium* is only distantly related to *Agrobacterium* (and *Rhizobium*) and is not allied to known pathogens, suggesting that rhizobia do not necessarily evolve from lineages with a predisposition for pathogenic lifestyles. This conclusion is confirmed and extended by more general analyses of the phylogenetic relationships among pathogens and cooperative symbionts. Commonly, bacteria with the two lifestyles are phylogenetically distinct, with the implication that the cooperative symbionts and many pathogens may generally be highly specialized with little or no evolutionary "opportunity" for transition to (or from) the other lifestyle.

Section Summary
- Metabolic cooperation (also known as syntrophy) is promoted by close proximity between the participating organisms.

- The degradation of recalcitrant organic compounds and the efficiency of energy production in oxygen-free environments are promoted by metabolic cooperation among microorganisms with different metabolic capabilities.
- Higher organisms gain access to complex metabolic capabilities (e.g., nitrogen fixation and photosynthesis) by forming symbioses with cooperative microorganisms.

GENOMIC CONSEQUENCES OF COOPERATION FOR MICROORGANISMS

Genomic Decay in Vertically Transmitted Microorganisms

Organisms that are persistently vertically transmitted are predicted to display cooperative traits because their fitness depends on the fitness of their partner. Recent molecular analyses, however, have revealed that persistent vertical transmission can also have dramatic genomic consequences for microorganisms.

Much of the research has been conducted on bacteria of the genus *Buchnera*, a member of the gammaproteobacteria allied to *E. coli*, which occurs exclusively in aphids. The *Buchnera* cells reside in specialized cells, called mycetocytes, in the aphid body cavity; a small number of bacterial cells are transferred from this location to each egg in the ovaries of the female insects, and they become incorporated into the mycetocytes of the offspring aphid as these cells differentiate. As a consequence, *Buchnera* cells never encounter any environment apart from the aphid.

Genes with functions required only in the free-living condition or duplicated by host functions are predicted to be lost by relaxed selection, resulting in a reduced genome size and evolution of dependence on the host. Consistent with this prediction, the genome size of *Buchnera* is 0.45 to 0.64 Mb, less than 20% of the genome size of its relative *E. coli*.

The impact of relaxed selection on the genome of *Buchnera* and other vertically transmitted microorganisms is compounded by a second evolutionary process, called Müller's ratchet (Moran, 1996). This is best explained by considering the small inoculum of bacterial cells acquired by each host offspring. The cells are closely related to each other by descent and, therefore, very likely to be genetically uniform. If they bear the same mildly deleterious mutation, then that fixed mutation cannot be eliminated from the *Buchnera* population in the insect by selection because there is no alternative source of *Buchnera* cells lacking the mutation. As a result, the deleterious mutations accumulate inexorably, until such time that the host lineage bearing the *Buchnera* is selected against. In summary, Müller's ratchet refers to the progressive accumulation of mildly deleterious mutations in small asexual populations.

Buchnera and other obligately vertically transmitted microorganisms display two traits characteristic of genomes subject to Müller's ratchet. First, the rates of sequence evolution are elevated, e.g., the substitution rate for the 16S rRNA gene is 0.007 to 0.018 substitutions per site every 100 million years for *E. coli*, but 0.019 to 0.054 substitutions per site every 100 million years for *Buchnera*. Second, protein-coding genes have elevated rates of nonsynonymous substitutions, i.e., substitutions that change the amino acid in the protein coded by the gene, such that the ratio of synonymous to nonsynonymous (K_s/K_a) is invariably lower for *Buchnera* than for related free-living bacteria (Fig. 8).

Genomic decay of vertically transmitted microorganisms is countered by selection, especially for cooperative function. This can also be illustrated by the *Buchnera*-aphid association, in which *Buchnera* spp. provide the aphid with essential amino acids, nutrients in short supply in the aphid diet of plant phloem sap. Any deleterious mutation in *Buchnera* genes involved in essential amino acid synthesis would reduce the fitness of the aphid and, consequently, the *Buchnera* (because the number of aphid offspring available for colonization by *Buchnera* spp. is lowered). The gene inventory for *Buchnera* provides a vivid demonstration of the importance of selection in shaping the

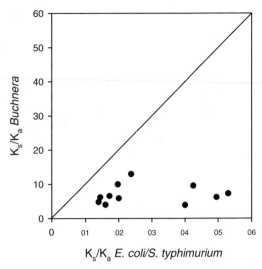

FIGURE 8 Ratio of synonymous to nonsynonymous substitutions per site (K_s/K_a) for protein-coding genes between *Buchnera* in two aphid species (*Schizaphis graminum* and *Diuraphis noxia*) and between the enteric bacteria (*E. coli* and *Salmonella enterica* serovar Typhimurium) to which *Buchnera* is closely allied. Redrawn from Clark et al. (1999).

genome of vertically transmitted symbionts. Virtually all the genes in *Buchnera* have unambiguous sequence similarity to *E. coli* genes, and, in most functional categories, *Buchnera* has 20% to 50% of the genes present in *E. coli*. Exceptionally, 53 (95%) of the 56 *E. coli* genes mediating essential amino acid synthesis have homologues in *Buchnera*, indicative of the strong selection pressure to maintain essential amino acid synthesis, despite the evolutionary tendency toward genomic deterioration.

The Evolutionary Transition from Microorganisms to Organelles

There is close to universal acceptance that mitochondria evolved from alphaproteobacteria and plastids from cyanobacteria in symbiosis with early eukaryotic cells. Mitochondria and plastids are, thus, bacterial-derived organelles, which cooperate with their eukaryotic hosts by providing aerobic respiration and photosynthetic carbon fixation, respectively.

Mitochondria and plastids have very small genomes, as a result of both genomic decay (as described in "Genomic decay in vertically transmitted microorganisms") and also the transfer of genes to the nucleus. Genes transferred to the nucleus are "protected" from genomic deterioration through Müller's ratchet such that the selection pressure on the organelle/bacterium to retain function is relaxed. In other words, gene transfer both compensates for and promotes genomic decay of the organelle and results in the progressive genomic "assimilation" of the microorganism into the eukaryotic host. The transfer to the nucleus of genes whose products are targeted back to the organelle has been adopted as a defining feature of the evolutionary transition from a cooperative microorganism to an organelle.

It can be argued that microorganisms are much more likely to evolve into organelles in unicellular than in multicellular eukaryotic hosts. In unicellular hosts, the cell bearing the microorganism is also the cell giving rise to offspring host, and, therefore, any genes transferred from the microorganism/organelle to the cell nucleus are transmitted, together with the microorganism/organelle, to the next host generation. In multicellular hosts, the cells bearing the microorganisms are usually specialized somatic cells (e.g., the aphid mycetocytes bearing *Buchnera*). These host cells, which are the most likely cell to receive DNA from the microorganism, die with the host individual in which the DNA transfer event occurred. Because the microbial DNA does not have ready access to the host gametes, microorganisms are unlikely to evolve into organelles in multicellular hosts. In other words, the genomic integrity and function of cooperative microorganisms is more likely to be retained in multicellular hosts with defined division between the germ line and soma than in unicellular eukaryotic hosts.

Section Summary

- Cooperation is promoted by partner fidelity through persistent vertical transmission.
- Relaxed selection and Müller's ratchet are significant evolutionary processes for

obligately vertically transmitted microorganisms, resulting in genomic decay and gene deletion.
- The defining feature of the evolutionary transition from a microorganism to an organelle is the transfer to the host nucleus of genes whose products are targeted back to the organelle. This process accelerates the genomic decay of an organelle.

SUMMARY OF THE ANTAGONISTIC AND COOPERATIVE INTERACTIONS INVOLVING BACTERIA

It is the natural condition for microorganisms to interact with other microbial cells, including genetically different cells. These interactions may be antagonistic or cooperative, and the outcome of certain interactions is ecologically variable, depending on environmental conditions, and evolutionarily labile.

A diverse array of antibiotics are produced by microorganisms, and many bacteria also produce proteinaceous toxins called bacteriocins. Antibiotics are widely accepted to mediate interspecific competition and bacteriocins to mediate competition within species and between closely related species, and the coexistence of antibiotic/bacteriocin-producing and sensitive cells is promoted in spatially structured environments, where the conditions are heterogeneous and microorganisms are not uniformly distributed. However, the significance of antibiotic and bacteriocin production as a factor shaping the composition of natural microbial communities is still uncertain.

The relationship between bacteria and virulent bacteriophage is unambiguously antagonistic and, in the absence of disturbance, subject to persistent antagonistic coevolution, i.e., reciprocal selection for increased bacterial resistance and phage infectivity. However, it is to the advantage of lysogenic phage (which are passively replicated with the bacterial host genes) to minimize the costs to the host, an illustration of the generality that vertical transmission tends to reduce antagonism and promote cooperation.

Metabolic cooperation, also known as syntrophy, among bacteria is promoted in spatially structured environments and by close proximity between the participating cells (a condition known as a consortium). Syntrophy involving interspecies transfer of hydrogen and sulfur is believed to be very widespread in low-oxygen environments, where metabolic efficiency is at a selective advantage, linked to the low-energy yield of anaerobic metabolism. However, the taxonomic identity of the organisms and nature of the metabolic interactions have not been elucidated for many consortia.

Many microorganisms cooperate with higher organisms (hosts), by the net microbe-to-host transfer of the products of metabolic capabilities possessed by the microorganisms but not their hosts (e.g., nitrogen fixation and photosynthesis). Further research is required to establish the genetic and metabolic bases of nutrient transfer in these associations.

Cooperation is promoted by partner fidelity, as achieved, for example, by vertical transmission. Persistent vertical transmission can, however, lead to genomic deterioration of the cooperating microorganisms through relaxed selection and Müller's ratchet, resulting, in certain conditions, in the evolutionary transformation of a microorganism into an organelle. However, the relative importance of selection and deleterious evolutionary processes in shaping the genomic organization of vertically transmitted microorganisms has yet to be quantified.

QUESTIONS

1. Why can antibiotic-producing bacteria be isolated from soils particularly readily? Can you think of other habitats with features similar to soils and likely to bear antibiotic-producing bacteria?

2. (a) What do you understand by the term "antagonistic coevolution"? (b) How prevalent is antagonistic coevolution in natural environments? What factors may limit the incidence of antagonistic coevolution?

3. Vertical transmission is widely accepted to reduce antagonism and promote mutualistic interactions.

(a) Provide a verbal description of the evolutionary argument on which this view is based, and some empirical data that support this view.

(b) Can you think of any factors that may promote mutualistic traits in associations that are horizontally transmitted and, contrariwise, promote antagonistic traits in vertically transmitted organisms? (The biology of the reproductive parasites, such as *Wolbachia*, can help you answer this question.)

4. (a) What does the term "microbial consortium" mean? (b) A culturable consortium between two bacteria has been shown to degrade a toxic organic compound to harmless products. Describe the approach you would adopt to investigate the contribution of the two bacteria to the degradative metabolism.

5. Unculturable bacteria have recently been described in blood-sucking leeches. What experiments would you conduct to investigate whether the bacteria are (a) vertically transmitted and (b) subject to genomic deterioration?

REFERENCES

Buckling, A., and P. B. Rainey. 2001. Antagonistic coevolution between a bacterium and a bacteriophage. *Proc. R. Soc. London, Ser. B* **269:** 931–936.

Christensen, B. B., J. A. J. Haagensen, A. Heydorn, and S. Molin. 2002. Metabolic commensalisms and competition in a two-species microbial consortium. *Appl. Environ. Microbiol.* **68:**2495–2502.

Clark, M. A., N. A. Moran, and P. Baumann. 1999. *Mol. Biol. Evol.* **16:**1586–1598.

Currie, C. R., J. A. Scott, R. C. Summerbell, and D. Malloch. 1999. Fungus-growing ants use antibiotic-producing bacteria to control garden parasites. *Nature* **398:**701–704.

de Souza, M. L., D. Newcombe, S. Alvey, D. E. Crowley, A. Hay, M. J. Sadowsky, and L. P. Wackett. 1998. Molecular basis of a bacterial consortium: interspecies catabolism of atrazine. *Appl. Environ. Microbiol.* **64:**178–184.

Dohlen, C. D. von, S. Kohler, S. T. Alsop, and W. R. McManus. 2001. Mealybug β-proteobacterial endosymbionts contain γ-proteobacterial symbionts. *Nature* **212:**433–436.

Gordon, D., M. A. Riley, and T. Pinou. 1998. Temporal changes in the frequency of coligenicity in *Escherichia coli* from house mice. *Microbiology* **145:**2233–2240.

Huber, H., M. J. Hohn, R. Rachel, T. Fuchs, V. C. Wimmer, and K. O. Stetter. 2002. A new phylum of Archaea represented by a nanosized hyperthermophilic symbiont. *Nature* **417:**63–67.

Kusch, J., L. Czubatinski, S. Wegmann, M. Hubner, M. Alter, and P. Albrecht. 2002. Competitive advantages of *Caedibacter*-infected paramecia. *Protist* **153:**47–58.

Moran, N. A. 1996. Accelerated evolution and Muller's ratchet in endosymbiotic bacteria. *Proc. Natl. Acad. Sci. USA* **93:**2873–2878.

Nealson, T., M. Schmidt, and B. Bleakley. 1988. Luminescent bacteria: symbionts of nematodes and pathogens of insects. *NATO ASI Series* **H17:**101–113.

Overmann, J., and H. van Gemerden. 2000. Microbial interactions involving sulfur bacteria: implications for the ecology and evolution of bacterial communities. *FEMS Microbiol. Rev.* **24:**591–599.

Pagie, L., and P. Hogeweg. 1999. Colicin diversity: a result of eco-evolutionary dynamics. *J. Theor. Biol.* **196:**251–261.

Thiele, J. H., M. Chartrain, and J. G. Zeikus. 1988. Control of interspecies electron flow during anaerobic digestion: role of floc formation in syntrophic methanogenesis. *Appl. Environ. Microbiol.* **54:**10–19.

Wiener, P. 2000. Antibiotic production in a spatially structured environment. *Ecol. Lett.* **3:**122–130.

FURTHER READING

Buchner, P. 1965. *Endosymbiosis of Animals with Plant Micro-organisms.* John Wiley & Sons, London, United Kingdom.

Bull, J. J., I. J. Molineux, and W. R. Rice. 1991. Selection of benevolence in a host-parasite system. *Evolution (Lawrence, Kans.)* **45:**875–882.

Douglas, A. E. 1994. *Symbiotic Interactions.* Oxford University Press, New York, N.Y.

Fenchel, T., and B. J. Finlay. 1995. *Ecology and Evolution in Anoxic Worlds.* Oxford University Press, New York, N.Y.

Ochman, H., and N. A. Moran. 2001. Genes lost and genes found: evolution of bacterial pathogenesis and symbiosis. *Science* **292:**1096–1098.

Riley, M. A. 1998. Molecular mechanisms of bacteriocin evolution. *Ann. Rev. Genet.* **32:**255–278.

Riley, A. D., and D. Gordon. 1999. The ecological role of bacteriocins. *Trends Microbiol.* **7:**129–133.

Russell, A. D., and I. Chopra. 1996. *Understanding Antibacterial Action and Resistance,* 2nd ed. Ellis Horwood Publishers, Chichester, United Kingdom.

Shigenobu, S., H. Watanabe, M. Hattori, Y. Sakaki, and H. Ishikawa. 2000. Genome sequence of the endocellular bacterial symbiont of aphids *Buchnera* sp. APS. *Nature* **407:**81–86.

Torsvik, V., J. Goksor, and F. L. Daae. 1990. High diversity in DNA of soil bacteria. *Appl. Environ. Microbiol.* **56:**782–787.

WHY ARE GENES LOST?
WHY DO GENES PERSIST?

Martin J. Day and Robert V. Miller

18

WHY WOULD CELLS HAVE A STRATEGY FOR GENE ESTABLISHMENT AND PERSISTENCE?

WHY DO GENES AGGREGATE AND PERSIST?

WHY DO GENE SEQUENCES INTERACT?

It is worth referring back to the summary chapters 6 and 11 when reflecting on the issues raised in those chapters.

WHY WOULD CELLS HAVE A STRATEGY FOR GENE ESTABLISHMENT AND PERSISTENCE?

Previously, we showed prokaryotic genomes are compact and size-restricted compared with eukaryotes. Thus, gene inactivation and deletion are just as important to genome evolution as are duplication and other mutational events. Overlay all the genomic modification events with the capacity to exchange genes (see chapter 11) and the opportunities for "rapid" evolution of strains, e.g., for pathogenicity, become clear. In reality, this is only rapid evolution of the strain, it is not rapid evolution of the pathogenicity gene islands. These will have taken just as long to evolve as other important islands of associated gene functions, such as catabolic pathways and antibiotic resistance clusters.

Martin J. Day, Cardiff School of Biosciences, Cardiff University, Cardiff, Wales CF10 3TL, and *Robert V. Miller*, Department of Microbiology, Oklahoma State University, Stillwater, OK 74078.

This leads us to an important concept, that of core and accessory DNA. It differentiates the effectively nonpathogenic *Escherichia coli* (strain K-12) genome with core sequences, from those *E. coli* (e.g. strain O157) that have acquired pathogenic (accessory) sequences. Kaper pointed out in 1998 that these *E. coli* O157 virulence characteristics enable it to occupy a niche not accessible to the normal strain. The environment and phenotype (delivered by accessory gene sequences) act selectively in concert to enable *E. coli* to diversify both ecologically and genetically. Having strategies for acquiring genetic information of all sorts of sizes is helpful to cells in rapidly adapting to new environmental niches.

WHY DO GENES AGGREGATE AND PERSIST?

If sequences provide selective advantage to the cell then they become established and persist. When a cell has ceased to grow in what we consider a normal way, in nature or in laboratory experiments, it then is posed a serious physiological problem. It has to have a survival strategy and, if it fails, it will certainly die. So, as bacteria pass back and forth through periods of plenty and periods of famine, they have evolved characteristics that enable them to survive through change with some degree of success. These processes do not always ensure total genomic safety, however. So a cell entering stationary phase develops a state of physiological stress, relative to the state prior to this. The genome of most bacteria is limited in size and, unlike higher organisms, it consists almost entirely of useful information. Logically, it will contain information to "get through" some of the physiologically stressful demands encountered, but some environments will require more "survival" information than can be comfortably stored in the genome. Thus, alternative strategies to survive have been evolved. For example, some of the gram-positive bacilli have evolved a complex set of pathways enabling them to sporulate and persist through adverse conditions. These cells have opted for a state of suspended animation. To do this requires not a small set of genes, but a large number. Why is it not a feature of all bacteria? Why is it not a feature of all gram-positive bacteria? What do others do?

They enact alternative survival strategies. Sporulation is just one escape strategy; other bacteria initiate a cycle of mutagenesis, producing an immediate spectrum of mutants. Any mutations allowing a suitable metabolic capability will raise the cell's survivability. Finally, we know that cells "talk" to one another. This is termed quorum sensing and enables a population of cells to function in unison or individual cells to modify their activity in response to the size or performance of the colony (Davies et al., 1998). Bacteria are able to titrate these population functions, enabling a single cell to sense the density of bacteria around it. In natural environments many different bacteria live together by using various classes of signaling molecules. These are effectively different languages and confine "conversations" to those that can understand them. For example, quorum sensing enables individuals to recognize that appropriate numbers are around in the right physiological state for transformation to occur. This could be envisaged as a means to share operons, encoding a diverse set of metabolic properties. If competence for transformation were integrated with other events as cells passed through the hypermutable state, it could provide environmentally relevant gene sequences to other transformable species present in the neighborhood. Uptake and integration of exogenous DNA can also overcome the effects of lethal mutations produced in the hypermutable state too. The integration of quorum sensing, mutation, and gene exchange provide for a very effective evolutionary strategy.

Most survival studies are done in laboratory model systems as in situ work is still technically difficult. When we attempt to extrapolate these studies to environments like the deep subsurface, we are really looking into a dark abyss! Why? In some instances bacteria have been there for extremely long periods, thousands if not hundreds of thousands of years. Their persistence can almost be considered geological! However, despite being isolated for such prolonged periods, bacteria should have evolved, as did their laboratory cousins! They have the same basic genetic organization and biochemistry and may only really differ from their laboratory cousins in that they will have been growing very, very slowly and have been physically isolated, within strata, from other bacteria. Eventually, we will determine whether the bacteria living in such isolation have different physiological characteristics. What is the effect on evolution of bacteria growing at best with generation times of months to years compared with those with generation times in hours? Can we yet conceive what metabolic and genetic problems such organisms have had to solve?

It is easier to understand life at another extreme, in aquatic culture. Here bacteria have two major choices to obtain nutrients. They can remain in the water column and be free living, or they can attach to a surface through the production of exopolysaccharides and let the nutrients come to them! In either case, there is a predictable outcome. Successfully growing cells will scavenge nutrients and grow. In the water column small flocs, aggregates of cells, will form. On surfaces a biofilm will develop, and soon the film will consist of more than one layer and a diverse community of cells that will differentiate metabolically. A biofilm is a dynamic structure. Biofilms causing medical problems are less taxonomically diverse, such as in the infection of medical implants, typically by staphylococci, and of lungs in cystic fibrosis by *Pseudomonas aeruginosa*. Here the exopolysaccharides in the biofilm act as a barrier to antibiotic treatment and the immune system. The immune system may induce an autoimmune response that can extensively damage healthy tissues. In nature, biofilms consist of many species of bacteria. These form wherever there is a solid-water interface, such as on riverbeds, water pipes, ship hulls, and teeth or in the gut. The types of species present will describe the ecological state of the biofilm; for example, anaerobes to the bottom and aerobes to the top. The same will happen in the flock. This "zoological garden" represents a community of species, capable of a range of interactions, in which the associations change with time as the biofilm matures. In 2001, Hallet showed that an individual species may be present both within the biofilm and planktonically and yet display very different phenotypes due to phase variation. Others compete successfully on one phase and not in the other. It is clear from this simple analysis that the physiological demands on a cell in any site within an environment determine its life success or failure and this is, in part, determined by its ability to interact with other species! It is true that in the environment no individual cell is ever alone; it is always competing for space and nutrients! So what benefits are there to the bacteria of living in a biofilm? There are several. It offers mutual protection and enables survival in a wide range of environments hostile to individual planktonic cells. There is exclusion of harmful compounds and behavioral shifts that enhance resistance. A biofilm is an organized structure, and quorum sensing is involved. Quorum sensing allows cells that are separated by as much as 500 μm to communicate. Many species modify their behavior as population densities increase; for example, dental plaque formation in *Streptococcus gordonii*, pathogenicity in *Aeromonas hydrophila*, bioluminescence in *Vibrio fischeri*, motility and swarming in *Serratia liquefaciens*. Quorum sensing is energetically costly and so it is not appropriate for low cell densities; thus, bacteria need to monitor population densities effectively.

However, whether a cell is in a biofilm or deep underground it is going to be interacting biochemically with others, and so we return to quorum sensing. If cells interact biochemically (syntrophy), then they can exchange genes also. In a structured community, like a biofilm, the potential is likely to be higher as the adjacent cells are close and in similar physiological states. To interact successfully signifies a coevolved characteristic enabling successful DNA exchange in the past and thus the potential for exchange between "like-minded" organisms has a probability of being effective. As in many microbiological investigations, the advent of easy sequencing, the applications of bioinformatic analysis, and array technology are going to make a huge contribution to our understanding of microbial interactions.

WHY DO GENE SEQUENCES INTERACT?

Consider the mobile genetic elements (transposons, plasmids, phage, etc.) as entities independent of their host-cell genomes (Mira et al., 2002). To "anthropomorphize" for a moment, imagine you are such a "selfish" sequence and are entirely dependent on the host genome you currently occupy. To ensure your persistence, since cells die, you should adopt strategies to spread your sequence as far and wide as is possible. What could they be? You could become highly transmissible, "hitch a lift" on every element moving through and spread yourself through genome after genome. This is the strategy adopted by IS elements. The down side is that if you are lost (e.g., by deletion) the host cell may be physiologically better off without having to devote the energy necessary for your replication and maintenance. So you might decide to become more adventurous and acquire a gene or two of selective value to your host to ensure a better chance for your persistence. So you move up the evolutionary ladder to become a transposon and on up to gain a catabolic pathway or even become a pathogenicity island! You are still reliant on the host, are more costly but have something extra to offer now. Maybe this is good enough, but

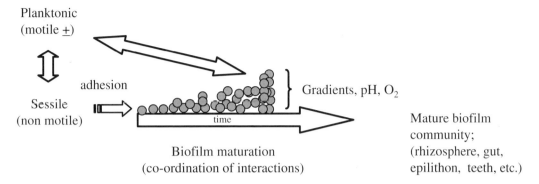

FIGURE 1 Interactions in a developing biofilm community. Events occurring are predation, quorum sensing, syntrophy, gene exchange, antagonisms, cheating, nutrient uptake, excretion, etc.

could you go further? Well you could "hitch up" to a replicon and become a plasmid or become associated with a phage genome. This would allow you to be more organized at moving around between cells. As long as you offer some phenotypic advantage to the cell, or add positively to the energy balance, then your survival potential would be enhanced. But now you are growing in size, if not stature! You are now becoming part of the community of genes, as opposed to being a migrant and an opportunist. If you are not careful the next step is to become a chromosome and join the establishment! This scenario provides a hypothesis for how genes, genomes, and mobile elements might interact over time. It provides a rationale for why we have some elements (IS sequences) which appear to be relatively unevolved and an array of ever more complex ones. It provides a basis for the idea of a continuum of sequence

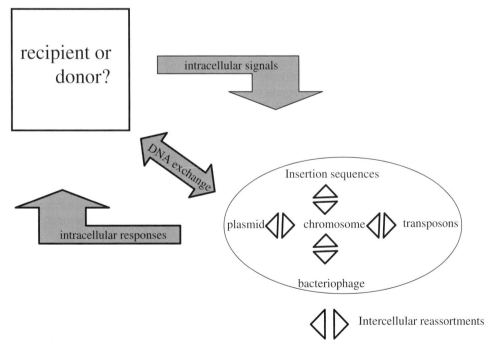

FIGURE 2 Summary of possible inter- and intracellular interactions.

interaction through time and space. Figure 1 summarizes the principal components of interactions in a biofilm.

So we know some of the things that are happening in microbial populations, but we do not comprehensively understand the interactions (Fig. 2). Group activities are effectively promoted by interactions between different species and genera. Because there are opportunists whose presence is not for the benefit of the interacting group, a diverse range of inhibitory agents targeted against specific or general competitors seems to have evolved. If this hypothesis is correct, then the agents used range from bacteriophage and bacteriocins, which generally target related strains, to general antimicrobials (antibiotics, acids, alkalis, etc.). So we have hypotheses but lack some of the technology and the time to do the analyses. When we do we will be able to appreciate better how natural populations survive in the environment, evolve, and interact.

REFERENCES

Davies, D. G., M. R. Parsek, J. P. Pearson, B. H. Iglewski, J. W. Costerton, and E. P. Greenberg. 1998. The involvement of cell-to-cell signals in the development of a bacterial biofilm. *Science* **280:**226–227.

Hallet, B. 2001. Playing Dr. Jekyll and Mr. Hyde: combined mechanisms of phase variation in bacteria. *Curr. Opin. Microbiol.* **4:**570–581.

Kaper, J. B. 1998. Enterohemorrhagic *Escherichia coli. Curr. Opin. Microbiol.* **1:**103–108.

Mira, A., L. Klasson, and S. G. E. Andersson. 2002. Microbial genome evolution: sources of variability. *Curr. Opin. Microbiol.* **5:**506–512.

MECHANISMS FOR DETECTING GENOMIC DIVERSITY

HORIZONTAL TRANSFER, GENOMIC DIVERSITY, AND GENOMIC DIFFERENTIATION

Roger Milkman

19

INDIVIDUAL GENOMES

CLONES AND MEROCLONES

COMPARATIVE SEQUENCING
Preliminary Survey Techniques
Primer Extension in Sequencing and Amplification
Subjects of Analysis

GENOMIC DIVERSITY

HORIZONTAL TRANSFER
Mechanisms and Vectors
Integrons
Recombination
Restriction

KINDS OF SELECTION

SELECTION AND RANDOM
GENETIC DRIFT

GENOMIC DIFFERENTIATION

PERIODIC SELECTION

TARGETS OF SELECTION REVISITED

PROKARYOTIC HORIZONTAL GENE
TRANSFER VS. EUKARYOTIC
RECOMBINATION

THE BIOLOGICAL SPECIES CONCEPT IS
NOT APPLICABLE TO BACTERIA, SO
WHERE DO WE GO FROM HERE?

Roger Milkman, Professor Emeritus, The University of Iowa, Iowa City, Iowa.

The sequencing of the genome of *Escherichia coli* K-12 MG1655 has brought a new stage of explicit and precise detail to microbiology. This unambiguous information provides both enhanced comprehension and utility to the study of this important bacterium. A peak has been reached, and at once another has come into view: the definition of one representative *individual genome* calls for a description of the *species genome* (Lan and Reeves, 2000). As a subject, the species genome of *E. coli* must be more structured and coherent than *E. coli*'s population genetics, which has existed for decades as an interesting sample of information on the state and dynamics of the distribution (and occasionally association) of alleles of particular genes. Population genetics data support inferences about population structure. The species genome will eventually become quite explicit, embodying both the general *diversity* of individual genomes and some *differentiation* common to the genomes of all members of the species and some properties found in one or another important constituency of the species, but not in all of its members. It is likely that, while the individual genome has a single dimension, the species genome will be conceived as having two.

In *E. coli*, *genomic diversity* comprises nucleotide sequence variation, which often occurs in a mosaic pattern of disparities among isolates (Milkman, 1996; Perna et al., 2001; Kudva et al., 2002). *Genomic differentiation* in *E. coli* is exemplified by two common diametrically opposed regions of hypervariability amid the

Microbial Evolution: Gene Establishment, Survival, and Exchange
Edited by Robert V. Miller and Martin J. Day, ©2004 ASM Press, Washington, D.C.

far more orderly, though not uniform, remaining 85% of the chromosome of most members of the species studied to date. Other departures from the ordinary, including the largely non-homologous *pathogenicity islands*, are variously distributed in the genomes of particular groups of pathogens. Also, integrated bacteriophages, fragmentary or complete, and various elements containing insertion sequences are commonly encountered in *E. coli* chromosomes; finally, various *plasmids*, replicating independently, have stable genomic tenures in many lines of descent. As we shall see, mutation, horizontal gene transfer, and particular modes of selection are responsible for these occasional features of the species genome.

INDIVIDUAL GENOMES

The genome of an individual cellular organism or a DNA virus is best defined as a single example of a set of unique and uniquely placed DNA sequences. A given DNA sequence may appear more than once within a bacterial chromosome, and all the repeats are counted. Only one copy of a plasmid is counted, as is only one copy of a chromosome. The genome may consist of a single, continuous, and usually circular chromosome (as is common in prokaryotes), alone or together with independently replicating plasmids. Some prokaryotes do have two or even more chromosomes. A frequent misconception of the term "genome" is "the entire amount of DNA within a cell," and this leads to serious misunderstanding, especially when applied to eukaryotic cells with a variable number of copies of DNA strands (as in *Drosophila* salivary glands, whose polytene chromosomes are famous). Less obvious sources of confusion are, for example, dinoflagellate cells, each of which often contains many copies of its genome.

Section Summary

- The genome is the set of instructions for developing and operating an organism.

CLONES AND MEROCLONES

A clone comprises all the descendants of a single ancestor and nothing else. Any consideration of a clone requires an implicit reference time, namely, a time of origin. For example, since $2^{69} \approx 6 \times 10^{20}$, in theory, a single *E. coli* cell could give rise, in 69 generations of binary fission, to a progeny of about 6×10^{20} cells (a millimole of cells!). That happens to be one estimate of the number of *E. coli* cells in the world. Clearly clones this large are hypothetical: death is too frequent, and so is horizontal transfer. But in any clone of the same age, given a nucleotide substitution rate of 3×10^{-10} per site per generation, each nucleotide in the genome would have an average chance of $69 \times 3 \times 10^{-10}$ of having undergone a substitution since the origin of the clone. With a genome like that of strain K-12 MG1655 (4.6×10^6 nucleotide pairs [Blattner et al., 1997]), the surviving cells would average $69 \times 3 \times 10^{-10} \times 4.6 \times 10^6 = 0.095$, or about 0.1 nucleotide substitution per genome. In pairwise comparisons this would result in an average divergence of 0.2 nucleotide (nt) per genome, which represents an average proportion of about 4×10^{-8} of the genome. On the other hand, an average pairwise difference of 1% (4.6×10^4 bp), which is often encountered or exceeded in the comparison of natural isolates of *E. coli*, would take $(4.6 \times 10^4/2) \times 69$ = about 16 million generations of divergence from a common ancestor, or about 80,000 years at 200 generations per year. Actually, since the species has been around for some 100 million years, it appears that its clones have grown, diversified, diverged, and often departed, either by extinction or by being recognized as distinct from *E. coli*.

The hypothetical examples above, in which divergence is attributed solely to mutation, are definitive clones in that they represent a group of cells totally descended from a single common ancestor that existed at time t_1. Mutation does not compromise clonality, since the line of descent, involving *vertical transfer*, is not altered in the course of a nucleotide substitution. For that matter, a linear duplication of a gene or a nucleotide does not compromise clonality either; both duplicates are descended from the same ancestor. The members of a clone are *identical by descent* whether or not they are *identical in sequence*. On the other hand, *horizontal transfer* of DNA results in a local change of ancestry if the source of the inserted or replacement DNA is a donor cell outside of the clone.

Over evolutionary time, horizontal gene transfer has been frequent enough in *E. coli* to result in easily detectable mosaicism in sequence comparisons. Natural isolates (maintained as "strains") are seen to differ in local stretches of various sizes. In nature, as opposed to isolated culture, a clone acquires these replacements with time and is no longer a true clone. At first the vast majority of its DNA is identical by descent. This DNA constitutes the *clonal frame* and continues to share a common tree. A replacement is continued as a set of *clonal segments* in cells descended vertically from the recombinant parent. Each clonal segment thus has its own local tree originating at the time of horizontal transfer (sporadic mutations notwithstanding). With respect to the common ancestor that existed at time t_1, the cells no longer compose a true clone, but the common ancestry of most of their genome, the clonal frame, is significant and justifies the term *meroclone* ("partial clone"; Milkman, 1996). With considerable passage of time and further horizontal transfer, no single tree is supreme, and the diverse strains are said to have a variety of chromosomally local "gene trees" (Avise, 1989). Comparative analysis of *E. coli* isolates has revealed large groups of closely to moderately similar strains that can be placed in larger, looser groups in a tree-like arrangement (Herzer et al., 1990) that is not a strict phylogenetic tree. (Note that what is commonly called a phylogenetic tree shares a property of mobiles, in that any pair of branches may revolve about their common node, before coming to rest on the page.) In any event, the Herzer diagram is not a consensus but rather a sample of many distinct trees, and it is better referred to as a phenogram, since it is based on observed similarities in various parts of the genome. Phenotypic characters are observable attributes. With modern technology, even DNA is an observable attribute; phenotypic characters and genes are no longer mutually exclusive!

Section Summary
- A clone is all the descendants of a single cell. A meroclone is a clone compromised by a modest amount of horizontal transfer.

COMPARATIVE SEQUENCING

Preliminary Survey Techniques
Two techniques that have provided detailed bases of comparison among *E. coli* strains, as well as among those of many other bacterial species, are multilocus enzyme electrophoresis (MLEE) and restriction fragment length polymorphism (RFLP, also referred to as restriction analysis). Both reveal sequence-related variation. MLEE was applied to bacteria in the early 1970s (Milkman, 1973), following its dramatic success in the analysis of genetic variation in a variety of eukaryotes. Entire organisms or particular parts are ground up or sonicated in buffer and applied to a solid support medium: gels of starch, agarose, or polyacrylamide, or else buffer-moistened sheets of paper or cellulose acetate. The variously charged macromolecules in the extracts (notably proteins) move differentially in an applied electric field, and afterwards individual enzymes on the support medium can be stained specifically in appropriate mixtures of substrate, cofactor(s), and product-specific stains. Since proteins are encoded by DNA, nonsynonymous nucleotide substitutions lead to amino acid substitutions, and these influence electrophoretic mobility by changing the charge distribution on the protein. An amino acid substitution that introduces (or eliminates) an ionizable group is an obvious example, but changes in protein conformation often cause important charge changes as well, due to new interactions among the amino acid residues, which may result in changes in the ionization dynamics of existing groups.

The wealth of enzyme data could not at first be mapped easily to the corresponding genomes, but the simplicity of bacterial genomes made the connection easier, although even today only 21 of 37 MLEE enzymes have been mapped to specific gene loci. This is because 14 catalytic activities described in terms of experimental staining conditions have not been assigned a specific Enzyme Classification (EC) number, so that mapping is more difficult than merely testing candidate loci from the extensive genetic map. Also, a defined set of catalytic properties as in alcohol dehydrogenase (EC 1.1.1.1) is occasionally found in two or more distinct enzymes (*isozymes*), each coded by its own gene (Whittam, T. S., http://foodsafe .msu.edu/whittam/ecor/enzymab.html).

RFLP provided genetic markers to characterize specific strains in bacterial populations growing in chemostats, but its use as a preliminary survey tool for nucleotide sequencing emerged in 1986 when Dykhuizen and Green (1991) established the evolutionary significance of horizontal transfer in *E. coli* by a small but convincing sequencing experiment that showed *gnd* to have a tree different from *trp*, which had previously been sequenced in the same set of strains. Since RFLP reveals strain differences in the distribution of restriction sites (these are mainly 4- to 6-bp sequences), this simple precursor to nucleotide sequencing became an attractive technique. This led to RFLP-based physical mapping of entire genomes, notably that of *E. coli* by Kohara et al. (1987), its elaboration (Rudd, 1998) and use with the *E. coli* linkage map (Berlyn, 1998), and its important connection with the subsequent spectacular industry of genome sequencing that we know today.

A recent derivative of MLEE is multilocus sequence typing (MLST), in which "alleles at multiple housekeeping loci are assigned directly by nucleotide sequencing" (Spratt, 1999, p. 312), with obvious advantages over electrophoretic mobilities in data processing. This technique has been particularly useful in epidemiological studies of pathogenic bacteria such as *Campylobacter*, *Helicobacter*, *Neisseria*, and *Streptococcus*.

Primer Extension in Sequencing and Amplification

Today's automated nucleotide sequencing is a remarkable metamorphosis of Fred Sanger's remarkable invention. The Sanger technique begins with *primer extension*, the progressive addition by DNA polymerase of nucleotides to a short existing sequence (primer) bound to a complementary part of a long strand of DNA, the exposed portion of which acts as a template that determines the sequence of nucleotides added to the primer. The Sanger "dideoxy" sequencing method runs primer extension in only one direction and exploits the special use of added tiny amounts of *dideoxy*nucleotides, one type per lane. These infrequently land at the growing 3′ end of the strand (having just paired with a complementary template nucleotide), and they terminate strand growth right there. This happens because dideoxynucleotides lack the 3′-hydroxyl group necessary to hook on to the next nucleotide presented by the DNA polymerase. Bypassing the (highly productive) radioisotope methods of the recent past, each of four reaction mixes contains a specific primer labeled with a dye molecule fluorescing in a distinct part of the visual spectrum, as well as one type of dideoxynucleotide (A, C, G, or T). The four reaction mixes are combined and electrophoresed; each band's dye is recognized by a scanner, which in turn instructs a graphing needle to squirt one of four colors of ink as it draws its record. There is no intrinsic similarity between the color of the dye attached to each primer and the color of the ink. This method, outlined on pages 291 to 292 of Smith et al. (1987), includes the electrophoretic process that produces discrete bands from the mix of growing polynucleotides; the template's color identifies the specific dideoxynucleotide at the other end of the chain. Thus sequencing no longer requires four parallel lanes, each containing a reaction doped with one of four dideoxynucleotides that are labeled in common with one radioisotope.

To produce the amount of uniform template DNA sufficient for sequencing, PCR is used. Primer extension itself is automated in the *amplification* of DNA, in which a pair of primers, designed to recognize specific sequences, land facing one another at some distance (e.g., 1 kb) on complementary DNA template strands, which have been separated in the cycle of temperatures produced by a thermocycler. In this process, primer extension proceeds in *both* directions; each primer is extended until it has reached the end of its template and is now a template itself, binding its complementary primer. In a representative example, the temperature is raised briefly to about 95°C to separate the two complementary strands; next, it declines to about 56°C, where new primers can land; and then it rises to 72°C, where primer extension resumes. PCR easily amplifies tiny, really tiny, amounts of a stretch of DNA, recognized specifically by the primers used (typically 24 bp in length), to a sufficient amount of DNA pure enough to use as a sequencing template. PCR thus bypasses much of the need for an earlier miracle of genetic engineering, gene cloning, and has earned a prominent place in elementary biology texts along with bee dances and the *lac* operon.

Subjects of Analysis

Another exciting application of the PCR/ comparative nucleotide sequencing procedure is *culture-independent* characterization of organisms that don't individually contain enough DNA for identification as part of a known phyletic group. For example, deep sea sediments contain certain microbes that can't multiply in surface (e.g., laboratory) conditions, but which can be captured dead or alive, subjected to PCR in a microtube containing the appropriate ingredients, and placed in a thermocycler. Primers complementary to certain *conserved sequences* that are *common to a phylogenetically broad range of ribosomal RNA genes* are used in pairs that amplify *intervening* DNA that is specific and diagnostic of its place on the tree of life. There are real limitations on the proportion of living things that we can learn about through comparing their DNA with that of known organisms, but the barriers are currently being reduced in this way.

In terms of evolutionary analysis, comparative sequencing is used in two major ways. The first, as indicated, is the assignment of phylogenetic position where more obvious criteria such as bone structure are not applicable. Here, complex, highly conserved sequences are extremely useful, as Carl Woese (1987) had found when microbial DNA coding for ribosomal RNA fell into two clearly distinct categories, leading to the splitting of bacteria into the domains *Eubacteria* (now just *Bacteria*) and *Archaebacteria* (now *Archaea*). Only then did their cell membranes attract interest, and it turned out that although bacterial membrane lipids are like those of eukaryotes in containing *ester* linkages, the *Archaea* typically have lengthy *ether* molecules. More recently, explorations of new environments and advances in analytic lipid chemistry have revealed a variety of distinct and often informative lipids, including one type of molecule with parts formerly diagnostic of each of the two respective domains (Hayes, 2000)! The second major evolutionary application of comparative sequencing is the general analysis of genomic variation, an outgrowth of population genetics that has now proceeded to comparisons of very long stretches of DNA. These stretches were never accessible to MLEE, whose phenotypic units correspond to small, scattered, discrete components of the genome; nor are they efficiently described by gene mapping, which is based on the recombinational analysis of various selected phenotypes (morphological or chemical), each related specifically to a gene at one of a variety of chromosomal locations.

Instead, all genomic variation, whether neutral or functionally critical, is important to the analysis of horizontal transfer. DNA sequences may be compared at any phylogenetic level and at any level of genomic detail. The analysis of the comparisons may focus on a variety of properties and serve, directly or indirectly, a variety of purposes.

Phylogenetic divisions at the highest level emerge from the comparison of the most conserved traits. The large (23S and 16S) RNA molecules in bacterial ribosomes fit closely (and of course functionally) to some 50 ribosomal proteins. The tightly integrated ribosomal structure makes acceptable nucleotide substitutions rare. Also noteworthy is the virtual impossibility, under everyday circumstances, of the horizontal transfer of ribosomal RNA genes, whose retention would require an improbably strong adaptive advantage for an incoming 16S or 23S RNA gene over the resident homologue. This prevents false implications of membership switches between major phylogenetic groups, bacterial or otherwise. As already mentioned, unculturable microorganisms can be assigned to a major phylogenetic group by the use of PCR and sequencing, exploiting appropriate primers in highly conserved regions. Rare lipid molecules, also mentioned earlier, can now be identified increasingly often, providing positive evidence as well as raising new questions. At the lowest phylogenetic levels, bacterial isolates cloned from a single cell and cultured as named or numbered strains can be identified as members of a species and grouped according to their similarity in relatively changeable nucleotide sequences or sequence-dependent properties such as RFLP. Recent advances in technology, however, have made sequencing actually quicker than the less informative preliminary methods of indirect sequence sampling.

Complete genome sequences have provided a genetic description, in ultimate detail, of individual representatives of a growing number of species. So far, comparisons above the species level predominate, but the existence of two complete *E. coli* sequences (Blattner et al., 1997; Perna et al., 2001; Kudva et al., 2002) points the way to a vast range of detail in intraspecific differences. In both publications and electronic records, detailed annotation going far beyond the mere sequence of nucleotides adds tremendous value to each achievement. Shorter continuous sequences and local or genomewide sets of intermittent sequences have been compared among considerable numbers of individuals, with a variety of interesting targets. These range from analysis of number and type of nucleotide differences, through nucleotide and codon bias, to striking chromosome-positional differences both in degree of variation and in signs of horizontal transfer.

Section Summary
- Sequencing is a technical triumph enlisted in the effort to analyze and describe the workings and relationships of living beings.

GENOMIC DIVERSITY
The central framework for the consideration of genome diversity is the species. Unfortunately, a commonly accepted rigorous and universal definition of bacterial species does not yet exist, and the possibility of such a definition is not widely taken for granted. Nevertheless, a tentative empirical notion does emerge with reference to the immediately evident characteristics of eukaryotic species, including commonality of conserved phenotypes within species and clear differences between the species. This pragmatic concept is supported, though not decisively, by the presently limited genomic comparisons among eukaryotic species. In bacteria, generalizations relating to higher tentative taxonomic groups can also be quite informative. The evidence of important ongoing horizontal gene transfer across great phylogenetic distances in bacteria, however, is clearly at odds with the Biological Species Concept currently accepted for eukaryotes (Mayr, 1963, 2001), in which interspecific gene transfer is understood to be (depending on phylogenetic distance) infrequent, minimal, or the result of a series of steps in a very complex trajectory. A constructive view of this discordance is that its resolution will lead to important conceptual advances in biology.

The *E. coli* Reference (ECOR) collection of 72 strains derived from wild isolates (Ochman and Selander, 1984) has been studied extensively, together with a growing number of other members of the species. The best known laboratory strain, K-12, is closely related to ten of the ECOR strains, as evidenced by an average pairwise difference of less than 0.1% between homologous nucleotides and a small scattering of recombinational replacements of various lengths (Milkman and McBride, manuscript in preparation; Milkman et al., manuscript in preparation) differing by greater percentages. This set of 11 strains, known as the *Big Ten* (Milkman, 1999; Milkman et al., 1999), provides a useful baseline against which to measure nucleotide variation in *E. coli*, as we shall see. Among the ECOR collection as a whole, a mosaic of local pairwise homologous differences with nucleotide substitution frequencies ranging up to 4% or 5% is common; and rarer observations include 24% over a stretch of 283 bp with 15 at the first codon position, 7 at the second, and 46 at the third (Milkman, 1996). In addition there are various presence/absence differences: occasional bacteriophages, complete or incomplete, occurring as infrequent insertions of the order of 50 kb; other large assemblies of genes, including pathogenicity islands (see "Individual genomes"); and integrons (see "Mechanisms and vectors"), which serve as ports of entry for migratory genes. Comparisons at higher taxonomic levels often also reveal differences in the organization of genes coding for the enzymes of a particular pathway. For example, the *trp* genes, relating to tryptophan synthesis, form a single 7-kb operon in *E. coli* and other enterobacteria and in some actinobacteria, but occur in other taxa as small groups or even individual genes scattered in various chromosomal locations (Crawford, 1989). Clearly, an operon is not only a unit of regulation, but also a potential unit of efficient horizontal transmission of a complex set of functions. (Numerous taxonomic advances are accessible through the National Center for Biotechnology Information [www.ncbi.nlm.nih.gov/Taxonomy/taxonomyhome.html]. Note, for example, that *Brevibacterium lactofermentum* is now *Corynebacterium glutamicum*.)

Section Summary

• Comparisons lead toward a satisfying understanding of the big picture.

HORIZONTAL TRANSFER

Mechanisms and Vectors

Bacterial horizontal transfer mechanisms are diverse, to say the least, even within a species. In *homologous* gene transfer, a stretch of incoming DNA replaces largely similar resident DNA. (*Homology* is similarity due to common ancestry.) *Nonhomologous* gene transfer introduces DNA that is novel to its incorporation site (and usually to the entire recipient genome). The most familiar mechanisms include *conjugation*, mediated by *F* (for fertility) factors located on a free *plasmid* either without (*F*) or with (*F'*) some chromosomal DNA aboard, or integrated into the chromosome (*Hfr*). Thus anything from the *F* factors themselves to an entire chromosome can be transmitted in the conjugation process, which is described clearly and extensively by Salyers and Whitt (2001, p. 140–145) and in chapter 11. The phylogenetic range of transmission is occasionally impressive: bacterial conjugative plasmids are capable, with the aid of a plasmid that can replicate and be selected in yeast, of delivering DNA to yeast (Heinemann and Sprague, 1989).

Transduction (see chapter 9), mediated by bacteriophages, comes in two forms: specialized and generalized. Specialized transduction involves transmission of some or all of an integrated or episomal phage together with some flanking host-chromosomal DNA. Generalized transduction is occasionally incidental to the disintegration of a bacterial host's DNA by multiplying bacteriophages: a newly formed phage head mistakenly incorporates a fragment of host DNA about the same size as the normal phage genome. The mechanically functional phage particle subsequently introduces this DNA fragment into another host, where homologous recombination can take place. While most familiar transducing phages have genomes of less than 100 kb, a variant of bacteriophage T4 (not ordinarily capable of transduction) has been found to do it (Masters, 1996, p. 2435). This raises the size limit for transduced fragments to about 175 kb, in theory, and thus affects the inference of the transfer mechanism responsible for observed replacements.

Transformation (see chapter 10), in which a naked DNA fragment is incorporated into a cell, is the simplest form of gene transfer. Some species prefer linear fragments; *E. coli* prefers circles. Some species, like *Bacillus subtilis*, sporadically engage in rolling-circle replication and extrude single-stranded DNA into their environment, where it is taken up and made double-stranded by other members of their species. As one contemplates this process, it is tempting to regard conjugation as a form of facilitated transformation.

DNA movement may also be intragenomic. An *insertion sequence* (IS element) contains a gene encoding a *transposase*, an enzyme that allows it to enter DNA at a random or specific position and to excise itself. Two insertion sequences together with some coding DNA between them can often move as a functional unit; they qualify as a *transposon*. Such intragenomic changes of position may interrupt a resident unit (disable a gene, for example) or modify its control (by placing a gene next to a different promoter). The event, though obviously more complex than a nucleotide substitution, is effectively a form of mutation. It is not in itself transfer, either horizontal or vertical.

The incorporation of horizontally transferred DNA falls into three patterns. First, simple homologous recombination involves the recognition, by the incoming DNA, of similar resident DNA, wherever it lies. The result can be the entire or fractional replacement of a functional coding DNA, producing an allelic difference, for example. Alternatively noncoding or even nonfunctional DNA may be replaced. Second, a conjugative transposon can move between cells and introduce a nonhomologous change at a random site. (Other types of incoming DNA, for example, the bacteriophage Mu genome, integrate in any of a very large number of acceptable landing spots, whose chromosomal distribution is thus virtually random. Phage lambda, on the other hand, has a main preferred integration site.) Third, another type of IS element or transposon uses sequence recognition when it integrates (after immigrating, for example); its transposase targets a specific site. This occasionally leads to the formation of a complex agent of bidirectional integration called an *integron*, whose mechanisms facilitate both immigration

and importation. An integron contains a gene for a second enzyme, *integrase*, which recognizes and incorporates certain *gene cassettes*, often leading in turn to the formation of an operon-like sequence in which several genes have a polar orientation and may thus be driven by a single promoter (Salyers and Whitt, 2001, p. 145–149). So in this system transposase integrates, and integrase incorporates! The component features of this *integron* help explain the structural and functional organization of the O-antigen region (see "Genomic differentiation"), including coordinate control and the repeated changes upon which circumstantial selection (see "Kinds of selection") is based. For example, an O antigen's immunological shape is dictated by the particular sugars that adorn it and by the way they are attached. How are various genes, capable of altering the form of an O antigen, generally introduced where they can replace an analogous alternative? Their vectors evidently recognize a stretch of target DNA. Thus, a gene codes for an enzyme whose product, a saccharide different from that made by the enzyme it replaces, is introduced in a functionally appropriate place.

Integrons

The mechanism of site-specific recombination of genes such as those in the O-antigen region has long implied some form of functional recognition mechanism resident in the chromosome. During the past few years, some specific sequences have been implicated, such as the "JUMPstart" sequence found in such enteric bacteria as *E. coli*, *Salmonella*, *Vibrio*, and *Yersinia* (Hobbs and Reeves, 1994). Integrons (Hall and Collis, 1995) begin with an *Int* family integrase gene enclosing a promoter oriented toward a few promoterless gene cassettes separated by DNA elements with recognition sites for integrases. In the words of Vaisvila et al. (2001), "The discovery of integrons and antibiotic-resistance gene cassettes . . . illustrates that bacteria have a natural genetic engineering tool to acquire, mix and match potentially useful genes." Superintegrons, containing up to and over 100 gene cassettes, have been found in *Vibrio cholerae* and now in *Pseudomonas alcaligenes*, a widely distributed, ecologically diverse nonpathogen, suggesting that a major common natural mechanism of horizontal gene transfer is available for use in the genomic arsenal (Vaisvila et al., 2001).

Recombination

The most frequent result of horizontal gene transfer is intraspecific recombination, in which fragments of DNA from different lineages within a species are transferred to other lineages, and the succession of overlapping transfers of diverse origin breaks down large sequence differences into a mosaic of diverse, smaller stretches. Recombination of this sort is far less frequent than the crossing over of eukaryotic homologues, but it leads to similar results in opposing the tendency of lineages to diverge in isolation. By definition, the frequencies of the recombining alleles do not change dramatically as an immediate result of the transfer events. A major difference between prokaryotic and eukaryotic recombination processes lies in the great phylogenetic distances crossed (very infrequently, to be sure, but with evolutionarily important effects) in prokaryotic horizontal transfer.

Restriction

The demonstration that incoming DNA molecules are frequently cut by restriction endonucleases and evidently nibbled subsequently by exonucleases (McKane and Milkman, 1995; Milkman et al., 1999) has important implications for homologous replacement. While restriction can incapacitate invading phage genomes and presumably some potentially useful gene transfers as well, the frequent cutting and shortening of transferred DNA that is homologous to parts of the genome can, on balance, be quite advantageous. The resulting increase in the number of recombinogenic fragments is likely to increase the frequency of successful horizontal transfers, especially since small fragments are less likely than large stretches to contain both deleterious and beneficial DNA, so that the good may be saved by

selection and the bad lost. Also, since bacterial chromosomes are generally somewhere in the process of replicating, the nascent chromosomal copies often provide the small fragments with two or more alternative targets for integration, so that a single transfer can actually result in a mosaic lineage.

Section Summary

- Horizontal transfer expands an organism's horizons and holds the living world together.

KINDS OF SELECTION

Within a species, selection favors some organisms over others in terms of the number of offspring they leave per generation, their *fitness*. In prokaryotes, this is simpler than in sexual eukaryotes, because only one ancestor is involved in producing offspring each generation. (When a given gene or allele confers an increase in fitness on an organism, the gene is sometimes described, for convenience, as having the fitness itself. This practice is inaccurate and often leads to confused calculations.)

Selection may be classified by its effect (purifying, directional, stabilizing, and diversifying) or by its stable dependence on another parameter (frequency-dependent). *Purifying selection* is the ongoing removal of harmful mutant alleles by their effects (death and debilitation); it balances ongoing mutation by opposing the accumulation of inferior individuals. *Directional selection* is the rise in frequency of a favorable allele, or the progressive change in a quantitative phenotypic parameter value (like height or generation time). Since quantitative parameters are generally influenced by alleles at numerous loci, quantitative directional selection ordinarily becomes *balancing* because of the assembly of genotypes on both sides of the optimum phenotypic value. *Balancing selection* is thus a form of *stabilizing* selection in acting eventually to preserve the status quo. *Diversifying selection* in its simplest form is merely directional selection operating divergently within a single population: two alternative values of an environmental property (like temperature or humidity) in different parts of the population's habitat might cause diversifying selection.

Frequency dependent selection can be classified as direct (or positive: the higher the frequency, the higher the fitness) or inverse (the higher the frequency, the lower the fitness). In bacteria, both kinds of frequency-dependent selection are well known. For example, inverse frequency-dependent selection is often incidental to the extensive work of Richard Lenski's group at Michigan State University in East Lansing. In experiments involving successive transfers in batch culture, the newly inoculated bacteria grow fast, and as the food supply per cell diminishes with increased numbers, the division rate progressively declines to a crawl. Also in chemostats, where population density is controlled at any desired level, division rate is systematically lower at high density.

There are numerous cases of direct frequency-dependent selection where bacteria gain from synergistic activity. Luminescent species like *Vibrio fischeri* live in aquatic organisms and protect them from being detected as dark objects against a light sky by predators lurking below. Their luminescence is useful to bacterial species in other contexts as well, and it is promoted by *quorum sensing* (see chapter 21), the detection of a high population density of members of the species (Madigan et al., 2002).

A type of habitat leading to frequency-dependent selection might be one evenly divided into two areas favoring different plant species such as rye and wheat. An insect species containing two freely interbreeding genotypes, a common one doing best on rye and a rare one doing best on wheat, could move into the habitat; the rare genotype preferring wheat would at first multiply faster than the rye eaters, and as its frequency rose its food supply would be shared among more individuals and its fitness would decline. At low frequency, the rye-eating genotype would have high fitness, but with higher frequency comes lower fitness. This is *inverse frequency-dependent selection*, but the term is useful only if the relationship is stable. For example, if

more of the habitat changed to favor wheat, the fitness of the wheat eaters would go up and that of the rye eaters would decline. Sporadic events or independent circumstances may coincidentally result in false inferences of frequency dependence; such inferences emerged early in the analysis of the horizontal transfer of O antigens. As will be discussed later, selection may operate briefly and then cease, or resume cyclically with repeated recombination and related events. These patterns reflect the important episodic influence of the environment on the process of selection, and the term *circumstantial selection* has been proposed in this context (Milkman et al., 2003).

Section Summary

- Selection is a powerful force for improvement. It works in diverse ways.

SELECTION AND RANDOM GENETIC DRIFT

In striking contrast to the frequent exchanges in intraspecific recombination, rare imports from long-separated lineages make dramatic primary contributions and, because of hitchhiking, result in impressive secondary phenomena as well. How is it that genes horizontally transferred from a phyletically distant source, at a rate far lower than intraspecific horizontal transfer, can nevertheless have strikingly important evolutionary effects on the recipient genome? Favorable selection is a major part of the answer.

The fortunes of a rare transferred gene begin with its entry into a recipient cell, followed by its integration into the chromosome. At this point it is not certain whether it will enter the evolutionary stream or, often after any number of indecisive generations, disappear. The initial arrival of the single transferred gene entails that, unless it is strongly favored by selection, it will be swept away by random genetic drift. Indeed, the interplay of selection and drift is nowhere so compelling as in the battle between their respective champions, namely, extremely high fitness and very small numbers.

To follow this interplay, it is important to know quantitatively how it works. The size of the *E. coli* population of a human colon is more-or-less constant, and we can consider that it contains two kinds of cells. One kind, initially, is the single cell whose novel O antigen is unrecognized by the host's immune system, leaving it free to grow and multiply at a higher rate than the other kind. Although the numerous other cells are likely to contain a good deal of genetic diversity, their critical distinction here lies in their recognition and resulting moderate restraint by the host. Each of these cells contains one of what may be several different O antigens present, but the host knows them all.

A single genetic difference between (initially) one cell and all the others accounts for the selective advantage of the novel undetected cell. Cells of this type multiply faster than the others and, as they become numerous, their fitness declines due to strong competition from others of the same genotype. It is still greater than that of the detected cells, which also suffer from the stronger competition, and so the proportion of undetected cells continues to increase.

Fitness in bacteria can be defined as "the average number of cells left in each new generation by each cell genotype *under a given set of conditions*, including the properties of its competitors." From this definition, it has to be clear that *fitness is a result, not a cause*. Also the "properties of its competitors" are clearly a cause, not a result. Since fitness has the dimension *time*$^{-1}$, it is a rate. Because the population size is essentially constant, the *mean fitness* of all its cells has a constant value of 1: the undetected cells have a declining fitness greater than 1 and a rising frequency; the detected cells have a declining fitness less than 1 and a decreasing frequency.

The constant size of populations is not limited to the colon. Although some may experience sustained growth at the expense of other populations, or in expanding habitats, populations are generally taken to have a constant size in the long run. This usual constancy contradicts the fallacy that natural selection leads inevitably to a steady improvement in fitness.

What, then, does natural selection inevitably

improve? Natural selection improves a general function that bears on growth rate in the same way a golfer's skill bears on his/her wealth. If the toughest competitors increase in number, the others' winnings decline. Between genotype and the resulting fitness lies the intermediate cause, *fitness potential*. Fitness is a nondecreasing function of fitness potential. *In a population of constant size, mean fitness potential increases as the result of selection, and mean fitness does not.* There are only 100 members of the 100 best milers in the world (fitness), but the 100 best times (fitness potential) improve consistently. In aggregate, fitness potential consists of all the various features of an organism that combine to influence its fitness (Milkman, 1978, p. 399).

Returning to the interplay of selection and drift (introduced in the second paragraph of this section), one way of following its course is to simulate the change of frequency of an allele or gene as influenced by selection and random genetic drift operating at the same time. The need for simulation stems from the random nature of random genetic drift. While the effects of selection can be calculated precisely for a given set of parameter values, with results that are identical on repetition, random processes have no single specific trajectory.

One practical setting, for example, is a population whose size, N, is 10,000, containing a single newly arrived gene (frequency, $p = 1/10,000 = 0.0001$) with initial fitness, w_a, of 1.5. Because the population size is constant, the mean fitness of all its cells is 1, and the prior residents (whose frequency is $q = 1 - p = 0.9999$) now have a common initial fitness, w_b, of 0.99995 [$= (10,000 - 1.5)/9,999$]. Recall that fitness has the dimension *time*$^{-1}$ and is thus a rate. As the number of individuals with fitness w_a rises, they compete with one another and with the individuals whose fitness is w_b.

If the new gene rises in frequency (as is likely here), the number of competitors with an *initial* fitness of 1.5 increases in the population, whose mean fitness, W, remains at 1 and equals $pw_a + qw_b$. The average selective advantage of a cell with gene *a* over all of its competitors is given by its selection coefficient, namely, $s_a = w_a - 1$. With each new generation, the change in p due to selection is $\Delta p_{sel,i+1} = s_a pq/W = s_a pq/1$. At any value of p, the value of w_a can be calculated from the mean fitness equation, $pw_a + qw_b = 1$, and the approximate relationship $w_a - 0.5 = w_b$ (from their initial values). Thus when p is 0.04 $(0.04 \times w_a) + 0.96 (w_a - 0.5) = 1$, so $w_a - (0.96 \times 0.5) = w_a - 0.48 = 1$; and $w_a = 1.48$. So as p has risen from 0.0001 to 0.04, w_a has declined from 1.5 to 1.48, and w_b has declined from 1 to 0.98. These numbers fit the mean fitness equation perfectly.

At the same time, random sampling is taking place. The colon contents move through in a day or two. During this time, the bacterial cells, including *E. coli*, grow and divide fast enough to keep the population size constant. Half the cells have been lost, and the other cells have doubled in number, so that this period is effectively one generation in a population of constant size. The cells are lost at random and replaced selectively (competitively). The succession of gene frequency changes due to random loss constitutes random genetic drift. Both selective and random processes can be quantified on a per-generation basis, and so in each generation the changes in number of each of the two genotypes represent the sum of selective and random effects, which are, in this case, Δp_{sel} and Δp_{rgd}. The selection process operates deterministically and, in the present example, directionally. The net random effects per generation are about half positive and half negative.

To simulate the random effects, one begins with a series of random numbers. One series that has the essential properties of random numbers can be obtained by taking the final four digits of a phone number expressed as a decimal fraction, add π, and then raise the sum to the fifth power. Eliminate the integers, leaving a new decimal fraction, which is the next "random" number. Iteration produces the desired series, which is, in truth, only a series of *pseudorandom* numbers, because starting over with the same phone number will produce an identical series each time. Instead, keep a phone book handy for a source of "seed" numbers. The "random" numbers, u_i, are distributed uniformly between 0 and 1 (there are about as many between 0.100 and 0.200 as there are, for example, between 0.623 and 0.723). They thus have what is called a *uniform*, or *rectangular*, distribution.

For our present purposes we need "random" numbers that have a *normal* (Gaussian) distribution. Recall that a binomial distribution approaches a normal distribution as N approaches infinity, just as a polygon approaches a circle. One straightforward method of converting uniformly distributed random numbers into random normal numbers uses the Box-Muller equations (Box and Muller, 1958), which use uniform random numbers between 0 and 1. Starting with u_1 and u_2 from a series of uniform random numbers (and proceeding with successive pairs), one obtains the random normal numbers, x_i, in pairs as follows:

$$x_1 = (-2 \ln u_2)^{1/2} [\cos(2\pi u_1)]$$
$$x_2 = (-2 \ln u_2)^{1/2} [\sin(2\pi u_1)].$$

In fact, the present simulation uses only one of the two equations. This is easier to set up, and it is an efficient use of the uniformly distributed random numbers (which are available essentially without limit).

The variance of p, which is binomially distributed in any finite population, is pq/N, and the standard deviation, σ, is $(pq/N)^{1/2}$. Further, $\Delta p_{i,rgd} = x_i (p_0 q_0/N)^{1/2}$, where x_i is the random normal number for the ith generation. So $p_1 = p_0 + \Delta p_{1,sel} + \Delta p_{1,rgd}$, calculated as indicated above. Iterations lead to assured fixation of the new gene or to its loss, as indicated by rows of zeroes, from which there is no return for successive generations. Simulations of this sort with various fitness values can easily be set up in Microsoft Excel. An example is described in "Instructions for simulation," an appendix to this chapter.

Section Summary

- Random genetic drift causes unpredictable quantitative differences in the composition of samples of a genetically heterogeneous population.

GENOMIC DIFFERENTIATION

Individual nucleotide differences may be analyzed collectively in terms of nucleotide frequencies, frequencies of linear pairs of nucleotides, and codon usage. These analyses may be done locally or on a genomewide basis; their interpretations, to be discussed subsequently, are of considerable interest. Some chromosomally local differences, including features already mentioned, can be regarded, together with even broader features of genomes, as examples of genomic differentiation.

Plasmids are an obvious example of genomic differentiation, in that they are physically distinct units of coding, regulatory function, and replication. Many types of antibiotic resistance are coded by plasmids, and *conjugative* plasmids facilitate horizontal gene transfer. Indeed, antibiotic resistance is often carried on a conjugative plasmid. Some conjugative plasmids can integrate into, and excise from, specific (or not so specific) regions of the chromosome.

Another form of genomic differentiation, discussed here in relation to the prevalent, nonpathogenic forms of *E. coli*, is exemplified by the two diametrically opposed *hypervariable* regions of the chromosome, each 7′ to 8′ in length (Milkman et al., 2003). These regions center on minute 45 and minute 99 of the chromosome: each contains a functionally distinct core and flanking regions that are characterized by a high level of nucleotide variation.

The core near minute 45 codes for the O antigen, which covers the entire flat surface of the cell. Numerous restriction endonuclease genes (but not necessarily all that are present in the genome) are located near minute 99. Both cores feature nonhomologous variation (Amor et al., 2000; Li and Reeves, 2000; Barcus et al., 1995). Each variant results in a distinct change either in a single overall product (the O antigen) or in a distinct array of defenses (restriction endonucleases). The O antigen consists of a lipopolysaccharide integrated into the outer cell membrane (Madigan et al., 2002, p. 79–81), which ordinarily is recognized by host antibodies; the resulting moderate restraint of the growth, even of these benign bacteria, is a host adaptation. The introduction of a novel O antigen, which is not recognized by the host, enables evasion of this restraint and confers a great increase in fitness on its possessor. Its novelty results from a new analogous replacement due initially to horizontal gene transfer over a great phyletic distance, ordinarily via a plasmid. The replacement is a gene whose product results either in the synthesis of a novel saccharide or in a novel attachment of a saccharide already produced in the cell. The probability of retention of such a single replacement is great when strong favorable selection is involved. This high fitness lasts while the frequency of the antigen in the population within a host rises, until it is sufficiently numerous to be recognized and responded to by the host's immune system. The response constitutes the production and release of a specific soluble *secretory immunoglobulin A* (sIgA), which binds both to the cells and to the network of microfibers in the colon's mucin layer. As a result, the cells of the new clone will now be trapped frequently in the colon's mucin layer just like the resident cells bearing other, already recognized O antigens. Thus, the new clone's multiplication rate falls to the level of these other cells, but not below. Its numbers do remain high. This phenomenon suggests inverse frequency-dependent selection, since the new clone's fitness declined after its frequency increased. The present case, however, illustrates an inconstant relationship between frequency and fitness; the high frequency led to the immunological recognition of the new clone, whose fitness underwent a resultant change. A decrease in frequency (due, for example, to the arrival and multiplication of another novel serotype) would not have brought a corresponding increase in fitness. This is an example of circumstantial selection.

Now two processes evidently occur that cause the observed hypervariability in the flanking region. First, ordinary intraspecific horizontal transfer mechanisms such as conjugation and transduction continue throughout the chromosome among all the diverse genotypes in the colon, and second, the cell bearing the new O-antigen gene is spread to other hosts via egestion and subsequent ingestion. In each naive host, its fitness is immediately high because its serotype is not recognized by the host's immune system. Thus, the *effective replacement rate* is even higher than it is in horizontal transfer over a great phylogenetic distance, because the intraspecific transfer frequency is much higher and the O-antigen gene's retention rate is once again high. The rates of horizontal transfer among the various bacteria in the same colon have not been investigated to any important extent; they would be worth knowing.

Both conjugation and transduction can carry along substantial stretches of DNA flanking the core. Here, since the effective replacement rate is extremely high due to high fitness conferred by the new O antigen, horizontal transfer among strains whose sequences vary will result in local hypervariability. The sequencing of the very closely related Big Ten strains (see "Genomic diversity," above) reveals a level of nucleotide variation near the O antigen region about 50 times as high as that encountered in the nonhypervariable 85% of the chromosome (Milkman et al., 2003) including, notably, a mosaic of small samples of sequences resembling a broad variety of ECOR strains (Milkman et al., unpublished data). This latter observation confirms the role of the relayed transfer of flanking DNA and demonstrates that the 72 ECOR strains are representative of the *E. coli* species as a whole, from which these flanking replacements presumably are drawn.

The cycle is repeated in every host and advances as a geographical wave. Behind this wave, where all O antigens present are now recognized and equally favored, the stage is set once again for the retention of any new, rare, horizontally transferred O-antigen gene. The result, on a relatively fast evolutionary timescale, is repeated selection for novelty. Thus, the *effective* individual replacement rate (the probability of retention × the *basic* individual replacement rate) is exceptionally high for nonhomologous

replacements of this type and location, in contrast to the tiny retention probability of any replacement that is merely mildly favorable.

In the restriction endonucleases coded near minute 99, the initial inverse frequency-dependent selection stems from an arms race between *E. coli* and the bacteriophages that prey on it. Restriction endonucleases inactivate invading phage DNA by cutting it up; natural selection favors phages that do not have the sequences that are cut by the restriction enzymes, as well as phages that modify such sites beyond recognition by methylating a given nitrogen base. The bacterial species occasionally acquire, ultimately from phyletically distant sources, new restriction endonucleases that the phages cannot frustrate, and the running battle continues. This type of arms race attests to the effectiveness of capitalizing on the sum of evolutionary processes taking place in many different lines of descent, rather than just one, by means of horizontal transfer. A dramatic laboratory demonstration of this general opportunity was achieved recently by the use of *gene-shuffling* of DNA from *Citrobacter*, *Enterobacter*, *Yersinia*, and *Klebsiella* to improve moxalactamase (a cephalosporinase) activity selectively up to 540-fold (Crameri et al., 1998).

The two sizeable hypervariable regions constitute genomic differentiation both in their site-specific dynamics of horizontal transfer and the resulting distinct local patterns of high variation.

Section Summary
- Genomic differentiation embodies the results of diverse genetic mechanisms, from mutation to horizontal transfer to selection.

PERIODIC SELECTION
Although traditional population genetics operates in fairly large populations between relative allele frequencies of, say, 0.10 and 0.90, it turns out that the case of an absolute frequency of one novel allele or gene in a population of hundreds to billions of cells is very important also. The probability of this single allele taking

over the entire population (or of never being eliminated) is about twice its selective advantage, *s* (Crow and Kimura, 1970, p. 422). A grasp of this situation has applications to two major subjects of present interest: the clonal origins of bacterial species structure and the generation of local chromosomal regions of hypervariability.

The Periodic Selection Model (Atwood et al., 1951) emerged from studies of batch (and later chemostat) cultures of *E. coli* containing equal frequencies of two neutral alleles that could be distinguished under special conditions. Over numerous generations at constant population size, occasionally one marker allele or the other would show a dramatic rise in frequency at the expense of the other. This was correctly interpreted as resulting from a favorable mutation at a different site (a single random event), which deterministically carried along its entire chromosome, including the particular marker allele that happened to be present. No gene transfer occurred in these cultures.

Applied to a natural situation, this phenomenon came to be called a *clonal sweep*, in which a vast number of identical genotypes were now present in the species. In nature, subsequent horizontal transfer would now occasionally compromise the clone, as previously indicated (see "Clones and meroclones," above), but the existence of *extensive* nucleotide polymorphism in two regions of the chromosome presents a real paradox. How could chromosomally local centers of great genetic variation exist in defiance of a sweep of uniformity? The explanation of such "bastions of polymorphism" (Milkman, 1997, 1999) emerges from the interpretation of Reeves (1993) built on a repeated process of selection for new genes (not alleles) transferred to the O-antigen region from phyletically distant sources and rising to high frequency, as described previously. Were this the only event, it would be just another clonal sweep, but two other processes join to produce great variation. First, the new gene is transferred to other strains of the species by normal and relatively frequent *intraspecific* horizontal transfer

via conjugation or transduction, which usually takes along some flanking DNA. This introduces essentially neutral nucleotide variation of the type by which various strains differ. Also, as the most recently arriving gene rises to a sufficiently high frequency for recognition by the host's immune system, its fitness declines to the average level, and the process is ready to be repeated. Note that transfer to a naive host, however, results in a new burst of high fitness until the new host's immune system responds.

Section Summary
- Periodic selection makes vast clones, and horizontal transfer diversifies them.

TARGETS OF SELECTION REVISITED
The unit target of selection may be a population, whose components (down to a given individual) may vary quantitatively in their *response*, which is measured by their *fitness*. Discussion of this subject is simpler in bacteria than in sexual unicellular eukaryotes, and it is much simpler than in promiscuous multicellular eukaryotes. In bacteria, the *absolute fitness* of a given component of a population can be defined as the ratio of its number of survivors plus descendants at a given time, t_{0+g}, to that number at an earlier time, t_0, where g stands for a unit generation time interval. This absolute fitness value can be normalized to a reference value (for example, that of the population), yielding a *relative fitness* (still with the dimension t^{-1}). *Random genetic drift*, the random fluctuation of allele frequencies in a population, is also a rate, and the average time between generations (*generation time*) is a convenient time unit to use when fitness and random genetic drift are considered together as their additive action modifies allele frequencies. Whether positive selection or random genetic drift will prevail is critical to the retention of a horizontally transferred gene.

The preceding summary may be useful for a clear understanding of the role of selection in the *rate of effective replacement*, which, in turn, is essential to understanding horizontal gene transfer over great phyletic distances, which though far less frequent than horizontal gene transfer within a species, is capable of playing a dramatically more powerful role in certain cases. Two of these cases, as we have seen, are those of the hypervariable regions of the *E. coli* chromosome centering, respectively, on the O-antigen region at minute 45 and the restriction endonuclease region near minute 99.

Section Summary
- *Effective replacement rate* is a concept that integrates the consequences of phylogenetic distance, horizontal transfer, and favorable selection.

PROKARYOTIC HORIZONTAL GENE TRANSFER VERSUS EUKARYOTIC RECOMBINATION

Crossing over in *Drosophila*, together with that in maize and in yeast, has long been a familiar example of recombination in eukaryotes. Heterozygosity in a diploid chromosome permits the reciprocal crossing over of particular sets of alleles in meiotic prophase. This is followed by the independent assortment of homologous chromosomes in metaphase and anaphase, leading to the formation of haploid gametes with combinations of alleles that differ from those in the "parental" gametes that united to form the diploid organism. Mating joins two populations of gametes, from which opposite mating types or sexes pair to form diploid zygotes. In each generation, each new diploid genome is thus formed by two half-genomes, each of which is a sample containing half the alleles (one at each locus) in a parental diploid. A great deal of genetic variation is generated in this way by a ritual of common types of events.

The diverse mechanisms of horizontal transfer in bacteria, as described earlier, lead to the incorporation of a relatively small amount of DNA into a recipient genome, generally but not always replacing (i) a similar amount of (ii) homologous DNA. Unlike recombination in eukaryotes, this process is not tied to repro-duction, and there are other major differences as well. Horizontal transfer can cover great phylogenetic distances, though intraspecific transfers are far more frequent, due to more frequent opportunities and specific mechanisms that facilitate the transfer.

The success of an allele in a eukaryotic population where recombination is connected to sexual reproduction depends on the average fitness of a vast number of combinations containing the allele. Numerous unicellular eukaryotes, like yeast; metazoans, like many jellyfish, aphids, and Daphnia; and metaphytes, like *Castilleja* (Indian paintbrush), have an alternation of sexual and asexual generations, so that newly recombinant genotypes are tested as clones rather than as individuals. In this, they are like bacteria, in which strongly favorable transferred genes or alleles are given the opportunity to multiply to a frequency that significantly reduces random genetic drift.

Section Summary
- Genetic exchange in prokaryotes and in eukaryotes differs so greatly as to promise a synthesis of understanding at a higher level. Thank you, Hegel!

THE BIOLOGICAL SPECIES CONCEPT IS NOT APPLICABLE TO BACTERIA, SO WHERE DO WE GO FROM HERE?

In the classification of animals, and plants and other eukaryotes, a critical level of organization is the species. In arriving at conclusions, taxonomists were hindered, to say the least, by the absence of a general and objective species criterion. Such a criterion, the biological species concept, was developed by Ernst Mayr in 1940 (Mayr, 1963, p. 19). Species are, said Mayr in a discussion of species definitions, "groups of actually or potentially interbreeding natural populations which are reproductively isolated from other such groups." Various mechanisms, including behavioral mechanisms, promote intraspecific breeding and discourage interspecific mating. Natural selection operates in favoring the avoidance by an individual of those possible mates that would produce weak, unhealthy, or sterile offspring. Thus, the wastage of gametes would be prevented by effective mating preference, and this would obviously benefit the species. More recently, Mayr (2001) discussed the biological species concept in relation to molecular evolution. Important features of molecular evolution, such as horizontal transfer, present a moving target, however, due to the amazing rapidity of major tangible advances in the description of these features. We have not yet reached the point where we know enough to incorporate molecular evolution into a species concept, but the prospect is exciting.

If bacteria could laugh, they would go into hysterics at the notion of being forced to mate only with their own close relatives or of belonging to one vast domain, *Bacteria*.

The fact, though, is that objectively ascertainable physical similarities and discontinuities exist in bacteria, so that classification is now broadly practical (after some false steps in the past). How have bacteria avoided losing their specific identities? Or perhaps the question is, "Why don't they follow the rules of (eukaryotic) biology?"

This brings us to tree diagrams and to the many current deduced contradictions to a single "Tree of Life," which is a modern expression of Darwin's Doctrine of Common Descent (Darwin, 1859), quoted by Woese (2002) as follows, "[P]robably all of the organic beings which have ever lived on this earth have descended from one primordial form . . . ."

Now, to characterize the foregoing Woese article as presenting "important conceptual advances in biology" (see "Genomic diversity," above) turns out to be an understatement; to read it is heartily recommended. It summarizes Woese's view that horizontal transfer preceded vertical transfer and that horizontal transfer was the engine of evolution in supramolecular aggregates, which preceded cells, whose eventual achievement of sufficient complexity led to heredity and Darwinian evolution. This was the crossing of the "Darwinian Threshold," between what may eventually be seen as several melanges of prebiotic activity and a cell-based evolutionary trajectory that looks like a tree.

Section Summary
- In the next 10 years, the influence of Ernst Mayr may contribute decisively to the incorporation of molecular biology into a lifewide perspective.

QUESTIONS
1. Of what use is growth in pure culture to the characterization of a newly isolated bacterium (as opposed to a fish)? Why are culture-independent methods important in the classification of bacteria? What is a basic example of such a method? How does it work, and how is it applied?

2. Can Mayr's Biological Species Concept be applied to bacteria? What would it take to do this, and where might it lead?

3. What can we conclude from the striking similarity of the Big Ten strains, including K-12? How do they serve as a standard of comparison in the chromosomal distribution of sequence variation?

4. How do small regions of specific non-homologous variation arise in *E. coli* chromosomes?

5. Use Entrez at the National Center for Biotechnology Information (www.ncbi.nlm.nih.gov/) to acquire and apply certain information of your choice.

6. Describe the interplay of selection and random genetic drift in a colon population of *E. coli*.

REFERENCES
Amor, K., D. E. Heinrichs, E. Frirdich, K. Ziebell, R. Johnson, and C. Whitfield. 2000. Distribution of core oligosaccharide types in lipopolysaccharides from *Escherichia coli. Infect. Immun.* **68:**1116–1124.

Barcus, V. A., J. B. Titheradge, and N. E. Murray. 1995. The diversity of alleles at the *hsd* locus in natural populations of *Escherichia coli. Genetics* **140:**1187–1197.

Berlyn, M. K. B. 1998. Linkage Map of *Escherichia coli* K-12, edition 10: the traditional map. *Microbiol. Mol. Biol. Rev.* **62:**814–984.

Box, G. E. P., and M. E. Muller. 1958. A note on the generation of random normal deviates. *Ann. Math. Stat.* **29:**610–611.

Crameri, A., S.-A. Raillard, E. Bermudez, and W. P. C. Stemmer. 1998. DNA shuffling of a family of genes from diverse species accelerates directed evolution. *Nature* **391:**288–291.

Crow, J. F., and M. Kimura. 1970. *An Introduction to Population Genetics Theory.* Harper & Row Publishers, New York, N.Y.

Hayes, J. M. 2000. Lipids as a common interest of microorganisms and geochemists. *Proc. Natl. Acad. Sci. USA* **97:**14033–14034.

Kudva, I. T., P. S. Evans, N. T. Perna, T. J. Barrett, F. M. Ausubel, F. R. Blattner, and S. B. Calderwood. 2002. Strains of *Escherichia coli* O157:H7 differ primarily by insertions or deletions. *J. Bacteriol.* **184:**1873–1879.

Lan, R., and P. R. Reeves. 2000. Intraspecies variation in bacterial genomes: the need for a species genome concept. *Trends Microbiol.* **8:**396–401.

Li, Q., and P. R. Reeves. 2000. Genetic variation of dTDP-l-rhamnose pathway genes in *Salmonella enterica*. *Microbiology* **146:**2291–2307.

Milkman, R. 1978. Selection differentials and selection coefficients. *Genetics* **88:**391–403.

Milkman, R. 1999. Gene transfer in *Escherichia coli*, p. 291–309. *In* R. L. Charlebois (ed.), *Organization of the Prokaryotic Genome*. American Society for Microbiology, Washington, D.C.

Milkman, R., and M. McKane Bridges. 1990. Molecular evolution of the *Escherichia coli* chromosome. III. Clonal frames. *Genetics* **126:**505–517.

Milkman, R., E. A. Raleigh, M. McKane, D. Cryderman, P. Bilodeau, and K. McWeeny. 1999. Molecular evolution of the *Escherichia coli* chromosome. V. Recombination patterns among strains of diverse origin. *Genetics* **153:**539–554.

Milkman, R., E. B. Jaeger, and R. D. McBride. 2003. Molecular evolution of the *Escherichia coli* chromosome. VI. Two regions of high effective recombination. *Genetics* **163:**475–483.

Ochman, H., and R. K. Selander. 1984. Standard reference strains of *E. coli* from natural populations. *J. Bacteriol.* **157:**690–693.

Perna, N. T., G. Plunkett III, V. Burland, B. Mau, J. D. Glasner, D. J. Rose, G. F. Mayhew, P. S. Evans, J. Gregor, H. A. Kirkpatrick, G. Posfai, J. Hackett, S. Klink, A. Boutin, Y. Shao, L. Miller, E. J. Grotbeck, N. W. Davis, A. Lim, E. T. Dimalanta, K. D. Potamousis, J. Apodaca, T. S. Anantharaman, J. Lin, G. Yen, D. C. Schwartz, R. A. Welch, and F. R Blattner. 2001. Genome sequence of enterohaemorrhagic *Escherichia coli* O157:H7. *Nature* **409:**529–533.

Reeves, P. 1993. Evolution of *Salmonella* O antigen variation by interspecific gene transfer on a large scale. *Trends Genet.* **9:**17–22.

Rudd, K. E. 1998. Linkage map of *Escherichia coli* K-12, edition 10: the physical map. *Microbiol. Mol. Biol. Rev.* **62:**985–1019.

Spratt, B. G. 1999. Multilocus sequence typing: molecular typing of bacterial pathogens in an era of rapid DNA sequencing and the internet. *Curr. Opin. Microbiol.* **2:**312–316.

Vaisvila, R., R. D. Morgan, J. Postal, and E. A. Raleigh. 2001. Discovery and distribution of superintegrons among pseudomonads. *Mol. Microbiol.* **42:**587–601.

Woese, C. R. 1987. Bacterial evolution. *Microbiol. Rev.* **51:**221–271.

Woese, C. R. 2002. On the evolution of cells. *Proc. Natl. Acad. Sci. USA* **99:**8742–8747.

FURTHER READING

Atwood, K. C., L. K. Schneider, and F. J. Ryan. 1951. Selective mechanisms in bacteria. *Cold Spring Harbor Symp. Quant. Biol.* **16:**345–355.

Avise, J. 1989. Gene trees and organismal histories: a phylogenetic approach to population biology. *Evolution* **43:**1192–1208.

Blattner, F. R., G. Plunkett III, C. A. Bloch, N. T. Perna, V. Burland, M. Riley, J. Collado-Vides, J. D. Glasner, C. K. Rode, G. F. Mayhew, J. Gregor, N. W. Davis, H. A. Kirkpatrick, M. A. Goeden, D. J. Rose, B. Mau, and Y. Shao. 1997. The complete genome sequence of *Escherichia coli*. *Science* **277:**1453–1474.

Crawford, I. P. 1989. Evolution of a biosynthetic pathway: the tryptophan paradigm. *Annu. Rev. Microbiol.* **43:**567–600.

Darwin, C. 1859. *On the Origin of Species*, p. 484. Harvard University Press, Cambridge, Mass.

Dykhuizen, D. E., and L. Green. 1991. Recombination in *Escherichia coli* and the definition of biological species. *J. Bacteriol.* **173:**7257–7268.

Hall, R. M., and C. M. Collis. 1995. Mobile genetic cassettes and integrons: capture and spread of genes by site-specific recombination. *Mol. Microbiol.* **15:**593–600.

Heinemann, J., and G. Sprague. 1989. Bacterial conjugative plasmids mobilize DNA transfer between bacteria and yeast. *Nature* **340:**205–209.

Herzer, P. J., S. Inouye, M. Inouye, and T. Whittam. 1990. Phylogenetic distribution of branched RNA-linked multicopy single-stranded DNA among natural isolates of *Escherichia coli*. *J. Bacteriol.* **172:**6175–6181.

Hobbs, M., and P. R. Reeves. 1994. The JUMPstart sequence: a 39 bp element common to several polysaccharide gene clusters. *Mol. Microbiol.* **12:**855–856.

Kohara, Y., K. Akiyama, and K. Isono. 1987. The physical map of the whole *E. coli* chromosome: application of a new strategy for rapid analysis and sorting of a large genomic library. *Cell* **50**:495–508.

Madigan, M. T., J. M. Martinko, and J. Parker. 2002. *Brock's Biology of Microorganisms*, 10th ed. Prentice-Hall, Upper Saddle River, N.J.

Masters, M. 1996. Generalized transduction, p. 2421–2441. *In* F. C. Neidhardt (ed.), *Escherichia coli and Salmonella: Cellular and Molecular Biology*. American Society for Microbiology, Washington, D.C.

Mayr, E. 1963. *Animal Species and Evolution*, p. 19. The Belknap Press of Harvard University Press, Cambridge, Mass.

Mayr, E. 2001. *What Evolution IS*. Basic Books, New York, N.Y.

McKane, M., and R. Milkman. 1995. Transduction, restriction and recombination patterns in *Escherichia coli*. *Genetics* **139**:35–43.

Milkman, R. 1973. Electrophoretic variation in *Escherichia coli* from diverse natural sources. *Science* **182**:1024–1026.

Milkman, R. 1996. Recombinational exchange among clonal populations, p. 2663–2684. *In* F. C. Neidhardt (ed.), *Escherichia coli and Salmonella: Cellular and Molecular Biology*. American Society for Microbiology, Washington, D.C.

Milkman, R. 1997. Recombination and population structure in *Escherichia coli*. *Genetics* **146**:745–750.

Salyers, A. A., and D. D. Whitt. 2001. *Microbiology*. Fitzgerald Science Press, Bethesda, Md.

Selander, R. K., J. Li, and K. Nelson. 1996. Evolutionary genetics of *Salmonella enterica*, p. 2691–2707. *In* F. C. Neidhardt (ed.), *Escherichia coli and Salmonella: Cellular and Molecular Biology*. American Society for Microbiology, Washington, D.C.

Smith, L. M., R. J. Kaiser, J. Z. Sanders, and L. E. Hood. 1987. The synthesis and use of fluorescent oligonucleotides DNA sequence analysis. *Methods Enzymol.* **155**:260–301.

Whittam, T. S. 1996. Genetic variation and evolutionary processes in natural populations of *Escherichia coli*, p. 2708–2720. *In* F. C. Neidhardt (ed.), *Escherichia coli and Salmonella: Cellular and Molecular Biology*. American Society for Microbiology, Washington, D.C.

APPENDIX

Instructions for Simulation

First, go to the website (http://jbpc.mbl.edu/Pages/milkman/book.html). You can use the text and download an Excel table to work on. The text appearing below can be used in parallel with the website. Note, for example, that the next paragraph uses an initial fitness, w_a, of 1.5 for simplicity, while the downloaded table starts with $w_a = 1.25$.

This is a Microsoft Excel simulation of the change in the frequency of a gene, a, in a population, beginning with its arrival in a single horizontal transfer and the replacement, to some extent, of an analogous gene that makes an alternate component of the O antigen. As a first example, the *E. coli* cell bearing the new, immunologically unrecognized gene is assigned an initial fitness, w_a, of 1.5. The population's size (here set at 10,000) is constant, and, thus, the mean population fitness, $W = pw_a + qw_b$, has the value 1. The new gene, a, starts with a relative frequency, p, of $1/10,000 = 0.0001$, and the population's mean fitness value remains at 1. The initial selective advantage, s_a, of a is $w_a - W = 1.5 - [(0.0001)1.5 + (0.9999)1] = 1.5 - (0.00015 + 0.99990) = 1.5 - 1.00005 = 0.49995$. Between generation 0 and generation 1, p changes by the sum of Δp_{sel} (due to selection) and Δp_{rgd} (due to random genetic drift). The value of p changes due to both processes in each successive generation, and the simulation can be set to operate iteratively for up to 50 generations or more, for example. By changing the current initial random normal number (each drawn as a simple random number by Excel and then converted each time into a random normal number with Box-Muller equations [see Spratt, 1999]), you can repeat the simulation with a different 50-generation trajectory several times (say, 100). Now, changing the initial value of w_a, you can get such sets of 100 simulations each for fitnesses of 2.00, 1.75, 1.60, 1.50, 1.40, 1.25, 1.20, 1.10, 1.05, 1.03, and 1.01, for example. We ask, "What are the respective frequencies of retention of gene a in each set?" As you do these simulations, it will become clear that at the high values of w_a, random genetic drift must strike fast to eliminate the newcomers, otherwise their relative frequency, p, will rise inexorably to 1, because of powerful selection. At the lower values of w_a that still exceed 1.00, the contest between selection and random genetic drift can go on for a long time; even in a population as small as 10,000, the *effective retention* of gene *a* (i.e., a very small probability of its eventual extinction) does not require *p* to get very high.

Setting Up the Active Table ("Worksheet") in Windows

Begin by opening Excel (double-click the "green X" icon on the desktop or in the task bar at the bottom of the screen), so that an arrangement of rows and columns appears. In the File menu use Page Setup to get Landscape orientation. Set the top and bottom margins at 0.75″ and 0.5″, respectively; set the left and right margins at 0.5″ each to accommodate columns A through M at their necessary widths. For this, set the cursor on the border between column heads (e.g., E and F), and drag the border to the right or left. (It's useful to have an Excel user at hand. This is a situation where the instant voice of experience is eminently superior to the available guidebooks.)

To type the headings in their respective cells, click on Format, then Font. In the appropriate boxes, select "Arial," "regular," and "10." Use italics or not, as you like. Columns B, C, and D require no special treatment, but columns E, F, G, H, and M contain subscripts. These will appear in the title cells as w_a in E2, but *wa* in the formula bar. Column J has σ (but s in the formula bar). The headings in I2 (Δp_{sel}) and L2 (Δp_{rgd}) have both symbols and subscripts. For L2, click on Format, then Font. Select Arial and dive down to Symbol, which you select also. Click OK. Type "D." Return to Format and Font, dive down to Symbol and deselect it. Click OK, and type "p." Finally, return to Format and Font, check "subscript," click OK and type "rgd." The formula bar will say "Dprgd."

In row 1 of column A (i.e., in cell A1), enter *N* for population size. In row 2 of column A (cell A2), enter 10,000. This value remains constant over the entire simulation, and it is used in the formula bar as $A$2, which denotes the constant value in cell A2.

The remaining columns, B through M, contain descriptive headings in row 2 and values in the subsequent rows. These headings are

as follows: under B, *g* (for generation); under C, *p* (frequency of gene *a*); under D, *q* (= 1 − *p*), the frequency of gene *b*; under E, w_a (fitness of *E. coli* cells containing gene *a*); under F, w_b (fitness of *E. coli* cells containing the various immunologically recognized analogs of gene *a*); under G, $pw_a + qw_b$ (= mean fitness); under H, $w_a - w_b$ (= selective advantage of the *E. coli* cells containing gene *a* relative to those containing the analogs of gene *a*); under I, Δp_{sel} (change in the value of *p* due to selection); under J, σ (standard deviation ≡ square root of the binomial variance of *p*); under K, *norand* (random number drawn from a *normal distribution*); under L, Δp_{rgd} (change in the value of *p* due to random genetic drift); and, finally, under M, p_{i+1} (*p* as of one generation later than that in the same row of column B).

The initial values in row 3 are either set (as 0 for *g* in column B) or calculated. Using the Formula Bar, one simply clicks on (selects for) cell B3, enters 0 in the Formula Bar space, and clicks on the green check mark ("Enter"). For cell B4, one enters "=1+B3". *The entered equal sign directs the calculation of a value*, in this case "1 plus the value in cell B3", or 1 + 0. To prepare for the completion of column B, click on cell B4 and move the cursor to its lower right corner, where it will produce a plus sign. Drag the plus sign straight down through cell B53; this results in a progression of values (generation numbers), each greater by 1 than the last. For column C, begin row 3 by typing "=1/$A$2" and hit the check. This will produce the value 0.0001. Now subsequent *p* values are obtained by typing "=ROUND(M3,4)" (for cell C4) and extending the progression by dragging. The value in cell M3 is rounded to four decimal places in cell C4 (and the M4 value will be similarly rounded in C5, and so on).

(Rounding is critical to accurate simulation of changes in gene frequency. The unit variables are whole genes, not mathematical artifacts [fragments], and failure to round distorts the results, as you can easily show by typing "=M3" for cell C4, and so on. The same principle applies to changes in *allele* frequencies, the far more common focus of population genetics.)

You can also set the population size at one billion instead by entering 1,000,000,000 in cell A2 and round to one-bacterium units by directing A4 with "=ROUND(M3,9)".

So far we have set column C to utilize M values that will result from operations on values in columns C through L of the same generation, but the intermediate directions have yet to be given. For column D, the Formula Bar begins with "=1−C3", and this gives the current value of q, the relative frequency of E. coli cells not containing gene a. For cell E3 we use the formula "=1+D3*(0.50)". (Note the arrival of a new symbol: "*" is the Excel multiplication sign.) This formula results in an initial value for w_a just under 1.50. It will decrease progressively in column E as the fitter cells, which bear gene a, become more numerous and make the competition tougher. The C column numbers change progressively with dragging, as do the D and E column numbers. Only $'s keep the cell numbers from changing, and in the present simulation the $'s are used only in cell A2.

To set other initial fitness values in cell E3 for respective simulation programs, use 1+D3*(1) (= 1+D3) for 2.00; similarly, 1+D3*(0.75) for 1.75; 1+D3*(0.60) for 1.60; 1+D3*(0.40) for 1.40; 1+D3*(0.25) for 1.25; 1+D3*(0.20) for 1.20; 1+D3*(0.10) for 1.10; 1+D3*(0.05) for 1.05; 1+D3*(0.03) for 1.03; and 1+D3*(0.01) for 1.01. These can be labeled and stored separately, as you like, for example as RS-1.75.xls. (R for round and S for small [here 10,000]). Proceeding from one such program file, you can make and save a family of program files differing only by w_a values by resetting column E. To change the initial value of w_a to make a new set, click on cell E3 and enter a new value once in the formula bar, and drag from cell E4 down through cell E53.

Column F utilizes the constancy of the population fitness, which is held to a value of 1 by the constancy of the population's size. Thus $pw_a + qw_b = 1$, and we can solve this equation for w_b, as follows: $1 − pw_a = qw_b$, and so $w_b = (1 − pw_a)/q$. This requires the following entry for cell F3 in the Formula Bar: =(1−C3*E3)/D3.

The Formula Bar entry for cell G3 is "=C3*E3+D3*F3". We enter the value by clicking on the green check mark and drag down to row 53 as before. The constant value 1 is reassuring, as is the constancy in column H. Cell H3's value is directed by "=E3 − F3". Cell I3, the change in p during generation 1 due to selection, is determined by $pw_a − p$ and expressed in the Formula Bar space by "=C3*E3−C3". It may be useful at this point to make explicit the fact that the association of 3's with generation 0 is merely the correspondence between row number and generation number as the table of values has been assembled.

To determine the change in p due to random genetic drift requires the product of a random normal number and the standard deviation of p as it is sampled randomly. This standard deviation, represented by σ, is calculated on the basis of a binomial distribution, which approximates a normal distribution when the population size, N, is large. The initial value of σ is estimated as $(p_0 q_0/N)^{\frac{1}{2}}$ and directed by the Formula Bar as (C3*D3/$A$2)^0.5. The product of σ and a random normal number is the change in p due to random genetic drift, Δp_{rgd}.

To set the value of Δp_{rgd} during the first generation, obtain an initial random normal number by clicking on the top cell below the title in column K. The box, now outlined heavily, is the active cell. Above, in the formula window, type: =((−2*LN(RAND()))^0.5)*(COS(2*PI()*RAND())). In Excel, PI stands for π, and the empty closed parentheses indicate that PI is a number, not a function; that is, it does not need an "argument" to operate on. RAND is also followed by a closed pair of parentheses, signifying the same thing. In this case, Excel reaches into its vast hoard of ordinary (uniformly distributed) random numbers (designated u) and picks one. The other parentheses are used as in normal algebra. Note that LN and COS are functions: we use "log of something" or "cosine of something". The entire expression is: $(−2 \ln u_2)^{\frac{1}{2}} [\cos (2\pi u_1)]$, which equals x_1, a random normal number.

New random numbers are automatically served up by Excel in successive cells of the column. To change the column of random normal numbers, click on a cell above the column K heading, *norand*, and then double click on the same cell. Then click on the top cell containing a number in column K: all the numbers in the column will change. Subsequent replacements are easier in most setups: just press your F9 button for each new replacement. All the numbers in column L, under Δp_{rgd}, are the products of the random normal numbers that are multiplied by σ (the binomial standard deviation of p). Now 100 sets of, say, 50 cumulative changes each, will provide 100 trajectories. Use a special notebook to write these records. Each time you make a new trajectory, you can enter a Retention (p is still >0 at generation 50) or a Loss (p stops the process by reaching a value at or below 0 in column C at the generation displayed in column B). For convenience, record Retentions in groups of five in a line as they occur, as R: | +++++ +++++ | +++++ +++ . . . . , and record the number of generations to each Loss, as L: | 42617 − 13312 |

23(17)16 − 421 . . . . [Note that "(17)" is "seventeen."] Count the total cases of R and L in each set, and calculate %R and the average number of generations in each L trajectory of each set. In a set of 100 trajectories with 82 retentions and 18 losses, "Av. gen. (L)" = 66/18 = 3.67 (based on the 18 numbers above).

(For a slightly different approach to the typography, to get symbols such as Δ and σ in Excel, type D or s in the appropriate cells [D in top cell of column I, s in top cell of column J, and later D in top cell of L], each time highlighting the same letter in the Formula Box. To get Δ, highlight D, click on "Format" and choose "Cells". Then scroll way down in the "Font" window and choose "Symbol". Δ will now appear in the top cell of column I [or L]; similarly, starting with s, σ will appear in the top cell of column J. Also, you can make a subscript or superscript after having chosen "Cells" by clicking on the appropriate box instead of scrolling down in the Font window.)

HORIZONTAL GENE TRANSFER AND PROKARYOTIC GENOME EVOLUTION

I. King Jordan and Eugene V. Koonin

20

DETECTION OF HORIZONTAL TRANSFER
Phylogenetic and Distance-Based Methods
Surrogate Methods Based on Nucleotide Composition

**IMPLICATIONS OF
HORIZONTAL TRANSFER**
Quantification and Classification
Functional Impact

**SUMMARY OF THE CONTRIBUTION
OF GENE TRANSFER TO GENOME
EVOLUTION**

Horizontal (lateral) gene transfer can be defined as the transmission of genes across species boundaries. In the pregenomics era, the extent and evolutionary significance of horizontal gene transfer were the subject of an intense debate. Most contentious was the question of whether evidence of horizontal gene transfer pointed to a major evolutionary phenomenon or whether it simply represented a collection of inconsequential anecdotes, exceptions that proved the rule of vertical transmission. The answer to this question has fundamental implications for the understanding of the pattern and process of evolution. If it proves to be the case that a substantial fraction of the genes in each genome has been acquired via horizontal gene transfer, the traditional, tree-based view of the evolution of life must be considered incomplete or even dubious. Prior to the genome-sequencing era, striking anecdotal examples of horizontal gene

transfer were described and prescient speculation on the potential major evolutionary impact of such events was published. However, the prevailing view seemed to hold that horizontal transfer events were rare enough not to affect our general understanding of evolution. The only exception to this was the acceptance of the endosymbiotic theory of the origin of organelles. In this case, the impact of horizontal gene transfer was underscored by the apparent massive flow of genes from the genomes of endosymbiotic organelles, mitochondria in all eukaryotes and chloroplasts in plants, to the eukaryotic nuclear genome (Gray, 1999).

More recently, comparative analyses of complete prokaryotic genome sequences have shown that horizontal gene transfers have been too common to be dismissed as inconsequential. An early indication of the extent of prokaryotic horizontal gene transfer came from the multifactorial analysis of codon frequencies in portions of the *Escherichia coli* genome that revealed significant deviations from the general pattern of codon usage in approximately 15% of this bacterium's genes. It was proposed that all the deviant genes, many of which had clear affinities to bacteriophage genes, were "alien" to the *E. coli* genome and had been acquired horizontally from various sources. This observation seemed to strongly support substantial and relatively recent horizontal gene flow among bacteria. In addition, the possibility of numerous more ancient horizontal transfers has been suggested by the lack of congruence between phylogenetic trees for different sets of orthologous genes

I. King Jordan and Eugene V. Koonin, National Center for Biotechnology Information, National Library of Medicine, National Institutes of Health, Bethesda, MD 20894.

Microbial Evolution: Gene Establishment, Survival, and Exchange
Edited by Robert V. Miller and Martin J. Day, ©2004 ASM Press, Washington, D.C.

from a wide range of organisms. For example, it was found that some archaeal genes showed a clear affinity to their eukaryotic counterparts, in agreement with the topology of the tree of life derived from ribosomal RNA sequences, whereas others clustered with bacterial homologs (Golding and Gupta, 1995).

A new age of "lateral genomics" was ushered in through the comparison of multiple, complete prokaryotic genomes. Such studies revealed major differences in gene repertoires even among bacteria that belong to the same evolutionary lineage, such as *E. coli* and *Haemophilus influenzae*. This clearly indicated that prokaryotic genome evolution could not be reasonably described in terms of vertical descent alone and much of the difference between genomes is attributable to lineage-specific gene loss and horizontal gene transfer. Comparative analyses of archaeal genomes presented particularly striking evidence that strongly suggests massive horizontal gene exchange with bacteria. Consistent with earlier phylogenetic studies, but now on the whole-genome scale, most archaeal proteins could be classified into two groups based on their phylogenetic affinities. The majority of archaeal proteins was most similar to bacterial homologs, while a minority was more similar to eukaryotic homologs (Koonin et al., 1997; Makarova et al., 1999). The "bacterial" and "eukaryotic" proteins in archaea were, for the most part, divided along functional lines. Archaeal proteins involved in information processing (translation, transcription, and replication) showed a eukaryotic affinity, while archaeal metabolic enzymes, structural components, and a variety of uncharacterized proteins clustered with bacterial proteins. Because the informational components generally appear to be less subject to horizontal gene transfer (however, some important exceptions are discussed below), these observations have been tentatively explained by massive horizontal exchange of operational genes between *Archaea* and *Bacteria*. This hypothesis is in accord with the "standard model" of early evolution, whereby eukaryotes share a common ancestor with *Archaea*, and was further supported by the result of genome analysis of two hyperthermophilic bacteria, *Aquifex aeolicus* and *Thermotoga maritima*. Each of these genomes was found to contain a significantly greater proportion of "archaeal" genes than any of the other bacterial genomes. Thus, a startling correlation exists between the similarity in the lifestyles of very distantly related organisms (bacterial and archaeal hyperthermophiles) and the apparent rate of horizontal gene exchange between them (Aravind et al., 1998). These findings also raised an interesting question with respect to the mode of horizontal transfer, namely whether it is adaptive or simply opportunistic in nature. Did the genes that were horizontally acquired from archaea facilitate the adaptation of *Aquifex* and *Thermotoga* to hyperthermal conditions, or have these particular bacteria acquired more archaeal genes than others simply because they share an environmental niche with many thermophilic archaea?

Genomic era revelations of the widespread contributions of horizontal gene transfer to the gene composition of prokaryotic genomes amounted to a major shift in our understanding of evolution. Indeed, it became apparent that, in many cases, genuine differences in the evolutionary histories of genes caused by horizontal transfer, as opposed to artifacts of tree-construction methods, were responsible for the commonly occurring reconstruction of different phylogenetic trees for different genes. Thus, in principle, it may simply be impossible to reconstruct a true bifurcating tree of life that accurately represents prokaryotic species evolution. It may, however, be possible to generate a consensus tree that depicts the evolution of a core of information processing genes that is conserved in all or the majority of species and is not subject to horizontal gene transfer. But the very existence of such a resilient core of genes, and more so its actual delineation, remain questionable.

Perhaps the high incidence of horizontal gene transfer in prokaryotes should not have come as a complete surprise. As demonstrated by the classic Avery-McLeod-McCarthy experiment of 1943 that proved the hereditary role of DNA, microbes have the ability to absorb DNA from the environment and integrate it into the genome. Indeed, high transformability has been demonstrated for many microbial species. In addition, bacteriophages and plasmids, some of which readily cross species barriers, provide highly effective potential vectors for horizontal gene transfer. Given that microbes typically coexist in tightly knit communities, such as microbial mats and the microflora of animal guts, opportunities should abound for DNA transfer by various means between diverse prokaryotes.

Despite recent findings based on the comparative genomics approach, the question of whether horizontal gene transfer is a major evolutionary force remains highly contested. One reason for this is the fact that if horizontal transfer were as widespread as suggested by comparative genomic studies, this would demand, on some level, a paradigm shift in evolutionary biology. Such a radical change in the biological world-view will necessarily be greeted with a healthy dose of skepticism from the scientific community. Another, perhaps more daunting problem, lies with the methodology for inferring horizontal transfer events. While the general significance of horizontal transfer seems to ensue from whole-genome comparisons, proving many individual cases beyond reasonable doubt often proves to be difficult if not impossible.

In this chapter, we discuss the methodology used to infer horizontal gene transfer events and discuss examples of such transfers among prokaryotes with an emphasis on adaptive and biological implications that they entail.

DETECTION OF HORIZONTAL TRANSFER

The various methods used for the detection of horizontal transfer events all rely on the characterization of some anomalous or unusual genomic subset of genes. This subset of horizontally transferred genes is distinguished from the majority of genes in a genome that are presumed to be derived from vertical descent. There are two generic classes of horizontal transfer detection methods: (i) phylogenetic or distance-based methods and (ii) the so-called "surrogate" methods based on the nucleotide composition of genes. Methods of each of these classes have their own strengths and liabilities. What's more, the application of different methods to a particular set of genomic data can yield quite different results (Lawrence and Ochman, 2002). Thus, it is often exceedingly difficult to obtain unequivocal proof of horizontal transfer for any particular gene. The inference of horizontal transfer is an essentially probabilistic endeavor. The goal of the detection mechanisms described below is to maximize the likelihood of correct identification of horizontal transfer events. An understanding of the rationale on which these methods are based, as well as the factors that can confound their application and interpretation, is critical to this end.

Phylogenetic and Distance-Based Methods

PHYLETIC PATTERNS

Orthologous genes are direct evolutionary counterparts (homologs) related by vertical descent. Comparative analysis of sequence data from completely sequenced genomes allows for the delineation of clusters of orthologs (Tatusov et al., 1997). Given a set of species and an orthologous cluster of genes, one can determine a phyletic (phylogenetic) pattern, which is simply the pattern of species that are present or missing in the cluster. Using this approach with the Clusters of Orthologous Groups of proteins (COGs) database revealed an extreme diversity of phyletic patterns. That the vast majority of phyletic patterns include only a small number of genomes (Fig. 1) indicates that horizontal gene transfer, as well as lineage-specific gene loss, has contributed extensively to the evolution of prokaryotic genomes. Certain types of phyletic patterns are strongly suggestive of horizontal transfer (Table 1). For example, when a set of orthologs shows

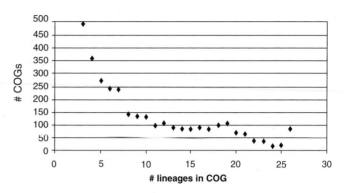

FIGURE 1 Distribution of the Clusters of Orthologous Groups of proteins (COGs) by the number of represented species. Each COG includes predicted orthologs from at least three genomes that belong to 26 distinct lineages. The number of COGs (y axis) is plotted against the number of phylogenetic lineages represented per COG (x axis). Adapted from Koonin et al. (2001).

the presence of a typical "archaeal-eukaryotic" protein in a single bacterial lineage, the odds for horizontal gene transfer underlying this pattern seem to be high. The B-family DNA polymerase (*E. coli* DNA polymerase II) is a clear-cut case of such obvious horizontal gene acquisition. There is additional evidence that bolsters this case, including the similarity between gammaproteobacterial Pol II sequences and their archaeal and eukaryotic orthologs and a plausible implication of bacteriophage and/or viruses as vectors that could have facilitated the interdomain horizontal transfer.

SEQUENCE SIMILARITY AMONG HOMOLOGS

Whenever a gene from one species encodes a protein with the strongest similarity to a homolog from a distantly related species, horizontal transfer can be suspected. For example, when all protein sequences encoded in a bacterial genome are compared (e.g., using the BLAST program) to the entire protein database, a certain fraction of proteins will show the greatest similarity to eukaryotic homologs, rather than to those from other bacteria. These genes are likely candidates for horizontal gene exchange between the given bacterium (or its evolutionary lineage) and eukaryotes. The strength of the claim depends on the difference between the levels of sequence similarity for the homologs from the relatively distantly related (eukaryotes) and more closely related (bacterial) taxa. Evidence from sequence comparisons is indispensable for obtaining a genomewide picture of probable horizontal transfers. However, such results are preliminary and need to be interpreted with extreme caution. Additional phylogenetic analysis is necessary for the validation of each individual case.

PHYLOGENETIC TREE TOPOLOGY

Analysis of phylogenetic tree topologies is the most traditional and perhaps most robust way to infer horizontal gene transfer events. For example, if a bacterial protein groups (in a well-supported group) with eukaryotic homologs, to the exclusion of homologs from other bacteria, and shows a reliable affinity with a particular eukaryotic lineage, the evidence of horizontal gene transfer seems to be strong. Unfortunately, phylogenetic analysis does not offer such clear-cut solutions in many suspected cases of horizontal gene transfer. In addition, there are many sources of potential error in phylogenetic analysis. The problem of long-branch attraction is particularly relevant for the detection of horizontal transfers because these events may be accompanied by accelerated evolution, resulting in long branches in phylogenetic trees. It is also critical that the tree topology is strongly supported statistically by bootstrap analysis and/or likelihood estimates for different topologies. Unfortunately, many gene families seem to have undergone "star-like" evolution, with very short internal branches. In such cases, phylogenetic analysis is useless for verifying the candidate horizontal transfer events because of the uncertainty in the tree. From a pragmatic

TABLE 1 Probable horizontal gene transfers detected using phyletic patterns[a, b]

COG no.	Function	Occurrence in complete bacterial genomes	Occurrence in complete archaeal and eukaryotic genomes
From archaea or eukaryotes to bacteria			
0417	DNA polymerase, B family	Ec, Pa, Vc	All (also many viruses and bacteriophages)
0430	RNA phosphate cyclase	Ec, Pa, Aa	All
0467	KaiC-like ATPase of RecA superfamily (implicated in signal transduction)	Ssp, Aa, Tm	All
0615	Predicted cytidylyltransferase	Aa, Bs	All
1257	Hydroxymethylglutaryl-CoA reductase	Vc, Bb	All
1577	Mevalonate kinase	Bb	All
2519	Predicted S-adenosylmethionine-dependent methyltransferase involved in tRNA^{Met} maturation	Aa, Mtu	All
From bacteria to archaea and eukaryotes			
0847	DNA polymerase III, epsilon subunit/domain (3′–5′ exonuclease)	All except Bb, mycoplasmas	Af, Sc, Ce, Dm
0566	rRNA methylase	All	Af, Sc, Ce, Dm
0188	DNA gyrase (topoisomerase II) A subunit	All except Af, Sc, Ce, Dm	
0187	DNA gyrase (topoisomerase II) B subunit	All	Af, Sc, Ce, Dm
0138	AICAR transformylase/IMP cyclohydrolase (PurH)	All except Hp, Rp, spirochetes, chlamydia, mycoplasmas	Af, Sc, Ce, Dm
0807	GTP cyclohydrolase II (riboflavin biosynthesis)	All except Rp, spirochetes, mycoplasmas	Af, Sc

[a] Species name abbreviations in this and subsequent tables: Aa, *Aquifex aeolicus*; Af, *Archaeoglobus fulgidus*; Ap, *Aeropyrum pernix*; Bb, *Borrelia burgdorferi*; Bh, *Bacillus halodurans*; Bs, *Bacillus subtilis*; Ce, *Caenorhabditis elegans*; Cj, *Campylobacter jejuni*; Cp, *Chlamydia pneumoniae*; Ct, *Chlamydia trachomatis*; Dm, *Drosophila melanogaster*; Dr, *Deinococcus radiodurans*; Ec, *Escherichia coli*; Hi, *Haemophilus influenzae*; Hp, *Helicobacter pylori*; Mth, *Methanobacterium thermoautotrophicum*; Mtu, *Mycobacterium tuberculosis*; Nm, *Neisseria meningitidis*; Pa, *Pseudomonas aeruginosa*; Rp, *Rickettsia prowazekii*; Sc, *Saccharomyces cerevisiae*; Ssp, *Synechocystis* sp.; Tm, *Thermotoga maritima*; Tp, *Treponema pallidum*; Vc, *Vibrio cholerae*; Uu, *Ureaplasma urealyticum*; Xf, *Xylella fastidiosa*. AICAR, 5-aminoimidazole-4-carboxamide-1-β-4-ribofuranoside.

[b] Adapted from Koonin et al. (2001).

perspective, phylogenetic analysis is time- and labor-consuming because it depends critically on correct sequence alignments and is hard to automate without compromising the robustness of the results. However, an attempt was made to automatically construct the complete sets of phylogenetic trees for seven prokaryotic genomes and to systematically compare the topologies in search of horizontal transfer events. The results seem to be coherent with those produced by analysis of unexpected distribution of sequence similarity and reveal significant amounts of horizontal transfer (see also below). More recent attempts at high-throughput phylogenetic analysis using distance and likelihood methods have gone even further in revealing genomewide patterns of horizontal transfer. Despite the inherent challenges in automating phylogenetic tree reconstruction, this approach appears to be suitable for uncovering preliminary evidence and revealing major trends in horizontal transfer.

CONSERVATION OF GENE ORDER BETWEEN DISTANT TAXA

Comparative genomic analyses have revealed very little conservation of gene order between distantly related genomes; apparently, evolution of bacterial and archaeal genomes involved extensive genome rearrangements. From this, it is taken that the presence of three or more genes in the same order in distant genomes is extremely unlikely unless these genes form an operon (Wolf et al., 2001). In addition, any given operon typically emerges only once in the course of evolution and is subsequently maintained by selection. Based on the principle of parsimony, horizontal transfer is invoked to explain the distribution of an operon when it is found only in a few distantly related genomes. An even stronger indication of horizontal transfer can be made if phylogenetic analysis confirms that genes comprising the operon form monophyletic groups in their respective phylogenetic trees. For example, there is extremely compelling evidence that points to a role for horizontal transfer in distributing operons that encode restriction-modification systems (Naito et al., 1995), and a number of other functionally distinct operons also appear likely to have been spread by horizontal transfer. The archaeal-type H^+-ATPase operon is a well-characterized example of such horizontal dissemination of an operon among bacteria.

Surrogate Methods Based on Nucleotide Composition

The term "surrogate methods" was coined to describe methods of detecting horizontally transferred genes based on their anomalous nucleotide composition (Ragan 2001). This class of methods is both well known and widely applied despite the fact that it has a relatively narrow utility (i.e., it is applicable only to recent horizontal transfers) and is fraught with difficulties (see discussion below). In general, these approaches are based on the idea that each genome has a characteristic pattern of GC content and codon usage that can be taken as a "signature" of the genome. Genes whose nucleotide or codon composition differ significantly from the mean signature for a given genome are considered to be probable horizontal acquisitions. Invariably, when these methods are applied to prokaryotic genomes, a significant fraction, up to 15% to 20% of the genes, is inferred to be derived from recent horizontal transfer. Not surprisingly, many of the horizontally transferred genes revealed by this approach are prophages, transposons, and other mobile elements, for which this mode of evolution is to be expected.

Recently, close inspection of results of the application of surrogate methods to genome sequences has fundamentally challenged their accuracy (Wang, 2001). Different surrogate methods often fail to detect a common set of putatively horizontally transferred genes for a given genome. For example, when four different surrogate methods were applied to the same genomes, they each detected vastly different sets of genes that in some cases overlapped even less than expected by chance. This may result, in part, from the identification of distinct sets of genes that have been horizontally transferred at different times. However, reinspection of E. coli genes classified as horizontally transferred based on nucleotide composition revealed numerous cases of vertically inherited genes with atypical nucleotide compositions and horizontally transferred genes with typical nucleotide compositions (Koski et al., 2001). Thus, some of these surrogate tests appear to suffer from a debilitating combination of low sensitivity and lack of selectivity. Clearly, results based on surrogate methods need to be interpreted with caution.

Section Summary

- It can be difficult or even impossible to obtain unequivocal proof of horizontal transfer for any particular gene. Inferences of horizontal transfer are probabilistic, and, as such, the goal of horizontal transfer detection mechanisms is to maximize the likelihood of correct identification of horizontal transfer events.

- The different methods used for the detection of horizontal transfer events all rely on the characterization of some anomalous or unusual genomic subset of genes.
- Phylogenetic and distance-based methods used for the detection of horizontal transfer are based on the discordant evolutionary patterns of horizontally transferred genes.
- Surrogate methods used for the detection of horizontal transfer are based on the anomalous nucleotide composition of horizontally transferred genes.

IMPLICATIONS OF HORIZONTAL TRANSFER

Quantification and Classification

A key question regarding horizontal transfer is the extent to which prokaryotic genomes are made up of horizontally transferred genes. Below, we report on an attempt to quantify this effect of horizontal transfer on prokaryotic genomes using one of the detection methods described previously. In addition to measuring the extent of horizontal transfer events, classification of different types of horizontal transfer events was also attempted. The relationships between horizontally acquired genes and homologous genes (if any) preexisting in the recipient lineage can be used to classify horizontal gene transfer events into at least three distinct categories: (i) acquisition of a new gene missing in other members of a given clade; (ii) acquisition of a paralog of the given gene with a distinct evolutionary ancestry; and (iii) acquisition of a phylogenetically distant ortholog followed by xenologous gene displacement, that is, elimination of the ancestral gene (xenology has been defined as homology of genes that is incongruent with the species tree and so implies horizontal gene transfer).

The first two classes of events in some cases may result in nonorthologous gene displacement, that is, acquisition of an unrelated (or distantly related) gene with the same function as an essential ancestral gene typical of the

given clade, with subsequent elimination of the latter.

As discussed above, there are a number of different ways to infer horizontal transfer, each with its own advantages and liabilities. On a genomic scale, currently the most reliable and practicable approach falls into the phylogenetic and distance-based class of methods and consists of the taxonomic classification of database hits based on sequence similarity. This approach was applied to completely sequenced bacterial and archaeal genomes in an attempt to quantitatively assess the amount of horizontal gene transfer in prokaryotic genomes and to tentatively classify the transfer events with respect to the three categories described above. The approach is based on the identification of proteins that are significantly more similar to homologs from other evolutionary lineages than to homologs from those lineages to which the species being evaluated belongs. Best database hits to taxa from distantly related lineages are referred to as paradoxical best hits, while hits to taxa from the same lineage being considered are referred to as being similar to reference taxa. Protein sets from 31 complete prokaryotic genomes (9 archaeal and 22 bacterial) available at the time of the analysis were extracted from the Genome division of the Entrez retrieval system and used as queries to search the nonredundant (NR) protein sequence database at the National Center for Biotechnology Information (National Institutes of Health, Bethesda, Md.) with the gapped BLASTP program. From the results of these searches, three sets of paradoxical best hits were identified using the Tax_Collector program of the SEALS package. The first set was designed to include candidate horizontal transfers between most distantly related organisms. For nine archaeal species, all proteins were detected whose best hits to bacterial or eukaryotic proteins had expectation (E) values significantly lower than the E value of the best hit to an archaeal protein (see legend to Table 2). Similarly, for 22 bacterial proteomes, the paradoxical best hits to archaeal and eukaryotic proteins were collected. A separate subset of

TABLE 2 Probable interdomain horizontal transfers[a, b]

Species	Reference lineage	Paralog acquisition or xenologous displacement, n (%) of the genes	Acquisition of new genes, n (%) of the genes
Aeropyrum pernix	Archaea	34 (1.8)	47 (2.5)
Archaeoglobus fulgidus	Archaea	103 (4.3)	100 (4.2)
Methanobacterium thermoautotrophicum	Archaea	100 (5.3)	61 (3.3)
Methanococcus jannaschii	Archaea	43 (2.5)	39 (2.3)
Pyrococcus horikoshii	Archaea	55 (2.7)	39 (1.9)
Pyrococcus abyssi	Archaea	72 (4.1)	39 (2.2)
Thermoplasma acidophilum	Archaea	112 (7.8)	54 (3.7)
Halobacterium sp.	Archaea	204 (8.4)	174 (7.2)
Aquifex aeolicus	Bacteria	87 (5.7)	45 (3.0)
Thermotoga maritima	Bacteria	207 (11.1)	53 (2.9)
Deinococcus radiodurans	Bacteria	47 (1.5)	45 (1.5)
Bacillus subtilis	Bacteria	71 (1.7)	28 (0.7)
Bacillus halodurans	Bacteria	79 (1.9)	40 (1.0)
Mycobacterium tuberculosis	Bacteria	50 (1.3)	62 (1.7)
Escherichia coli	Bacteria	26 (0.6)	13 (0.3)
Haemophilus influenzae	Bacteria	3 (0.2)	3 (0.2)
Rickettsia prowazekii	Bacteria	23 (2.8)	7 (0.8)
Pseudomonas aeruginosa	Bacteria	66 (1.2)	39 (0.7)
Neisseria meningitidis	Bacteria	6 (0.3)	5 (0.2)
Vibrio cholerae	Bacteria	12 (0.3)	16 (0.4)
Xylella fastidiosa	Bacteria	22 (0.8)	8 (0.3)
Buchnera sp.	Bacteria	0 (0.0)	0 (0.0)
Treponema pallidum	Bacteria	10 (1.0)	4 (0.4)
Borrelia burgdorferi	Bacteria	3 (0.4)	6 (0.7)
Synechocystis strain PCC6803	Bacteria	219 (6.9)	115 (3.6)
Chlamydia pneumoniae	Bacteria	23 (2.2)	9 (0.9)
Mycoplasma pneumoniae	Bacteria	0 (0.0)	1 (0.1)
Ureaplasma urealyticum	Bacteria	1 (0.2)	1 (0.2)
Helicobacter pylori	Bacteria	5 (0.3)	3 (0.2)
Campylobacter jejuni	Bacteria	5 (0.3)	4 (0.2)

[a] All protein sequences from each genome were compared to the NR database using the BLASTP program [expect (E) value cutoff 0.001, no filtering for low complexity] and the results were searched for paradoxical best hits, i.e., those that either had a hit to a homolog from a nonreference lineage with an E value 10 orders of magnitude lower (more significant) than that of the best hit to a homolog from the reference taxon, or had statistically significant hits to homologs from nonreference taxa only. All automatically detected paradoxical best hits were manually checked to eliminate possible false positives.
[b] Adapted from Koonin et al. (2001).

the paradoxical best hits was formed by proteins with a significant hit (E < 0.001) detected only outside the reference taxon.

The second group of paradoxical best hits was to include candidate gene-exchange events between major bacterial lineages. With the current state of genome sequencing, this type of analysis is best applicable to small genomes of parasitic bacteria when at least one larger genome sequence of a related species is available.

Thus, for *H. influenzae* and *Rickettsia prowazekii*, hits to Proteobacteria (two large proteobacterial genomes, those of *E. coli* and *Pseudomonas aeruginosa*, have been sequenced) were compared with hits to all other bacteria, and paradoxical hits were collected. To isolate likely horizontal transfer events between distantly related bacteria, the paradoxical hits to the archaea and the eukaryotes included in the first set were subtracted. Similarly, for *Mycoplasma genitalium, M.*

pneumoniae, and *Ureaplasma urealyticum*, the reference taxon was *Firmicutes* (gram-positive bacteria; available large genomes, *Bacillus subtilis* and *Mycobacterium tuberculosis*). In the case of the spirochetes *Treponema pallidum* and *Borrelia burgdorferi*, the reference taxon was *Spirochaetales* (no large genome available for this lineage).

The third set was to include paradoxical best hits due to relatively recent horizontal transfers. For this purpose, two closely related species pairs, namely *Chlamydia pneumoniae*-*Chlamydia trachomatis* and *M. genitalium*-*M. pneumoniae* were compared. The conservative criteria used to register a paradoxical best hit were the same for all three sets. Database hits were considered to be paradoxical best hits if they (i) hit a homolog from a nonreference taxon with an E value 10 orders of magnitude lower (more significant) than that of the best hit to a homolog from the reference taxon, or (ii) had statistically significant hits to homologs from nonreference taxa only.

Even with the conservative threshold used in this analysis, paradoxical best hits provide only a first-approximation estimate of horizontal gene transfer events. To deal with this issue, a detailed analysis of all candidates was performed for four selected genomes, the bacteria *H. influenzae*, *Vibrio cholerae*, and *A. aeolicus* and the archaeon *Methanobacterium thermoautotrophicum*. In these cases, all paradoxical best hits were examined individually via the construction of phylogenetic trees and examination of the phyletic distribution of the corresponding protein family. Here, multiple protein sequence alignments were constructed with the ClustalW program, checked for the conservation of salient sequence motifs, and used for constructing phylogenetic trees, with the neighbor-joining method as implemented in the NEIGHBOR program of the PHYLIP program package.

A quantification of putative interdomain horizontal transfers for all prokaryotic genomes analyzed is shown in Table 2. Interdomain transfers seemed to involve ~3% of the genes in most free-living bacteria. This fraction was significantly lower in parasitic bacteria, with the exception of *Chlamydia* and *Rickettsia*. In contrast, archaea had a greater fraction of candidate interdomain gene transfers, typically between 4% and 8%. It should be emphasized that the protocol used to obtain these numbers is a conservative one that will detect primarily relatively recent horizontal transfer events. Ancient transfer events, for example, those that could have occurred prior to the divergence of the analyzed archaeal species, will be obscured by interarchaeal hits. In addition, the fraction of transfers recorded for each of the gammaproteobacterial species is likely to be an underestimate given that this lineage is represented by several genomes, including two large ones, *E. coli* and *P. aeruginosa*. Given these limitations in detection of paradoxical best hits and the conservative cutoff values used, the fraction of each genome that has been horizontally transferred between domains is quite likely to be a substantial underestimate. Clearly, the level of gene exchange between different domains of life seems to have been quite substantial.

In terms of the number of genes that appear to have been horizontally acquired from a different domain of life, there were several outlier organisms. For example, *Synechocystis* sp. (a cyanobacterium, the progenitor of chloroplasts) and *R. prowazekii* (an alphaproteobacterium, the group of bacteria to which the progenitor of the mitochondria is thought to belong) both show relatively high levels of interdomain transfers. In these cases, the reported high numbers probably reflect interdomain horizontal gene transfers between chloroplasts and/or mitochondria and eukaryotic nuclear genomes. In addition, as noticed previously, the hyperthermophilic bacteria, *A. aeolicus* and *T. maritima*, are significantly enriched in genes that appear to have been horizontally transferred from archaea. Conversely, the archaea *Thermoplasma acidophilum* and *Halobacterium* sp. appear to possess a much greater number of acquired bacterial genes than other archaeal species. This could be because these organisms are moderate thermophiles that share their habitats with multiple bacterial species. Consistent with this notion, the difference in the number of acquired genes between two close bacterial species, *B. subtilis* and *B. halodurans*, is notable. Specifically, there is an excess of archaeal genes in *B. halodurans* that may be explained by its environmental coexistence with halophilic archaeal species.

Because of the nature of this genomewide analytical approach, apparent acquisition of new genes (cases when a given protein simply has no detectable homologs in the reference taxon) could be automatically distinguished from paralog acquisition/xenologous displacement. However, differentiating between paralog acquisition and xenologous displacement was not possible without additional detailed study. The number of probable horizontal transfer events that resulted in the acquisition of new genes was generally similar to the number of paralog acquisition/xenologous displacement events (Table 2).

The data in Table 3 provide a more complete estimate of probable horizontal gene transfer events (not just ancient events) by using the respective bacterial lineage as the reference taxon for each bacterial genome and accordingly including gene exchange between major lineages. These estimates are expected to be particularly reliable for small genomes because, with the exception of the spirochetes, larger genomes of related bacteria are available, making it unlikely that the paradoxical best hits are due to an insufficient representation of the given taxon in the sequence database. The estimates of horizontal transfer rate obtained by this approach widely differ, from 1.6% of the genes for *M. genitalium* to 32.6% in *T. pallidum*. The spirochete data could be an overestimate due to differential gene loss in the two parasitic spirochetes with very different lifestyles, but, in general, a substantial amount of relatively recent horizontal gene transfer seems to be supported by the data. The number of even more recent horizontal transfer events was estimated by collecting the paradoxical best hits for two pairs of closely related bacterial species, *M. genitalium*/*M. pneumoniae* and *C. trachomatis*/*C. pneumoniae*. As may be expected, this analysis revealed a smaller number of candidate horizontal transfers, with none at all seen in *M. genitalium* (Table 3). These last results stand in stark contrast to the much

TABLE 3 Probable horizontal gene transfers between major bacterial lineages[a, b]

Species	Reference lineage	Paralog acquisition or xenologous displacement, *n* (%) of the genes	Acquisition of a new gene, *n* (%) of the genes
Mycoplasma genitalium	Firmicutes	6 (1.2)	2 (0.4)
Mycoplasma pneumoniae	Firmicutes	9 (0.9)	8 (1.2)
Bacillus subtilis	Firmicutes	685 (16.7)	383 (9.3)
Bacillus halodurans	Firmicutes	772 (19.0)	400 (9.8)
Treponema pallidum	Spirochaetales	132 (12.8)	204 (19.8)
Borrelia burgdorferi	Spirochaetales	109 (12.8)	141 (16.6)
Haemophilus influenzae	Proteobacteria	32 (1.9)	21 (1.2)
Rickettsia prowazekii	Proteobacteria	49 (5.9)	32 (3.8)
Escherichia coli	Proteobacteria	223 (5.2)	102 (2.4)
Pseudomonas aeruginosa	Proteobacteria	448 (8.1)	275 (5.0)
Neisseria meningitidis	Proteobacteria	55 (2.7)	34 (1.7)
Vibrio cholerae	Proteobacteria	130 (3.4)	85 (2.2)
Xylella fastidiosa	Proteobacteria	88 (3.2)	83 (3.0)
Buchnera sp.	Proteobacteria	0 (0.0)	0 (0.0)
Mycoplasma genitalium	Mycoplasma	0 (0.0)	0 (0.0)
Chlamydia pneumoniae	Chlamydiales	4 (0.4)	25 (2.4)

[a] The schema for detection of candidate horizontal transfers was the same as in Table 2.
[b] Adapted from Koonin et al. (2001).

TABLE 4 Classification of candidate horizontal gene transfer events in selected genomes[a, b]

Species	Reference taxon	Xenologous gene displacement	Acquisition of paralog	Uncertain
Haemophilus influenzae	Proteobacteria	6	6	20
Vibrio cholerae	Proteobacteria	5	40	85
Aquifex aeolicus	*Bacteria*	31	8	48
Methanobacterium thermoautotrophicum	*Archaea*	19	17	

[a] Paradoxical best hits representing the "paralog acquisition or displacement" category (Table 3, third column) were examined case-by-case, which involved establishing the phylogenetic distribution of the corresponding protein family and constructing phylogenetic trees. Trees were generated from multiple alignments, constructed using the ClustalW program, with the neighbor-joining method as implemented in the NEIGHBOR program of the PHYLIP program package. If a protein sequence showed a clear phylogenetic affinity (with an at least 60% bootstrap probability) to a nonreference taxon, whereas another member of the same protein family from the same species (a paralog of the protein in question) belonged to the reference-taxon branch, a case of probable acquisition of a paralog was recorded. When the given species was represented by a single ortholog in a family and that sequence displayed a clear affinity to a non-reference taxon, this was considered a case of xenologous gene displacement. Other complicated cases and those that were not supported by bootstrap analysis were classified as "uncertainty."

[b] Adapted from Koonin et al. (2001).

higher numbers of candidate recent horizontal transfers detected using surrogate methods based on anomalous nucleotide composition.

In an attempt to distinguish between xenologous gene displacement (i.e., acquisition of a phylogenetically distant ortholog followed by elimination of the ancestral gene) and acquisition of a paralog of a preexisting gene, we performed a phylogenetic analysis of the candidate horizontally transferred genes for three bacterial genomes and one archaeal genome. While the phylogenetic tree topology was too complex to make the distinction in many cases, multiple clear-cut cases of both types were identified in each of the analyzed genomes (Table 4 and data not shown). The relative contributions of xenologous gene displacement and paralog acquisition appeared to differ significantly for the compared genomes. Xenologous gene displacement is far more prevalent in the hyperthermophilic bacterium *A. aeolicus* and the archaeon *M. thermoautotrophicum*, while paralog acquisition predominates in the parasitic bacterium *V. cholerae* (Table 4).

Functional Impact

One of the most striking examples of horizontal transfer is the acquisition of eukaryotic genes by prokaryotes. Eukaryotic derived genes are particularly abundant in certain functional classes of bacterial genes. Below, we describe

several examples of this phenomenon and their potential functional impact.

The acquisition of eukaryotic genes by bacteria may be of particular interest because of the possible role of such horizontally transferred genes in bacterial pathogenicity. Conservative estimates of the number of acquired eukaryotic genes in completely sequenced prokaryotic genomes were produced by using the "paradoxical best hit" approach described above (Table 5). All prokaryotic proteins that showed significantly greater similarity to eukaryotic homologs than to bacterial ones (with a possible exception for closely related bacterial species) were considered likely to be derived from interdomain horizontal transfer. The number of apparent eukaryotic acquisitions for most of the prokaryotic genomes tends to be relatively small, typically on the order of 1% of the genes. The high number of eukaryotic (plant-derived) genes in *Synechocystis* is for the most part an artifact caused by the direct evolutionary relationship between cyanobacteria and chloroplasts. However, *Synechocystis* also has an unusual excess of animal-specific "eukaryotic" genes including several proteins that share conserved domains with cadherins and other extracellular receptors. A possible explanation for this could be that, much as with the extant cyanobacterial symbionts of sponges, the ancestors of *Synechocystis* have passed through a phase of animal symbiosis

TABLE 5 Apparent phylogenetic affinities of eukaryotic best hits in bacteria and archaea[a]

Species	Metazoa	Plants	Fungi	Other
Aeropyrum pernix	1	2	0	0
Archaeoglobus fulgidus	5	6	2	2
Methanobacterium thermoautotrophicum	5	3	2	1
Methanococcus jannaschii	2	2	0	0
Pyrococcus horikoshii	3	3	2	1
Pyrococcus abyssi	5	1	2	1
Thermoplasma acidophilum	7	2	2	2
Halobacterium sp.	9	10	5	4
Aquifex aeolicus	0	2	1	0
Thermotoga maritima	2	4	1	1
Deinococcus radiodurans	11	11	8	1
Bacillus subtilis	10	10	6	3
Bacillus halodurans	9	5	3	2
Mycobacterium tuberculosis	16	5	0	2
Escherichia coli	7	2	2	1
Haemophilus influenzae	0	0	0	1
Rickettsia prowazekii	15	6	3	3
Pseudomonas aeruginosa	26	14	16	7
Neisseria meningitidis	1	2	2	0
Vibrio cholerae	4	5	3	0
Xylella fastidiosa	5	10	4	2
Treponema pallidum	3	3	0	2
Borrelia burgdorferi	1	1	1	1
Synechocystis strain PCC6803	22	167	4	69
Chlamydia pneumoniae	9	12	2	0
Helicobacter pylori	8	2	0	0
Campylobacter jejuni	1	0	1	1
Mycoplasma pneumoniae	0	0	1	0
Ureaplasma urealyticum	0	1	0	0

[a] Adapted from Koonin et al. (2001).

in their evolution. There also seems to be a modest but consistent excess of acquired eukaryotic genes in at least some parasites, such as *M. tuberculosis*, *P. aeruginosa*, *Xylella fastidiosa*, and *C. pneumoniae* (Table 5). However, only a very small number of probable acquired eukaryotic genes was detected for other parasitic bacteria such as spirochetes. A more detailed phyletic breakdown of the eukaryotic acquisitions shows some limited correlation with parasite-host affinities. For example, *P. aeruginosa* has an excess of "animal" genes, whereas the plant pathogen *Xylella fastidiosa* seems to have acquired a greater number of "plant" genes (Table 5). *Chlamydia* is an unusual case in that this animal pathogen seems to have acquired a greater number of genes from plants than from animals. Chlamydiae and their close relatives had a long history of parasitic or symbiotic relationships with eukaryotes and, at some stages of their evolution, could have been parasites of plants or their relatives. Most unexpected was the observation that genes apparently acquired from eukaryotes are seen in each of the archaeal genomes, with the greatest number detected in *Halobacterium* spp., the archaeon that also appears to have acquired the greatest number of bacterial genes (Table 5 and discussion above).

It should be emphasized that the approach used here probably underestimates, perhaps considerably, the true number of eukaryotic genes transferred to prokaryotes. For more ancient transfers, some of the proteins encoded by transferred genes may not show highly significant similarity to their eukaryotic ancestors and are therefore missed. This is the case for most of the signaling proteins discussed below. For the most part, detection of such subtle cases requires a more detailed and not easily automated analysis.

AMINOACYL-tRNA SYNTHETASES

In general, genes that encode components of the translation machinery appear to belong to a conserved genomic core of information-processing genes that is less prone to horizontal transfer than other groups of genes (Jain et al., 1999). However, recent work points to a role for horizontal transfer in the evolution of some ribosomal proteins. One example of this phenomenon is the ribosomal protein S14, for which several probable horizontal transfer events have been revealed by phylogenetic analysis. Furthermore, the evolution of one group of essential components of the translation machinery, aminoacyl-tRNA synthetases (aaRS), appears to have involved numerous horizontal transfers. In fact, phylogenetic analysis shows that varying amounts of horizontal transfer probably have contributed to

the evolution of aaRS of all 20 specificities (Wolf et al., 1999). This may reflect the relative functional autonomy of these enzymes when compared with ribosomal proteins that function as subunits of a tight complex. On many independent occasions, eukaryotic aaRS appear to have displaced the resident bacterial ones (Table 6). Acquisition of eukaryotic aaRS genes is most prevalent in two groups of parasitic bacteria, chlamydiae and spirochetes. It may not be particularly surprising that parasites have acquired more aaRS genes than free-living bacteria, but why spirochetes are so unusual in this respect, remains unclear.

A highly unexpected, albeit very well supported, case of probable horizontal transfer from eukaryotes is the TrpRS from the genus *Pyrococcus* that includes free-living, hyperthermophilic archaea. Given its environmental niche, the presence of an aaRS in the pyrococci appears extremely surprising. It is most likely that *Pyrococcus* acquired a eukaryotic TrpRS gene from one of the few eukaryotic species that have been found in hyperthermophilic habitats, such as a polychaete annelid.

In most cases, the apparent horizontal transfer of eukaryotic aaRS genes into prokaryotes

TABLE 6 Horizontal transfer of eukaryotic aaRS genes into different bacterial lineages[a, b]

Bacterial group	Horizontally transferred aaRS genes	Comment
Spirochetes	Pro (*Borrelia* only), Ile, Met, Arg, His, Asn	
	Ser, Glu	Apparent acquisition of the mitochondrial gene
Chlamydia	Ile, Met, Arg, Asn??	
	Glu	Apparent acquisition of the mitochondrial gene
Bacillus	Asn	
Mycobacteria	Ile, Asn??	
Mycoplasma	Pro, Asn	
Gammaproteobacteria	Asn, Gln	
Helicobacter	His	
Deinococcus	Gln, Asn	
Cyanobacteria	Arg, Asn	
Pyrococcus (archaeon)	Trp	

[a] An update of the results presented in Wolf et al. (1999).
[b] Adapted from Koonin et al. (2001).

involves xenologous gene displacement of the corresponding ancestral prokaryotic aaRS. A special case involves the GlnRS that apparently first emerged in eukaryotes through a duplication of the GluRS gene and subsequently has been horizontally acquired by gammaproteobacteria. In most of the other prokaryotes, glutamine incorporation into protein is mediated by a completely different mechanism, namely transamidation, whereby glutamine is formed from Glu-tRNA^{Gln} in a reaction catalyzed by the specific transamidation complex, GatABC. The GatABC complex genes are missing in gammaproteobacteria, indicating that horizontal gene transfer, followed by nonorthologous gene displacement, has resulted in a switch to a completely different pathway for an essential biochemical process. In several other Proteobacteria and in *Deinococcus radiodurans*, GlnRS and the GatABC complex coexist. This suggests that the elimination of the transamidation system has been a relatively recent event, specific to the gammaproteobacteria lineage, compared with the acquisition of the eukaryotic GlnRS gene. The finding that *Mycobacterium leprae* encodes a eukaryotic-type ProRR, in contrast to the closely related *M. tuberculosis* that has a typical bacterial form, supports the notion that xenologous displacement of aaRS is an active, ongoing process. This case represents a particularly compelling case of xenologous gene displacement of an individual gene since the displacement occurred relatively recently and without disrupting the conserved order of surrounding genes.

The topology of some of the aaRS trees, particularly that for IleRS, suggests a single horizontal gene transfer from eukaryotes, with subsequent lateral dissemination among bacteria. For example, all bacterial species that are suspected to have acquired eukaryotic IleRS and HisRS cluster together in their respective trees. However, the topologies of the trees for MetRS, ArgRS, and Asp-AsnRS do not show such clustering and suggest instead multiple cases of acquisition of the respective eukaryotic genes by different bacterial lineages.

The gene for the eukaryotic-type IleRS is spread through bacterial populations on plasmids, conferring resistance to the antibiotic mupirocin. Thus, in this case, both the vector of horizontal gene transfer and the nature of the selective pressure that results in the fixation of the transferred gene in the bacterial population are known.

SIGNAL TRANSDUCTION SYSTEMS

Eukaryotes have vastly more complex signal transduction systems than most bacteria and archaea. However, some bacteria, such as cyanobacteria, myxobacteria, and actinomycetes, also have remarkably versatile signal-transduction systems. Eukaryotic derived protein domains involved in various forms of signaling are frequently present, sometimes in highly divergent forms, in prokaryotes (Ponting et al., 1999). These domains can be classified into those that probably have been inherited from the Last Universal Common Ancestor and those that have evolved in eukaryotes and then were subsequently horizontally transferred to bacteria, and less commonly, archaea (Table 7).

The evidence for horizontal transfer appears most compelling when a domain is present in all eukaryotes but found in only one or two bacterial lineages. The SWIB domain, which is present in subunits of the SWI/SNF chromatin-associated complex in all eukaryotes, was detected in only one bacterial lineage, the *Chlamydia*. The chlamydial SWIB domain, one copy of which is fused to topoisomerase I, might participate in chromatin condensation, a distinguishing feature of this group of intracellular bacterial parasites. *Chlamydia* and *X. fastidiosa* also carry the SET domain, a characteristic eukaryotic histone methylase. The functions of these SET methylases in bacteria remain unclear, and it is possible that their elucidation will uncover novel prokaryotic regulatory mechanisms.

Some of the "eukaryotic" signaling domains acquired by bacteria probably perform functions that are mechanistically similar to their functions in eukaryotic systems. An example of such probable functional conservation is the

TABLE 7 Examples of eukaryotic-bacterial transfer of genes coding for proteins and domains involved in signal transduction[a, b]

Domain	Occurrence in eukaryotes	Occurrence in prokaryotes	Prevalent protein context		Functions in eukaryotes
			Eukaryotes	Prokaryotes	
WD40	All	Ssp, Hi, Dr, Tm, Streptomyces, *Cenarchium symbiosum* (archaeon)	Multiple repeats in various regulatory proteins including G protein subunits	Stand-alone multiple repeats, Ser/Thr protein kinases	Various nuclear and cytoplasmic regulatory protein-protein interactions
Ankyrin	All	Ec, Ssp, Bb, Tp, Nm, Rp, Dr, Streptomyces	Multiple repeats in various cytoplasmic and regulatory proteins	Stand-alone, multiple repeats	Various forms of signal transduction including regulation, cell cycle control, and PCD[c]
TIR	Animals, plants	Bs, Ssp, Streptomyces, Rhizobium	Toll/interleukin receptors, PCD adaptors	Stand-alone, AP-ATPases, WD40	Programmed cell death, interleukin signaling
EF-hand	All	Ssp, Streptomyces	Calmodulins, calcineurin phosphatase regulatory subunits, other Ca-binding proteins	Predicted Ca-binding proteins; transaldolase	Various forms of Ca-dependent regulation
FHA (Forkhead-associated)	All	Ct, Cp, Mtu, Ssp, Xf, Streptomyces, Myxococcus	Protein Ser/Thr kinases and phosphatase; nuclear regulatory proteins	Stand-alone, adenylate cyclase, histidine kinase	Phosphoserine binding, protein kinase-mediated nuclear signaling
SET	All	Ct, Cp, Xf	Various multidomain chromatin proteins	Stand-alone	Histone methyl-transferase
SWIB	All	Ct, Cp	SWI/SNF complex subunits	Stand-alone, Topoisomerase type I	Chromatin remodeling, transcription regulation
Sec7	All	Rp	Guanine nucleotide exchange factors; protein transport system components	Stand-alone (protein transport?)	Protein transport, regulation
Kelch	All	Ec, Hi, Af	Multiple repeats, fusions with POZ and other signaling domains; actin-binding proteins, transcription regulators	Stand alone, multiple repeats	Transcription regulation; cytoskeleton assembly

(continues)

TABLE 7 (*continued*)

| Domain | Occurrence in eukaryotes | Occurrence in prokaryotes | Prevalent protein context | | Functions in eukaryotes |
			Eukaryotes	Prokaryotes	
Adenovirus type protease	All; DNA-viruses	Ct, Ec	Stand-alone; ubiquitin-like protein hydrolases	Stand-alone; membrane-associated (Ct)	Regulation of ubiquitin-like, protein-dependent protein degradation
OUT-family protease	All; RNA- and DNA-viruses	Cp	Stand-alone or fused to ubiquitin hydrolase	Large protein with non-globular domains	Unknown; possible role in ubiquitin-mediated protein degradation
START domains	Animals and plants	Pa	Stand-alone or fused to homeodomains and GTPase regulatory domains	Stand-alone	Lipid binding

[a] An update of the results presented in Ponting et al. (1999).
[b] Adapted from Koonin et al. (2001).
[c] PCD, programmed cell death.

phosphoserine-peptide-binding FHA domain. This domain's partners in signal transduction, protein kinases and phosphatases, are present in both eukaryotes and prokaryotes, suggesting that it functions in similar phosphorylation-based signaling pathways in both.

Other examples point to the possible exaptation of signaling domains of eukaryotic origin for distinct functions in bacteria. This phenomenon is exemplified by predicted cysteine proteases of two distinct families that have been detected in *Chlamydia* (Table 4). In eukaryotes, both these protease families are thought to participate in ubiquitin-dependent protein degradation. However, the ubiquitin system does not exist in bacteria; this seems to rule out functional conservation for these proteins between eukaryotes and prokaryotes. In *Chlamydia*, it seems that these proteases contribute to the pathogen-host-cell interaction as is suggested by the predicted membrane localization of the adenovirus-family proteases in *C. trachomatis*.

MISCELLANEOUS EUKARYOTIC GENES ACQUIRED BY PROKARYOTES

Many genes of very diverse functions also seem to have been horizontally transferred from eukaryotes to bacteria or archaea (Table 8). For most cases, the selective advantage that may have been conferred on the prokaryote by the acquired eukaryotic gene remains unclear. However, there are a few exceptions where biologically feasible inferences can be made about the adaptive benefits that such transfers may confer on prokaryotic recipients. One clear-cut example is the chloroplast-type ATP/ADP translocase that is present in the intracellular parasites *Chlamydia* and *Rickettsia* and the plant pathogen *X. fastidiosa*. For *Chlamydia* and *Rickettsia*, the advantage of having this enzyme is obvious because it allows them to scavenge ATP from the host, consistent with their status as "energy parasites." In contrast, the presence of the ATP/ADP translocase in *X. fastidiosa* is unexpected and might indicate that such use of the energy-producing facilities of the host is also used by extracellular parasites. Another case when the adaptive value of the horizontally transferred gene seems clear is the sodium/phosphate cotransporter that was detected in *V. cholerae*, but so far not in any other bacterium. This transporter probably facilitates the survival of the bacterium in the host intestines and could be involved in pathogenicity.

TABLE 8 Examples of eukaryotic-prokaryotic transfer of functionally diverse genes[a]

Gene function	Representative (gene name_species)	Occurrence in prokaryotes	Occurrence in eukaryotes	Apparent source	Type of horizontal transfer
Hydrolase, possibly RNase	AF2335_Af	Af (archaeon) only	Animals, Leishmania	Animal?	Acquisition of paralog
Heme-binding protein	MTH115_Mth	Mth only	Plants, animals	Plant?	Acquisition of a new function
Glutamate-cysteine ligase	XF1428_Xf	Xf, Zymomonas, Bradyrhizobium	Plants	Plant	Xenologous gene displacement
Fructose-bisphosphate aldolase (FBA)	XF0826_Xf	Xf, Cyanobacteria	Plants, animals	Plant	Nonorthologous gene displacement (of the typical bacterial FBA)
ATP/ADP translocase	XF1738_Xf	Xf, Chlamydia, Rickettsia	Plants	Plant	Acquisition of a new function
Sulfotransferase	BH3370_Bh	Bh	Animals	Animal	Acquisition of a new function
Gamma-D-glutamyl-L-diamino acid endopeptidase I	ENP1_Bs	Bacillus	Animals	Animal	Acquisition of a new function
General stress protein	GsiB_Bs	Bs only	Plants	Plant	Acquisition of a new function
Superfamily I helicase	Cj0945c_Cj	Cj only	Fungi, animals, plants	Eukaryotic	Acquisition of paralog
Guanylate cyclase	Rv1625c_Mtu	Mtu only	Animals, slime mold	Animal	Acquisition of paralog
Purple acid phosphatase	Rv2577_Mtu	Mtu only	Plants, fungi, animals	Plant?	Acquisition of paralog
α/β hydrolase (possible cutinase or related esterase)	Rv1984, Rv3451, Rv2301, Rv1758 (Mtu)	Mycobacteria; multiple paralogs	Fungi	Fungal	Acquisition of a new gene
C-5 sterol desaturase	Slr0224_Ssp	Synechocystis, Mtu, Vibrio	Fungi, animals, plants	Eukaryotic	Acquisition of a new gene
Carnitine O-palmitoyltransferase	MPN114_Mp	Mp only	Fungi, animals	Eukaryotic	Acquisition of a new gene
Arylsulfatase	B1498_Ec	Ec only	Animals	Animal	Acquisition of paralog
Cation transport system component	ChaC-Ec	Ec only	Fungi, animals, plants	Eukaryotic	Acquisition of a new gene
Thiamine pyrophosphokinase	NMB2041_Nm	Nm only	Fungi, animals	Fungal?	Nonorthologous gene displacement?
Phospholipase A2	VC0178_Vc	Vc only	Plants, animals	Plant	Acquisition of paralog
Sodium/phosphate cotransporter	VC0676_Vc	Vc only	Animals	Animal	Acquisition of a new gene
Topoisomerase 1B	DR0690_Dr	Dr only	All eukaryotes	Eukaryotic	Acquisition of a new gene
RNA-binding protein Ro	DR1262	Dr, Streptomyces	Animals	Animal	Acquisition of a new gene

[a] Adapted from Koonin et al. (2001).

Many enzymes that have apparently been acquired by bacteria from eukaryotes seem to have been exapted for interactions with their eukaryotic hosts. For example, a hemoglobinase-like protease that is present in *Pseudomonas* (PA4016) might function as a virulence factor that degrades host proteins. The fukutin-like enzymes have been acquired horizontally from eukaryotes by certain pathogenic bacteria, including *Haemophilus* and *Streptococcus*, and probably have been adapted by these bacteria for the modification of their own surface molecules.

One surprising corollary from studies of interdomain horizontal transfer is the fundamental functional plasticity of many cellular systems. Apparently enzymatic components that have evolved in very distant related organisms retain functional compatibility. A striking example of this is demonstrated by topoisomerase IB, an enzyme that is ubiquitous in eukaryotes, but among prokaryotes is found encoded only in the genome of the extreme radioresistant bacterium *D. radiodurans*. There are major differences in the repertoires of the involved proteins and molecular mechanisms between the bacterial and eukaryotic repair systems. However, this typical eukaryotic topoisomerase contributes to the UV resistance of *D. radiodurans* and thus apparently maintains a functional role in repair.

Section Summary

- Horizontal transfers involving prokaryotes have been rather common in evolution. Indeed, prokaryotic genomes are made up of a substantial fraction of horizontally transferred genes.
- Genomewide analyses reveal that at least 3% of the genes of most free-living bacteria have been acquired by interdomain (between different domains of life, e.g., *Archaea* to *Bacteria*) horizontal transfer.
- More recent horizontal transfers, within and between bacterial lineages, have been even more common and contribute greater numbers of genes to bacterial genomes.
- Typically only ~1% of the genes in a bacterial genome have been acquired

from eukaryotes via horizontal transfer. However, certain functional classes of bacterial genes (e.g., aminoacyl-tRNA synthetases and signal-transduction system genes) are enriched with genes that have been transferred from eukaryotes.

SUMMARY OF THE CONTRIBUTION OF GENE TRANSFER TO GENOME EVOLUTION

A number of different methods exist to infer past horizontal transfer events. While each of these methods has its own scope, advantages, and liabilities, it remains exceedingly difficult to unequivocally demonstrate horizontal transfer for many cases. However, whole-genome comparisons do strongly suggest that the extent of horizontal transfer involving prokaryotic genomes has been more substantial than previously imagined. For example, estimates based on the analysis of taxon-specific best hits suggest a significant amount of horizontal gene transfer between the three primary domains of life as well as between major bacterial lineages. For selected genomes that have been analyzed in detail, this conclusion is largely supported by more detailed phylogenetic analysis.

Acquisition of eukaryotic genes by bacteria, particularly parasites and symbionts, is a surprisingly prevalent and functionally significant direction of horizontal gene flow. Gene transfer events of this class are particularly characteristic of certain functional categories of genes such as aminoacyl-tRNA synthetases and signal-transduction systems.

Horizontal gene transfer, particularly between eukaryotes and bacteria, emphasizes the remarkable unity of molecular-biological mechanisms in all life forms. This is underscored by the compatibility of recently united eukaryotic and bacterial proteins that have evolved in their distinct lineages for hundreds of millions of years prior to the horizontal gene transfers that brought them into the same genome. To be successfully fixed and retained in a bacterial population after horizontal transfer, an acquired eukaryotic gene must confer a tangible selective advantage on the recipient bacteria. In cases of xenologous gene displacement, the acquired version of a gene should immediately become adaptively superior to the resident version of the recipient species. This is demonstrated by the eukaryotic isoleucyl-tRNA synthetase displacement of the original gene in some bacteria that is due to the antibiotic resistance that is conferred by the transferred gene. Acquisition of a new gene, for example, the ATP/ADP translocases in the intracellular parasitic bacteria, *Chlamydia* and *Rickettsia*, would seem to lead to a more obvious selective advantage. Most often, comparative genomics can only identify the genes that probably have entered a particular genome by horizontal transfer. A fuller understanding of the biological significance of horizontal gene transfer will require direct experimental analysis of these genes.

QUESTIONS

1. The issue of horizontal transfer has been a contentious one in the biological community. Describe some of the reasons for this and some of the most debated aspects of the subject.

2. Describe the two major classes of methods used to detect horizontal transfer.

3. What are the advantages and disadvantages associated with these methods?

4. List the three distinct categories of horizontal transfer events in terms of the relationships between horizontally acquired genes and homologous genes (if any) preexisting in the recipient lineage.

5. List some specific examples of horizontal transfers that have conferred functional and/or adaptive benefits on the recipient genomes.

REFERENCES

Aravind, L., R. L. Tatusov, Y. I. Wolf, D. R. Walker, and E. V. Koonin. 1998. Evidence for massive gene exchange between archaeal and bacterial hyperthermophiles. *Trends Genet.* **14:**442–444.

Golding, G. B., and R. S. Gupta. 1995. Protein-based phylogenies support a chimeric origin for the eukaryotic genome. *Mol. Biol. Evol.* **12:**1–6.

Gray, M. W. 1999. Evolution of organellar genomes. *Curr. Opin. Genet. Dev.* **9:**678–687.

Jain, R., M. C. Rivera, and J. A. Lake. 1999. Horizontal gene transfer among genomes: the complexity hypothesis. *Proc. Natl. Acad. Sci. USA* **96:**3801–3806.

Koonin, E. V., A. R. Mushegian, M. Y. Galperin, and D. R. Walker. 1997. Comparison of archaeal and bacterial genomes: computer analysis of protein sequences predicts novel functions and suggests a chimeric origin for the archaea. *Mol. Microbiol.* **25:**619–637.

Koski, L. B., R. A. Morton, and G. B. Golding. 2001. Codon bias and base composition are poor indicators of horizontally transferred genes. *Mol. Biol. Evol.* **18:**404–412.

Lawrence, J. G., and H. Ochman. 2002. Reconciling the many faces of lateral gene transfer. *Trends Microbiol.* **10:**1–4.

Makarova, K. S., L. Aravind, M. Y. Galperin, N. V. Grishin, R. L. Tatusov, Y. I. Wolf, and E. V. Koonin. 1999. Comparative genomics of the Archaea (Euryarchaeota): evolution of conserved protein families, the stable core, and the variable shell. *Genome Res.* **9:**608–628.

Naito, T., K. Kusano, and I. Kobayashi. 1995. Selfish behavior of restriction-modification systems. *Science* **267:**897–899.

Ponting, C. P., L. Aravind, J. Schultz, P. Bork, and E. V. Koonin. 1999. Eukaryotic signalling domain homologues in archaea and bacteria. Ancient ancestry and horizontal gene transfer. *J. Mol. Biol.* **289:**729–745.

Ragan, M. A. 2001. On surrogate methods for detecting lateral gene transfer. *FEMS Microbiol. Lett.* **201:**187–191.

Tatusov, R. L., E. V. Koonin, and D. J. Lipman. 1997. A genomic perspective on protein families. *Science* **278:**631–637.

Wang, B. 2001. Limitations of compositional approach to identifying horizontally transferred genes. *J. Mol. Evol.* **53:**244–250.

Wolf, Y. I., L. Aravind, N. V. Grishin, and E. V. Koonin. 1999. Evolution of aminoacyl-tRNA synthetases: analysis of unique domain architectures and phylogenetic trees reveals a complex history of horizontal gene transfer events. *Genome Res.* **9:**689–710.

Wolf, Y. I., I. B. Rogozin, A. S. Kondrashov, and E. V. Koonin. 2001. Genome alignment, evolution of prokaryotic genome organization, and prediction of gene function using genomic context. *Genome Res.* **11:**356–372.

FURTHER READING

Avery, O. T., C. M. MacLeod, and M. McCarty. 1944. Studies on the chemical nature of the substance inducing transformation of pneumococcal types. *J. Exp. Med.* **79:**137–158.

Doolittle, W. F. 1999. Phylogenetic classification and the universal tree. *Science* **284:**2124–2129.

Doolittle, W. F. 2000. Uprooting the tree of life. *Sci. Am.* **282:**90–95.

Koonin, E. V., K. S. Makarova, and L. Aravind. 2001. Horizontal gene transfer in prokaryotes: quantification and classification. *Annu. Rev. Microbiol.* **55:**709–742.

Koonin, E. V., and M. Y. Galperin. 2002. *Sequence-Evolution-Function: Computational Approaches in Comparative Genomics.* Kluwer Academic Publisher, Boston, Mass.

Medigue, C., T. Rouxel, P. Vigier, A. Henaut, and A. Danchin. 1991. Evidence for horizontal gene transfer in Escherichia coli speciation. *J. Mol. Biol.* **222:**851–856.

Smith, M. W., D. F. Feng, and R. F. Doolittle. 1992. Evolution by acquisition: the case for horizontal gene transfers. *Trends Biochem. Sci.* **17:**489–493.

Sprague, G. F., Jr. 1991. Genetic exchange between kingdoms. *Curr. Opin. Genet. Dev.* **1:**530–533.

Syvanen, M. 1985. Cross-species gene transfer; implications for a new theory of evolution. *J. Theor. Biol.* **112:**333–343.

Syvanen, M., and C. I. Kado (ed.). 2002. *Horizontal Gene Transfer.* Academic Press, New York, N.Y.

Tatusov, R. L., A. R. Mushegian, P. Bork, N. P. Brown, W. S. Hayes, M. Borodovsky, K. E. Rudd, and E. V. Koonin. 1996. Metabolism and evolution of *Haemophilus influenzae* deduced from a whole-genome comparison with *Escherichia coli. Curr. Biol.* **6:**279–291.

WHAT MAKES A BACTERIAL SPECIES? WHEN MOLECULAR SEQUENCE DATA ARE USED, IS rRNA ENOUGH?

Lorraine G. van Waasbergen

21

CURRENTLY RECOMMENDED "GOLD STANDARD" FOR BACTERIAL SPECIES CIRCUMSCRIPTION

TAXONOMIC METHODS: THE "EVOLUTION" OF BACTERIAL CLASSIFICATION
Phenotypic Methods
Genotypic Methods

PHYLOGENETIC ANALYSIS USING MOLECULAR SEQUENCE INFORMATION
Methods for Generating Phylogenetic Trees Based on Molecular Sequence Data
Choosing a Molecule for Sequence Analysis
rRNA as a Marker Molecule
Alternative Marker Molecules

SUMMARY OF THE CURRENT SPECIES CONCEPT FOR PROKARYOTES

What constitutes a species for prokaryotes? This has been and continues to be a difficult question to answer. A biological species has traditionally been considered to be a group of organisms that are able to interbreed and produce fertile offspring. This criterion cannot be directly applied to asexually reproducing organisms. Although under certain circumstances bacteria can exchange genetic material (see "Intercellular mechanisms for gene movement"), bacterial reproduction is fundamentally asexual in nature. Thus, there is debate as

to whether the concept of a species for bacteria can and should have such a natural basis. Bacteria have been found to naturally fall into discrete clusters based on various phenotypic and DNA-sequence properties. It is not clear what mechanisms generate these units; they may be a result of genetic isolation via barriers to recombination (Vulić et al., 1997) or a result of various ecological forces (Cohan, 2001, 2002). It is also not clear whether these clusters can be said to constitute bacterial species. Irrespective of whether the concept of a species for bacteria has a fundamental biological meaning or is simply an operational definition, there is a need to be able to determine the relationships among bacteria and to identify and classify individuals as belonging to specific groups. This is especially true given the tremendous diversity within the prokaryotes and their critical importance to the ecology of the environment, past and present.

For bacteria, a species has commonly been defined simply as a group of strains that are distinctly similar to one another. Often similarity is determined with reference to a designated type strain within the group, even though this strain may not necessarily be the most typical of the group. Just how the similarity within a group is to be determined and how similar these strains need to be to constitute the same species has proven to be a matter of controversy. Thus, the definition of a species for bacteria has been an evolving process. In this chapter I will begin by briefly introducing the

Lorraine G. van Waasbergen, Department of Biology, Box 19498, University of Texas at Arlington, Arlington, TX 76019.

Microbial Evolution: Gene Establishment, Survival, and Exchange
Edited by Robert V. Miller and Martin J. Day, ©2004 ASM Press, Washington, D.C.

presently recommended criteria for establishing species relationships for bacteria. I will then survey the methods that have been used for identifying and classifying prokaryotes, highlighting the ways in which these methods have been used to help construct a definition of a bacterial species. Since sequence analysis of molecules, in particular, 16S rRNA, is providing a powerful phylogenetic framework for the classification of organisms, we will discuss the methods involved in molecular sequence analysis. We will discuss the advantages and limitations in using 16S rRNA sequence analysis in taxonomic studies and explore the importance of including analyses of other gene sequences in bacterial species circumscription.

CURRENTLY RECOMMENDED "GOLD STANDARD" FOR BACTERIAL SPECIES CIRCUMSCRIPTION

In 1987, the Ad Hoc Committee for the Reconciliation of Approaches to Bacterial Systematics met to discuss the criteria by which the taxonomic divisions for bacteria might be defined. The results of this meeting were published in the *International Journal of Systematic Bacteriology*, renamed the *International Journal of Systematic and Evolutionary Microbiology*. This is the official journal of record for reporting on new bacterial taxa and other matters involved in bacterial systematics. At that time it was recommended that a bacterial species be defined as strains with approximately 70% DNA-DNA relatedness and 5°C or less ΔT_m (genotypic concepts that we will introduce below), since these criteria appeared to be descriptive for numerous bacterial species as defined by other methods. It was further suggested that phenotypic characteristics should be consistent with this definition, and only in special cases would phenotype be allowed to override the DNA-DNA similarity criteria. Most recently, in February 2002, the Ad Hoc Committee for the Re-Evaluation of the Species Definition in Bacteriology met to revisit the concept of what should define a bacterial species in light of newer methods and approaches now available to taxonomists. Their recommendation was to

adhere to the DNA-DNA similarity criterion, citing it as a practical and operational approach to species circumscription, while recognizing that newer methods and technologies are currently available and more are likely to be developed. They encouraged testing and application of these methods with a view to their agreement with the DNA-DNA similarity criterion. Moreover, the Committee recommended that species should be identifiable by a variety of currently available methods, both phenotypic and genotypic. In the next section we will review the various methods that have been and are currently being used in determining bacterial taxonomic relationships.

Section Summary

- A species cannot be defined for bacteria as it is for plants and animals, i.e., a group of organisms potentially capable of interbreeding. Therefore, a species of bacteria has been considered simply as a group of strains that are distinctly similar to one another.
- The current recommendation is that a bacterial species should consist of a group of strains with approximately 70% or greater DNA-DNA relatedness and a 5°C or less ΔT_m.

TAXONOMIC METHODS: THE "EVOLUTION" OF BACTERIAL CLASSIFICATION

Phenotypic Methods

CLASSICAL PHENOTYPIC CHARACTERIZATION

The oldest methods of classifying bacteria place them into groups based on observable (phenotypic) similarities between organisms. The traditional phenotypic methods are those that may be familiar to those who have had an introductory microbiology laboratory course. They are frequently used in diagnostic laboratories to determine the identities of unknown organisms. They involve surveying various characteristics, including colony and cell morphology, physiological and biochemical traits, chemical composition and serological reaction, and habitat. The analysis includes characteristics such as cell shape and arrangement; staining characteristics; the presence of endospores or other cellular inclusions; the presence of motility and flagellar orientation; the presence of various enzymes or fermentation products; substrate utilization; antimicrobial susceptibilities; growth preferences for pH, temperature, and salt concentration; and aerobic or anaerobic nature. In diagnostic and taxonomic laboratories these tests are traditionally used in an ordered series (from more general to more specific) to establish the identity of an unknown organism (Fig. 1). Two important resources are commonly used to aid in this process, *Bergey's Manual of Systematic Bacteriology* and *The Prokaryotes*. These works incorporate both traditional phenotypic data and newer genotypic information (most recently, the results of 16S rRNA sequence analysis), presenting detailed information on various bacterial groups in different sections and in diagnostic keys and tables to aid in the identification process.

The use of conventional phenotypic classification systems has a number of drawbacks. They are very time-consuming to perform. Moreover, most of the tests necessitate isolation of the organism in pure culture. Since we now know that less than 1% of bacteria in the environment can be cultured, these methods are obviously extremely limiting for characterizing the natural diversity of bacteria. In addition, in general, it is often difficult to determine which of the phenotypic tests are the most determinative for various groups. For some groups (e.g., pathogens), the set of tests to be used that will be most diagnostic have been better defined than they have been for other groups of organisms (e.g., environmental isolates). Thus, these studies can be limited by the tests or diagnostic keys available or be biased by the choice of the number and type of tests the investigator chooses to use in classifying an unknown organism. If too few tests are used or the incorrect ones inadvertently chosen, then those that are most diagnostic for that particular organism may be missed. Nevertheless, the traditional phenotypic methods have had a tremendous impact in shaping our pres-ent-day view of bacterial taxonomic structure, and phenotypic characteristics continue to be an important part of the description of various taxa.

NUMERICAL TAXONOMY USING PHENOTYPIC DATA

Individual or low numbers of phenotypic tests do not provide enough information to allow an accurate determination of relatedness among organisms, especially if the characteristics being utilized happen not to be those that are most

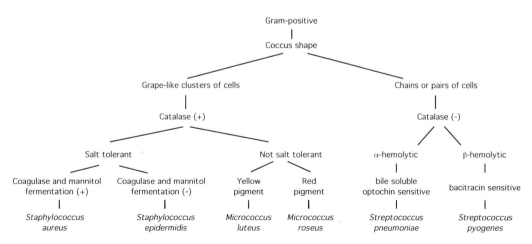

FIGURE 1 Example of the use of a series of classical phenotypic tests to identify an unknown organism. The example shown here shows the procedures that could be used to differentiate among several gram-positive cocci.

diagnostic for those particular strains. However, if enough unweighted phenotypic characteristics are taken into account, groups that are taxonomically relevant begin to appear. This is the basis of numerical taxonomy. Whereas traditional phenotypic methods use the information on various phenotypic traits simply qualitatively (as a description), numerical taxonomy uses a quantitative measurement of the character states of numerous phenotypic features to determine systematic relationships.

An overview of the method of numerical taxonomy is presented in Fig. 2. A set of strains to be analyzed is chosen; often carefully selected, recognized type strains are included in this set to establish hierarchical rankings within the resulting taxonomic clusters. Data

The chosen set of bacterial strains is assayed for a i
different phenotypic properties

The data is coded and scaled (e.g. 1 or 0 for pres
absence of character; no preferential weighting i:
to any particular property)

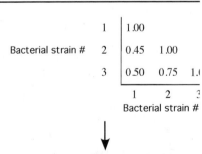

Matrix data is used to generate a dendrogr:

FIGURE 2 The method of numerical taxonomy. A simple example is given in which three organisms are compared.

on a multitude of different traits (i.e., several hundred) are collected from these strains, and these data are coded for computer analysis. A similarity coefficient that shows the overall degree of similarity is then calculated for each pair of organisms within the group based on the data. The similarity coefficients are then arranged in a matrix, and this information is used to generate dendrograms of strains that graphically show their relationships.

Numerical taxonomic methods survey many different characteristics and typically give equal weight to each characteristic, thus avoiding biases that give weight to some characteristics that may not be the most determinative for the particular group under study. However, the outcome still may be somewhat influenced by the choice of tests and by the reference type strains included in the analysis. Therefore, it is wise to evaluate the integrity of numerical taxonomic groupings by examining representative strains from each group using chemotaxonomic and molecular systematic methods (described below). Numerical taxonomic methods often fail to reliably compare distantly related organisms, partially because similar phenotypic features between distantly related organisms are more likely to be analogous rather than homologous traits, and partially because of the difficulty in finding enough similar features between dissimilar organisms for use in the comparisons. Nevertheless, numerical taxonomic analyses have proven to be extremely useful in helping to delineate bacterial groupings at the species level, and the databases of information generated from these analyses have been used to develop reliable identification systems for particular bacterial taxa.

ADVANCED PHENOTYPIC TECHNIQUES

The development of various analytical techniques has allowed the categorization of bacteria based on aspects of their chemical composition. Application of methods for separating and characterizing particular compounds from the cell has provided information for the classification of bacteria based on the composition of specific cellular components (chemotaxonomy). Such analyses include surveying the types of peptidoglycan or teichoic acids present in the cell wall (used for gram-positive organisms), various aspects of the lipid composition of the cell (e.g., fatty acid methyl ester content or FAME), the types of isoprenoid quinones present in the cell membrane, and the amount and character of polyamines present within the cell (used for discrimination at approximately the genus-species level and above). Phenotypic typing methods such as serotyping, phage typing, or comparisons of the electrophoretic profiles of either proteins (e.g., multienzyme electrophoresis) or lipopolysaccharides present within cells have proven useful for characterizing relationships only within closely related organisms (i.e., within a species). Some of the newest phenotypic methods, which have thus far been applied only to a limited number of bacterial groups, involve bulk cellular composition analysis through the use of various physical methods: Fourier-transform infrared spectroscopy (FTIR), pyrolysis mass spectrometry, matrix-assisted laser desorption/ionization time-of-flight mass spectrometry (Maldi/Tof), or spray-ionization mass spectrometry.

Genotypic Methods

DNA BASE COMPOSITION

The discovery of DNA as genetic material and the development of molecular techniques have allowed the development of techniques for classifying bacteria based on their genetic relatedness. One of the first genotypic methods to be developed for classifying bacteria was based on the determination of the overall nucleotide base composition of DNA. The ratio of guanine (G) and cytosine (C) relative to adenine (A) and thymine (T) naturally varies from organism to organism (genome to genome). The base composition is typically expressed as the relative abundance of G and C in the cell. This is referred to as the G+C content, the GC ratio, or moles percent G+C $[(G+C)/(A+T+C+G) \times 100]$. G+C content for prokaryotes ranges from about 20 to 80 mol% G+C. Strains

that are similar tend to have a similar G+C content (within approximately 5 mol% G+C for closely related species). Strains having differences in G+C content greater than 20 to 30 mol% appear to have essentially nothing in common, while those whose differences are greater than 10 mol% different are considered not to belong to the same genus (with differences of more than 15 mol% indicating heterogeneity within an established genus). However, because the same ratio of bases can be constructed out of very different base sequences, strains that are unrelated can have very similar or even identical G+C contents. Thus, having a similar G+C content does not necessarily ensure a close relationship between strains, but having a significantly different G+C content does indicate that the strains are not closely related. In this way, G+C content is a useful measure, and it is commonly included in the description of bacterial taxa.

DNA-DNA HYBRIDIZATION

Whereas the G+C content does not make use of the information present in similarities in the linear arrangement of bases in DNA, it is utilized in DNA-DNA hybridization analyses. This extra information, and the fact that it samples the whole genome, makes DNA-DNA hybridization methods extremely sensitive. DNA-DNA similarity methods (when performed carefully) can give a very accurate determination of whole-genome relatedness. These methods are based on the ability of DNA strands to hybridize with DNA strands of identical or similar linear sequence. The proportion of hybridization between two populations of denatured DNA is a relative indication of their degree of relatedness.

The typical steps involved in a DNA-DNA hybridization experiment are outlined in Fig. 3. DNAs isolated from two separate organisms are mixed and denatured to yield single-stranded DNA molecules. The mixture is allowed to renature and the amount of heteroduplex DNA subsequently formed is measured by one of several different methods. In the procedure, typically the reference DNA (the DNA from

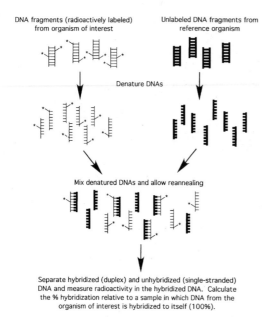

DNA fragments (radioactively labeled) from organism of interest

Unlabeled DNA fragments from reference organism

Denature DNAs

Mix denatured DNAs and allow reannealing

Separate hybridized (duplex) and unhybridized (single-stranded) DNA and measure radioactivity in the hybridized DNA. Calculate the % hybridization relative to a sample in which DNA from the organism of interest is hybridized to itself (100%).

FIGURE 3 Overview of the steps involved in a typical DNA-DNA hybridization assay.

a strain with which the DNA from the strain of interest is being compared) is included in excess to prevent DNA from the strain of interest from reassociating with itself. The amount of heteroduplex hybrid formed is commonly compared with a separate sample in which the strain's DNA is reannealed to itself (which would be 100%) to give the percent

DNA-DNA relatedness between the two strains. A related parameter that can be used to indicate the degree of relatedness of the hybrid is its thermal stability. The thermal stability can be calculated as the difference in the thermal denaturation midpoint (T_m, the temperature at which half of the DNA strands are denatured) of the heteroduplex DNA hybrid as compared with the T_m of DNA from the organism of interest hybridized to itself (homoduplex DNA), or the ΔT_m (Fig. 4). Thermal stabilities decrease 1 to 2.2% for each 1% mismatch in sequence. Both the DNA-DNA relatedness and ΔT_m parameters are used for delineation of bacterial species. The use of these methods is restricted to very closely related organisms such as species; DNA heteroduplexes do not form if the difference between their sequences is greater than 10 to 20% (approximately genus level).

As discussed above, the current standard for what constitutes a bacterial species is based on percentage of DNA-DNA hybridization and ΔT_m. For species circumscription it is recommended that species include strains that have approximately 70% or more DNA-DNA relatedness and 5°C or less ΔT_m. These somewhat arbitrary values arose based on observations of numerous studies using strains that had been well defined by other methods to constitute the same species.

FIGURE 4 Thermal denaturation profiles of two hypothetical duplex DNA samples, one a heteroduplex of DNA from two different organisms, the other a homoduplex of DNA from one of the organisms hybridized to itself. The T_m of both samples is indicated, as is the ΔT_m.

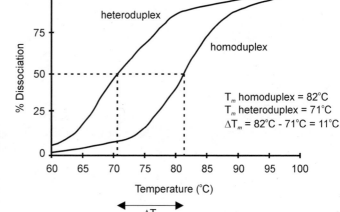

T_m homoduplex = 82°C
T_m heteroduplex = 71°C
ΔT_m = 82°C - 71°C = 11°C

Despite its sensitivity, there are some problems associated with the application of DNA-DNA reassociation methods. The techniques must be carefully performed to avoid significant differences in experimental outcome (typical errors are on the order of about 5 to 6%). The DNA must be of high quality and the experimental conditions must be carefully adjusted to take into account the mole percentage G+C of the organism. Another drawback to the method is that only two organisms can be compared at a time, which limits the number of analyses that can be done. Since typically only a limited number of comparisons can reasonably be done, and only DNA from very similar organisms will form duplexes, the reference organisms must be chosen with care to allow an accurate species identification.

DNA FINGERPRINTING

Like phenotypic typing methods (see above), DNA typing (also called DNA fingerprinting) methods generally allow determination of relationships only between members of the same species. How the resolving power of DNA-typing methods compares with that of some of the other methods of classification of bacteria discussed in this chapter is given in Fig. 5. However, there is some indication that certain fingerprinting methods may be useful for helping to delineate species, and, in general, their use gives support for placing strains within the same species. DNA-fingerprinting methods are based on the principle that naturally evolving changes in the sequence of bases in DNA between organisms can lead to the loss or gain of particular restriction enzyme recognition sites. The more differences there are between two organisms, the more often they will have changes that lead to the loss or gain of a site. There have been a plethora of various fingerprinting techniques developed, but the methods are basically of two types: the first-generation methods that are based on restriction enzyme digests of the whole genome, and later methods that use PCR to amplify particular regions of the genome. In the first-generation methods, DNA is extracted and subjected to restriction enzyme digestion, followed by separation of fragments by agarose gel electrophoresis, and visualization of the resulting fragment patterns. The complexity of the patterns can be reduced through the use of rare-cutting enzymes (low-frequency restriction fragment analysis [LFRFA]) followed by separation of the large fragments by pulsed-field gel electrophoresis. More frequently cutting enzymes could also be used, followed by probing a Southern blot of the DNA digest with a probe specific for a particular gene and examination of the pattern of hybridizing fragments. Probes specific for the highly conserved rRNA gene are often used for this purpose (in which case the DNA-based typing method is termed ribotyping). A variety of different PCR-based typing methods have been developed. For instance, there are those that involve the use of arbitrary primers to randomly amplify the genome, while others amplify highly repetitive elements that are scattered about the genome of different organisms. There are also methods that combine

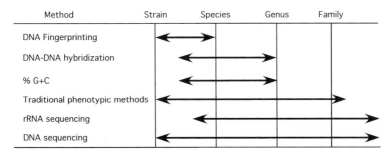

FIGURE 5 Efficacy of resolution by various taxonomic methods. Information is adapted from Vandamme et al. (1996).

PCR-based approaches with restriction enzyme analysis. Examples of these methods include restriction fragment length polymorphism (RFLP) analyses of amplicons of rRNA or other genes, as well as the amplified fragment length polymorphism (AFLP) method in which the DNA is first digested, followed by a selective amplification of certain fragments. In the various fingerprinting methods, similarities among the resulting fragment patterns can be analyzed by numerical analysis, resulting in a dendrogram showing relatedness among strains. They can also be compared with the fragment patterns obtained from known organisms to facilitate identification of unknown organisms.

DNA SEQUENCE ANALYSIS

DNA sequencing is the determination of the order of nucleotides in DNA. Use of sequence data in taxonomic studies involves the comparative analysis of homologous DNA sequences from extant organisms to infer their evolutionary relationships. There are a much larger number of definable characters present in sequence information than there can be in phenotypic analyses, making taxonomic analysis using sequence data a much more accurate measure of relatedness. Moreover, sequence analysis allows direct insight into the genetic relatedness of organisms.

Sequence analysis provides a phylogenetic framework for modern taxonomic classification. The development of molecular phylogenetic analysis, and sequence analysis of the 16S rRNA gene in particular, has allowed the determination of natural relationships among individuals and has in no small way revolutionized bacterial systematics. 16S rRNA studies led to the significant discovery that life evolved along the three main lines of descent of *Archaea*, *Bacteria*, and *Eukarya*, and they have revealed the rich diversity of prokaryotes present in the natural environment. Through the use of gene-cloning and PCR techniques, organisms need not be isolated in pure culture to obtain DNA sequence information from them, although, of course, not having the organism isolated in pure culture prohibits the use of other technologies such as those discussed

above to confirm a taxonomic position based on DNA-sequence analysis.

Because of the power behind the sequencing approach in helping to determine taxonomic relationships of organisms, and its widespread use, we will consider it here in some detail. An overview of the methods and considerations involved in molecular sequence analysis of the rRNA genes and other molecules for phylogenetic purposes is given in the next section with special reference to their use in helping to determine bacterial species relationships.

Section Summary

- Phenotypic methods

  Bacteria can be grouped based on similarities between their observable characteristics, collectively known as their phenotype.

  Numerical taxonomy is the multivariate analysis of numerous characteristics for the classification of organisms.

  Bacteria can be grouped using chemotaxonomy.

- Genotypic methods

  The nucleotide base composition (G+C content) can be used as part of the description of a species or genus.

  Percentage DNA-DNA hybridization and thermal stability serve as sensitive indicators of the relatedness between species.

  Intraspecies relationships can be determined using DNA-fingerprinting methods.

  DNA sequence data can be analyzed to determine the evolutionary relationships among organisms.

PHYLOGENETIC ANALYSIS USING MOLECULAR SEQUENCE INFORMATION

Methods for Generating Phylogenetic Trees Based on Molecular Sequence Data

Sequence information for the organisms of interest may be present in publicly accessible databases or it can be generated experimentally.

For organisms that have not been cultured, gene sequences can be generated from DNA extracted from an environmental sample containing the organisms of interest (e.g., soil, water, and tissue). The typical steps in obtaining sequence data from an isolate of interest or an environmental sample are as follows. DNA is extracted from a pure culture of an organism or from a natural community. The target gene may be cloned, detected in a library by hybridization, and then sequenced by standard methods. Alternatively, a more contemporary approach involves PCR amplification using primers designed to target conserved areas of the gene. It is desirable to obtain a large amount of the gene for sequencing; therefore, the clone libraries should be composed of large fragments of DNA or, in the case of PCR approaches, the primers must be designed to regions of the target gene that are a fair distance apart to maximize the amount of usable data present in the amplicon. In this regard, certain molecules may be limiting in the amount of usable information present between conserved regions. The PCR product is then cloned and sequenced or sequenced directly. In using the PCR-based approach there is always the risk of production of PCR-based artifacts. These arise from polymerase error and through chimera formation from the coamplification of mixed genomes from environmental samples.

A phylogenetic tree is a graphical representation of the predicted evolutionary relationships among extant organisms based on comparisons of molecular sequence data. Trees show the relative evolutionary relationship of organisms with one another and comparisons can be made concerning the differences in branching order and overall topology of trees based on the sequence of different marker molecules. Internal nodes connect groups of organisms with a common ancestor (monophyletic groups) and trees may be examined for the presence of monophyletic clusters of organisms that represent taxonomic groups. The taxonomic rank these clusters represent depends on the composition of the group. No clear cutoff for a particular unit (e.g., species) can typically be assigned to these clusters based on tree data alone. However, comparisons can be made of the relationships among representative individuals from the cluster based on other methods to assign a tentative level of relatedness.

The sequences from the set organisms of interest must be aligned before phylogenetic analysis can be performed to ensure that the characters (bases or amino acids) being compared with one another are homologous. A number of computer programs are available to assist in the alignment process, and it is prudent to also refine the alignment by eye, particularly when deciding whether a character difference is a result of a change in that character or a result of an insertion or deletion event. For protein-encoding sequences, conservative changes in the amino acid sequence can be taken into consideration to aid in the alignment of the characters. For rRNA molecules, there are regions of the primary sequence of the molecule that are more highly conserved than other regions (see Fig. 6). In addition, rRNAs play an important role in ribosome assembly and translation, and the higher order structure of rRNAs appears to be highly functionally conserved. The conserved regions of primary sequence can be used to initially align rRNA sequences. Then the more difficult task of aligning the more variable regions can be attempted, using the secondary structures for other known bacterial rRNAs as a guide, conserving the higher order sequence elements in alignments.

Once the sequences are aligned, they can be analyzed by a number of different methods to generate phylogenetic trees. Several of the most popular tree-reconstruction approaches are maximum parsimony, distance matrix, and maximum likelihood. The treeing methods operate using certain underlying models of evolution. Distance matrix methods compare the differences in the aligned sequences for every pair in the data set and construct a matrix of dissimilarity values. These dissimilarity values are transformed into phylogenetic distances

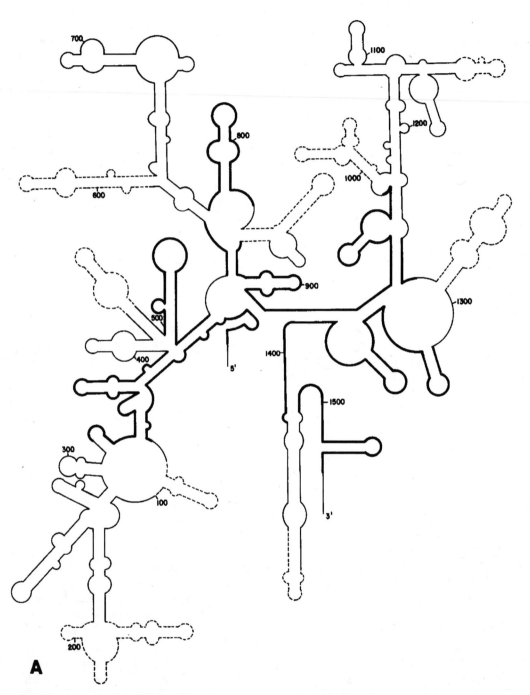

FIGURE 6 Conservation of regions in the 16S rRNA primary sequence and use of the conserved sequences and secondary structure to facilitate alignment of 16S rRNA sequences. Figure is reprinted from *Advances in Microbial Ecology* (Ward et al., 1992) with permission from the publisher. (A) Diagram of the secondary structure of the 16S rRNA molecule from *Escherichia coli*. Line thickness indicates the level of conservation of various regions of the primary sequence. The thick lines indicate regions of nearly universal conservation, thin lines indicate regions with an intermediate level of conservation, and dashed lines indicate hypervariable regions.

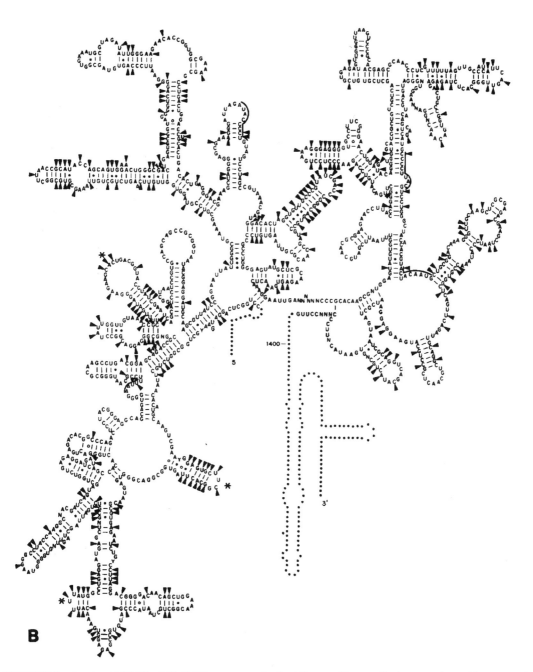

FIGURE 6 *(continued)* (B) The 16S rRNA sequence obtained from an Octopus Spring cyanobacterial mat sample superimposed on the *E. coli* 16S rRNA secondary structure. Arrowheads indicate the nucleotides that differ from the *E. coli* sequence. Features that characterize the sequence as being from a cyanobacterium are indicated with lines (primary structure features) or asterisks (secondary structure features).

in an attempt to correct for the possibility of multiple changes at any given site. The corrected distances are then used to generate the tree whose topology and branch lengths lead to distances that most closely approximate those in the data set. Maximum parsimony analysis examines each position and seeks to build the most parsimonious tree, one that requires the minimum number of mutational changes to have taken place. Maximum likelihood methods (the most sophisticated but most computationally intensive treeing method) examine individual sequence positions and attempt to generate a tree topology that best fits the observed changes to a particular model of evolution. Since no model of evolution can be perfect (take into account all the possible complexities involved in the actual evolution of the sequences), it is a good idea to generate trees based on different methods and compare them to give an idea of the robustness of any particular tree topology. Various filters and variable weighting methods can also be used to indicate that multiple substitutions have taken place and to reduce their effect. Since the input order of sequences can affect the outcome, various input orders should be tested when performing analyses. It is also important to apply some type of resampling analysis to the data (such as bootstrapping or jackknifing) to provide an estimate of the statistical support present for that particular tree topology. Rooting of trees can be done to give an order to the evolution of a group by including an outgroup in the study (a close relative that is outside of the group of interest).

Choosing a Molecule for Sequence Analysis

The idea that certain cellular macromolecules could be used as molecular chronometers or marker molecules to document the evolutionary history of organisms was first proposed by Zuckerkandl and Pauling in 1965. Carl Woese and his colleagues a decade later produced seminal studies in which they used the sequence of 16S rRNA to establish phylogenetic relationships of prokaryotes and the recognition of

the three-domain system of life. The 16S rRNA molecule has a number of attractive features for its use in molecular phylogenetic studies. However, it may not be the best marker molecule for use in all cases, particularly when trying to examine certain close relationships such as between species.

There are a number of considerations in choosing a molecule to sequence for a particular study (Table 1). Naturally the molecule must be present in all the strains of interest. Moreover, to ensure relatively smooth-running "clock-like" behavior, mutations should occur randomly, at a rate that will reflect the true evolutionary distance between organisms of interest. The ideal evolutionary chronometer would seem to be one that mutates randomly with no constrictions on functional sites that may mutate at different rates. The only sequences that would do this are functionless regions of the genome. Unfortunately, these regions accumulate multiple mutations very rapidly and are only useful for detecting extremely close relationships. Thus, for most studies it is necessary to use parts of the genome that are functional and mutate less rapidly. These sequences will naturally have sites that are more or less functionally constrained than other sites. If a molecule under consideration as a marker has evolved a different function between two organisms of interest, then the functionally important parts of the molecule would likely have changed, and, correspondingly, the selective pressures on different parts of the molecule will have become different. Thus, local nonrandom mutations could accumulate that would lead to an overestimation of phylogenetic distances and determination of improper branching orders

TABLE 1 Considerations in choosing a phylogenetic marker molecule

Wide distribution
Functional consistency
Genetic stability
Reasonable number of independent evolving regions
Reasonable level of conservation
High information content

among organisms. Therefore, another characteristic of a marker molecule is that it must have the same function in all organisms of interest.

The rate of evolution of the marker molecule should also be appropriate for the degree of divergence of the strains of interest. If the molecule is too conserved (highly functionally constrained), then it will not have recorded enough changes to be able to separate strains as being dissimilar. If too many changes have taken place, it is difficult to align sequences, and multiple mutations may have accumulated, obscuring the true relationships between more distantly related strains. A marker molecule should be large enough to contain a fair amount of phylogenetically useful information (for DNA sequence data, approximately 1,000 nucleotides or more is desirable, depending on the rate of evolution of the sequence and the level or relationship between strains). It is also helpful in a marker molecule to have variable and conserved regions, to allow different degrees of resolution among different groups of interest. Moreover, having more conserved regions interspersed with more variable regions facilitates PCR and sequencing primer design to the more conserved regions, and these more conserved areas facilitate alignment (the first step in the phylogenetic analysis), while the more variable regions still allow a degree of discrimination. Another helpful attribute is to have a number of independent domains within the molecule. If that is the case, then, if a large change affects one domain of the molecule, the rest of the molecule will have been unaffected, and the overall molecule may still be able to function as a good marker molecule. It is also important to choose a marker molecule that plays an important role in a complex process in the cell so that it is not likely to have been subject to lateral transfer during the course of the evolution of the group, which would confuse phylogenetic analyses. Finally, it is desirable that the marker molecule gene exists in a single copy per cell, thus avoiding having complicated phylogenetic analyses with an underlying duplication event.

rRNA as a Marker Molecule

The larger molecules of rRNA (i.e., 16S and 23S in bacteria) have many of the characteristics of a desirable marker molecule. For historical and technical reasons the rRNA from the small subunit (SSU) of the ribosome, the 16S rRNA molecule, has been more commonly used than the 23S rRNA (from the large ribosomal subunit), and therefore the data sets available are larger for this molecule. Ribosomal RNAs are ancient molecules that occur in all organisms. As important elements of the cell's translational machinery, they are highly structurally and functionally conserved. They have both variable regions and conserved regions (Fig. 6), allowing determination of more distant relationships (such as domain-level) as well as closer, sometimes species level relationships. The highly conserved regions and higher-order structure elements facilitate alignment of sequences from different individuals. There are also short stretches of the rRNA sequence that are unique to certain groups of organisms, and these signature sequence regions can be used as probes for the identification of particular organisms. The 16S rRNA and 23S rRNA molecules (approximately 1,600 and 3,000 nucleotides, respectively) are large enough to contain a fair amount of information. Bacterial cells have been found to contain more than one rRNA gene, but these alleles are typically identical or nearly so, although there are examples where this is not the case. In addition, rRNA molecules are thought to be largely unaffected by lateral transfer, as are other genes that are involved in the fundamentally involved processes of transcription, translation, or DNA replication. However, there are some cases in which it appears that horizontal transfer of rRNA genes may have taken place. Finally, rRNA molecules contain many independent domains so that local nonrandom rearrangements normally do not obscure relationships.

The 16S rRNA molecule of bacteria has been used extensively for phylogenetic analysis. It allows a wide range of discrimination among organisms, and, within the expected range of resolution for various methods, there is generally a good agreement between relationships obtained using other taxonomic methods and those obtained using 16S rRNA analysis. Moreover, phylogenetic trees for the major bacterial groups based on alternative marker molecules (see below) tend to give trees with generally similar topologies as those based on 16S rRNA. Indeed, the latest editions of the *Bergey's Manual of Systematic Bacteriology* and *The Prokaryotes* use 16S rRNA analysis as the phylogenetic framework upon which to base relationships among organisms.

The 16S rRNA molecule has proven extremely useful for determining the phylogenetic relationships of the prokaryotes, but it is highly conserved and this places limitations on its discriminatory power. Although the relationship is not linear, strains that show DNA-DNA relatedness values of greater than 70% (the current criterion for a species) tend to have very similar 16S rRNA identities (above 97%) (Stackebrandt and Goebel, 1994). Comparison studies using 16S rRNA homology and DNA-DNA reassociation methods indicate that the latter technique has better resolving power when comparing closely related strains; there are cases where organisms have been found to have almost identical 16S rRNA sequences but much less than 70% DNA-DNA relatedness values (see, for example, Fox et al., 1992). Nevertheless, 16S rRNA sequences are useful for helping to determine the coarse-scale phylogenetic positions of organisms. Moreover, the 16S rRNA sequence may also help to reposition strains that have been incorrectly classified as belonging to the same species; strains showing 16S rRNA sequence identities less than 97% similar are unlikely to have a DNA-DNA relatedness of greater than 60%, and thus would not fit the criterion of species. Recognizing the utility and widespread use of 16S rRNA sequence analyses in bacterial taxonomy, it has been recommended that 16S rRNA sequence information (greater than 1,300 nucleotides) be included in the description of a species (Stackebrandt et al., 2002). Since the small percentage of sequence differences that distinguish certain species tend to be clustered in different regions of the molecule, it is necessary to have nearly complete sequences to reliably differentiate between different groups.

Alternative Marker Molecules

Since there are cases where rRNA fails to discriminate between closely related strains, in order to confirm relationships among organisms based on rRNA analyses, investigators have been exploring the use of alternative marker molecules. Various protein-coding sequences that have many of the qualities of good molecular marker molecules have been proposed for use as alternatives to rRNA for determining evolutionary relationships. Some of the markers that have been used for this purpose are listed in Table 2. In some cases, use of these "alternative" markers allows better discrimination of species-level differences than does use of the 16S rRNA gene. For example, Fig. 7 shows the results of a study (Mollet et al., 1997) in which use of the gene for the β-subunit of RNA polymerase, *rpoB*, was compared with use of the 16S rRNA gene for genotypic and phylogenetic analyses of strains of the family *Enterobacteriaceae*. The results of this study indicated that relationships based on *rpoB* were generally congruent with those based on 16S rRNA. However, use of *rpoB* allowed a finer degree of resolution for distinguishing among strains, and trees based on *rpoB* were more compatible with the accepted view of relationships among species than were those based on 16S rRNA. The selective pressure for a protein-coding sequence is at the amino acid level, whereas for an rRNA it is directly at the nucleotide level. Due to degeneracy present in the genetic code, the rates of change in protein coding sequences at the nucleotide level are naturally faster than that for an rRNA mole-

TABLE 2 Alternative phylogenetic molecular markers

RNA polymerase
Elongation factor Tu
β-Subunit of F_1F_0-ATPase
RecA protein
Heatshock protein 60

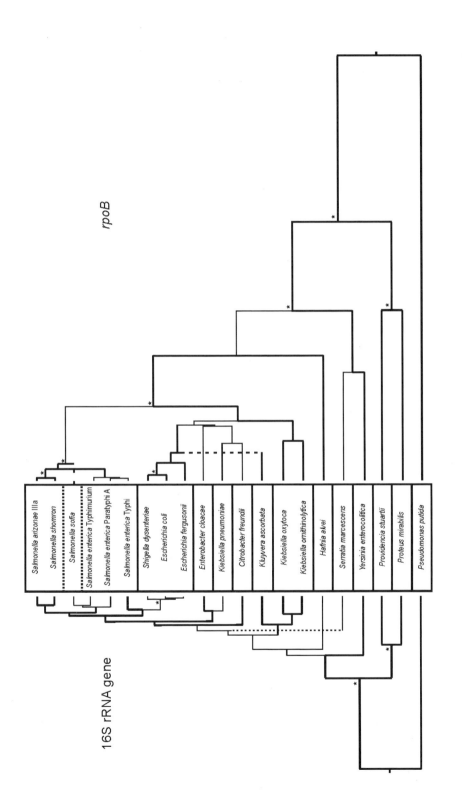

FIGURE 7 Phylogenetic trees based on 16S rRNA gene and partial *rpoB* gene sequences from 20 enteric strains. The figure is reproduced from *Molecular Microbiology* (Mollet et al., 1997) with permission of the publisher. The trees are those obtained from DNA maximum likelihood analyses with a molecular clock using *Pseudomonas putida* as an outgroup. Asterisks indicate nodes with bootstrap values >90%, and the thick lines indicate topologies that are conserved in analyses using neighbor joining from Jukes and Cantor DNA distance matrix and DNA parsimony.

cule, but just how much faster depends on the particular sequence. Thus, species that are not differentiated using 16S rRNA may be distinguished through analysis of the more rapidly evolving protein-encoding sequences.

There are some advantages in using sequences of protein molecules over rRNA sequences for phylogenetic analyses. Consideration of the amino acid translation of a protein-encoding gene facilitates the alignment of its nucleotide sequence, making it easier than rRNA alignments. Also, translation of the gene sequence into amino acid sequence also permits a verification of the accuracy of the nucleotide sequence. Moreover, comparisons can be done using either the amino acid or the nucleic acid sequence, allowing determination of relationships at different levels. In addition, for protein-coding genes, restrictions can be placed on how the nucleotides at the three different places in the codon are included in the analysis with respect to nucleotide compositional differences between individual sequences to reduce the chances of branching order artifacts.

There are some potential drawbacks and other considerations in using protein-encoding genes in phylogenetic studies. These genes may be more subject to lateral transfer than the rRNA genes. Moreover, in some cases there are multiple copies of genes in an individual organism or evidence of past duplication of protein-encoding genes. Also, not all genes will evolve at the same rate. Some protein sequences evolve too fast to allow discrimination between most any group. On the other hand, phylogenetic analyses of distantly related bacterial groups based on sequences of the alternative marker molecules listed in Table 2 give trees with overall similarity to those based on 16S rRNA, with only relatively minor differences in topology with low statistical significance. In general, 16S rRNA is useful for exploring deeper relationships, while protein-coding genes appear to better resolve closer relationships. The potential for using various metabolic protein-encoding (housekeeping) genes to help in delineating bacterial species

was recognized by the ad hoc committee that reevaluated the bacterial species definition (Stackebrandt et al., 2002). They acknowledge that analysis of such genes holds great promise for species circumscription and recommended that such analysis be further tested for its ability to delineate species as compared with other methods such as rRNA sequence analysis.

When comparing trees resulting from phylogenetic analysis of different genes, there are bound to be conflicts in aspects of the tree topologies. In some cases the inclusion of more sequence data may help resolve differences. Continued differences in gene trees may result from the lateral transfer of certain molecules, or they may reflect the differences in the resolution ability of different molecules. Therefore, it is wise to choose several different molecules to sequence for any particular phylogenetic study. This is the essence of the multilocus sequence typing (MLST) technique (Maiden et al., 1998), a method that has been used for determining interspecies variability, which involves classifying strains according to the internal sequences of a number (seven or more) of housekeeping genes.

Section Summary

- Methods for generating phylogenetic trees based on molecular sequence data
    Sequence data can be obtained from public databases or generated experimentally for cultured or noncultured organisms.
    From alignments of homologous sequences, phylogenetic trees can be generated based on various models of evolution.
- Choosing a molecule for sequence analysis
    Among the attributes of a good marker molecule for phylogenetic sequence analysis are universal distribution, functional consistency, and high information content.
    The molecular marker molecule chosen for a particular phylogenetic study should have a rate of evolution that is appropriate for the degree of diversity among the strains of interest.

- rRNA as a marker molecule

  The highly conserved rRNA molecules have many of the attributes of good molecular markers and are especially good for use in determining relationships among distantly related organisms. Bacterial strains of the same species tend to have 97% or greater sequence identity in their 16S rRNA genes.

- Alternative marker molecules

  The 16S rRNA gene can be too conserved to distinguish between organisms at the species level.

  Sequence analysis of certain housekeeping, protein-coding genes may help to distinguish bacterial species.

SUMMARY OF THE CURRENT SPECIES CONCEPT FOR PROKARYOTES

In lieu of a definitive biological concept of a bacterial species, most microbiologists embrace a practical, operational definition. The ad hoc committee for the reevaluation of the bacterial species definition (Stackebrandt et al., 2002) maintains that the 70% DNA-DNA similarity and 5°C or less ΔT_m criteria should continue to define a bacterial species. They recommend that species should be identifiable by a number of methods. Thus, the description of a species should include an almost complete 16S rRNA gene sequence. In addition, phenotypic characteristics (especially those most discriminatory, including chemotaxonomic markers) should continue to play an important role in helping individuals to delineate species. Also the $G+C$ content of DNA should be included in the description for the type strain of the type species of a new genus (but not necessarily for the type strain of a new species). It was recognized that there are some very promising new genomic technologies that should be further developed to help define inter- and intraspecies relationships, including sequence analysis of various housekeeping genes and DNA-typing methods, as well as DNA arrays and other technologies that take advantage of whole-genome sequence data.

When exploring prokaryotic taxonomy, due to variability in classifications resulting from the use of different methods, many feel that the best way to obtain a reliable classification of prokaryotes is to use as many techniques as possible, the so-called polyphasic approach (Fig. 8). Polyphasic taxonomy combines the results of a wide range of phenotypic, genotypic, and phylogenetic information to determine a consensus classification. The first step in delineating bacterial species is to determine the DNA relatedness. Information from phylogenetic analysis based on various marker molecules can be integrated into the analysis as well as the results of phenotypic studies. Thus, the polyphasic approach is not necessarily inconsistent with the present recommendation of a bacterial species as long as the DNA-relatedness criteria are adhered to in defining a species. As more information is gathered on the evolutionary processes that generate taxonomic diversity within prokaryotes, it will likely be necessary to revisit the possibility of adopting a less absolute and perhaps more natural definition of a bacterial species.

QUESTIONS

1. Why is it difficult to determine what constitutes a bacterial species?

2. List the various methods, phenotypic and genotypic, that have been used to identify and classify bacteria.

3. Briefly outline the steps involved in phylogenetic analysis using molecular sequence data.

4. What are the considerations in choosing a molecule for phylogenetic sequence analysis?

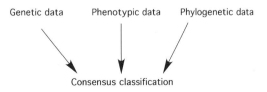

FIGURE 8 Diagram indicating the major categories of the many pieces of taxonomic data that are used to ultimately form a consensus classification using a polyphasic taxonomic approach.

Why might the use of certain protein-coding sequences be preferable to the use of rRNA for distinguishing bacteria at the species level?

5. What is the currently accepted definition for a bacterial species?

REFERENCES

Cohan, F. M. 2001. Bacterial species and speciation. *Syst. Biol.* **50:**513–524.

Fox, G. E., J. D. Wisotzkey, and J. P. Jurtshuk. 1992. How close is close: rRNA sequence identity may not be sufficient to guarantee species identity. *Int. J. Syst. Bacteriol.* **42:**166–170.

Maiden, M. C. J., J. A. Bygraves, E. Feil, G. Morelli, J. E. Russel, R. Urwin, Q. Zhang, J. Zhou, K. Zurth, D. A. Caugant, I. M. Feavers, M. Achtman, and B. G. Spratt. 1998. Multilocus sequence typing: a portable approach to the identification of clones within populations of pathogenic organisms. *Proc. Natl. Acad. Sci. USA* **95:**3140–3145.

Mollet, C., M. Drancourt, and D. Raoult. 1997. rpoB sequence analysis as a novel basis for bacterial identification. *Mol. Microbiol.* **26:**1005–1011.

Stackebrandt, E., and B. M. Goebel. 1994. Taxonomic note: a place for DNA-DNA reassociation and 16S rRNA sequence analysis in the present species definition in bacteriology. *Int. J. Syst. Bacteriol.* **44:**846–849.

Vandamme, P., B. Pot, M. Gillis, P. D. Vos, K. Kersters, and J. Swings. 1996. Polyphasic taxonomy, a consensus approach to bacterial systematics. *Micobiol. Rev.* **60:**407–438.

Vulíc, M., F. Dionisio, F. Taddei, and M. Radman. 1997. Molecular keys to speciation: DNA polymorphism and the control of genetic exchange in enterobacteria. *Proc. Natl. Acad. Sci. USA* **94:**9763–9767.

Ward, D. M., M. M. Bateson, and R. Weller. 1992. Ribosomal RNA analysis of microorganisms as they occur in nature, p. 219–286. *In* K. C.
Marshall (ed.), *Advances in Microbial Ecology*, vol. 12. Plenum Press, New York, N.Y.

FURTHER READING

Brenner, D. J., J. T. Staley, and N. K. Krieg. 2001. Classification of prokaryotic organisms and the concept of a bacterial species, p. 27–31. *In* D.R. Boone, R.W. Castenholz, and G.M. Garrity (ed.), *Bergey's Manual of Systematic Bacteriology*, 2nd ed. Springer, New York, N.Y.

Cohan, F. M. 2002. What are bacterial species? *Annu. Rev. Microbiol.* **56:**457–487.

Goodfellow, M., G. P. Manfio, and J. Chun. 1997. Towards a practical species concept for cultivable bacteria, p. 25–59. *In* M.F. Claridge, H.A. Dawah, and M.R. Wilson (ed.), *Species: The Units of Biodiversity*. Chapman & Hall, London, United Kingdom.

Rosselló-Mora, R., and R. Amann. 2001. The species concept for prokaryotes. *FEMS Microbiol. Rev.* **21:**39–67.

Stackebrandt, E., W. Frederiksen, G. M. Garrity, P. A. D. Grimont, P. Kämpfer, M. C. J. Maiden, X. Nesme, R. Rosselló-Mora, J. Swings, H. G. Trüper, L. Vauterin, A. C. Ward, and W. B. Whitman. 2002. Report of the ad hoc committee for the re-evaluation of the species definition in bacteriology. *Int. J. Syst. Evol. Microbiol.* **52:**1043–1047.

Vandamme, P., B. Pot, M. Gillis, P. de Vos, K. Kersters, and J. Swings. 1996. Polyphasic taxonomy, a consensus approach to bacterial systematics. *Microbiol. Rev.* **60:**407–438.

Wayne, L. G., D. J. Brenner, R. R. Colwell, P. A. D. Grimont, O. Kandler, M. I. Krichevsky, L. H. Moore, W. E. C. Moore, R. G. E. Murray, E. Stackebrandt, M. P. Starr, and H. G. Trüper. 1987. Report of the ad hoc committee on reconciliation of approaches to bacterial systematics. *Int. J. Syst. Bacteriol.* **37:**463–464.

Woese, C. R. 1987. Bacterial evolution. *Microbiol. Rev.* **51:**221–271.

CAN WE UNDERSTAND BACTERIAL PHYLOGENY, AND DOES IT MAKE ANY DIFFERENCE ANYWAY?

Robert V. Miller and Martin J. Day

22

MODELING GENETIC CHANGE

WHY DO WE NEED BACTERIAL
SPECIES ANYWAY?

In a 2002 article in *ASM News*, Radhey Gupta states:

> Understanding evolutionary relationships among prokaryotes constitutes one of the most fundamental challenges in biology. Because prokaryotes were the sole inhabitants for nearly the first 2 billion years of life on earth, they are key to understanding fundamental questions about the nature and origin of the first cell, metabolism, photosynthesis, information transfer processes and eukaryotic cells. Earlier efforts based on morphological, biochemical, and physiological characteristics met with limited success in describing how various groups of prokaryotes are evolutionarily related.

However, recent advantages in genomics and other molecular techniques have allowed us to make great headway in the classification of prokaryotes. The examination of 16S rRNA sequences by Carl Woese and others has allowed a new classification of prokaryotes and has led to the realization that they are divided into two domains, the *Bacteria* and the *Archaea*. These advances in molecular genetics have allowed us to gain true insights about the "universal ancestor," if in fact there was one,

for the first time. We have acquired much knowledge and vast amounts of data on the subject. However, two basic questions remain. First, "Can we ever truly understand the phylogeny of bacteria?" Second, "Does it make any difference anyway?" In this, the final chapter of this book, we will examine some of the current thinking on these two questions.

MODELING GENETIC CHANGE

We have seen that there is evidence for the alteration of bacterial genomes through evolutionary times and we've investigated some of the pressures and molecular mechanisms underlying these changes. In fact, given the change just in the atmosphere during this period, it would be astonishing if considerable evolutionary change did not occur. Consequently, some of the insights have been surprising and dramatic. Genomic studies have made it clear that many bacterial genes began as parts of bacteriophages (Villareal, 2001), that genes have been traded across generic and even higher taxonomic groups, and that mutation and selection have led to a vast diversity of microorganisms (Huynen and Bork, 1998). Since it is estimated that we are acquainted with less than 5% of the bacterial types on planet Earth, imagine what vast diversity and unique phenotypes remain to be discovered. Given this it seems a daunting task to ever hope to fully understand the processes that have led to this amazing diversity. But then reflect on what we have learned over the past 50 years and how new techno-

Robert V. Miller, Department of Microbiology, Oklahoma State University, Stillwater, OK 74078, and *Martin J. Day*, Cardiff School of Biosciences, Cardiff University, Cardiff CF10 3TL, Wales, United Kingdom.

Microbial Evolution: Gene Establishment, Survival, and Exchange
Edited by Robert V. Miller and Martin J. Day, ©2004 ASM Press, Washington, D.C.

logical advances have made what seemed an impossible task, like genome sequencing, easy.

Ever since Luria and Delbrück and Lederberg and Lederberg convinced us in the 1940s and early 1950s that bacteria did obey the rules of Mendel and Darwin, they have been proposed as models for the study of evolution. Bacteria are small and reproduce rapidly. In the time it takes to analyze the genetic change in one generation of oaks, we could analyze millions of generations of *Escherichia coli*. In the time it takes to analyze the F_1, F_2, and F_3 of Mendel's sweet peas, we could have studied and been amazed by the genetic and phenotypic variance in the $F_{100,000}$'th of a progenitor *Pseudomonas aeruginosa*. Clearly bacteria have many advantages as models for the study of genomic change, but can we ever truly model their evolution—much less ours? Probably not, but that shouldn't stop us from making every attempt to do so!

Many models designed to test many outcomes are being used by many microbiologists throughout the world. In this essay, we will discuss only one. In a paper that appeared in *Genetics* in 2002, Miriam Barlow and Barry Hall describe a model system that their laboratory has developed to emulate evolution of a gene in vitro. This model examines the processes of genetic change and selection in the TEM-1 β-lactamase gene of *E. coli*. Barlow and Hall mutated the gene in vitro and selected for increased resistance to the "extended-spectrum" penicillins: cefotaxime, cefuroxime, ceftazidime, and aztreonam.

Certainly their exercise was more than academic. Stuart Levy and many other microbiologists and physicians have warned that the misuse of antibiotics is rapidly reducing their effectiveness. And genes such as TEM-1 mutate in situ to allow the acquisition of resistance to new antibiotics almost as soon as they are put into use.

Barlow and Hall used a commercially available mutagenesis kit to produce an average of two mutations per copy of the TEM-coding sequence. They had shown that this mutagenesis system produced a suite of mutations

similar to the spectrum of mutations naturally found in *E. coli*. By producing an average of two mutations per coding sequence, these investigators ensured that all possible double- and triple-point mutations were present in the test populations but only at a very low frequency. This rate of mutation was chosen because in nature mutations occur one at a time and rarely in pairs. They then used the selective pressure of growth in the presence of cefotaxime, cefuroxime, ceftazidime, and aztreonam to evolve the TEM-1 gene. These antibiotics were chosen because they have been in use for a number of years and TEM-1 does not confer resistance to them. Natural variants of bacteria resistant to these antibiotics have been isolated and show alterations in their TEM-1 gene leading to the resistance phenotype.

Amino acid substitutions recovered in 10 independent in vitro "evolutions" were compared with the amino acid substitutions in the naturally occurring "extended-spectrum" TEM alleles. This phenotype is caused by nine different types of substitutions in nature. Barlow and Hall's in vitro-evolved populations contained seven of the changes multiple times. These data indicate that the model developed by these investigators does mimic the natural evolution processes.

Not only did this study help to validate their model, it allowed Barlow and Hall to predict that resistance to cefepime, a new "extended-spectrum" antibiotic, will develop. Armed with this knowledge, it may be possible to carefully control the use of this antibiotic to delay and hopefully prevent the conditions that select clones which develop resistance to it. An understanding of how the evolution of a resistance genotype occurs and with what level of likelihood is clearly clinically important. So if we could integrate our knowledge of intracellular and intercellular gene-mobility mechanisms, and the frequency with which they "move" sequences around, with selection pressure, then perhaps a predictive model for the "evolution" of resistant organisms can be made.

Models such as the Barlow-and-Hall system are proving very useful in understanding evolution and predicting change in natural populations of bacteria. Use of genomics, study of the fossil record, identification of environmental stresses, past, present, and future, and modeling are all important to increasing our knowledge and understanding of the mechanisms and consequences of bacterial genome diversification.

WHY DO WE NEED BACTERIAL SPECIES ANYWAY?

> What's in a name? That which we call a rose, by any other name would smell as sweet.
> —Shakespeare: Romeo and Juliet, ii. 2.

For the past 50 years nomenclature of bacteria has been in constant flux as new species and genera are added or removed for the standard systematics. These changes reflect the latest fashions in classification and the application of new molecular and physiological methods for characterizing bacteria. These changes have been confusing and have lead some to question the need for identifying bacterial species at all. An excellent commentary on the value and dangers of naming bacteria by Howard Gest can be found in the August 2003 issue of *Microbiology*.

Using the most modern molecular techniques to define bacteria taxa, the eubacteria are divided into 23 phyla (Gupta, 2002). Still the criteria used to identify the members of these phyla are ill defined. Some phyla contain only a few species, while only a few phyla account for more than 90% of the known species. These phyla are recognized by their branching on the 16S rRNA tree of Woese (1998). No other markers have been used to define them. Initially, this tree was based on limited sequence data and showed clear divisions between the phyla. As better sequence data based on a more complete sequence of the 16S gene in these bacteria have become available, the branches have blurred and the tree is not as well defined (Doolittle, 1999). Clearly, as we learned in chapter 21, other cri-

teria are needed to define a species and, for that matter, a phylum accurately.

Many criteria have been suggested and you have learned about them as you read this book. But why are we interested in developing a definition of these phyla and genera and species at all? The answer depends greatly on the question you are asking and those questions that you hope to ask in the future. The human mind has always wanted to categorize and "name" things and place relationships between them. It allows us to put order and structure and understanding into our interactions with the environment and to communicate them to others. Some say we can't really imagine something we cannot name. Certainly phylogenetic nomenclature has allowed us to differentiate and thus name bacteria. Naming allows us to better identify the relationships of bacteria among themselves, which may have a common ancestor, or how one may have evolved from another. This is an exciting and intellectually stimulating area of investigation and research. It has allowed and will allow us to ask very exciting questions. What are the relationships among the major phyla? Does the evolutionary tree have a trunk, rooted in a common ancestor, from which the phyla branch, or is the tree multitrunked with the major branches arising from, as Ford Doolittle (2000) suggests, "a common ancestral community of primitive cells?" These kinds of questions make it necessary to clearly define and name individual types of bacteria.

On the other hand, if our goal is to understand bacterial diversity (Hugenholtz and Pace, 1996), do we need to name bacteria and divide them into phylogenetic groups? In these studies we need to understand the richness of the microbial world and to protect this richness and diversity wherever possible. When we look at an ecosystem, we understand that the biochemical activities of bacteria form its foundation. They carry out the mineralization of many elements necessary for life. They destroy compounds toxic to themselves and to other forms of life. They form the basis for the food chain, providing energy and raw materials used by those organisms that feed on them and in turn are fed upon. However, for these studies, is the naming of species really necessary or even desirable? Understanding the relationships important to an ecosystem may only require identification of the functional groups of organisms that are necessary for the health of the ecosystem. To many, if the function is carried out, we needn't bother about how many "species" are carrying out the activity.

Is this dangerous? Potentially! Why? Because redundancy of a function, implicit in the diversity of the microbial populations that we place into different phylogenetic divisions within an ecosystem, may be necessary to ensure continuation of the activity during times of stress and change in the physical environment. We have certainly seen how these factors have caused elimination of many species in the past. Just consider the earth's atmospheric changes over the past billion or so years. If we allow a reduction in the genetic diversity of organisms that, for instance, fix nitrogen from a natural habitat, we may not have altered the apparent health of the ecosystem, but we have reduced the robustness of the system. Addition of another, and perhaps minor, stress to the system could eliminate the activity of fixing nitrogen altogether, and the ecosystem might die. In this case, we have unwittingly allowed the elimination of the functional redundancy of the system to its, and potentially our, ruin. An understanding and awareness of the taxonomic and evolutionary relationships among the groups that inhabit this ecosystem might have warned us that genomic diversity was being reduced and the ecosystem was not healthy at all.

But does this awareness depend on naming a species and identifying the fact that many species exist in an ecosystem or is it only necessary to measure genetic diversity among the organisms that carry out the function? This richness of genetic potential does not depend on our perception and certainly not on the naming of species. Our awareness of this richness, or lack thereof, does depend on our understanding of the diversity of genetic potential and the mechanisms that led to it.

Whether or not we feel the need to name and categorize bacteria, whether or not we wish to be bacterial genealogists, we need to understand the ways in which bacteria evolve and change. Insight into the richness of genetic potential in the world around us is essential if we are to protect our natural environmental heritage and leave to our posterity a world that is vital and healthy. To do otherwise is to doom future generations to evolutionary extinction. As Carl Sagan and Ann Druyan (1992) observed in their book *Shadows of Forgotten Ancestors*:

> By 3 billion years ago, life had changed the color of the inland seas; by 2 billion years ago, the gross composition of the atmosphere; by 1 billion years ago, the weather and the climate; by a third of a billion years ago, the geology of the soil; and in the past few hundred million years the close-up appearance of the planet. These profound changes, all brought about by forms of life we tend to consider 'primitive,' and of course by processes we describe as natural, mock the concerns of those who hold that humans, through their technology, have now achieved 'the end of nature.' We are rendering many species extinct; we may even succeed in destroying ourselves. But this is nothing new for the Earth. Humans would then be just the latest in a long sequence of upstart species that arrive on-stage, make some alterations in the scenery, kill off some of the cast, and then themselves exit stage-left forever. New players appear in the next act. The Earth abides. It has seen all this before.

Should we take this as a threat, a promise, or a challenge? You, the reader, must decide.

REFERENCES

Barlow, M., and B. G. Hall. 2002. Predicting evolutionary potential: *In vitro* evolution accurately reproduces natural evolution of the TEM β-lactamase. *Genetics* **160**:823–832.

Doolittle, W. F. 1999. Phylogenetic classification and the universal tree. *Science* **284**:2124–2128.

Doolittle, W. F. 2000. Uprooting the tree of life. *Sci. Am.* **282**(2):72–77.

Gupta, R. S. 2002. Phylogeny of *Bacteria*: Are we how close to understanding it? *ASM News* **68**:284–291.

Hugenholtz, P., and N. R. Pace. 1996. Identifying microbial diversity in the natural environment: a molecular phylogenetic approach. *TIB TECH* **14**:190-197.

Huynen, M. A., and P. Bork. 1998. Measuring genome evolution. *Proc. Natl. Acad. Sci. USA* **95:** 5849–5856.

Sagan, C., and A. Druyan. 1992. *Shadows of Forgotten Ancestors.* Ballantine Books, New York, N.Y.

Villarreal, L. P. 2001. Persisting viruses could play role in driving host evolution. *ASM News* **67:**501–507.

Woese, C. R. 1998. The university ancestor. *Proc. Natl. Acad. Sci. USA* **95:**6854–6859.

FURTHER READING

Arber, W. 2000. Genetic variation: molecular mechanisms and impact on microbial evolution. *FEMS Microb. Rev.* **24:**1–7.

Atwood, K. C., L. K. Schneider, and F. J. Ryan. 1951. Periodic selection in *Escherichia coli. Proc. Natl. Acad. Sci. USA* **37:**146–155.

Bricks, J. J., G. A. Logan, R. Biuck, and R. E. Summons. 1999. Archean molecular fossils and the early rise of eukaryotes. *Science* **285:**1033–1036.

Koch, A. L. 1998. How did bacteria come to be? *Adv. Microb. Physiol.* **40:**353–399.

Lawrence, J. G., and H. Ochman. 1997. Amelioration of bacterial genomes: Rates of change and exchange. *J. Mol. Evol.* **44:**383–397.

Levy, S. B. 2002. *The Antibiotic Paradox.* Perseus Publishers, Cambridge, Mass.

Posada, D., and K. A. Crandall. 2001. Intraspecific gene genealogies: Trees grafting into networks. *TIEE* **16:**37–45.

Woese, C. R. 1987. Bacterial evolution. *Microbiol. Rev.* **51:**221–271.

Woese, C. R. 1991. The use of ribosomal RNA in reconstructing evolutionary relationships among bacteria, p. 1–24. *In* R. K. Selander, A. G. Clark, and T. S. Whittam (ed.), *Evolution at the Molecular Level.* Sinauer Associates Inc., Sunderland, Mass.

INDEX

Microbial Evolution: Gene Establishment, Survival, and Exchange
Edited by Robert V. Miller and Martin J. Day, ©2004 ASM Press, Washington, D.C.